Dyl
dyc
am
gel
eite

This
the c
to b
reader
items.

Cyst Nematodes

NATO ASI Series

Advanced Science Institutes Series

A series presenting the results of activities sponsored by the NATO Science Committee, which aims at the dissemination of advanced scientific and technological knowledge, with a view to strengthening links between scientific communities.

The series is published by an international board of publishers in conjunction with the NATO Scientific Affairs Division

A Life Sciences **B Physics**	Plenum Publishing Corporation New York and London
C Mathematical and Physical Sciences	D. Reidel Publishing Company Dordrecht, Boston, and Lancaster
D Behavioral and Social Sciences **E Engineering and Materials Sciences**	Martinus Nijhoff Publishers The Hague, Boston, and Lancaster
F Computer and Systems Sciences **G Ecological Sciences**	Springer-Verlag Berlin, Heidelberg, New York, and Tokyo

Recent Volumes in this Series

Volume 114—Cardiorespiratory and Cardiosomatic Psychophysiology
edited by P. Grossman, K. H. Janssen, and D. Vaitl

Volume 115—Mechanisms of Secondary Brain Damage
edited by A. Baethmann, K. G. Go, and A. Unterberg

Volume 116—Enzymes of Lipid Metabolism II
edited by Louis Freysz, Henri Dreyfus, Raphaël Massarelli, and Shimon Gatt

Volume 117—Iron, Siderophores, and Plant Diseases
edited by T. R. Swinburne

Volume 118—Somites in Developing Embryos
edited by Ruth Bellairs, Donald A. Ede, and James W. Lash

Volume 119—Auditory Frequency Selectivity
edited by Brian C. J. Moore and Roy D. Patterson

Volume 120—New Experimental Modalities in the Control of Neoplasia
edited by Prakash Chandra

Volume 121—Cyst Nematodes
edited by F. Lamberti and C. E. Taylor

Volume 122—Methods for Mycological Examination of Food
edited by A. D. King, Jr., J. I. Pitt, and L. R. Beuchat

Series A: Life Sciences

Cyst Nematodes

Edited by

F. Lamberti

Institute of Agricultural Nematology of the CNR
Bari, Italy

and

C. E. Taylor

Scottish Crop Research Institute
Invergowrie, Dundee, Scotland

Plenum Press
New York and London
Published in cooperation with NATO Scientific Affairs Division

Based on the proceedings of a NATO Advanced Study Institute on
Cyst Nematodes,
held September 21-October 3, 1985,
in Martina Franca, Italy

Library of Congress Cataloging in Publication Data

NATO Advanced Study Institute on Cyst Nematodes (1985: Martina Franca, Italy)
Cyst nematodes.

(NATO ASI series. Series A, Life sciences; v. 121)
"Based on the proceedings of a NATO Advanced Study Institute on Cyst Nematodes, held September 21-October 3, 1985, in Martina Franca, Italy"—T.p. verso.
"Published in cooperation with NATO Scientific Affairs Division."
Includes bibliographical references and index.
1. Plant nematodes—Congresses. 2. Plant nematodes—Control—Congresses. I. Lamberti, F. (Franco), 1937- . II. Taylor, C. E. (Charles E.) III. Title. IV. Title: Cyst nematodes. V. Series.
QL391.N4N36 1985 632.6′5182 86-25381
ISBN 0-306-42475-4

A Division of Plenum Publishing Corporation
233 Spring Street, New York, N.Y. 10013

Printed in the United States of America

PREFACE

The cyst nematodes are a major group of plant pathogens of economic importance in many countries throughout the world. Considerable yield losses have been attributed to cyst nematodes attacking potatoes, sugar beet, soybean and cereals. Because of the protective cyst that is formed, which encloses the eggs, they pose special problems in their control and also in preventing their distribution in soil from infested areas.

There has been much research on many aspects of the biology of cyst nematodes and the large number of publications are testimony to their importance as crop pests and of the interest of nematologists in the group. Nevertheless, they continue to pose problems because of the complexity of their phylogeny and taxonomy and there is still much to be investigated on their biology, particularly with those aspects which relate to improving the effectiveness of control. In many respects the biology of the various cyst nematode species is similar, although with variations due to different interactions with hosts and geographical location. Thus, research on any one species has general applicability and this is particularly so with regard to evaluation of control measures, or to the methodology of soil sampling for the assessment of populations prior to growing crops.

The NATO Advanced Study Institute held at Martina Franca, Italy, during 21 September to 3 October 1985, provided an opportunity to assess current knowledge of cyst nematodes and a forum for critical discussion among nematologists who have been concerned with the biology of cyst nematodes in a research or advisory (extension) capacity. The major areas covered in the lectures included systematics and phylogeny, as a basis for understanding the evolution of the group and problems of identification; the relation between population densities and crop yield; the species associated with major crops throughout the world and an assessment of their economic importance; structure and physiology; and control by the chemical, cultural or legislative means. These published proceedings of the Institute are a comprehensive and authoritative account of particular aspects of the biology of cyst nematodes given by invited lecturers and supplemented by papers offered by other participants; there are also a few papers from authors who were unable to attend the Institute. An important element of the Institute was the discussion of the views presented in the lectures and it is hoped that some indication of the flavour and substance of these is conveyed in the summaries of discussions following the main papers.

Sixty five participants from twenty different countries were present and the organisers are grateful to them for their interest and enthusiasm which contributed greatly to the success of the Institute. Other than the participants the success of the Institute was dependent on several other individuals and organisations. We thank Mrs Angela Tosi of the Istituto de Nematologia Agraria who acted as Secretary of the Institute,

Mr F. Elia and Mr G. Zacheo, also of the Istituto de Nematologia Agraria, who were responsible for technical services, and the Municipality of Martina Franca and the Azienda Autonoma di Soggiorno e Turismo, Martina Franca, who provided facilities and generous hospitality. We are grateful to Mrs Loraine Galloway of the Scottish Crop Research Institute who organised and typed the final text for the book, and to Miss Maureen McMaster, Mr G. Menzies and Mr T. Geoghegan, also of the Scottish Crop Research Institute, who prepared the figures and photographs for publication.

F. Lamberti
C. E. Taylor

Alan R. Stone

While this book was in the final stages of preparation we learned of the tragic and untimely death of Alan Stone. Alan has had a distinguished career as a nematologist and his contribution to this volume on cyst nematodes is testimony to his academic authority as a taxonomist.

CONTENTS

TAXONOMY AND PHYLOGENY OF CYST NEMATODES

Alan R. Stone

Rothamsted Experimental Station
Harpenden, Herts, AL5 2JQ
England

INTRODUCTION

The major nematode pests in agriculture throughout the world are generally those forms showing greatest development of the parasitic habit. Of these, the cyst nematodes and their relatives are the most highly adapted. They are also probably the only large group of plant-feeding nematodes for which sufficient information has been accrued to permit rigorous phylogenitic or phenetic analyses and development of hypotheses describing how the nematode-plant host-parasite relationship has evolved. The group provides a paradigm for other truly plant parasitic nematodes. We require a systematic framework within which to order our names and descriptions of organisms. Such a framework, unless it is merely a catalogue, must have a logical basis if it is to be predictive and enable us to make assumptions about the biology of an organism from comparison with related forms. An understanding of the evolution of host-parasite relationships permits predictions to be made about nematode behaviour in relation to a host as well as contributing to the development of a rational systematics.

Studies of nematode taxonomy, phylogeny and host-parasite interactions are far from being satisfying intellectual exercises alone: they make possible a more complete understanding of the biology of nematodes as agricultural pests. This understanding is more important today than ever before as control measures turn increasingly from the use of broad action nematicides to resistant cultivars and, in the future, biological control agents. Both are specific in their action and require thorough understanding, in their development and application, of the biology of the nematodes we seek to control.

MORPHOLOGY

Cyst nematodes have a broadly similar morphology deriving their common name (and the name of the type genus of the Heteroderidae) from the endoparasitic swollen female, the sedentary habit of which commences with the second stage juvenile after invasion of the host root and finishes with the development of the mature female into an egg-filled cyst by tanning of the body wall. Because cysts are readily extracted from soil and second-stage juveniles readily obtained from the eggs they contain, the morphology of these two stages is of overriding importance in the taxonomy of the group; female morphology plays a lesser role; male morphology is remarkably consistent (or perhaps less intensively studied) throughout the genera and species of cyst nematodes; morphology of the intermediate juvenile stages plays no part because of their transient nature and inaccessibility within roots. Those genera containing species of importance to agriculture (*Heterodera, Cactodera, Globodera* and *Punctodera*, can be distinguished by female and cyst morphology. Characterisation of species within each genus requires both cysts and second-stage juveniles, in the majority of cases with use of quantitative characters with the attendant problems of character variation and overlap so that identification of some species presents considerable difficulty.

Mature females are swollen into a spherical, sub-spherical or lemon shape to accommodate the enormous development of the paired ovaries which lie coiled within the body cavity. The mature female cuticle is not annulated, except at the head where several prominent annules occur, and the lateral field is absent. The stylet and oesophagus are strongly developed with a prominent median bulb and lie on the anterior part of the body which forms the "neck". Posterior to the excretory pore the swelling of the body is greatly developed and the excretory pore thus lies at the base of the neck. The vulva lies at the opposite pole of the body and in lemon shape species (major genera: *Heterodera* and *Cactodera*) is raised on a greater or lesser projection, the vulval cone. The morphology of the vulval cone is of primary importance, recognised in the pioneering work of Cooper (1955) who described most of the features and established their terminology. The composite cyst cone diagrammatically illustrated in Fig. 1 shows these features and others subsequently found of value; such a combination of characters has not been found in nature!

The vulval slit runs transversely, i.e. at 90° to the anal vulval axis. Because the anus is posterior to the vulva in the unswollen immature female, it comes to lie dorsal to the vulva in mature females. Cysts retain the female shape. The surface of the cyst is covered by a pattern of ridges derived from the pattern on the female cuticle. A thin walled area surrounds the vulva and the cuticle is lost from this region in older

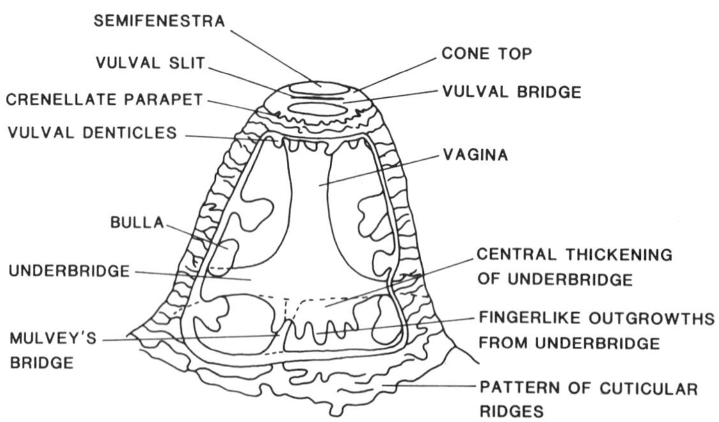

Fig. 1 Diagnostic features found in cyst nematode vulval cones (partly after J.J. Hesling, 1978, in "Plant Nematology", ed, J.F. Southey.

cysts, forming an opening, the fenestra. The fenestra may be a single, circular thin-walled region (circumfenestrate) as in *Cactodera* species but in most the vulva lies in a strip of thicker walled cuticle, the vulval bridge, which crosses the fenestral region and divides it into two semi-fenestrae. The semi-fenestrae are separated by a greater or lesser distance which is the width of the vulval bridge. Where this is narrow the opposed faces of the semi-fenestrae are flattened, imparting a D-shape to each semi-fenestral aperture but where the vulval bridge is wider (generally equal to or more than the semi-fenestral length) the semi-fenestrae are circular; the former condition is termed ambifenestrate and the latter bifenestrate. The vaginal wall may remain in the cyst cone and be connected to the sides of the cone by a transverse structure, the underbridge. A second lateral structure may be present between the vulval bridge and the under bridge, running from the vaginal remnant to the cone walls at 90° to the vulval bridge and underbridge; this is termed Mulvey's bridge. Thickening of the inner surface of the vulval cone may be present, forming variously shaped knobs termed bullae. Golden and Raski (1977) proposed the term vulval denticles for small, tooth-like pigmented structures occurring near the apex of the vulval cone which they distinguished from bullae; Mulvey and Golden (1983) described these structures in a number of species. The anus can usually be seen as a small opening on the side of the vulval cone.

In species forming round-ended cysts (major genera: *Globodera* and *Punctodera*) the vulva is not on a raised cone but lies in a shallow, circular depression, the vulval basin. The vulval slit runs transversely and is surrounded dorsally and ventrally by a crescent-shaped, papillated area within the vulval basin (with the exception of those *Globodera* species parasitising Compositae). Fenestration is always circumfenestrate and the area of the fenestra is always more or less equal (sometimes rather less) to that of the vulval basin. No underbridge is present. In *Punctodera* the anus is set within a second, circular fenestra more or less equal in size to the vulval fenestra.

Second-stage juveniles are vermiform, with an offset dome-shaped head and conical tail tapering to a point. On death the body assumes a gentle curve with the ventral surface concave so that the nematode lies on its side. The cuticle is regularly annulated with lateral fields running from near the head on to the tail; the number of incisures in the fields may reduce anteriorially and posteriorally, it is four in the majority of species and three in the exceptions [*Cactodera estonica* Kir'janova & Krall, 1968, was incorrectly described as having five; type specimens I have examined show four]. The oral disc and lip configurations are taxonomically useful features. The head skeleton, stylet and oesophagus are well-developed, the latter occupying approximately one-third of the body length. The median bulb is rounded in shape with a prominent valve. The oesophageal glands overlap the intestine ventrally and sub-ventrally; the single dorsal oesophageal gland nucleus is more prominent than, and anterior to, the two sub-ventral gland nuclei. The excretory pore is clearly visible on the ventral surface opposite the oesophageal glands with the hemizonid anterior. A hemizonion and cephalids may also be visible. The anus may be marked by a small notch in the cuticle and the tail has a clear tip, the hyaline terminus. Phasmids may be visible and the size has some correlation with genus. The second-stage juvenile is folded four times within the egg from which it hatches as the invasive stage.

Males are vermiform with an off-set, hemispherical head. The tail is characteristic: very short (about one-quarter maximum body width) and bluntly rounded without a bursa. The cloaca is extended into a short tubus. After death males frequently are twisted 90° along the longitudinal axis at a point two-thirds of the body length from the head, conferring an S-shape. The cuticle is annulated and the lateral fields terminate abruptly on the tail but often appear to continue around it. Morphology of the anterior end is similar to that of the second-stage juvenile but lip morphology is of less taxonomic value. The median bulb is more slender and the oesophagus occupies proportionally less of the body length. The single testis extends anteriorally for one-third to two-thirds of body length and is not reflexed. The paired spicules are curved, flask-shaped proximally and fit together distally to form a tapering tube. Spicule tips are single-pointed or bifid.

Genera other than *Heterodera*, *Cactodera*, *Globodera* and *Punctodera* differ in some respects from the above account (see below).

TAXONOMY

Family Group Level

Until 1978 the familial classification had demonstrated an unexceptional development but was criticised for its arbitrary basis and the proliferation of taxa within the family. Golden (1971) proposed the super family Heteroderoidea containing two families: Heteroderidae, with sub-families Heteroderinae, Meloidoderinae and Meloidogyninae; and Nacobbidae, with Nacobbinae and Rotylenchulinae. Wouts (1973) recognised two independent families: Heteroderidae with sub-families Heteroderinae, Meloidoderinae and Ataloderinae; and Meloidogynidae, both within the super family Tylenchoidea. Andrassy (1976) included the sub-families Heterodérinae, Meloidoderinae and Meloidogyninae in the Heteroderidae, within the super family Hoplolaimoidea. Krall and Krall (1978) proposed the families Meloidoderidae and Ataloderidae but these have not gained acceptance. Stone (1978) and Hooper (1978) recognised the separate families Meloidogynidae and Heteroderidae, with sub-families as in Wouts (1973); Hooper (1978) included them in the super family Heteroderoidea, from which he excluded Nacobbidae and Rotylenchulinae. Husain (1976) followed Golden (1971) but placed *Rotylenchulus* in the monogeneric family Rotylenchulidae within the Heteroderoidea and erected the sub-family Sarisoderinae for *Sarisodera sensu* Wouts (1973). Close linkage of nacobbids and rotylenchulids with heteroderids at the family level is unacceptable because of apparently fundamental differences in morphology, especially the bursate tail of nacobbid and rotylenchulid males and the vermiform shape of their young adult females. The heteroderids seem best limited to forms with abursate males with a short tail, usually less than half the body width long and with didelphic females in which enlargement usually begins by the third juvenile stage. Because Meloidogynidae and Heteroderidae share this unique combination of characters they are best placed on their own in the same super family, Heteroderoidea. Stone (1978) provides a detailed review of the taxonomy to 1977. Subsequently, Krall and Krall (1978) proposed a further sub-family, Punctoderinae, for *Punctodera* while for certain non-cyst-forming genera Cryphoderinae Coomans, 1978 and Verutinae Esser, 1981 were also proposed. At this time proliferation of sub-family taxa in the Heteroderidae was accelerating. Luc et al. (1978) had rightly sounded a warning when describing *Hylonema*, a form which in their view could not be accommodated in the purportedly highly artificial extant sub-families. Consequently, they proposed abandoning the Heteroderinae, Meloidoderinae and Ataloderinae with reduction of the Heteroderidae to an unordered assemblage of genera until further research provided a solution.

With the application of cladistic analysis a sound theoretical basis for heteroderid taxonomy is becoming established (see below: Phylogeny and Systematics). Such analyses have been published by Coomans (1979), Ferris (1979, 1985) and Wouts (1985) with much agreement among the phylogenies produced; those by Coomans and Wouts are used to justify family group classifications. Coomans (1979) designated tribes below the sub-family level, an approach which although limiting inflation of heteroderid family group taxa and over-valuation of relatively minor differential characters has not subsequently been adopted in classifications of the groups. Wouts (1985) provides the most recent classification of the Heteroderidae and, although now justified by a phylogenetic analysis, largely retains taxa established prior to Luc et al. (1978) with further additions; there is a large measure of congruence with the classification provided by Coomans.

A family group classification is not presented here, the reader being referred to that of Wouts (1985) much of which I accept despite some differences from the interpretation in this paper. Furthermore, classification of the Heteroderidae is likely to remain in flux as phylogenetic analyses of further data (see below) are undertaken and applied to the taxonomy. An important departure in Wouts (1985) is to place together in the Punctoderinae, the genera *Dolichodera, Globodera, Cactodera* and *Punctodera*.

Genus Group Level

At the time of writing nineteen genera available under the International Code of Zoological Nomenclature have been proposed in the Heteroderidae, ten from 1975 onwards of which five are not accepted by at least one authority on the systematics of the family (Table 1). Much of the taxonomy is unstable, with a tendency to erect new genera on trivial characters and, in the case of the less well understood taxa, changes are proposed as basic morphology is re-examined. Among the genera of agricultural importance the following proposals and revisions have occurred:-

The subgenera *Heterodera Heterodera)* Skarbilovich, 1959, and *Heterodera (Globodera)* Skarbilovich, 1959 became synonymous respectively with *Heterodera* Schmidt, 1871 and *Globodera* Skarbilovich, 1959 following the elevation of *Globodera* from subgeneric to generic rank by Behrens (1975). Mulvey and Stone (1976) made the same revision in ignorance of Behrens' action.

Punctodera proposed by Mulvey and Stone (1976) to contain species without a vulval cone, with a circumfenestrate vulval fenestra and a unique, second circular fenestra incorporating the anus.

Bidera proposed for bifenestrate *Heterodera* species with a short vulval slit up to 25 µm long and always less than the length of a vulval bridge. Krall and Krall's (1978) distinction is unsatisfactory because the bifenestrate condition is variable (specimens of *H. avenae*, the type species of *Bidera*, are commonly ambifenestrate) and is not unique to the proposed genus (*H. humuli* is bifenestrate but falls without *Bidera*, see Stone and Rowe, 1977) while a "short" vulval slit, varying from <10µ to 25µ is an inadequate character. Mulvey and Golden (1983) synonymised *Bidera* back to *Heterodera*, a revision with which I concur but Wouts (1985) retains the genus.

Cactodera also proposed by Krall and Krall (1978) and emended by Mulvey and Golden (1983) is an accepted genus erected for former *Heterodera* species with a vulval cone and circumfenestrate fenestration.

Table 1 Available genera of Heteroderidae Filip'jev & Schuurmans Stekhoven, 1941

Cyst forming	Non-cyst forming
Heterodera Schmidt, 1871	
	Meloidodera Chitwood, Hanon & Esser, 1956
Globodera Skarbilovich, 1959	
	Cryphodera Colbran, 1966
	Atalodera Wouts & Sher, 1971
	*Sarisodera*** Wouts & Sher, 1971
	*Sherodera** Wouts, 1973
	*Zelandodera** Wouts, 1973
Punctodera Mulvey & Stone, 1976	
*Bidera** Krall & Krall, 1978	*Thecavermiculatus* Robbins, 1978
Cactodera Krall & Krall, 1978	*Hylonema* Luc, Taylor & Cadet, 1978
Dolichodera Mulvey & Ebsary, 1980	
*Ephippiodera** Shagalina & Krall, 1981	*Verutus* Esser, 1981
	*Rhizonema** Cid del Prado Vera, Lownsbery & Maggenti, 1983
Afenestrata Baldwin & Bell, 1985	
	Bellodera Wouts, 1985

* Not accepted by at least one authority
**The non-cyst forming status of *Sarisodera* is disputed, see below.

The combination of vulval cone and circumfenestration is unique but cyst morphology within *Cactodera* is rather variable and the circumfenestrate condition is different from that within *Globodera*.

Ephippiodera was proposed by Shagalina and Krall (1981) for *Heterodera* of the avenae species group with circular semi-fenestrae widely separated by a vulval bridge >18μm wide, vulval slit <10μm and a saddle-shaped depression between the semi-fenestrae. This definition does not differentiate satisfactorily from other species of the avenae group. I do not accept *Ephippiodera* and Wouts (1985) synonymised *Ephippiodera* to *Bidera*, a de facto return of the species concerned to *Heterodera* if *Bidera* is also not accepted, as I believe is correct.

Genera of cyst-forming nematodes currently without species of importance to agriculture are *Dolichodera* and *Afenestrata*.

Dolichodera was erected by Mulvey and Ebsary (1980) to contain one newly described species *D. fluvialis*, based on only two females, twelve cysts and five juveniles. Females and cysts lack a vulval cone, show secondary annulation in the mid-body and are elongate. The genus is distinguished from *Globodera* by female/cyst shape and cuticular pattern, presence of bullae and absent of perineal tubercles (i.e. papillae in the vulval basin distributed dorsally and ventrally to the vulval slit). However, bullae occur in certain *Globodera* species (Mulvey and Stone, 1976) and the absence of perineal tubercles in *Dolichodera* cannot be determined with certainty because all specimens apparently were fenestrated or with details of the vulval basin unobservable. Furthermore, Mulvey and Ebsary state that *Dolichodera* second-stage juveniles are also distinguished by

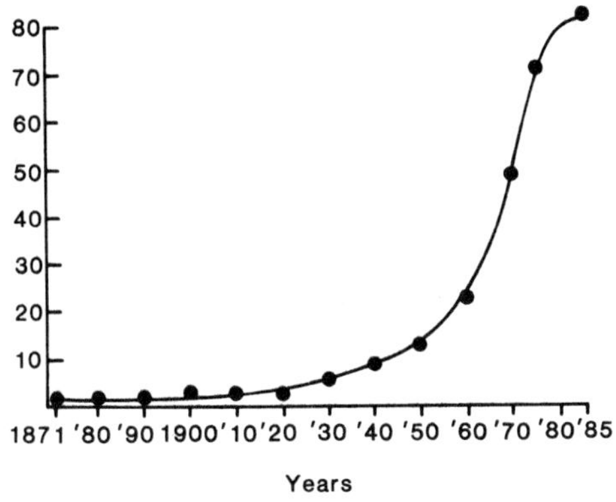

Fig. 2. Numbers of recognised species of cyst nematodes, to August 1985.

taxonomy undergoes revisions. Many of the nominal species are based on inadequate descriptions often of few specimens and with no congnizance of intraspecific variation, especially in the quantitative characters widely used at the species level. This is reflected by the large proportion (17%) of nominal species recognised as species inquirendae, i.e. ones for which the description does not permit identification without doubt. Unfortunately, the name of a species inquirenda remains fixed in the nomenclature as available because it meets the minimum requirements of the International Code of Zoological Nomenclature but in many cases improved descriptions are unlikely to be forthcoming.

Nominal species within each genus are listed below together with comments on their taxonomy; those regarded in the literature as species inquirendae are marked with an asterisk. As Stone (1978) pointed out, the great majority of species inquirendae have been described since Goodey (1959) laid down minimum standards for descriptions of plant parasitic nematodes, standards which apply as forcibly to cyst nematodes as to any other group.

AFENESTRATA

Type species: *Afenestrata africana* (Luc, Germani & Netscher, 1973)
See discussion of this species under Genus Group Level, above.

CACTODERA

Type species: *Cactodera cacti* (Filip'jev & Schuurmans Stekhoven, 1941)

Other Nominal Species:

Cactodera amaranthi (Stoyanov, 1972)
Cactodera aquatica (Kir'janova, 1971)*
Cactodera betulae (Hirschmann & Riggs, 1969)
Cactodera chaubattia (Gupta & Edward, 1973) n.comb.*
Cactodera eremica (Baldwin & Bell, 1985)
Cactodera estonica (Kir'janova & Krall, 1963)
Cactodera thornei (Golden & Raski, 1977)
Cactodera weissi (Steiner, 1949).

Krall and Krall (1978) transferred the listed species, formerly all *Heterodera*, to this genus with the following exceptions: Mulvey and Golden (1983) considered *C. betulae* to be distinct from *Cactodera* and *Heterodera* sensu lato. *Heterodera chaubattia* was described as a round cyst nematode by Gupta and Edward (1973) but their illustration of cysts clearly shows a low vulval cone. Consequently, Mulvey and Stone (1976) and Stone (1978) placed this species in *Heterodera* but Wouts (1984) included it in *Globodera*. Although Sharma and Swarup (1983) regarded *H. chaubattia* as a species inquirenda the description provides some basis for characterisation. Because the cysts are circumfenestrate with a low vulval cone *Cactodera* as emended by Mulvey and Golden (1983) is the only cyst nematode genus in which this taxon can be placed for the present and *H. chaubattia* is here regarded as a new combination in that genus. *Globodera mali* may be a junior synonym, of *C. chaubattia* both species being associated with apple and described from Asia (Krall, 1978). The new combination for *H. thorneii* was attributed to Krall and Krall (1978) by Mulvey and Golden (1983) who listed this species as *Cactodera* but Krall and Krall did not treat with *H. thornei*. The new combination therefore first appears in Mulvey and Golden (1983) and is attributable to these authors (International Code of Zoological Nomenclature, 3rd edition (1985) article 48). *C. eremica* was recently described as *Cactodera* by Baldwin and Bell (1985).

DOLICHODERA

Type species: *Dolichodera fluvialis* (Mulvey & Ebsary, 1980)
See discussion of this species under Genus Group Level, above.

GLOBODERA

Type species: *Globodera rostochiensis* (Wollenweber, 1923).

Other Nominal Species:

Globodera achilleae (Golden & Klindić, 1973)
Globodera artemisiae (Eroshenko & Kasachenko, 1972)
Globodera leptonepia (G.S. Cobb & Taylor, 1953)
Globodera mali (Kir'janova & Borisenko, 1975)
Globodera millefolii (Kir'janova, & Krall, 1965)
Globodera mirabile (Kir'janova, 1971)*
Globodera pallida (Stone, 1973)
Globodera pseudorostóchiensis (Kir'janova, 1963) — synonymised to *G. rostochiensis* by Krall (1978)
Globodera tabacum tabacum (Lownsbery & Lownsbery, 1954)
Globodera tabacum solanacearum (Miller & Gray, 1972)
Globodera tabacum virginiae (Miller & Gray, 1968)
Globodera zelandica Wouts, 1984

Globodera leptonepia is unique among Heteroderidae in the second-stage juvenile being folded seven times in the egg and having a stylet length of only 18 µm. Found in a ship-borne consignment of potatoes thought to originate from a Peruvian port, it is presumed to be a South American species parasitising potato. Extensive field collections of potato cyst-nematodes in the Andean Highlands (Evans et al., 1975; Stone and Evans, unpublished) have not resulted in its rediscovery. *G. mali* is probably synonymous with *C. chaubattia* (see above). *G. mirabile*, transferred from *Heterodera* by Mulvey and Stone (1976), is considered by Krall (1978) species et genus inquirendae with which I agree. *G. solanacearum* and *G. virginiae* are considered (Stone, 1983) sub-species of *G. tabacum* because full hybridisation with *G. tabacum* has been demonstrated and morphological differentiation is slight; known distributions of these taxa are adjacent but allopatric, if sympatric distribution is demonstrated as seems likely, the sub-species should be regarded as junior synonyms of *G. tabacum*. *G. pseudorostochiensis* was distinguished from *G. rostochiensis* by presence of punctations in the cyst wall but this character is inconsistently present in all *Globodera* species I have examined; Krall (1978) rightly synonymised this species to *G. rostochiensis*. Apart from the questionable *G. mali*, *G. mirabile* and *G. leptonepia* all *Globodera* species except *G. zelandica* are parasities of Solanaceae or Compositae. Those parasitising Compositae (*G. millefolii*, *G. achilleae*, *G. artemisiae*) are distinguished by absence of perineal tubercles in the vulval basin (Stone, 1984). Such differences in genitalia may have particular significance and the generic status of this group requires revision; *G. millefolii* is considered a species inquirenda.

HETERODERA

Type species: *Heterodera schachtii* Schmidt, 1871.

Other Nominal Species:

Heterodera acnidae Schuster & Brezina, 1979
Heterodera amygdali Kir'janova & Ivanova, 1975
Heterodera avenae Wollenweber, 1924
Heterodera bifenestra Cooper, 1955
Heterodera cajani Koshy, 1967
Heterodera canadensis Mulvey, 1979
Heterodera cardiolata Kir'janova & Ivanova, 1969*
Heterodera carotae Jones, 1950
Heterodera cruciferae Franklin, 1945
Heterodera cyperi Golden, Rau & Cobb, 1962
Heterodera daverti Wouts & Sturhan, 1978
Heterodera delvii Jairajpuri, Khan, Setty & Govindu, 1979
Heterodera elachista Oshima, 1974
Heterodera fici Kir'janova, 1954
Heterodera filipjevi (Madzhidov, 1981) n.comb.
Heterodera galeopsidis Goffart, 1936 — synonymised to *H. trifolii* by Maas et al. (1983)
Heterodera gambiensis Merny & Netscher, 1976
Heterodera glycines Ichinohe, 1952
Heterodera goettingiana Liebscher, 1892
Heterodera graduni Kir'janova in Kir'janova & Krall, 1971*
Heterodera graminis Stynes, 1971
Heterodera graminophila Golden & Birchfield, 1972
Heterodera hordecalis Andersson, 1975

Heterodera humuli Filip'jev, 1934
Heterodera indocyperi Husain & Khan, 1964 — Nomen nudum in Khan et al. (1964)
Heterodera iri Mathews, 1971
Heterodera latipons Franklin, 1969
Heterodera lespedezae Golden & Cobb, 1963
Heterodera leuceilyma Di Edwardo & Perry, 1964
Heterodera limonii Cooper, 1955*
Heterodera longicaudata Seidel, 1972 — synonymised to *H. bifenestra* by Andersson & Wouts (1983)
Heterodera longicolla Golden & Dickerson, 1973
Heterodera mani Mathews, 1971
Heterodera medicaginis Kir'janova in Kir'janova & Krall, 1971
Heterodera mediterranea Vovlas, Inserra & Stone, 1981
Heterodera menthae Kir'janova & Narbaev, 1971
Heterodera methwoldensis Cooper, 1955*
Heterodera mothi Khan & Husain, 1965
Heterodera oryzae Luc & Berdon-Brizuela, 1961
Heterodera oryzicola Rae & Jayaprakash, 1978
Heterodera oxiana Kir'janova, 1962*
Heterodera paratrifolii Kir'janova, 1963* — synonymised to *H. trifolii* by Krall (1978)
Heterodera polygoni Cooper, 1955*
Heterodera raskii Basnet & Jayaprakash, 1984
Heterodera rosii Duggan & Brenan, 1966
Heterodera rumicis Pogosyan, 1961
Heterodera sacchari Luc & Merny, 1963
Heterodera salixophila Kir'janova, 1969
Heterodera sonchophila Kir'janova, Krall & Krall, 1976 — nomen nudum in Kir'janova (1969)
Heterodera sorghi Jain, Sethi, Swarup & Srivastava, 1982
Heterodera tadshikistanica Kir'janova & Ivanova, 1966*
Heterodera trifolii Goffart, 1932
Heterodera turcomanica Kir'janova & Shagalina, 1965
Heterodera urticae Cooper, 1955
Heterodera ustinovi Kir'janova, 1969*
Heterodera uzbekistanica Narbaev, 1980
Heterodera vigni Edward & Misra, 1968
Heterodera zeae Koshy, Swarup & Sethi, 1971

Bidera filijevi was not included in those *Bidera* species transferred to *Heterodera* by Mulvey and Golden (1983), *H. filipjevi* is therefore here regarded as a new combination. *Heterodera scleranthii* Kaktyna, 1957 is not listed, being a nomen nudum Goodey (1963).

PUNCTODERA

Type species: *Punctodera punctata* (Thorne, 1928)

Nominal species:

Punctodera matadorensis Mulvey & Stone, 1976
Punctodera chalcoensis Stone, Sosa Moss & Mulvey, 1976

Groups within *Heterodera* sensu lato

Heterodera sensu lato includes the genera *Heterodera, Globodera, Punctodera* and *Cactodera.* It has been divided into sub-groups by Mathews (1971) and Mulvey (1972) on the basis of cyst, especially genital, morphology and by Stone (1975) on the basis of lip patterns of second-stage juveniles revealed by scanning electron microscopy. Green (1975) provided further evidence of grouping on the basis of development of vulval structures. There is much agreement between the six groups proposed by Mathews, the five proposed by Mulvey and the six proposed by Stone. The congruence of groupings based on two independent morphological character sets which are most unlikely to be functionally related is a powerful argument for the naturalness of the groups. Further evidence of group biological integrity is provided by limitation of host ranges of constituent species of each group to assemblages of related plant taxa demonstrated by Krall and Krall (1978) and Stone (1979) (see below: Coevolution and Phylogeny). Ferris (1979) synthesised the morphological data considered by Matthews, Mulvey, Green and Stone, included information on male spicule morphology, applied phylogenetic analysis (see below) and recognised six groups within *Heterodera* sensu lato which agree very substantially with the previously proposed groupings.

Three of the groups within *Heterodera* sensu lato have been recognised in the genera *Globodera, Punctodera* and *Cactodera. Bidera* is a contentious fourth genus which, although supported by phylogenetic arguments, currently lacks satisfactory definition as a taxon. Similar arguments apply to the remaining sub groups recognised by Ferris (1979) and Stone (1979) in *Heterodera* sensu stricto.

Group 4 of Stone (1975) based on common juvenile lip morphology is heterogenous and Stone (1979) divided it into four elements (named avenae, humuli, fici and oryzae groups) on the basis of cyst morphology and host range. Ferris (1979) also recognised this heterogeneity, placing Stone's juvenile lip-type 4 species in an avenae group corresponding exactly with the avenae group of Stone (1979) and the remainder in a group termed fici-humuli. Among this assemblage, *H. humuli, H. fici* and possibly other species have oral disc shapes subtly different from that typified by *H. avenae*, suggesting that second-stage juvenile lip shape can provide further data for phylogenetic analysis.

For convenient sub-division of the assembly of species in *Heterodera* sensu stricto without consideration of phylogenetic relationships, cyst characters alone can be used. Three groups of species are recognised by using criteria for Mulvey's (1972) group 3 (≡ avenae group), 4 (≡ schachtii group) and 5 (≡ goettingiana group) see Golden, this volume. However, divisions based on limited cyst characters work imperfectly. Sharma and Swarup (1983) attempted improvement of the group definitions, with sub-divisions of group 5 but without avoiding the need to use variably defined characters.

PHYLOGENY AND SYSTEMATICS

Coevolution and Phylogeny

Coevolution, a term coined to cover the mutual evolutionary influences of phytophagous insects and their food plants, has a variety of definitions (Futuyma and Slatkin, 1983a). The definition used here is the widely used one of Janzen (1980) meaning that, in a pair of coevolving species, a trait of one species has evolved in response to a trait in the other species, which has in turn evolved in response to a trait in the first. Such a pattern may be assumed to occur widely in interactions between parasites and their hosts. Parasite requires definition in our present context. A broad definition (e.g. Henderson and Henderson, 1979) "... an organism which at some stage of its life history makes connection with the tissues of an organism of a different species (the host) from which it derives food ..." includes all plant feeding nematodes. However, the vast majority of such forms browse on root systems, depleting cell contents and moving to feed elsewhere without restriction to a single plant individual or even species. Apart from their specialised feeding apparatus such nematodes may no more be true parasites than a cow grazing on a mixed pasture. Thompson (1982) provides a more restrictive and satisfactory definition: "A parasite in an organism that lives throughout a major period of its life in or on a single host individual, deriving its food from the host and causing lower survival or reproduction in the host." This definition restricts plant parasitic nematodes to those forms with sedentary females and other specialised types such as *Ditylenchus* and *Anguina*. Cyst nematodes show extreme adaptation to the parasitic habit.

Throughout the literature on parasite systematics the assumption is made that parasite phylogeny tracks host phylogeny. This is expressed in a number of "rules" reviewed, for example, by Mitter and Brooks (1983) of which the most important in our context is Fahrenholz' rule which states parasite phylogeny mirrors host phylogeny. However, plant nematode systematists have ignored almost entirely the arguments for derivation of parasite phylogenies from those of their host. Conversely, nematode-plant host-parasite relationships have been virtually ignored in the literature on coevolution, most evidence for reciprocal evolution between hosts and parasites coming from studies of plants and plant pathogens, plants and insects, and helminths and animals. A recent review with 23 contributors and several chapters specifically on host-parasite relationships contains not one mention of plant parasitic nematodes (Futuyma and Slatkin, 1983b). For a group like the Heteroderidae, especially the better understood genera of cyst-forming species, there is ample evidence of host specificity and sufficient knowledge of host ranges to justify attempts to construct nematode phylogenies taking into account that of their hosts. This has been done by Krall and Krall (1978) and Stone (1979) who produced phylogenies for the Heteroderidae and for cyst-forming species based on nematode morphology and host range. The congruence between cyst nematode groupings based on female/cyst characters, juvenile lip characters and host ranges has been discussed above.

Nevertheless, caution needs to be exercised in developing a phylogeny for cyst nematodes from host data. Alternatives exist to the hypothesis that parasite and hosts are associated by descent with congruent phylogenies (Fahrenholz' rule). "Colonisation" of a host group by a parasite or parasites after the hosts have undergone diversification may have occurred, and host parasite associations may even have formed by chance encounter between a pre-adapted parasite and a new host. Association by descent and colonisation are not mutually exclusive and both probably contribute to the processes which have resulted in apparently related host and parasite phylogenies. Formation of host parasite associations by chance encounter will provide no basis for congruence between host and parasite phylogenies but does provide a convenient explanation for parasites found in hosts unrelated to those of their relatives and vice versa.

An example of a host parasite interrelationship for which we can currently provide no coevolutionary explanation and hence derive no phylogenetic inference is *Globodera zelandica* parasitising New Zealand tree fuschia, a member of the angiosperm family Onagraceae (Wouts, 1984). This is the first *Globodera* species described from Australasia and the first reported parasitising this family. Other *Globoderas* are described from the New World parasitising Solanaceae, or from the Eurasian palearctic parasitising Compositae and for these association by descent can be postulated (Stone, 1979) because Solanaceae and Compositae are both placed in the highly natural dicotylendon sub-class Asteridae (Cronquist, 1968). Onagraceae are, however, considerably more distant and although Cronquist proposes a tentative link between Asteridae and the sub-class containing Onagraceae (the Rosidae) Dahlgren (1980) in a more detailed analysis separates them widely. Other examples of cyst nematodes whose hosts are best explained by chance colonisation are discussed in Stone (1979). Cases of similar cyst nematodes parasitising purportedly disparate plant taxa must be so explained unless wholesale extinction of intermediate forms parasitising the intermediate plant groups is accepted.

Although Cronquist (1968) is the most widely cited system of angiosperm systematics with a phylogenetic content and Dahlgren (1980) provides another widely accepted approach, neither is derived from objective phylogenetic analysis. Thus, although cladograms for the nematodes are available, apparently those for their hosts are not. This severely limits the use of host phylogeny in constructing or attempting to falsify heteroderid phylogenies. Henning's (1966) "parasitological rule" that parasite phylogeny can be determined from cladistic analysis of host phylogeny cannot be applied. Similarly, the most rigorous test of the hypothesis that these nematodes have indeed coevolved with their hosts (thus conforming with Fahrenholz' rule) is a comparison of nematode and host phylogenies both produced by cladistic methods (the nematode analysis excluding host data) and this cannot at present be done.

Heteroderid Systematics

Apart from innovation of host relationships, construction of heteroderid systematics must depend upon comparisons of extant forms because no fossil record of the nematodes exists. Using cladistic methods objectively derived phylogenies can be produced which are falsifiable (i.e. testable) by reinterpretation of the data on which they are based or by inclusion of new data sets. Ferris (1979, 1985) and Coomans (1983) provide accounts of cladistic methodology in a nematological context. The phylogenetic analyses of Coomans (1979), Ferris (1979, 1985) and Wouts (1985) discussed in earlier sections were produced by this method. Differences in the analyses result from differences in attributes assigned to particular taxa (e.g. whether cyst-forming or not) and whether particular characters are considered primitive (plesiomorphic) or derived (apomorphic). Construction of a cladogram requires the most parsimonious arrangement of taxa in relation to the distribution of plesiomorphic and apomorphic characters (e.g. an apomorphic character should, if possible, appear only once) and produces a hierarchy of monophyletic groups. Such analyses are widely (but not uniformly) held to have a high phylogenetic content and each point of division in the cladogram represents a speciation event (cladogenesis). The importance of cladistics is that it provides a rigorous way of producing phylogenies which are falsifiable, i.e. it is a truly scientific method. Cladistic analyses of Heteroderidae produced to date depend upon relatively few characters and it is not surprising that they differ because differing treatments of a single character can, where few characters are used, produce different results. The comparatively high level of agreement between the different phylogenies is encouraging and they are undoubtedly an advance on those produced before. The "truthfulness" and predictive value of a phylogenetic analysis depends in part upon the number of characters used. New data sets are being produced for Heteroderidae, e.g. cuticular ultrastructure (Shepherd and Clark, 1978); Baldwin, 1983: Cliff and Baldwin, 1985); feeding site type (Jones, 1981; Mundo-Ocampo and Baldwin, 1983); biochemical differentiation (numerous papers but not as yet surveying a substantial number of taxa within the family) and new information on morphology including lip structure. In my view there is no point in further phylogenetic analysis based on the data sets used by Ferris (1979) and Wouts (1985) until these can be tested against new information. When this position is achieved, with large data sets, use of computer based methods of cladogram construction will become a necessity.

The caveat about numbers of characters applies equally to numerical taxonomy, the other objective approach available to systematics and now coming to be applied to heteroderid systematics. Of course, numerical taxonomy is not a single methodology but an entire discipline within itself. Green (1971) used canonical analysis of variance to produce groupings of round cyst nematode populations based on vulval characters with results not dissimilar to those subsequently established; Roberts and Stone (1981) applied cluster and principal coordinate analyses to host range data for 35 populations of *Globodera* species, concluding that there was little phylogenetic information content; Stone and Hill (1982) used principal coordinate analysis of juvenile characters of 30 populations of species in the *Heterodera avenae* complex to demonstrate phenetic relationships within the group; Stone (1983) made use of the same method with juvenile and cyst characters from 57 populations of *Globodera* to produce groupings of species parasitising Solanaceae which resulted in a revision of their taxonomy; Fox and Atkinson (1984) analysed data from electrophoresis of proteins and enzymes of eight potato cyst-nematode populations to produce a cluster analysis of species and pathotypes which confirmed the separation of *G. pallida* from *G. rostochiensis*. None of these studies, being limited to a small number of species within a single genus, has contributed substantially to the elucidation of heteroderid phylogeny.

The two types of approach considered in this section can be argued to differ in their philosophy: the cladistic approach producing phylogenetic analyses (i.e. indicating genealogies), the numerical techniques used so far with cyst nematodes producing patterns of phenetic relationships (arrangements based on overall similarities between phenotypes, without necessary implications about evolutionary relationships or ancestry (Sneath and Sokal, 1973)). Arguments about the relative merits of different methods in systematics are unlikely to involve plant nematologists until the taxonomy of the group has advanced substantially.

COEVOLUTION, SPECIES COMPLEXES AND PATHOTYPES

Whatever the arguments for and against the role of coevolution in the phylogenetic history of heteroderids at the level of higher taxa there is powerful intuitive argument for its involvement in host parasite relations at the species level. Conventional wisdom holds that parasitic helminth evolution is slower than that of its hosts; however, anagenesis (evolutionary change of a species through time) may be rapid (Holmes 1983). The same argument can be applied to plant parasitic nematodes, where a single species may parasitise a series of phylogenetically linked hosts, with specialised host races resulting from adaptation within the species to particular hosts. Where coevolution occurs this may result in anagenic changes in both host and parasite, of which host resistance and parasite virulence are examples. Indeed, coevolutionary relationships should be closest at the local population level.

Models of host parasite coevolution are of three types (Holmes, 1983): (a) the arms race or mutual aggression model in which the parasite exploits the host to the full, a defence by the host is countered by a change in the parasite and this is in turn countered by a further change in the host ..., (b) the prudent parasite model in which damage is limited so that the parasite's substrate is less impaired, usually involving limitation of parasite reproduction, (c) the mutualism model in which host and parasite evolve together to promote each other. Of these, the arms race model predicts the appearance of plant phenotypes resistant to cyst nematodes and resistance-breaking nematode pathotypes (Stone, 1983). However, substantive damage to hosts seems generally observed only in agriculture where crops are grown to a stand while in my experience high cyst nematode population densities and evidence of damage are rarely seen in the wild. This perhaps suggests the prudent parasite model could be appropriate. A corollary of the arms race model is that gene-for-gene interactions governing resistance and virulence will evolve and this has indeed been demonstrated in the case of potato cyst-nematodes (Parrott, 1981) while most known examples of resistance to plant parasitic nematodes involve major genes (Sidhu and Webster, 1981; Stone, 1985). Barratt (1983) argued that major gene interactions are atypical products of host parasite (or pathogen) coevolution and that major genes, being more amenable to selection by plant breeders than polygenes, receive undue attention when developing hypotheses about coevolution. However, gene-for-gene interactions do provide the most parsimonious explanation for the appearance of resistance and virulence and it is not necessary to invoke such interactions only for major genes. Rapid selection for virulence in *G. pallida* against polygenic resistance in potatoes provides indirect evidence that gene-for-gene type interactions are involved in non-major gene resistance (Stone, 1985). On balance, the arms race hypothesis seems to describe best that which we observe.

Further evidence for coevolution is provided by the abundance of cyst nematode species complexes and coevolutionary arguments predict:

(i) that whenever resistance to cyst nematodes is bred into crop cultivars, resistance-breaking elements in field populations of nematode will be found

(ii) that such resistance, whether based on major or polygenes, is unlikely to be durable in the long term

(iii) that the species level taxonomy of the group must involve recognition of morphologically difficult to separate sibling species.

It is in focussing our attention on these problems that coevolutionary theory has practical value for us at the species level. Pathotypes pose an especially difficult problem, as shown by difficulties which have arisen with potato cyst-nematodes from failure to define pathotypes against identified genes for resistance. See Stone (1985) for a review of recent experience, and a discussion of pathotype definitions in the light of coevolutionary theory; also Sturhan (1985).

THE FUTURE

Significant developments in phylogenetic and phenetic analyses using cladistic methods and numerical taxonomy will come only with increased information from all aspects of heteroderid biology but a disproportionately large contribution may be expected from biochemical analyses, an until recently largely untapped source of information and one which can provide data in large quantities. Single dimension electrophoresis (including isoelectric focussing) of soluble proteins and isozymes will produce much of the information because the techniques are relatively low in cost, easily acquired and well suited to providing rapid characterisation by relatively few characters. Such methods seem effective down to about the sibling species level of genetic distance. They will become diagnostic tools which may well replace microscopy in the routine identification of taxa difficult to separate by morphology. Fig. 3 shows routine analyses of soluble proteins from 14 samples of potato-cyst nematodes. However, a warning must be sounded: differentiation of cyst nematode species or pathotypes by a small number of biochemical characters is likely to be no more dependable than using typical morphological features. The variation inherent in living organisms dictates that if only few characters are used in characterisation, inevitably individuals will be encountered in which the species typical features depart from the norm. Biochemical "finger-prints" must therefore be used with the caution normal in the use of type characters. Methods of analysis producing relatively few differences are unlikely to make major contributions to studies of heteroderid phylogeny. Here a major contribution is more likely to come from methods capable of distinguishing large numbers of chemical taxonomic characters, producing

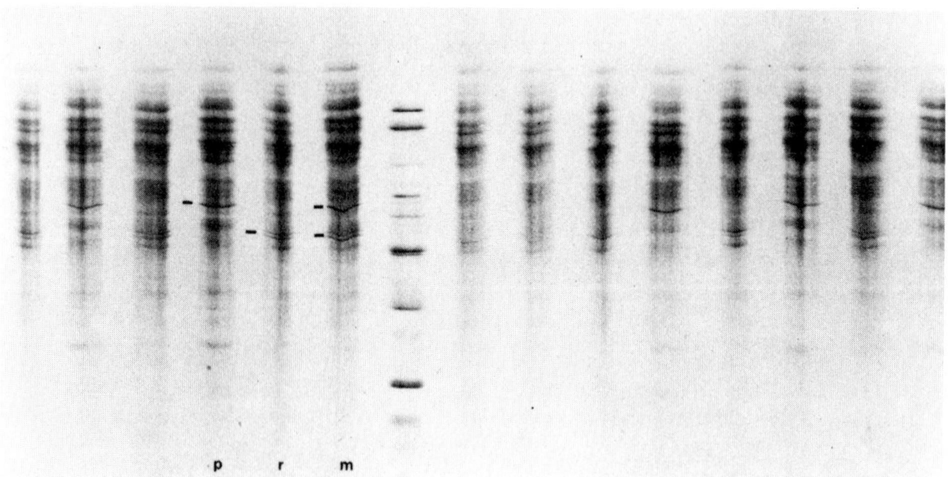

Fig. 3 Soluble protein patterns from 14 potato cyst-nematode samples from different populations, produced by isoelectric focussing of proteins from crushed cysts. The distinctly different groups of bands are pH indicators. Species typical bands (-) occur at pH 5.7 (*G. pallida*, p) and 5.9 (*G. rostochiensis*, r), m: species mixture.

the raw material for computer based analyses. Application of two-dimensional electrophoresis is already producing such data, e.g. Bakkers and Gommers (1982). Modern developments in serology and molecular biology have the potential to provide further methods of recognising components of closely similar taxa by, e.g. the application of ELISA or monoclonal antibodies. Restriction enzymes may be used to fragment DNA to produce characteristic electrophoretic fingerprints and DNA hybridisation could provide a wealth of data.

The future should see also a rationalisation of the species level taxonomy as the undoubtedly many synonymous descriptions are recognised, a process which has already begun in e.g. the *H. trifolii* complex (Maas et al., 1983) or the *Globodera* complex (Stone, 1983). Nevertheless increasing sophistication in the elucidation of species complexes will result in a splitting of what are currently recognised as good morpho-species.

Nematode taxonomy, including that of the Heteroderidae, is criticised for the too ready designation of new genera and higher taxa. But the diversity within the group cannot be ignored and must be accommodated in a systematic framework which, if it is to have predictive value, must have a rigorously derived basis. It is not so much the designation of taxa with only one or few components that is to be deplored, but the designation of taxa unjustified by a sound systematic theory. And although methods of phylogenetic or phenetic analysis may rationalise the systematics of the group, it seems likely that the already recognised diversity will require a highly structured taxonomy while increased sophistication in the detection of cryptic taxa will demand recognition of yet more.

ACKNOWLEDGEMENT

I am grateful to Dr W. M. Wouts for permission to cite extensively his paper Phylogenetic classification of the family Heteroderidae (Nematoda, Tylenchida), Wouts (1985) in press at the time of writing and to Mr P. R. Burrows for the electrophorogram reproduced in Fig. 3.

REFERENCES

Andrassy, A., 1976, "Evolution as a Basis for the Systematization of Nematodes," Pitman, London.

Bakkers, J. and Gommers, F.J., 1982, Differentiation of the potato cyst nematodes *Globodera rostochiensis* and *Globodera pallida* and of two *Globodera rostochiensis* pathotypes by means of two dimensional electrophoresis, Proc. of Koninklijke Nederl. Akad. Wetenschappen, Ser.C, 85:309.

Baldwin, J.G., 1983, Fine structure of body wall cuticle of females of *Meloidodera charis, Atalodera lonicerae* and *Sarisodera hydrophila* (Heteroderidae), J.Nematol., 15:370.

Baldwin, J.G. and Bell, A.H., 1985, *Cactodera eremica* n.sp., *Afenestrata africana* (Luc et al., 1973)n.gen., n.comb., and an emended diagnosis of *Sarisodera* Wouts and Sher, 1971 (Heteroderidae), J.Nematol., 17:187.

Barratt, J.A., 1983, Plant-fungus symbioses, in "Coevolution," D.J. Futuyma and M. Slatkin, eds., Sinauer Associates, Sunderland, MA.

Cid del Prado Vera, I., Lownsbery, B.F. and Maggenti, A.R., 1983, *Rhizonema sequoiae* n.gen., n.sp. from Coast Redwood *Sequoia sempervirens* (D. Don) Endb., J. Nematol., 15:460.

Cliff, G.M. and Baldwin, J.G., 1985, Fine structure of body wall cuticle of females of eight genera of Heteroderidae, J. Nematol., 17:282.

Coomans, A., 1979, General principles of systematics with particular reference to speciation, in "Root-knot Nematodes *(Meloidogyne* species) Systematics, Biology and Control," F. Lamberti and C.E. Taylor, eds, Academic Press, London.

Coomans, A., 1983, General principles for the phylogenetic systematics of nematodes, in "Concepts in Nematode Systematics," Systematics Association Special Volume No.22, A.R. Stone, H.M. Platt and L.F. Khalil, eds, Academic Press, London.

Cooper, B.A., 1955, A preliminary key to British species of *Heterodera* for use in soil examination, in "Soil Zoology: Proceedings of the University of Nottingham Second Easter School in Agricultural Science, 1955," D.K.M. Kevan, ed., Butterworths, London.

Cronquist, A., 1968, "The Evolution and Classification of Flowering Plants," Nelson, London.

Dahlgren, 1980, A revised system of classification of the angiosperms, Botn J. Linn. Soc., 80:91.

Evans, K., Franca, J. and de Scurrah, M.M., 1975, Distribution of species of potato cyst nematodes in South America, Nematologica, 21:365.

Ferris, V.R., 1979, Cladistic approaches in the study of soil and plant parasitic nematodes, Amer. Zool., 19:1195.

Ferris, V.R., 1985, Evolution and biogeography of cyst-forming nematodes, Bull. OEPP/EPPO, 15:123.

Fox, P.C. and Atkinson, A.J., 1984, Isoelectric focussing of general protein and specific enzymes from pathotypes of *Globodera rostochiensis* and *G. pallida,* Parasitology, 88:131.

Futuyma, D.J. and Slatkin, M., 1983a, Introduction in "Coevolution," D.J. Futuyma and M. Slatkin, eds., Sinauer Associates, Sunderland, MA.

Futuyma, D.J. and Slatkin, M., 1983b, in "Coevolution," eds., Sinauer Associates, Sunderland, MA.

Golden, A.M., 1971, Classification of the genera and the higher categories of the older Tylenchida in "Plant Parasitic Nematodes," Vol. 1., B.M. Zuckerman, W.F. Mai and R.A. Rhode, eds., Academic Press, New York.

Golden, A.M. and Raski, D.J., 1977, *Heterodera thornei* n.sp. (Nematoda: Heteroderidae) and a review of related species, J. Nematol., 9:93.

Goodey, J.B., 1959, Data to be considered, observed and, where possible, reported upon when presenting descriptions of new species, Nematologica, 4:211.

Goodey, J.B., 1963, "Soil and Freshwater Nematodes," 2nd edition, T. Goodey revised by J.B. Goodey, Methuen & Co. Ltd., London.

Green, C.D., 1971, The morphology of the terminal area of the round cyst nematodes *Heterodera rostochiensis* and allied species, Nematologica 17:34.

Green, C.D., 1975, The vulval cone and associated structures of some cyst nematodes (genus *Heterodera*), Nematologica, 21:134.

Gupta, P. and Edward, J.C., 1973, A new record of a cyst forming nematode (*Heterodera chaubattia* n.sp.) from the hills of Uttar Pradesh, Cur. Sci., 42:618.

Henderson, I.F. and Henderson, W.D., 1979, "Henderson's Dictionary of Biological Terms" 9th edition, S. Holmes, ed., Longman, London.

Henning, W., 1966, "Phylogenetic Systematics," Translated by D.D. Davis and R. Zangerl, Univ. Illinois Press, Urbana.

Holmes, J.C., 1983, Evolutionary relationships between parasitic helmiths and their hosts, in "Coevolution," D.J. Futuyma and M. Slatkin, eds., Sinauer Associates, Sunderland, MA.

Hooper, D.J., 1978, The structure and classification of nematodes, in "Plant Nematology," J.F. Southey, ed., H.M.S.O., London.

Husain, S.I., 1976, Phylogeny and inter-relationship of the superfamily Heteroderoidea (Skarbilooich, 1947) Gowen, 1971, Geobois, 3:9.

Janzen, D.H., 1980, What is coevolution?, Evolution, 34:611.

Jones, M.G.K., 1981, The development and function of plant cells modified by endoparasitic nematodes, in "Plant Parasitic Nematodes," Vol. III, B.M. Zuckerman and R.A. Rohde, eds., Academic Press, New York.

Krall, E., 1978, Compendium of cyst nematodes in the USSR, Nematologica, 23:311.

Krall, E. and Krall, H., 1978, [The revision of plant nematodes of the family Heteroderidae (Nematoda: Tylenchida) by using a comparative ecological method of studying their phylogeny] in "Printsipy i methody izucheniya vzaimootnoshenii mezhdu parazitichenskimi nematodami i rasteniyami," Akad. Nauk Estonoskoi SSR, Tartu.

Luc, M., Taylor, D.P. and Cadet, P., 1978, Description of a new tropical Heteroderidae, *Hylonema ivorense* n.gen., n.sp., and a new outlook on the family Heteroderidae (Nematoda: Tylenchida), Revue Nematol., 1:73.

Maas, P.W.T., du Bois, E. and Dede, J., 1983, Morphological and host range variation in the *Heterodera trifolii* complex, Nematologica, 28:263.

Mathews, H.J.P., 1971, Morphology of the nettle cyst nematode *Heterodera urticae* Cooper, 1955, Nematologica, 16:503.

Mitter, C. and Brooks, D.R., 1983, Phylogenetic aspects of coevolution, in "Coevolution," D.J. Futuyma and M. Slatkin, eds., Sinauer Associates, Sunderland, MA.

Mulvey, R.H., 1972, Identification of *Heterodera* cysts by terminal and cone top structures, Can. J. Zool., 50:1277.

Mulvey, R.H. and Ebsary, B.A., 1980, *Dolichodera fluvialis* n.gen., n.sp., (Nematoda: Heteroderidae) from Quebec, Canada, Can. J. Zool., 58:1697.

Mulvey, R.H. and Golden, A.M., 1983, An illustrated key to the cyst-forming genera and species of Heteroderidae in the Western Hemisphere with species morphometrics and distribution, J. Nematol., 15:1.

Mulvey, R.H. and Stone, A.R., 1976, Description of *Punctodera matadorensis* n.gen., n.sp., (Nematoda: Heteroderidae) from Saskatchewan with lists of species and generic diagnoses of *Globodera* (n. rank), *Heterodera* and *Sarisodera*, Can. J. Zool., 54:772.

Mundo-Ocampo, M. and Baldwin, J.G., 1983, Host-relationships of *Atalodera* spp. (Heteroderidae), J. Nematol., 15:234.

Parrott, D.M., 1981, Evidence for gene-for-gene relationships between resistance gene H1 from *Solanum tuberosum* ssp. *andigena* and a gene in *Globodera rostochiensis* and between H2 from *S. multidissectum* and a gene in *G. pallida*, Nematologica, 27:372.

Roberts, P.A. and Stone, A.R., 1981, Host ranges of *Globodera* species within *Solanum* subgenus *Leptostemonum*, Nematologica, 27:172.

Shagalina, L. and Krall, E., 1981, [A new genus of cyst nematodes, *Ephippiodera* n.g. and redescription of *Heterodera turcomanica* Kir'janova and Shagalina, 1965 (Nematoda: Heteroderidae),] Eesti NSV Teaduste Akad. Toimetised, Biol. Seer., 30:182.

Sharma, S.B. and Swarup, G., 1983, Identification scheme for *Heterodera* species in India, Indian J. Nematol., 13:171.

Shepherd, A.M. and Clark, S.A., 1978, Cuticle structure and 'cement' formation at the anterior end of female cyst-nematodes of the genera *Heterodera* and *Globodera* (Heteroderidae: Tylenchida), Nematologica, 24:201.

Sidhu, G.S. and Webster, J.M., 1981, The genetics of plant-nematode parasite systems, Bot. Rev., 47:387.

Sneath, P.H.A. and Sokal, R.R., 1973, "Numerical Taxonomy," W.H. Freeman, San Francisco.

Stone, A.R. and Hill. A.J., 1982, Some problems posed by the *Heterodera avenae* complex, Bull. OEPP/EPPO, 12:317.

Stone, A.R. and Rowe, J.A., 1977, *Heterodera humuli*, C.I.H. Descriptions of Plant Parasitic Nematodes Set 7, No. 105, 4pp.

Stone, A.R., 1975, Head morphology of second-stage juveniles of some Heteroderidae (Nematoda: Tylenchoidea), Nematologica, 21:81.

Stone, A.R., 1978, Recent developments and some problems in the taxonomy of cyst-nematodes, with a classification of the Heteroderoidea, Nematologica, 23:273.

Stone, A.R., 1979, Coevolution of nematodes and plants, in "Parasites as Plant Taxonomists," I. Hedberg, ed., Acta Universitatis Upsaliensis, Symbolae Botanicae Upsaliensis xxii, Uppsala.

Stone, A.R., 1983, Three approaches to the status of a species complex, with a revision of some species of *Globodera* (Nematoda: Heteroderidae) in "Concepts in Nematode Systematics," Systematics Association Special Volume No. 22, A.R. Stone, H.M. Platt and L.F. Khalil, eds., Academic Press, London.

Stone, A.R., 1984, Changing approaches in nematode taxonomy, Plant Dis., Vol. 68:551.

Stone, A.R., 1985, Coevolution of potato cyst nematodes and their hosts: implications for pathotypes and resistance, Bull. OEPP/EPPO, 15:131.

Sturhan, D., 1985, Species, subspecies, race and pathotype problems in nematodes, Bull. OEPP/EPPO, 15:139.

Thompson, J.N., 1982, "Interaction and Coevolution," Wiley, New York.

Wouts, W.M., 1973, A revision of the family Heteroderidae (Nematoda: Tylenchoidea). I. The family Heteroderidae and its subfamilies, Nematologica, 18:439.

Wouts, W.M., 1984, *Globodera zelandica* n.sp. (Nematoda: Heteroderidae) from New Zealand, with a key to the species of the genus *Globodera*, N.Z.J. of Zool., 11:129.

Wouts, W.M., 1985, Phylogenetic classification of the family Heteroderidae (Nematoda: Tylenchida), Syst. Parasitol., 8:2950.

DISCUSSION

Sikora referred to the situation in which certain genera were not accepted as valid by some taxonomists and suggested that there might be period meetings of those involved to arrive at a consensus of opinion and to set standards. Golden considered that such meetings would suggest an imposed censorship and that communication leads to agreement. Barnes suggested there was considerable utility in recognising variation in several populations and Stone agreed that broad analyses were of value. Wyss asked why the heteroderids induced the same transfer cell system in their host plants. Stone said this observation was true for a restricted group only but if the system proved to be successful there might be no selection pressure to change it. Stone agreed with Trudgill that environmental sex determination is an example of a prudent parasite.

MORPHOLOGY AND IDENTIFICATION OF CYST NEMATODES

A. Morgan Golden

Nematology Laboratory, PPI, ARS, US Department of Agriculture
Beltsville Agricultural Research Center, Beltsville, MD 20705

INTRODUCTION

Cyst nematodes represent a large group of plant parasites comprising 75 or more species. These species are presently included in six genera of the family Heteroderidae: *Heterodera, Globodera, Cactodera, Punctodera, Dolichodera,* and *Afenestrata*. In addition, *Sarisodera* was described and recognised as a cyst genus (Wouts and Sher, 1971) but Baldwin and Bell (1985) recently determined that cysts were not formed by its type species and placed the genus with related non-cyst forms. Whatever their origin, host, or economic status, accurate identification of cyst nematodes is of paramount importance for any decision making process. They may be revealed as a known species, an undescribed taxon, or possibly some species which is still so poorly described as to make identification impossible; but the key to their identity is their morphology.

In nematology, morphology has traditionally been the basis for identification and systematics. For a long time the light microscope provided the only means for detailed morphological observations; but in recent years, the scanning and transmission electron microscopes (SEM and TEM, respectively) have been utilised for ultrastructural examination; SEM has been especially useful in showing surface details and providing a sounder basis for identification. Generally, morphology is not useful or reliable for identification of races, pathotypes, biotypes or other populations below the species or subspecies level where biochemical and other approaches might be more appropriate.

In cyst nematodes, several life stages are important and useful in both morphology and identification. The cyst and its enclosed juveniles and eggs are of the greatest value and are the most widely used. Three factors perhaps account for this: first, the cyst in its entirety has been studied most over the years; second, the cyst with its contents is persistent, remaining viable and in relatively good condition for several years; and third, the cyst is the stage most commonly found in surveys, soil samples, and associated with infected plants. The male and female are more difficult to obtain than cysts and have been utilised far less in morphological studies and identification procedures. However, each has structures of value, and there is an increasing need and interest to develop better knowledge and understanding of the morphology of the adult stage of both sexes in cyst nematodes.

CYSTS

Shape

All cysts have an anterior protuberance representing the neck of the female; sometimes the head portion or more might be broken off but there is still a protuberance anteriorly on the cysts. A majority of cyst nematodes also have a posterior protuberance, the vulval cone. Cysts of the species of *Heterodera, Cactodera,* and *Afenestrata* are characteristically lemon shaped although there is much variation in exact shape and symmetry. Cysts of species of *Globodera* and *Punctodera* are basically round, without a posterior protuberance, but the exact shape may be spherical, ovoid or pear-shaped according to species. The elongate-ovoid cyst, without a posterior protuberance, is represented by the monotypic genus *Dolichodera.* Cyst shapes are well illustrated by Mulvey (1972) and Mulvey and Golden (1983).

Cyst shape is useful in placing a species in one of the above major groups and the shape should always be noted before preparing the specimens for study. However, cyst shape alone is not generally a reliable character for species identification. In reviewing reports of several workers, Hesling (1978) noted that "no lemon-shaped cyst can be positively identified by its shape alone ...". He also mentioned that efforts by some workers to measure cyst shape differences in certain species met with only limited success and were highly laborious. For highly experienced workers, cyst shape can be interpreted as indicative, though not definitive, for certain species. For examples, the heart-shaped cysts of *Heterodera cardiolata* Kirjanova & Krall, 1969 (and the species name itself) suggest uniqueness but half of the six cysts figured in the description are typical lemon shape; also *Cactodera estonica* (Kirjanova & Krall, 1963) Krall & Krall, 1978 is distinct from other known species of the genus by having elongate cysts with a length/width ratio of 2.3, compared with a ratio of 1.5-1.7 for the other species. Nevertheless, it is preferable to further examine the cysts and contents for final identification because of the ever present possibility of a related but undescribed form.

Size

This gross morphological parameter is properly included in species description to define the length, width, and length/width ratio of the cysts. The total length of the cysts varies from 210 µm, the minimum size in *H. carotae* Jones, 1950, to 1178 µm, the maximum size in *C. estonica.* The average length ranges from 485 µm for *H. cardiolata* Kirjanova & Ivanova, 1969 (smallest); about 500 µm for *H. cruciferae* Franklin, 1945 and *H. zeae* Koshy, Swarup & Sethi, 1971; 710 µm for *H. avenae* Wollenweber, 1924; to 922 µm for *C. estonica* (longest). Cyst widith varies in a similar manner. Also, variation in size within a species is generally great. For example, in measurements of 50 cysts of *G. rostochiensis* (Wollenweber, 1923) Behrens, 1975, Golden and Ellington (1972) reported an average length of 680 µm (450-990) and a width of 540 µm (250-810). In species with quite large cysts, such as *H. avenae* and *H. trifolii* Goffart, 1932, small cysts are present also. Thus, in general, cyst size alone would not be useful and reliable for species identification although with caution it might be helpful to experienced workers in separating certain species at the upper and lower ends of the size range.

Colour

Mature cysts range in colour from light tan, brown, dark brown, reddish brown to dark reddish brown or almost black. Cysts of some species, such as *H. zeae*, are lighter than those of other species and this characteristic can be helpful in delimiting populations of such species. However, it should be used with caution because of the inherent variation of colour within a species and because other factors can influence colour, including age, size, content and condition of the cysts (see Hesling (1978) for further comments). Also, the subcrystalline layer (described below) might mask the colour of new cysts.

Subcrystalline Layer (SCL)

This structure forms on the outer surface of females and can generally be found on new cysts. It probably occurs in most, if not all of the lemon-shaped cyst species (Brown et al., 1971). As pointed out by Hesling (1978), it varies from a short-lived white bloom to a thick, chalk-like deposit as can be seen on young *H. avenae* cysts. In a detailed study of four *Heterodera* species (*avenae, iri, trifolii* and *latipons*), Brown et al. (1971) found the SCL to consist of two layers: a thin, inner layer pressed to and replicating the cuticle pattern; and a thicker outer layer fractured into pyramidal blocks giving an irregular polygonal appearance. These layers were found to be composed of a fatty acid (tetracosanoic and derivatives) and a calcium salt (calcium tetracosanate), the latter being the main constituent of the outer layer. The fatty acid seems to be produced by an unidentified fungus that feeds upon and metabolises products excreted by the nematode. Brown et al. (1971) suggest that there is a symbiotic relationship between the fungus and nematode in which the fungus, in feeding, converts "copious sugary waste" from the nematode to inert fatty acid, preventing pollution of the immediate rhizosphere; and they also suggest that the layers may give protection against pathogens and predators. However, Zunke (1986) reports that *H. schachtii* grown in aseptic culture developed SCL, indicating that a fungus is not necessary for its presence. Because of its variability and brief existence, the SCL cannot at present be regarded as of diagnostic value. However, once it is known for certain which species have the SCL, and its origin, nature and chemical composition for each species have been established, then it may prove to be of significant value in identification and indicating relationships of species.

Cyst Wall Structure

The cyst wall is variable in thickness, and consists of two basic layers, the exo-and endo-cuticles (Ferris and Siegel, 1957); Weiser, 1953). These are now recognised to consist of four major layers (A, B, C, D) which are further divided into several sublayers as seen with transmission electron microscopy (TEM) (Shepherd et al., 1972; Cliff and Baldwin, 1985). With TEM, Ferris and Siegel (1957) earlier had recognised and assigned numbers to five "distinct bands" in the endocuticle of *G. rostochiensis*. In their comprehensive studies Shepherd et al. (1972) found differences in the layering of the cuticle of females of several cyst species. Of particular significance were their findings that round cyst species, including *G. rostochiensis*, possess a D layer but the several lemon-shaped species examined did not have this layer. This supported the validity of *Globodera* (Skarbilovich, 1959) Behrens, 1975 as a separate genus (Behrens, 1975); and it also contributed justification for the genus *Punctodera* Mulvey & Stone, 1976 since *P. punctata* (Thorne, 1928)

Mulvey & Stone, 1976 was found to have a D layer, although smaller than in the *Globodera* species. Cliff and Baldwin (1985) have shown the presence of the D layer in *Cactodera* Krall & Krall, 1978; and Baldwin (1983) described the fine structural details of the body wall of three cyst-related genera and discussed its phylogenetic significance. Thus, an understanding of the structure and nature of the cyst wall can provide deeper and more accurate insight into relationships of species and genera; but until comparative details are available on many species, the structure of the cyst wall will not be of practical value in routine identification of individual cyst species.

Cyst Wall Pattern

Surface markings on the cyst have been noted by several workers. In describing *Punctodera punctata*, Thorne (1928) stated with reference to the female that the "cuticle is distinctly striated and these striae are resolvable into rows of dot-like marking that have the appearance of being minute pores". Franklin (1939) described and illustrated "pits or dents" (punctations) in several cyst species which formed two groups: one being round cyst forms (now known to have been *G. rostochiensis* and *P. punctata*) with rows of punctations encircling the body; the other being several lemon shaped cyst species with irregularly arranged punctation (*H. schachtii*, *H. avenae*). She also discussed the "ridges" on the surface which formed a network and clearly illustrated a "reticulate pattern" of which is now *H. avenae*. In his description of *Cactodera weissi* (Steiner, 1949) Krall & Krall, 1978, Steiner (1949) briefly but concisely described the nature of the wall pattern; and Cobb and Taylor (1953) described the external cyst surface of *G. leptonepia* (Cobb & Taylor, 1953) Behrens, 1975, and noted a lower layer which had punctation in rows, later confirmed and further illustrated by Golden and Ellington (1972). Taylor (1957) was one of the first to group cysts on the basis of the outer wall pattern. Hesling (1978) discusses the efforts of various workers to determine how the outer wall pattern is derived, although this still remains unclear.

Earlier workers faced the problem of lack of details about the cyst wall pattern for individual species and also the fact that a known species might consist of one or more other species, such as was the case with *H. schachtii*. In more recent years species have been described in greater detail, the cysts have been better characterised, and more genera established to accommodate groups of related species. Thus, it now seems evident that the species of *Heterodera* have a wall pattern ranging from zig-zag to network (or reticulate) and varying from fine to coarse, depending upon the species. The pattern is best observed in the mid-portion of the cyst because the anterior and posterior portions commonly have modified or even different patterns. Punctation is often present, arranged in an irregular manner, and varies from inconspicuous to prominent as can be seen in *H. avenae*. Cysts of the monotypic genus *Afenestrata* Baldwin & Bell, 1985 have a wall pattern similar to *Heterodera* and with punctation irregularly arranged. The cyst wall pattern of *Cactodera* species is different from these genera and was characterised by Golden and Raski (1977) as "... pattern at midbody consisting of basically straight to wavy lines extending around cyst at right angles to long axis of cysts and frequently broken by short oblique or vertical lines ...". Punctation is commonly present, randomly arranged, and may be fine to coarse. Another type of wall pattern is seen in the monotypic genus *Dolichodera* Mulvey and Ebsary, 1980 with the cyst encircled latitudinally by fine, closely spaced, irregular striae; punctation is evidently absent. In *Globodera* the wall pattern varies from wavy lines to lacelike especially in midportion, and as rugose in the terminal area in some species. In *G. rostochiensis*, and perhaps some other species, the pattern in midportion extends latitudinally around the cyst as illustrated by Golden and Ellington (1972). However, in *G. leptonepia* (Cobb & Taylor, 1953) Behrens 1975 and *G. achilleae* (Golden & Klindic, 1973) Behrens, 1975 wavy lines extend longitudinally from the neck to the terminal area and these lines overlay rows of distinct punctation extending latitudinally around the body. Punctation generally is present in other species of the genus, is rather fine, and arranged irregularly (except in *tabacum* where it is in rows). The wall pattern of *Punctodera* species varies from wavy lines to lacelike, and extends longitudinally on much of the cyst. In the three known species of the genus, punctation is distinct, arranged in rows, and extends latitudinally around the cyst.

The cyst wall pattern has been of value in the grouping of species, as a generic characteristic, and to a limited extent in species identification (e.g. Stone (1973), Mulvey and Golden (1983)). However, more detailed information is needed on the wall pattern of cysts for each species. Also, a more precise and descriptive terminology is needed to better characterise the various kinds of patterns which have been observed. Such information might well give more insight into relationships between species or groups and provide a useful tool in identification.

Cyst Posterior

The nature of the posterior region of cysts is important in both classification and identification of cyst species. This region encompasses the portion of cyst from the anus to its terminus and comprises a complex of external and internal structures. It seems that Franklin (1939) was the first to examine end-on the posterior regions of cysts. She examined lemon shaped cysts from several hosts, representing what is now a number of different cyst species; she noted the vulval and anal openings, and "knob-like projections on the inside of the cyst wall" (now designated bullae). In round cyst species she differentiated the "potato strain" with "one posterior aperture" from a parasite on grass (probably *P. punctata*) having two apertures. In 1951, Franklin published photomicrographs of the posterior region of *G. rostochiensis* and *P. punctata*. In a conference discussion (Anon, 1952) Oostenbrink stated that the "vulval aperture" in *H. avenae*, being double, was useful in separating this species from *H. schachtii* and some other species, which have a "single" aperture. In 1952 Taylor (personal communication) was already using the knob-like structures (bullae) as diagnostic features in some species. Cobb and Taylor (1953) were apparently the first to include in the original description of a cyst nematode details and end on view of the posterior region. Oostenbrink and den Ouden (1954) gave data and illustrations of taxonomic value of the cone top of four cyst species. Cooper (1955) published details of cyst posterior regions which have served as a basis for subsequent studies and use in species descriptions. Hesling (1978) has more recently given an extensive account of the posterior region of the cyst nematodes.

Vulval Cone (Fig. 1)

Lemon-shaped cysts (*Heterodera, Cactodera*, and *Afenestrata*) have a protuberance or prominence at the posterior end commonly called the vulval cone. The size and shape of the cone is variable, ranging from prominent and rather narrow in *H. schachtii* and *H. trifolii* Goffart, 1932, short and nipple-like in several *Cactodera* species, to insignificant and button-like in *H. betulae* Hirschmann & Riggs, 1969. This last species as well as

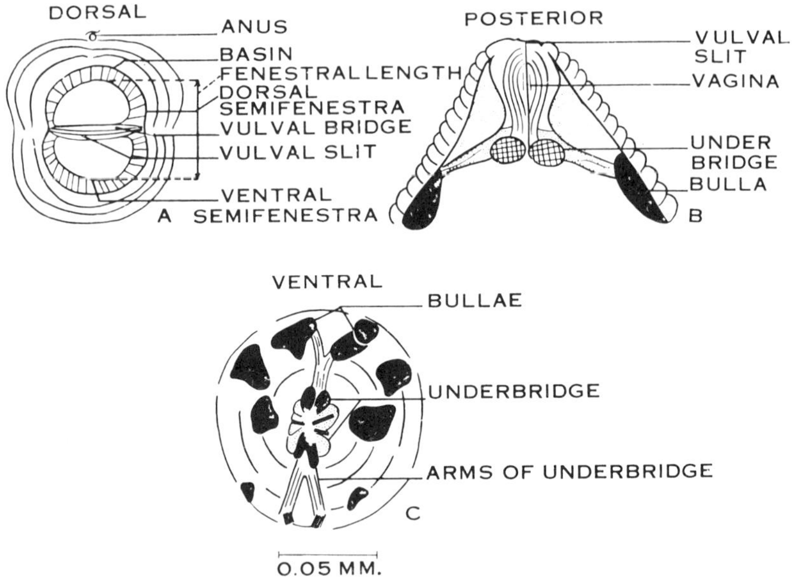

Fig. 1 Drawings of *Heterodera trifolii* cyst vulval cone, indicating various structures. A) Posterior view of cone top. B) Side view (dorso-ventral aspect) of cone. C) Anterior view of cone at underbridge level. (From Mulvey, 1959. Courtesy E. J. Brill, Leiden).

specimens of certain other species can sometimes be misidentified as not having a vulval cone unless the cysts are examined carefully and in appropriate position.

At the terminus of the cone is located the vulval slit which is bordered by two vulval lips, forming the vulval bridge; the bridge extends across the fenestra (the thin-walled area of the cone terminus) and divides the fenestra into two semifenestrae. However, in the monotypic genus *Afenestrata*, fenestrae are not present; and in cysts of *H. betulae* and *Cactodera* species, the fenestra disintegrates and leaves a round opening, thus forming a circumfenestrate cyst. In all *Heterodera* species (except *H. betulae*), there are two types of fenestration: ambifenestrate and bifenstrate. In the first, the semifenstrae on each side of the vulval slit and bridge disintegrate, leaving two openings each of about one half circle or less, separated by a narrow to medium vulval bridge. In certain species, e.g. *H. carotae* and in some old cysts, the vulval bridge can disappear along with the semifenestrae, giving a false appearance of a circumfenestrate cyst. In the second type (bifenestrate) the semifenestrae also disintegrate, leaving two openings each a full or more than half circle on each side of a strong, wide vulval bridge.

On cysts of many *Heterodera* species the fenestra is encircled by a band of cuticle which extends to the cuticular surface of the cyst and was designated by Cooper (1955) as the basin. He noted lines or wrinkles on the basin which he called rivulets. The basin is especially distinct in some species, e.g. *H. glycines* Ichinohe, 1952, and in *H. geottingiana*

Liebscher, 1982 it consists of a brickwork-like pattern. In some species, e.g. *H. schachtii*, the vagina or its remnants persist in the vulval cone and ridges or folds give it a striated appearance, like a sheaf of grain, and because of this Taylor (1957) called it "sheaf-like organ". Proximally, the vagina is thought to thicken to form the "basal bulb" (Cooper, 1955) or "cervix" (Thorne, 1961) which although initially hollow, later becomes solid forming two arms extending across the vulval cone and becoming linked to the cyst wall; this is the underbridge. The arms lie beneath and parallel to the vulval bridge and their ends are forked or furcated. The origin, nature and function of the underbridge have not been fully elucidated although different workers have given views about it (Cooper, 1955; Thorne, 1961; Green, 1975). In young cysts of *H. trifolii*, Mulvey (1959) reported another bridge-like structure (Mulvey's bridge) located just below the level of the fenestra and at right angles to the underbridge; he suggested this might be hardened muscle tissue which in the white female controlled the opening or closing of the vagina. Thorne (1961) described and illustrated in the vulval cone of *H. avenae* a spheroid vaginal chamber and associated complex musculature; he suggested the chamber forced the bullae into the terminal area of the cone but the structure is not detectable in older cysts of this species. Bullae (Cooper, 1955) are prominent, single or branched knob-like or finger-like structures attached to the cyst wall in the vulval cone. They may be clustered in the distal end of the cone, as in *H. avenae*, or rather scattered in the vicinity of the underbridge, as in *H. schachtii*. In general, species with a strong underbridge have conspicuous bullae while those with a weak (or absent) underbridge, have only rare or inconspicuous bullae. Some species have no bullae. In a redescription of *H. zeae* Koshy, Swarup & Sethi, 1971), Golden and Mulvey (1983) reported four conspicuous finger-like bullae always in a distinct arrangement immediately below (anteriad) the underbridge, followed by a number of randomised bullae. These four bullae were not observed in any of the many other *Heterodera* species examined (Mulvey and Golden, 1983). Vulval denticles were named and described by Golden and Raski (1977) as "small tooth-like structures seen below the fenestral surface and within the vulval cone". They are attached to the cone wall, may occur in small clusters or as a partial ring, and were seen in the five *Cactodera* species examined and in *H. schachtii*. Externally and located on the dorsal surface of the vulval cone is the anus; it may be small and inconspicuous to fairly prominent, and in some species, e.g. *H. cyperi* Golden, Rau & Cobb, 1962, it is surrounded by a perianal pattern. A structure often seen attached to or surrounding the vulval cone of females and young cysts is the egg sac or gelatinous matrix. This is a jelly-like substance apparently secreted through the vulva but its origin in cyst nematodes is not known. In those species with a long vulval slit, e.g. *H. schachtii*, the sac may contain a few to a large number of eggs; in species with a very short vulval slit, e.g. *H. avenae*, the sac may be present but without eggs. The sac size is highly variable and is not yet of proven diagnostic value. As pointed out by Hesling (1978), this structure probably occurs in most, possibly all, cyst nematode species and also is known in several other groups of nematodes, including *Meloidogyne* Goeldi, 1887.

The above described structures of the vulval cone have much value in species identification although some may be of greater use than others. The nature of fenestration (ambi-, bi-, circumfenstrate or absent) can group species and serve as a generic basis (*Heterodera* vs *Cactodera*); the size and shape of the semifenestrae can separate some species. The length of the vulva slit is very useful, ranging in *Heterodera* from 6 µm

H. betulae) to 70 µm *(H. salixophila* Kirjanova, 1969), from 13 µm to 25 µm in *Cactodera*, and with the slit in individual species being within certain limits. Width and length of the vulval bridge are diagnostic for certain species, e.g. *H. latipons* Franklin, 1969. Presence or absence of an underbridge, and its length and structure if present, are commonly used in identification as are the nature and location of bullae if present. Presence or absence of the vaginal sheath and perianal pattern are also useful diagnostic structures. Vulval denticles were recently used by Mulvey and Golden (1983) in a key for species identification. No doubt a fuller understanding of the vulval cone and its structures would not only contribute to improved identification of these cyst species but also to their systematics.

Terminal Region (Fig. 2)

This term is used here to refer to the vulval-anal area and associated structures of all the round cyst species (*Globodera, Punctodera* and *Dolichodera*) and is less restricted than the "terminal area" as defined by Green (1971). All of these species are rounded posteriorly,

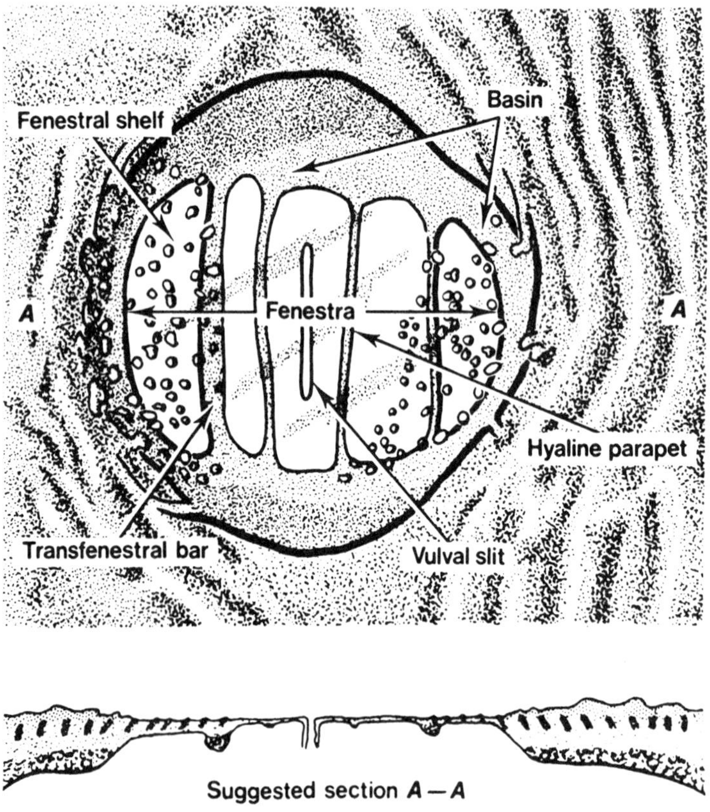

Fig. 2 Diagram of plan and cross-section A-A of vulval region of *Globodera pallida*. (From Hesling and Ellis, 1974. Courtesy E. J. Brill, Leiden).

circumfenestrate, and have a vulval fenestra and small anal pore, except for *Punctodera* species which have an anal fenestra about the same size as the vulval fenestra. Some of the structures discussed below are best seen in females or young cysts as they would not be present in older cysts after fenestration.

The vulva is the female genital opening at the external surface and its immediate margin of cuticle, the lips. In *Globodera* species the vulval slit is 4 µm to 11 µm in length, being fairly constant for a given species and therefore of diagnostic value (Mulvey, 1973); in *Punctodera* species it averages 4 µm, and has not been observed in the monotypic genus *Dolichodera*. The fenestra is the central, thin, translucent sheet of tissue with the vulva in its centre; fenestrae may be circular or oval and differ in size and shape, which are of diagnostic use. The fenestral margin is the junction of the fenestra and the cyst wall and can be readily delimited using transmitted light. Adjacent to and outside the dorsal and ventral margins of the fenestra of some species are regions containing external oval or rounded protuberances; these regions were designated as anterior and posterior vulval crescents by (Green 1971), and as tubercular crescents by Mulvey (1973); the protuberances, approximately 1 µm in width, were named vulval papillae by the former author and perineal tubercles by the latter. Mulvey (1973) suggested that the tubercles (or papillae) are knob-like modifications of the cuticular ridges (of the cyst surface) and Hesling (1978) discussed this further and agreed with Mulvey, although he preferred the term "vulval papillae". These structures also are useful in identification. Fenestral shelves (Hesling and Ellis, 1974) are two punctate layers of tissue, each on the outer area of the fenestra and are bounded on the inner side by the transfenestral bar. The shelves and bars are especially prominent on *G. pallida* (Stone, 1973) Behrens, 1975 and appear as "D"-shaped shelves of tissue (Hesling and Ellis, 1974). Near the vulva and parallel to it are two very slender bars called hyaline parapets which resemble structures of the same name associated with the vulval bridge of lemon-shaped species as described by Cooper (1955). Cysts of some species have a circumfenestral zone or basin (Cooper, 1955) which consists of less pigmented, more translucent tissue than the cyst wall and which may wholly or partially encircle the fenestral area. As defined by Green (1975), the perineum, or perineal area, is the surface between the vulva and anus; and its length and number of ridges in the cyst pattern on the perineum are useful in identification of certain round cyst species. Also, the ridges and furrows of the cyst surface on the perineum form a perineal pattern of two basic types: a maze-like or reticulate pattern; and a pattern of parallel ridges. These patterns too are of diagnostic value (Green, 1971; Hesling and Ellis, 1974). Granek's ratio (Granek, 1955), as redefined by Hesling (1973), is derived by dividing the distance from the anus to the nearest edge of the fenestra by the length of the fenestra. This ratio is of much value in separating certain species where there are rather wide differences in the mean ratio for each species, e.g. *G. rostochiensis* (4.5) (Golden and Ellington, 1972) and *G. achilleae* (1.6) (Golden and Klindic, 1973); but in some other species there is considerable overlap where the mean ratios are close. Bullae have been reported in two of the three species of *Punctodera* and the single species of *Dolichodera*, but perineal tubercles have not been observed in species of these two genera (Mulvey and Stone, 1976; Stone et al., 1976; Mulvey and Ebsary, 1980).

FEMALES

Many of the morphological details of females, including shape, size, and posterior region, are essentially similar to those discussed for cysts. However, some additional comments on certain facets of females seem appropriate.

As the female matures and changes to a cyst, the colour changes from white through yellow to brown in some species but some others do not have a distinct yellow phase. These sequences were designated as chromogenesis and used in differentiating potato cyst nematodes (Guile, 1966, 1967, 1970). Earlier, Ellenby (1946) had shown this tanning (or browning) and hardening process to be due to the action of polyphenol oxidase in the body wall; further comments on this are given by Shepherd et al. (1972) and Hesling (1978).

Vulval bodies were first named and described in *G. rostochiensis* and three other species by Wilson (1968), the name being given to "endo-cuticular bodies" closely associated with the vulva itself and located in the hypodermis. Similar structures were observed and described in *G. rostochiensis* females by Golden and Ellington (1972)who suggested that the vulval bodies might in time have some taxonomic value. In a study of the female posterior region of *G. rostochiensis* and three related species, Mulvey (1973) reported vulval bodies present in the four species and considered them to be of diagnostic use. The function of vulval bodies is unknown.

Length of stylet can be utilised in differentiation of species, e.g. Golden and Ellington (1972) found the stylet length in *G. rostochiensis* to average 23.0 µm and Stone (1973) reported the length of *G. pallida* as 27.4 µm. The width and shape of stylet knobs and the dorsal gland outlet are other structures which should be compared between species; and the nature of the head might prove to be of value when specimens of many different species have been thoroughly examined, especially en face with SEM.

MALES

In general, males have been used relatively little in systematics and identification of cyst nematodes. Among the reasons for this are two important contributing factors: the absence of males in some species, and the difficulty of obtaining males in the common procedures used in soil samplings and surveys. Nevertheless, there are some male structures which are of morphological value.

<u>Body length</u> - usually variable within a species but the means for some species differ significantly.

<u>Head</u> - offset or not; number of annules; size and nature of labial disc and labial region.

<u>Stylet</u> - length; size and shape of stylet knobs.

<u>Dorsal gland opening</u> - distance from base of stylet.

<u>Lateral field</u> - width; size and centre band; presence or absence of areolation.

<u>Spicules</u> - length; nature of tips (rounded, dentate) (note Clark et al., 1973).

Tail - length and shape.

Details of these and other structures of males are now generally given in descriptions of new species. However, observations with SEM, especially en face views, are rather limited and probably would be of systematic and diagnostic value once many species have been examined in such manner.

SECOND-STAGE JUVENILES (J2's)

Juveniles are second only to cysts in their importance in identification, and perhaps in systematics also. Many morphological features have been studied in juveniles and are usually included in recent and current descriptions. Unfortunately, for many species of cyst nematodes details of juveniles are often missing from descriptions. The stylet is perhaps the most stable structure, varying very little within a species if measured accurately, and a difference in length of as little as 2 µm between two species can be of diagnostic value (Mulvey and Golden, 1983). The number of incisures in the lateral field is of particular use as is the length of the tail and terminus. Other structures in juveniles and other stages that are of importance are mentioned by Goodey (1959). However, variations occur, as pointed out by Hesling (1978) and experience is needed to evaluate the variation in juvenile features.

Length - rather variable within a species but the mean is of diagnostic value for a number of species.

Width (at greatest diameter) - avoid measuring specimens which are flattened.

Ratio "a" - length divided by width.

Head - offset or not; number of annules; height, width and their ratios; size and nature of labial region (en face and especially with SEM); amount of sclerotisation of cephalic framework.

Stylet - length and degree of stoutness; size and shape of stylet knobs; nature of "stylet guide".

Dorsal gland outlet - distance from base of stylet.

Excretory pore and hemizonid - distance from anterior and relation to each other.

Median bulb valve - distance from its centre to anterior end.

Genital primordium - number of cells and location from posterior end.

Lateral field - width at midbody; number of incisures; presence or absence of areolation; its nature on the tail.

Phasmids - size and location.

Tail - length (from anus to tip); shape; width at anus.

Tail terminus (hyaline region) - length; shape; nature of its tip (rounded, acute, etc.); caudal ratio A (length of tail terminus divided by its width at its beginning).

These, and other structures also, will probably be of increasing diagnostic and systematic value in the future, particularly with the increased use of SEM and TEM. In a study of the juvenile head morphology of 37 cyst species using SEM, Stone (1975) placed the species in six groups based on lip patterns. His groupings generally accorded with the subdivisions of cyst nematodes at that time, demonstrating the value of SEM en face examinations.

EGGS

Eggs may be recovered from the egg sac of many species and from cysts on roots or in soil of all species. They are variable in size, shape, and in length/width ratio and this variation occurs within and between species. As pointed out by Hesling (1978) egg shape is unreliable in species identification but the morphometrics might be useful in some species as supplemental or confirmatory characters. The egg shell in most species is hyaline and without visible markings; however, three *Cactodera* species are known to have egg shells with punctations. These markings were useful in diagnosis of two species by Golden and Raski (1977) and recently Baldwin and Bell (1985) described a third species with punctations on the egg shell. Perhaps additional species will be found with similar, or even different egg shell markings through critical examination of eggs by both light microscopy and SEM.

IDENTIFICATION

Morphology is the essential basis for identification of cyst nematodes. However, new and different approaches are being explored for differentiating forms which are not amenable to morphological distinction, such as races, pathotypes and other subspecific populations. Hesling (1978) discussed possible roles of cytology, host ranges, and serological and electrophoretic methods in distinguishing cyst nematode forms. Hussey (1979) presented an extensive review of gel electrophoresis and serology as applied to differentiation of cyst and other plant and free-living forms, and suggested that a biochemical approach has potential for assisting in identification and characterisation of these forms as well as establishing phylogenetic relationships. Fox and Atkinson (1984) further discussed electrophoresis in nematode taxonomy; Curran (1984) presented details on population identification by biochemical analysis of genomic DNA; and McClure and Lucas (1984) discussed the use and specificity of monoclonal antibodies in nematology. Thus, biochemical approaches in taxonomy will no doubt be far more widely used in the future and will supplement and extend our morphological knowledge and provide new understanding of the identity and systematics of population and species.

In preparing the key given below, as well as the general text, type or other specimens of most species were examined; original descriptons of many or most were reviewed; and extensive literature was utilised. The papers by Mulvey (1972) and Mulvey and Golden (1983) were particularly useful and segments of the latter are incorporated into this key. The present key encompasses 59 species and the six cyst nematode genera (*Globodera, Puntodera, Dolichodera, Heterodera, Cactodera* and *Afenestrata)*. The genus *Ephippiodera* Shagalina & Krall, 1981 was established to accommodate two previous *Heterodera* species, *H. turcomanica* Kirjanova & Shagalina, 1965 and *H. latipons* Franklin, 1969. It seems preferable to recognise these two species as belonging in the genus *Heterodera* and it is proposed that *Ephippiodera* Shagalina & Krall, 1981 be placed as a new synonym under the genus *Heterodera* Schmidt, 1871.

KEY TO 6 GENERA AND 59 SPECIES OF CYST FORMING NEMATODES (*HETERODERIDAE*)
(Based mainly on cysts and second-stage juveniles)

1. Cysts (and females) without posterior protuberance 2
   Cysts (and females) with distinct posterior protuberance 11

2. Vulval fenestra large; anal area small and without fenestration; cysts spherical, subspherical or elongate ovoid 5
   Vulval and anal fenestra large, both of equal size; cysts ovoid or pear-shaped .... *PUNCTODERA* Mulvey & Stone, 1976, 3

3. Cysts pear-shaped; bullae absent; hosts certain grasses and wheat, oats, barley .... *P. punctata* (Thorne, 1928) Mulvey & Stone, 1976
   Cysts spherical, subspherical or ovoid; bullae usually present 4

4. Massive bullae between vulval and anal fenestra; second-stage juveniles with oesophageal gland lobe extending to about 50% of body length; stylet knobs strongly concave anteriorly; hosts unidentified permanent grasses .... *P. matadorensis* Mulvey & Stone, 1976
   Bullae small, scattered or absent; second-stage juvenile oesophageal gland lobe extending to only about 30% of body length; stylet knobs flattish to slightly concave anteriorly; host maize (corn) .... *P. chalcoensis* Stone, Sosa Moss & Mulvey, 1976

5. Cysts spherical or subspherical; bullae absent; second-stage juveniles with four incisures in lateral field .... *GLOBODERA* (Skarbilovich, 1959) Behrens, 1975 6
   Cysts elongate ovoid; bullae present; second-stage juveniles with three incisures in lateral field .... *DOLICHODERA* Mulvey & Ebsary, 1980
   .... *D. fluvialis* Mulvey & Ebsary, 1980

6. Cyst wall pattern with rugose striae at midbody, primarily extending latitudinally; second-stage juveniles with ratio "a" = about 18-25; stylet length 20 μm or more; outlet of dorsal oesophageal gland about one-third or less stylet length from its base 7
   Cyst wall pattern with prominent striae extending longitudinally from fenestra area to near base of neck; body of second-stage juveniles very thin, with "a" = 39; stylet length 18 μm; outlet of dorsal oesophageal gland about two-thirds of stylet length from its base .... *G. leptonepia* (Cobb & Taylor, 1953) Behrens, 1975

7. Granek's ratio, as modified by Hesling (1973), averages 2.0 or less 8
   Granek's ratio averages 2.5 or more 9

8. Cuticle between anus and vulval fenestra deeply etched, network-like; distinct circumfenestral area present in yellow and mature cysts; distance from anus to edge of fenestra averaged 48 μm; female stylet mean length 26.2 μm .... *G. tabacum solanacearum* (Miller & Gray, 1972) Behrens, 1975
   Cuticle between anus and vulval fenestra moderately etched,

with some network; circumfenestral area obscure in females; mean distance from anus to edge of fenestra 41 µm; female stylet mean length 22.4 µm; ....
.... *G. tabacum tabacum* Lownsbery & Lownsbery, 1954) Behrens, 1975
Cuticle between anus and vulval fenestra distinct, consisting of lines or ridges 7 (4-11) in number; circumfenestral area distinct and often showing an arched pattern of lines extending above anus; distance from anus to edge of fenestra, mean 27 µm; female stylet 26 µm ....
.... *G. achilleae* (Golden & Klindic, 1973) Behrens, 1975

9. Cuticular pattern maze-like between anus and vulval fenestra ...
.... *G. tabacum virginae* (Miller & Gray, 1968) Behrens, 1975
Cuticular pattern not maze-like between anus and vulval fenestra, being more or less straight lines or ridges 10

10. Number of cuticular ridges between anus and vulval fenestra 12.2 (8-20); Granek's ratio averages 2.5; distance from anus to edge of fenestra 43.9 µm (22-67); second-stage juveniles stylet mean length 23.6 µm; perineal tubercles small and discrete ....
.... *G. pallida* (Stone, 1972) Behrens, 1975
Number of cuticular ridges between anus and vulval fenestra 21.6 (16-31); Granek's ratio averages 4.5; distance from anus to edge of fenestra 68 µm (29-116); second-stage juvenile stylet mean length 22 µm; perineal tubercles large and clustered ......
.... *G. rostochiensis* (Wollenweber, 1923) Behrens, 1975

11. Cyst ovoid shaped and having "hunched backed" appearance; vulva sunken into a terminal cone, lacking fenestration or bullae .... *AFENESTRATA* Baldwin & Bell, 1985
..... *A. africana* (Luc, Germani & Netscher 1973) Baldwin & Bell, 1985
Cyst generally lemon shaped, rarely spherical and then with button-like protrusion; vulva on a terminal cone, with fenestration; bullae present or absent; circumfenestrate, bifenestrate, or ambifenestrate ....
.... *HETERODERA* Schmidt, 1871 and *CACTODERA* 12

12. Cysts spherical; vulval denticles absent; anal-vulval distance long (190-120 µm); vulval cone short or button-like and with distinct three-layered core; lateral field of second-stage juvenile with three incisures ....
.... *Heterodera betulae* Hirschmann & Riggs, 1969
Cysts basically lemon shaped; vulval denticles present or absent; anal-vulval distance generally short (30-70 µm); vulval cone generally prominent and without core structure; lateral field of second-stage juveniles with either three or four incisures 13

13. Vulval cone bifenestrate or ambifenestrate; bullae present or absent; cyst wall pattern at midbody zig-zag or network; vulval slit very short (6-24 µm) to very long (>30 µm) 18
Vulval cone circumfenestrate; cyst wall pattern at midbody straight to wavy lines encircling body, broken by short oblique or vertical lines; bullae absent; vulval slit short (12-25 µm); second-stage juvenile mean body length 440 µm (368-554) with four incisures in lateral field ....
.... *CACTODERA* Krall & Krall, 1978 14

14. Cysts very elongate, generally twice or more as long as wide; L/W ratio averaging 2.3; second-stage juvenile mean stylet length 22.8 µm; egg shell hyaline .... *C. estonica* (Kirjanova & Krall, 1963) Krall & Krall, 1978
Cysts stout, commonly much less than twice as long as wide; average L/W ratio ranging from 1.3 to 1.7; egg shell with or without punctation 15

15. Egg shells distinctly punctate; second-stage juvenile stylet length 23 µm or more 16
Egg shells hyaline, without visible markings; second-stage juvenile stylet length less than 21 µm 17

16. Stylet length of second-stage juvenile, mean 23.5 µm; tail mean 46 µm; hyaline tail terminus short, blunt and measuring 12-20 µm ....
C. cacti (Filipjev & Schuurmans Stekhoven, 1941) Krall & Krall, 1978
Stylet length of second-stage juvenile, mean 26.5 µm; tail, mean 40.4 µm; hyaline tail terminus with finely rounded tip, 17-24 µm .... *C. eremica* Baldwin & Bell, 1985
Stylet length of second-stage juvenile, mean 27 µm; tail, mean 56 µm; hyaline tail terminus tapering to finely rounded terminus, 23-28 µm .... *C. thornei* (Golden & Raski, 1977) Krall & Krall, 1978

17. Tail of second-stage juvenile, mean length 46 µm; hyaline tail terminus about the same length as stylet, mean length 20.4 µm. .... *C. weissi* (Steiner, 1949) Krall & Krall, 1978
Tail of second-stage juvenile mean length 35 µm; hyaline tail terminus shorter than stylet length (mean 20.7 µm) and mean length 15 µm .... *C. amaranthi* (Stoyanov, 1972) Krall & Krall, 1978

18. Vulval slit long to very long (>30 µm); vulval cone commonly ambifenestrate, occasionally bifenestrate 25
Vulval slit very short (6-24 µm); vulval cone always bifenestrate; vulval denticles present or absent; bullae few (or absent) to many .... *avenae* group 19

19. Vulval cone top deeply indented, appearing saddle-shaped; tail of second-stage juvenile short, mean length 35 µm .... *H. turcomanica* Kirjanova & Shagalina, 1965
Vulval cone top smooth or rounded; tail of second-stage juvenile, 48 µm or greater 20

20. Semifenestra widely separated; vulval bridge 18-39 µm; underbridge prominent, strongly developed, bullae, if present, few and scattered 21
Semifenestra not widely separated; vulval bridge less than 18 µm wide; bullae strongly developed (except commonly absent in one species); underbridge, if present, weakly developed 22

21. Vulval slit 7 µm (6-9); second-stage juvenile mean tail length 48 µm, hyaline tail terminus 27 µm, and stylet knobs lightly concave anteriorly .... *H. latipons* Franklin, 1969
Vulval slit 20.3 µm (17-23.5); second-stage juvenile mean tail length 52 µm, hyaline tail terminus 34 µm, and stylet knobs strongly projecting forward, often looking like hooks .... *H. hordecalis* Andersson, 1974

22. Bullae most commonly absent; second-stage juvenile mean length 426 µm and with three incisures in lateral field .... *H. bifenestra* Cooper, 1955
Bullae strongly developed, always present; second-stage juvenile mean length 575 µm or more and lateral field with four incisures (inner two may be indistinct) 23

23. Cysts mostly creamy brown in colour; underbridge present; second-stage juvenile tail 94µ m (83-103); hyaline tail terminus 62 µm (51-75) .... *H. iri* Matthews, 1971
Cysts mostly dark brown to black in colour; underbridge rarely present; second-stage juvenile mean tail length 68 µm (45-77); hyaline tail terminus about 40 µm (28-48) 24

24. Cysts under low power of microscope appear dark brown to black in colour (when 10 or more viewed); second-stage juvenile stylet knobs shallowly concave anteriorly; vulval slit mean length 9.6 µm; cereals host .... *H. avenae* Wollenweber, 1924
Cysts under low power and in number appear brown in colour (when 10 or more viewed); second-stage juvenile stylet knobs deeply concave anteriorly; vulval slit mean length 6.5 µm; grass hosts .... *H. mani* Matthews, 1971

25. Bullae present in vulval cone of all species, small to large in size; underbridge well developed; vulva cone ambifenestrate ... . *schachtii* group 26
Bullae generally absent, if present, few, scattered; underbridge either absent, slender or strongly developed; vulval cone ambifenestrate or bifenestrate .... *goettingiana* group 39

26. Underbridge massive, with dorso-ventral fingerlike projections; bullae few on peripheral wall of cone; second-stage juveniles with three incisures in lateral field 27
Underbridge weakly to strongly developed, without fingerlike projections; bullae few to many on peripheral wall of cone; second-stage juveniles with three or four incisures in lateral field 28

27. Second-stage juvenile length 550 µm (520-580); stylet length 26 µm (23-28); mean tail length 61 µm; hyaline tail terminus length 38 µm (31-41) .... *H. leuceilyma* Di Edwardo & Perry, 1964
Second-stage juvenile length 480 µm (420-530); stylet length 22 µm (21-24); tail length 55 µm (49-60); hyaline tail terminus length 26 µm (20-30) .... *H. sacchari* Luc & Merny, 1963

28. Bullae located at two levels in vulval cone; level one immediately below (anterior) the underbridge and having four fingerlike bullae projecting inward; level two a mass of randomly located long, heavy bullae; second-stage juvenile stylet mean length 19.9 µm; four incisures in lateral field .... *H. zeae* Koshy, Swarup & Sethi, 1971
Bullae at one level, randomly located; second-stage juvenile mean stylet length more than 19.9 µm; three or four incisures in lateral field 29

29. Second-stage juveniles with only three lines in lateral field; mean stylet length range 21-21.4 µm 30

Second-stage juveniles with four distinct lines in lateral field; mean stylet length range 23-31.4 µm 31

30. Vulval slit 53.7 µm (49-60); underbridge length 138.4 µm (108-163); second-stage juvenile tail 57.5 µm (50-66); hyaline tail terminus 33.4 µm (27-38) ...... *H. gambiensis* Merny & Netscher, 1976
Vulval slit 40 µm (38-44); underbridge length 40-50 µm; second-stage juvenile tail 44 µm; hyaline tail terminus 22 µm .... *H. vigni* Edwards & Misra, 1968
Vulval slit 31.8 µm (28-39); underbridge length 79.3 µm (70-115); second-stage juvenile tail 52 µm (42-60); hyaline tail terminus 31 µm (28-35) .... *H. sorghi* Jain, Sethi, Swarup & Srivastava, 1982

31. Anal bullae typically molar shaped; fenestra length 32.5 µm (24-38); second-stage juvenile stylet mean length 25.6 µm .... *H. Schachtii* Schmidt, 1871
Anal bullae not molar shaped, average fenestra length 36 µm or more 32

32. Bullae, small scattered; fenestra length 45-68 µm; underbridge weakly developed; second-stage juvenile mean stylet length 23µm .... *H. fici* Kirjanova, 1954
Bullae strongly developed, many; underbridge fairly well to very strongly developed (except weak in *H. medicaginis*) 33

33. Stylet length of second-stage juvenile, mean 23 µm; underbridge fairly well developed 34
Stylet length of second-stage juvenile, 25 µm or more 35

34. Vulval slit 45-60 µm; underbridge depth 35-37 µm; fenestra length 38-62 µm; cyst wall pattern fine to medium zig-zag .... *H. glycines* Ichinohe, 1952
Vulval slit 36-48 µm; underbridge depth 22-30 µm; fenestra length 25-42 µm; cyst wall pattern very large coarse, prominent zig-zag .... *H. cajani* Koshy, 1967

35. Stylet length of second-stage juvenile mean 25 µm 36
Stylet length of second-stage juvenile, 27.5 µm or greater 38

36. Underbridge weakly developed, only 40-60 µm in length; second-stage juvenile tail 51.8 µm (41-60); hyaline tail terminus 28.5 µm (22-33) .... *H. medicaginis* Kirjanova in Kirjanova & Krall, 1971
Underbridge fairly to heavily developed, 70 µm or more in length 37

37. Underbridge fairly well developed, 90-115 µm in length with a depth of 30-35 µm; mean length of second-stage juvenile 457 µm; mean tail length 56 µm; mean length of hyaline tail terminus 30 µm; males absent .... *H. lespedezae* Golden & Cobb, 1963
Underbridge heavily developed, 80-100 µm in length, with a depth of 37-40 µm; mean length of second-stage juveniles 503 µm; mean tail length 60 µm; mean length of hyaline tail terminus 35 µm; males absent .... *H. galeopsidis* Goffart, 1936

Underbridge heavily developed, 70-102 μm in length; mean length of second-stage juveniles 457 μm; mean tail length 55 μm; mean length of hyaline tail terminus 33 μm; males present .... *H. daverti* Wouts & Sturhan, 1978

38. Vulval slit length 39-50 μm; underbridge length 80-105 μm, depth 33-38 μm; second-stage juvenile stylet length 28 μm (26.5-29.0); outlet of dorsal oesophageal gland mean 7.3 μm ... . *H. trifolii* Goffart, 1932
Vulval slit, mean length 59 μm; mean length of underbridge 153 μm, depth 70 m; second-stage juvenile stylet length 31.4 μm (26.7-34.0); outlet of dorsal oesophageal gland mean 4.9 μm .... *H. rosii* Duggan & Brennan, 1966

39. Vulval cone bifenestrate; semifenestra separated by a wide vulval bridge; fenestral length generally twice width .... *H. humuli* Filipjev, 1934
Vulval cone ambifenestrate; semifenestra separated by a narrow to medium width vulval bridge; fenestral length about the same or generally slightly more than width 40

40. Cysts very elongate, ovoid; L/W ratio mean 3.0; bullae absent; lateral field of second-stage juvenile with only three incisures .... *H. raskii* Basnet & Jayaprakash, 1984
Cysts commonly lemon-shaped, sometimes spheroid or heart-shaped; L/W ratio mean less than 2.0, usually 1.2-1.7; bullae, if present, generally few and scattered; lateral field of second-stage juvenile with either three or four incisures 41

41. Lateral field of second-stage juvenile with four incisures stylet length mean more than 22.7 μm (except *H. cardiolata*) ... 42
Lateral field of second-stage juvenile with three incisures; stylet length mean 22.7 μm, commonly much less (except *H. canadensis*) 48

42. Cysts often heart-shaped; fenestra length 56 μm (54-60); second-stage juvenile stylet mean 19.2 μm; mean tail length 21 μm .... *H. cardiolata* Kirjanova & Ivanova, 1969
Cysts lemon-shaped to spheroid; fenestra length mean 42 μm or less; second-stage juvenile stylet 23 μm or more; tail length 40 μm or more 43

43. Some bullae in vulval cone, irregularly distributed on periphery or close to cone top surface; underbridge strong or absent 44
Bullae absent; underbridge narrow, weakly developed, often lost in preparation 45

44. Underbridge well developed, length 124 μm (120-130); second-stage juvenile body length 405 μm (360-430); tail length 40 μm (38-45); hyaline tail terminus length 22 μm (19-26) .... *H. mediterranea* Volvas, Inserra & Stone, 1981
Underbridge absent; second-stage juvenile body length 500 μm (405-540); tail length 52-64 μm; hyaline tail terminus 30 μm (26-32) .... *H. amygdali* Kirjanova & Ivanova, 1975

45. Fenestra length (mean 38 μm) less than width (mean 44 μm); second-stage juvenile stylet mean length 27 μm; outlet of

dorsal oesophageal gland (DGO) average 8 µm from stylet knobs; two outer lines of lateral field crenate, two outer bands areolated .... *H. urticae* Cooper, 1955
Fenestra length and width about equal; second-stage juvenile stylet mean length 23.4 - 24.6 µm; DGO about 5 µm; two outer lines of lateral field smooth, two outer bands not areolated 46

46. Circumfenestral area of cyst cone with brick work-like pattern; vulval slit mean 39 µm; second-stage juvenile mean body length 486 µm; mean tail length 60.1 µm; hyaline tail terminus mean length 37 µm .... *H. geottingiana* Leibscher, 1892
Circumfenestral area without brick-work like pattern; vulval slit mean 45.5-47.0 µm; second-stage juvenile mean body length 422-431 µm; mean tail length 50 µm; hyaline tail terminus length usually less than 30 µm 47

47. Cysts rather large, stout, appearing almost spherical but with distinct vulval cone; second-stage juvenile stylet strong, robust, with massive knobs, flattened or slightly concave anteriorly; hyaline tail terminus mean 25.2 µm .... *H. cruciferae* Franklin, 1945
Cysts fairly small, lemon-shaped; second-stage juvenile stylet rather delicate, with concave knobs; hyaline tail terminus, mean 28.3 µm .... *H. carotae* Jones, 1950

48. Cysts essentially spherical but with distinct vulval cone; L/W ratio 1.05; strongly developed underbridge; second-stage juvenile mean stylet length 24 µm; tail length 115 µm (110-120); hyaline tail terminus length 60 µm (55-65) .... *H. canadensis* Mulvey, 1979
Cysts lemon-shaped to elongate, mean L/W ratio 1.3-1.9; underbridge present but weak or rare in some species; second-stage juvenile mean stylet length 22.7 µm, usually much less; tail length less than 70 µm; hyaline tail terminus mean length 40 µm 49

49. Bullae may be present (few, scattered) or absent; underbridge weak or rare, without furcations; mean fenestral length 36 µm (except *H. longicolla*- 48 µm); second-stage juvenile stylet mean length 20.8 µm 50
Bullae absent; underbridge very strongly developed, 115-140 µm long, with furcate ends; fenestral length 52-58 µm; perianal pattern absent; second-stage juvenile stylet mean length 22.7 µm; mean tail length 62 µm .... *H. graminophila* Golden & Birchfield, 1972
No bullae; underbridge prominent, 70-107 µm long, ends furcate; fenestral mean length 50 µm; conspicuous perianal pattern present; second-stage juvenile stylet mean length 22 µm; mean tail length 56 µm .... *H. graminis* Stynes, 1971

50. Few scattered bullae in vulval cone; second-stage juvenile hyaline tail terminus length 31-40 µm 51
Bullae absent; cysts elongate, L/W ratio 1.9; prominent perianal pattern present; second-stage juvenile stylet mean length 20 µm; tail length 60 µm (56-63); hyaline tail terminus mean 24.7 µm ....*H. cyperi* Golden, Rau & Cobb, 1972

Cysts stouter, L/W ratio 1.4; perianal pattern absent; second-stage juvenile stylet mean length 17.5 μm; tail length 50μm (45-54); hyaline tail terminus mean length 27 μm .... *H. longicolla* Golden & Dickerson, 1973

51. Cysts elongate, L/W ratio 1.7; distinct periannal pattern present; second-stage juvenile body length 380-430 μm; stylet length 16-17 μm; tail mean length 67 μm; hyaline tail terminus length 30-40 μm .... *H. mothi* Khan & Husain, 1965
Cysts rather plump, L/W ratio about 1.3; without perianal pattern; underbridge 75-90 μm; second-stage juvenile mean body length 367 μm; stylet length 18.6 μm (18-19.5); tail length 53 μm (44-57); length of hyaline tail terminus 31.4 μm (26-36) .... *H. elachista* Ohshima, 1974
Cysts fairly stout, L/W ratio 1.3; perianal pattern absent; underbridge 52-70 μm; second-stage juvenile mean body length 440 μm; mean stylet length 20.8 μm; tail length 67-69 μm; hyaline tail terminus 39.5 μm (35-45) .... *H. oryzae* Luc & Berdon-Brizuela, 1961

REFERENCES

Anon, 1952, Remarks by Dr Oostenbrink, in "Proc.Int.Nemat.Symposium, Harpenden, 1951, Rome (FAO), 55.

Baldwin, J.G., 1983, Fine structure of body wall cuticle of females of *Meloidodera charis, Atalodera lonicerae,* and *Sarisodera hydrophila* (Heteroderidae), J.Nematol., 15:370.

Baldwin, J.G. and Bell, A.H., 1985, *Cactodera eremica* n.sp.,*Afenestrata africana* (Luc et al., 1973) n.gen., n.comb., and an emended diagnosis of *Sarisodera* Wouts and Sher, 1971 (Heteroderidae), J.Nematol., 17:187.

Brown, G., Callow, R.K., Green, C.D., Jones, F.G.W., Rayner, J.H., Shepherd, A.M. and Williams, T.D., 1971, The structure, composition and origin of the subcrystalline layer in some species of the genus *Heterodera*, Nematologica, 17:591.

Clark, S.A., Shepherd, A.M. and Kempton, A., 1973, Spicule structure in some *Heterodera* spp., Nematologica, 19:242.

Cliff, G. and Baldwin, J.G., 1985, Fine structure of body wall cuticle of females of eight genera of Heteroderidae, J.Nematol., 17:286.

Cobb, G.S. and Taylor, A.L., 1953, *Heterodera leptonepia*, n.sp., a cyst-forming nematode found in soil with stored potatoes, Proc.Helminthol. Soc.Wash., 20:13.

Cooper, B.A., 1955, A preliminary key to British species of *Heterodera* for use in soil examination, in "Soil Zoology", D.K.McE. Kevan, ed., Butterworths, London.

Curran, J., 1984, Identification of nematode populations by analysis of genomic DNA, Proc.First Int.Cong.Nematol., 102, (Abstract).

Ellenby, C, 1946, Nature of the cyst wall of the potato-root eelworm *Heterodera rostochiensis*, Wollenweber, and its permeability to water, Nature, 157:302.

Ferris, V.R. and Siegel, B.M., 1957, Electron microscopy of golden nematode cyst wall, Nematologica, 2:16.

Fox, P.C. and Atkinson, H.J., 1984, Electrophoresis and the biochemical taxonomy of nematodes, Proc.First Int.Congr.Nematol., 103, (Abstract).

Franklin, M.T., 1939, On the structure of the cyst wall of *Heterodera schachtii* (Schmidt), J.Helminthol., 17:127.

Franklin, M.T., 1951, The cyst-forming species of *Heterodera*, Wilding & Son, England.

Golden, A.M. and Ellington, D.M.S., 1972, Redescription of *Heterodera rostochiensis* (Nematoda: Heteroderidae) with key and notes on related species, Proc.Helminthol.Soc.Wash., 39:64.

Golden, A.M. and Klindic, O., 1973, *Heterodera achilleae* n.sp. (Nematoda: Heteroderidae) from yarrow in Yugoslavia, J.Nematol., 5:196.

Golden, A.M. and Mulvey, R.H., 1983, Redescription of *Heterodera zeae*, the corn cyst nematode, with SEM observations, J.Nematol., 15:60.

Golden, A.M. and Raski, D.J., 1977, *Heterodera thornei* n.sp. (Nematoda: Heteroderidae) and a review of related species, J.Nematol., 9:93.

Goodey, J.B., 1959, Data to be considered, observed and, where possible, reported upon when presenting descriptions of new species, Nematologica, 4:211.

Granek, I., 1955, Additional morphological differences between the cysts of *Heterodera rostochiensis* and *Heterodera tabacum*, Plant Dis.Rept., 39:716.

Green, C.D., 1971, The morphology of the terminal area of the round-cyst nematodes, S.G. *Heterodera rostochiensis* and allied species, Nematologica, 17:34.

Green, C.D., 1975, The vulval cone and associated structures of some cyst nematodes (Genus *Heterodera*), Nematologica, 21:134.

Guile, C.T., 1966, Cyst chromogenesis in potato cyst eelworm pathotypes, Plant Pathol., 15:125.

Guile, C.T., 1967, On cyst colour changes, bionomics and distribution of potato cyst-eelworm (*Heterodera rostochiensis* Woll.) pathotypes in the East Midlands, Ann.appl.Biol., 60:411.

Guile, C.T., 1970, Further observations on cyst colour changes in potato cyst eelworm pathotypes, Plant Pathol., 19:1.

Hesling, J.J., 1973, The estimation of Granek's ratio in round-cyst *Heteroderas*, Nematologica, 19:119.

Hesling, J.J., 1978, Cyst nematodes: morphology and identification of *Heterodera, Globodera* and *Punctodera*, 125-155, in "Plant Nematology", J.F. Southey, ed., London.

Hesling, J.J. and Ellis, P.R., 1974, The fenestralia of *Heterodera rostochiensis* and *H. pallida*, Nematologica, 20:43.

Hussey, R.S., 1979, Biochemical systematics of nematodes - a review, Helminthol.Abstr., 48:141.

McClure, M.A., and Lucas, D.O., 1984, Monoclonal hybridoma antibodies in plant nematology, Proc. First Int.Cong.Nematol., 106, (Abstract).

Mulvey, R.H., 1959, Investigations on the clover cyst nematode, *Heterodera trifolii* (Nematoda: Heteroderidae), Nematologica, 4:147.

Mulvey, R.H., 1972, Identification of *Heterodera* cysts by terminal and cone top structures, Can.J.Zool., 50:1277.

Mulvey, R.H., 1973, Morphology of the terminal areas of white females and cysts of the genus *Heterodera* (s.g. *Globodera*), J.Nematol., 5:303.

Mulvey, R.H. and Ebsary, B.A., 1980, *Dolichodera fluvialis* n.gen., n.sp. (Nematoda: Heteroderidae) from Quebec, Canada, Can.J.Zool., 58:1697.

Mulvey, R.H. and Golden, A.M., 1983, An illustrated key to the cyst-forming genera and species of Heteroderidae in the western hemisphere with species morphometrics and distribution, J.Nematol., 15:1.

Mulvey, R.H. and Stone, A.R., 1976, Description of *Puntodera matadorensis* n.gen., n.sp. (Nematoda: Heteroderidae) from Saskatchewan with lists of species and generic diagnoses of *Globodera* (n. rank) *Heterodera* and *Sarisodera*, Can.J.Zool., 54:772.

Oostenbrink, M. and den Ouden, H., 1954, De structuur van de kegeltop als taxonomisch Kenmerk by *Heterodera* - soorten met citroenvormige cysten, Tijdschr.Plantenziekt, 60:146.

Shepherd, A.M., Clark, S.A. and Dart, P.J., 1972, Cuticle structure in the genus *Heterodera*, Nematologica, 18:1.

Steiner, G., 1949, Plant nematodes the grower should know, Proc.Soil Sci.Soc.Fl.Proc., 4-B:72.

Stone, A.R., 1973, *Heterodera pallida* n.sp. (Nematoda: Heteroderidae), a second species of potato cyst nematode, Nematologica, 18:591.

Stone, A.R., 1975, Head morphology of second-stage juveniles of some Heteroderidae (Nematoda: Tylenchoidea), Nematologica, 21:81.

Stone, A.R., Sosa Moss, C. and Mulvey, R.H., 1976, *Punctodera chalcoensis* n.sp. (Nematoda: Heteroderidae) a cyst nematode from Mexico parasitizing *Zea mays*, Nematologica, 22:381.

Taylor, A.L., 1957, *Heterodera* taxonomy, Proc. 5-19 Workshop in Phytonematology, 1957, Univ.Tennessee, 1.

Thorne, G., 1928, *Heterodera punctata* n.sp. a nematode parasite on wheat roots from Saskatchewan, Sci.Agric., 8:707.

Thorne, G., 1961, Principles of Nematology, McGraw-Hill, New York.

Weiser, W., 1953, On the structure of the cyst wall in four species of *Heterodera* Schmidt, Medd.Vaxtskyddsanst.Stockh., 65:1.

Wilson, E.M., 1968, Vulval bodies in certain species of *Heterodera*, Nature, London, 217:879.

Wouts, W.M. and Sher, S.A., 1971, The genera of the subfamily Heteroderinae (Nematoda: Tylenchoidea) with description of two new genera, J.Nematol., 3:129.

Zunke, U., 1986, Zur Bildung der subkristallinen Schicht bei *Heterodera schachtii* unter aseptischen Bedingungen, Nematologica, 31: (in press).

DISCUSSION

Stone asked if keys can provide an assured method of species identification. Golden replied that he looked upon keys as a means of getting into the general area within a genus but that reference to original descriptions of species was necessary to confirm identification.

ULTRASTRUCTURE AND FUNCTION OF CYST NEMATODES

Burton Y. Endo

US Department of Agriculture, Agricultural Research Service
Plant Protection Institute, Beltsville, Maryland 20705

INTRODUCTION

Investigations of the structure of the cyst nematodes and other Tylenchid species have been stimulated by the detailed electron microscopy (TEM) observations of the cuticle of *Globodera (Heterodera) rostochiensis* (Woll.) Behr. by Günther and Kämpfe (1966) and Shepherd et al. (1972). Ultrastructure of the root knot species *Meloidogyne javanica* (Treub) Chitw. (Bird and Rogers, 1965a, b; Bird, 1968) was described in a previous symposium (Lamberti and Taylor, 1979). The extensive investigation by De Grisse (1977) of the ultrastructure of 22 genera of the Tylenchids showed the common structural base of these genera and from which future studies can be extended.

Although scanning electron microscopy (SEM) is an integral part of ultrastructural observations, the emphasis in this review will be primarily on TEM studies where tissue structure can be related to morphology and function. Nevertheless, SEM has played a significant role in permitting observations of obscure surface features that can be utilized in taxonomic studies (Wergin, 1982).

CUTICLE

The exoskeleton serves as the interface between the nematode and its external environment (Maggenti, 1981). The combination of a noncellular elastic cuticle and an underlying hypodermis makes the exoskeleton one of the most complex histological and chemical structures of the nematode body. The cuticle is the product of the hypodermis, which is derived from the ectoderm. Maggenti (1981) considers that the cuticle comprises four fundamental strata: epicuticle, exocuticle, mesocuticle and endocuticle.

Structural changes of the cuticle have been observed during the developmental stages of *Heterodera schachtii* Schmidt (Günther and Kämpfe, 1966). Although independent juvenile formation is suppressed in the egg this stage is designated as the J1. The J2 has a mean length of 440 μm and is 20.6 μm in diameter. The annulations are 1.4-1.7 μm broad while the lateral field consists of three heavy cuticularized ridges. Observations of the J2 show that the exocuticle is surrounded by a thin, intensely stained epicuticle. The exocuticle proper is dense and homogeneous in structure. The endocuticle is a striated structure when viewed in longitudinal and cross

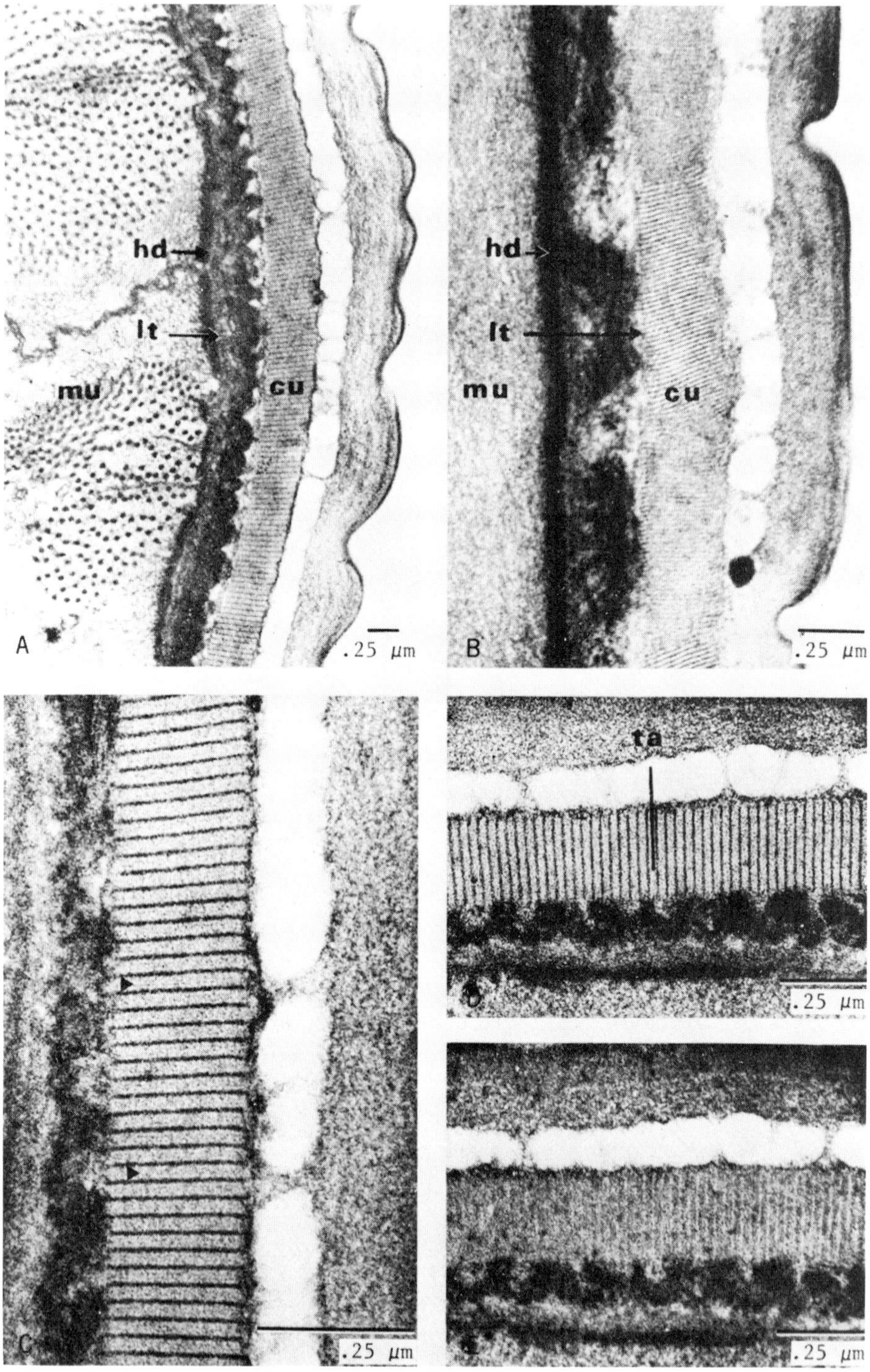
hd
lt
mu
cu
A
.25 µm
hd
lt
mu
cu
B
.25 µm
C
.25 µm
ta
.25 µm
.25 µm

sections but its basal boundary is homogeneous. The two inner layers are about 0.3 µm thick. In contrast, the J3 cuticle increases in thickness so that the exocuticle and endocuticle are each 0.9-1.2 µm. Correspondingly, the hypodermis during the moulting phase of J2 to J3 is 3-5 µm thick.

The epicuticle of *H. schachtii* shows characteristic changes in each juvenile stage. In the J3 the surface structure consists of fine, densely arranged longitudinal stripes that cover the whole body except at the posterior blunt end where they change to a zig-zag pattern. The exocuticle and endocuticle of the J3 show no structural detail.

The hypodermis forms a new layer about 3-4 µm thick prior to the moult of the J4 to the adult female. The basal portion of this layer becomes the epicuticle while the distal part becomes detached from the old cuticle. The hypodermis then secretes the future exocuticle that is characterised by a pattern of fine radial stripes; these stripes are assumed to be fine canals that transport to the periphery hypodermal secretions which dissolve the old cuticle and tan the new one.

Subsequently the hypodermis forms the new endocuticle. The deposition does not appear to be continuous but can occur in two or more stages that may account for the layered pattern of the cuticle of various cyst nematodes. The basic structure of the cuticle of the mature female corresponds to the endo-, exo- and epicuticles of the juvenile stages. However, in the adult the epicuticle is very dense and firm and has a fine outer boundary that stains intensely. The exocuticle is homogeneous and except for the stripe-like canals does not show any specific features. The endocuticle, which is distinct from the exocuticle, has a loose squamous structure. The inner boundary is marked by the basal membrane. The total thickness of the cuticle, which ranges from 14 to 20 µm, varies according to the overall size of the nematode.

Wisse and Daems (1968) observed a striated structure of the endocuticle or the basal layer in *H. rostochiensis*. Bird (1971) noted that a striated region of the endocuticle occurs in a wide range of plant- and animal-parasitic nematode species. In the transverse sections of the cuticle of J2

Fig. 1 ⟵ A, B: Special structure of the hypodermis between cuticle (cu) and muscle cells (mu) in a transverse (A) and a longitudinal (B) section of *Heterodera rostochiensis*. The local thickenings of the hypodermis (lt), which have a layered appearance, and the hemidesmosome (hd) are cut transversely in A and longitudinally in B. In both electronmicrographs the fine fibrous material of the muscle cells is located at the inner side of the hemidesmosomes. C. Transverse section through the cuticle. At some places, two thin lines are interposed between the normal periodicity (arrows). D. Part of the fourth layer of the cuticle in transverse section revealing the characteristic striated pattern of this layer. Tilting axis (ta). E. The same section tilted 12° over the axis indicated in D. Even at this low tilting angle the characteristic sharp striation of the fourth layer seen in D is diminished, an observation that supports the assumption that the sharp striation is the result of the projection of several rods to a single image. (Reprinted from E. Wisse and W. Th. Daems, 1968).

H. schachtii (Wisse and Daems, 1968; Fig. 1A), the striated fourth layer or endocuticle shows a periodicity of 24 µm while the periodicity in longitudinal sections is about 18 µm (Fig. 1B). The authors conclude that the striated fourth layer or endocuticle in the J2 of *G. rostochiensis* consist of rods that are arranged in a geometrical pattern with interconnecting membranes. Wisse and Daems (1968) did not observe fibres crossing the hypodermis that would attach the muscles to the cuticle. They further commented that the transverse canals described by Günther and Kämpfe (1966) may be spaces between two periodic thickenings in the hypodermis. The attachment of the cuticle to the somatic muscles via hemidesmosomes is clearly shown in the J2 of *H. rostochiensis* (Fig. 1B) as well as in the J2 and male of *H. glycines* Ichinohe (Baldwin and Hirschmann, 1975b; Endo, 1983).

Although the designation for cuticle components differs, a close resemblance exists between the cuticle of the adult male and the J2 of *G. rostochiensis* (Fig. 2). Shepherd et al. (1972) noted that the cuticle of the male *G. rostochiensis* has two main regions which they designated the outer (A) and the inner (B) or base region. In comparison with Günther and Kämpfe's (1966) description, the A layer is comparable with the epicuticle and the exocuticle; whereas the B layer is the striated endocuticle. The A layer of Shepherd et al. (1972) terminology has subscripts of A_1, A_2, and A_3 with A_1 being equivalent to the epicuticle. A_2 and A_3 correspond to the exocuticle; A_2 consists of a homogeneous, moderately electron dense layer and A_3 sometimes appears vacuolated and has moderately dense material among fibrous strands. This region is interspersed with globular electron dense deposits.

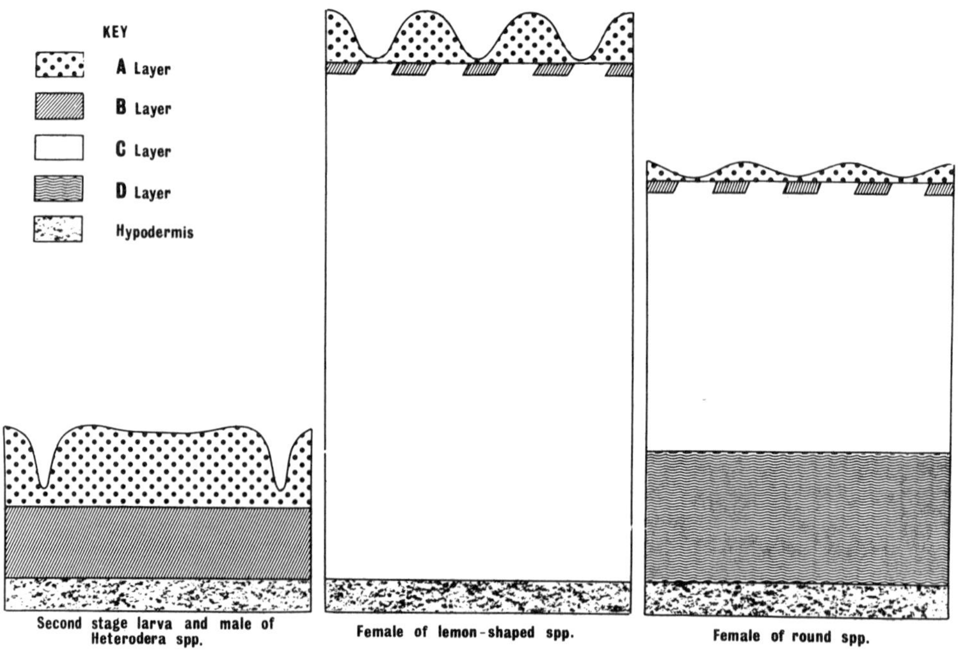

Fig. 2 Diagrammatic representation of the cuticular layering in the 2nd-stage juvenile and male, and in the mid-body region of females of lemon-shaped and round species of *Heterodera*. (Reprinted from A.M. Shepherd, S.A. Clark and P.J. Dart, 1972).

Shepherd et al. (1972) compared the cuticle of the lemon shaped and round cyst nematode species (Fig. 2). The cuticle of the female lemon-shaped cyst nematodes has an A layer comparable to that found in the male cuticle. However, only remnants of the B layer (or endocuticle) are retained after the underlying C layer is deposited. This layer (10-25 µm thick) is further subdivided into three layers: C_1, fibres with no preferred orientation; C_2 (10 µm), coarse, randomly orientated fibres dispersed through an electron dense matrix; and C_3, a fine fibrillar zone that occurs, in some specimens, next to the hypodermis. Shepherd et al. (1972) reported that a fourth D layer occurred only in the round cysts and its presence supported placing the round cyst species [of *Heterodera*] into a separate genus *Globodera*. The D layer is 6-8 µm thick and consists of very clearly defined fibres that are arranged in distinct and apparently parabolic patterns that occur in sublayers within the region. Shepherd et al. (1972) suggested that the D layer present in *Globodera* may contribute to its ability to withstand desiccation and similarly Baldwin (1983) considered that the non-cyst forming genera *Sarisodera* and *Atalodera*, both with the D layer, are well adapted to semi-arid conditions.

HYPODERMIS

The hypodermis underlies the cuticle. Although it is thin in regions between the somatic muscles and cuticle, it is quite extensive at the dorsal, ventral and lateral regions of males of *H. glycines* (Baldwin and Hirschmann, 1975b; Fig. 3).

In males of *H. glycines*, the hypodermis is a thin protoplasmic layer between the cuticle and somatic muscles. The hypodermis expands at the intermuscular regions to form the lateral, dorsal, and ventral cords that project into the pseudocoelom. The hypodermis is described as pervading the cephalic framework region and lining the stomatal cavity and stylet shaft. In the anterior region, the hypodermis has longitudinal radii of tissues that surrounds the stoma and are located between the anterior portions of the stylet protractor muscles (Baldwin and Hirschmann, 1975b). This hypodermal layer is described in juvenile *H. glycines* as being penetrated by numerous hemidesmosomes that attach the protractor muscles to the stomatal wall (Endo, 1983).

Posteriorly in males of *H. glycines*, the hypodermal cords are generally reduced in size to accommodate the amphidial glands. Below the amphids, the cords enlarge to partially enclose the lateral nerve bundles and to make contact with the subdorsal and subventral nerve bundles. In this region, the tissues are predominantly hypodermis and form a ring around the oesophagus.

In the ganglionic region which lies posteriad to the procorpus and beyond the nerve ring, the hypodermal cords are greatly expanded and enclose many perikaryons of the six cephalic nerve bundles. The hypodermis of *H. glycines* males is considered syncytial although occasional cells are enclosed.

In addition to the syncytial nuclei, the hypodermal cords contain numerous organelles that include large lipid globules, lysosomes, myelin-like products, smooth and rough ER, free ribosomes, beta-glycogen, Golgi bodies, and spindle-shaped structures. Microtubules occur within the hypodermis just below the cephalic framework (Baldwin and Hirschmann, 1975a).

The lamellae of parallel membranes in the cephalic region are considered unique to the hypodermis in males of *H. glycines*. Subsequent observations by others show that these membranes are part of ciliary processes related to accessory or supplementary nerves of various nematode species (Baldwin and Hirschmann, 1975a; Wergin and Endo, 1976; Natasasmita, 1979; Endo, 1980).

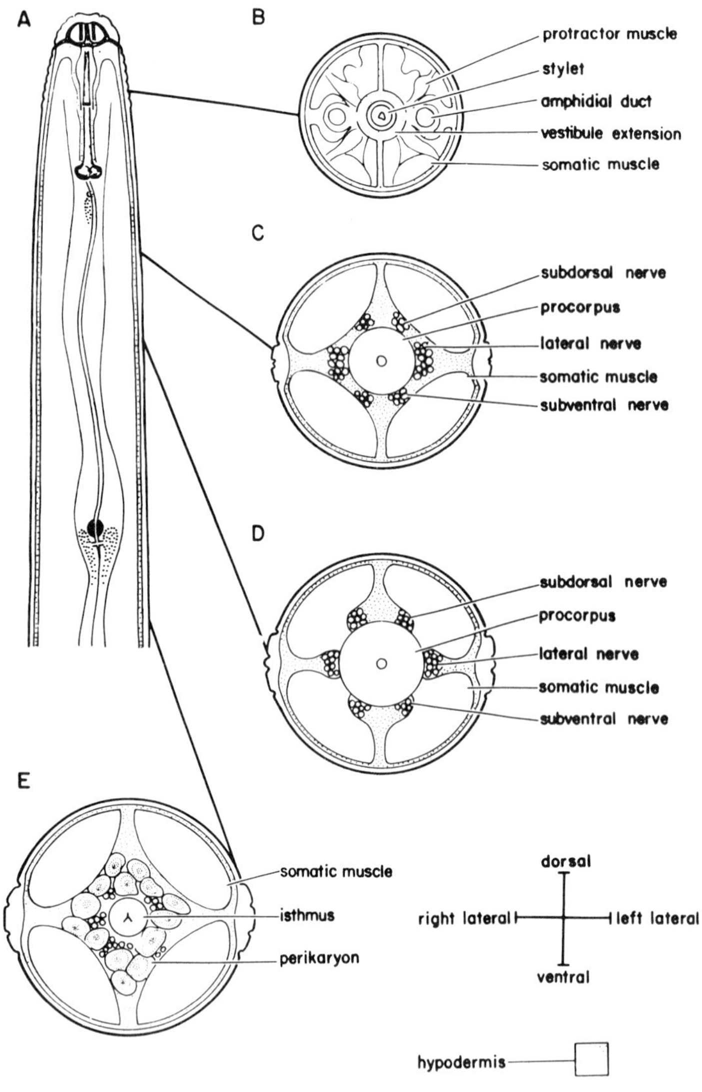

Fig. 3 Diagram illustrating relationship of hypodermis to other structures in anterior portion of *Heterodera glycines*. A. Ventral view. B. Cross section through region of stylet cone. C. Cross section through region of anterior procorpus. D. Cross section through region of middle procorpus. E. Cross section through region of isthmus. (Reprinted from J.G. Baldwin and H. Hirschmann, 1976).

Generally, the intercordal hypodermis is void of organelles, but at certain sites it encloses fine peripheral nerve processes and provides points of attachment for the cuticle and somatic muscles (Baldwin and Hirschmann, 1975). In the J2 and male of *H. glycines*, hemidesmosomes are arranged in circular arrays as paired units under each cuticle annulation. Less uniform but very extensive hemidesmosomes occur between the basal lamina of the somatic or protractor muscles and under the surface of the cephalic framework. Similar hemidesmosomal contacts occur between the protractor muscles and the stomatal wall (Endo, 1980; Fig. 4).

SOMATIC MUSCLES

The somatic musculature of *Heterodera glycines* males has platymyarian cells with a meromyarian arrangement. At the anterior of the nematode are two muscle cells per intercordal region; at least three muscle cells occur in the region of the stylet and four to five throughout most of the oesophageal region. The sarcoplasm, which extends into the pseudocoelom, supports the striated muscle. It contains a large nucleus with a distinct spherical nucleolus and dense peripheral chromatin, as well as mitochondria, beta-glycogen, and smooth ER. Innervation processes extend from the broadest part of the sarcoplasm to the nerve processes that are located in the major nerve bundles. A cross section of the contractile portion of the somatic muscles reveals thick and thin myofilaments, 24 mμ or 6 mμ in diameter (Baldwin and Hirschmann, 1975b).

The specialised musculature at the anterior of nematodes varies according to the morphology and function of the stomatal region. The arrangement of protractor muscles of the cyst nematodes is similar to that described in other Tylenchid species (Yuen, 1967; De Grisse, 1977: Endo, 1983).

In males and juveniles of *H. glycines*, the ten branches of the protractor muscles merge posteriorly to form three units, which attach to the dorsal and two subventral knob surfaces. Cross sections through the striated portion show the thin and thick myofilaments that typify contractile muscles (Baldwin and Hirschmann, 1976; Endo, 1983). Although Yuen (1967) reported that the stylet muscles do not attach laterally to the stomatal wall surface, hemidesmosomes are present at these junctures in juveniles and adults of *H. glycines*, emphasises the supporting role of the stomatal wall during stylet projection (Baldwin and Hirschmann, 1976; Endo, 1983).

In describing the noncontractile region of the protractor muscles in males of *H. glycines*, Baldwin and Hirschmann (1976) noted that the portion immediately posterior and surrounding the knobs is filled with fine filaments and various forms of sarcoplasmic reticulum. Similarly, the muscle components adjacent to the stylet knob are further accentuated in juveniles of *H. glycines* with spaces between the contact regions of the stylet knob surfaces and the thin myofilament terminals of the protractor muscles (Endo, 1983).

STOMATAL REGION

In a study of the stomatal region of males of *Heterodera glycines*, Baldwin and Hirschmann (1976) report that the cephalic framework is radially symmetrical. Six arched blades extend from the hexagonal vestibule to the basal ring. In juvenile *H. glycines* (Endo, 1983), the six blades of the framework are radially symmetrical anteriorly, but take on a bilateral

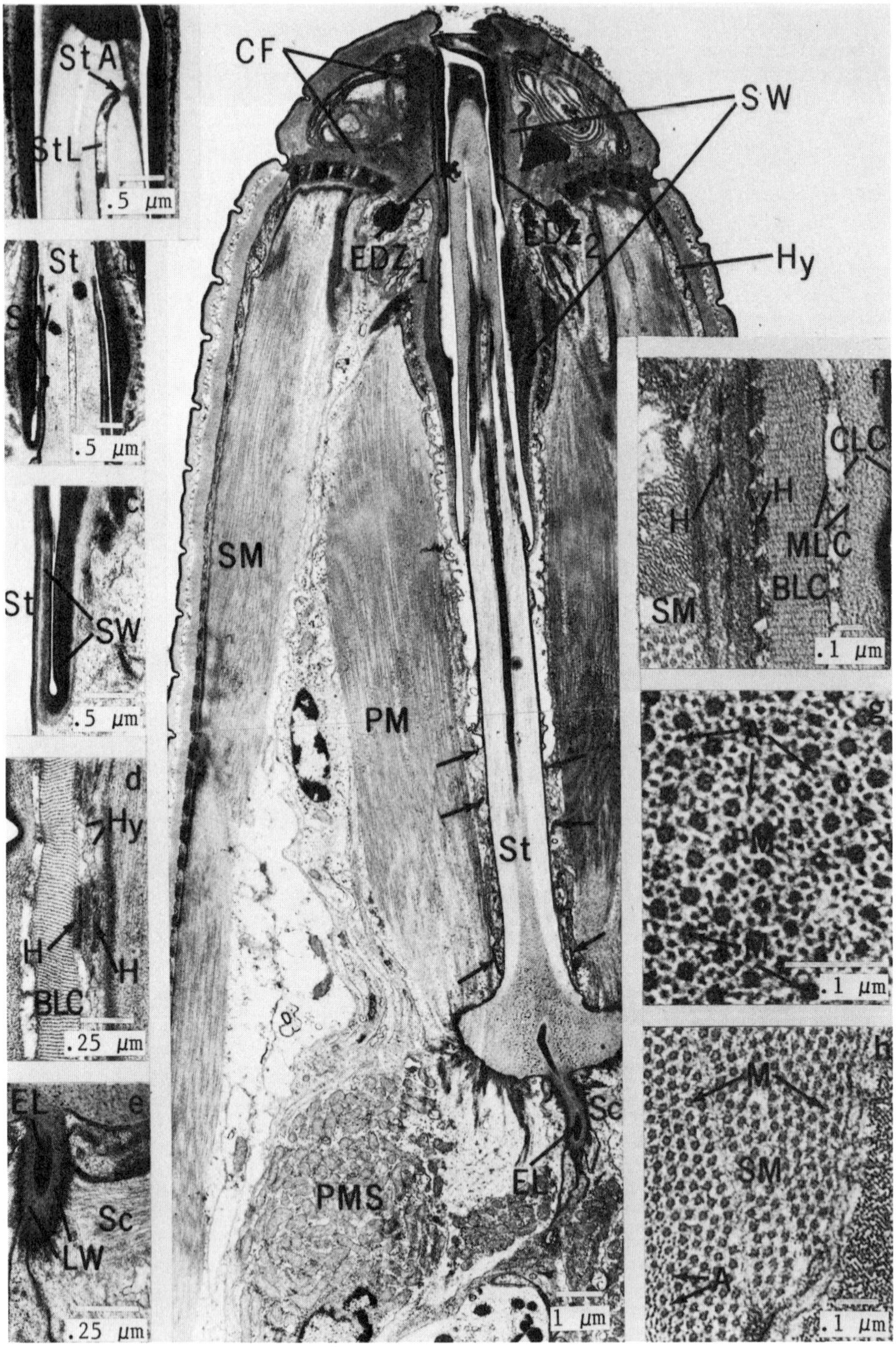
StA
StL
.5 μm
St
SW
.5 μm
c
St
SW
.5 μm
d
Hy
H
H
BLC
.25 μm
EL
e
Sc
LW
.25 μm
CF
SW
EDZ1
EDZ2
Hy
SM
PM
St
PMS
Sc
EL
1 μm
f
CLC
H
H
MLC
BLC
SM
.1 μm
g
A
PM
M
.1 μm
h
M
SM
A
.1 μm

symmetry as the dorsal and ventral blades broaden to form the basal plate. The sublateral blades are narrow and curved. The broad basal plate provides the zones of contact between the cephalic framework and the subjacent elements of the somatic and protractor muscles. These observations are consistent with similar observations of males of *Meloidogyne incognita* (Treub) Chitw., (Baldwin and Hirschmann, 1976), females of *Rotylenchus robustus* (de Man) Filipjev (De Grisse, et al., 1974) and males and females of *Tylenchulus semipenetrans* Cobb (Natasasmita, 1979). In *H. glycines* juveniles the bilateral symmetry is consistent with the arrangement of the termini of the somatic and protractor muscles and their attachment to the basal plate of the cephalic framework (Endo, 1983).

The fibrillar part of the cephalic framework of *H. glycines* juveniles (Endo, 1983) and males (Baldwin and Hirschmann, 1976) is similar to the moderately dense component of the labial framework of another Tylenchid, *Tylenchorhynchus dubius* (Bütschli) Filipjev (Byers and Anderson, 1982). In *H. glycines*, this cuticular component extends from the outer margins of the cephalic framework and comprises a major part of the stomatal wall. The junction of the fibrillar cuticle at the periphery of the basal plate with the striated layer of the body cuticle signifies a difference in either the composition or origin of these cuticular regions. It is at this juncture of the lip region annulation that the striated layer of the body cuticle terminates. Recent observations of stylet regeneration in *H. glycines* show some of the developmental pattern and tissues that are involved in the deposition of various components of the stomatal region (Endo, 1985).

Baldwin and Hirschmann (1976) compared the stylet morphology of males of *M. incognita* and *H. glycines* and noted similarity in the morphology of the stylet cone, shaft, and knobs. The cone portion is somewhat smoothly surfaced and interacts with the shaft through the shaft material that extends into the cone. This material is divided by radiating strands of

Fig. 4 ← Lateral longitudinal section through the anterior region of a second-stage juvenile of *H. glycines*. The cephalic framework (CF), stomatal wall (SW), and protractor muscles (PM) provide the guidance required for stylet protraction. A. Nematode salivation and food ingestion occur through the stylet lumen (StL), which has a ventrally located aperture (StA). B, C. Posterior portion of stomatal wall (SW) is invaginated during stylet (St) protraction. D. longitudinal section of a nematode showing hemidesmosomes (H) within the hypodermis (Hy) that interact with the striated basal layer of the cuticle (BLC) and with the basal lamina of the somatic muscle. E. Oblique view of a secondary muscle (Sc) attachment to the oesophageal lumen wall (LW). F. Cross section of a nematode with a linear orientation of hemidesmosomes (H) that show the interaction between the cuticle and somatic muscle (SM) basal lamina with the hypodermis as an intermediary. a, actin; Arrows, membrane junctions of tissue related to stylet shaft; CLC, cortical layer of cuticle; EDZ_1, particulate electron-dense zone; EDZ_2, amorphous electron-dense zone; EL, oesophageal lumen; M. myosin; MLC, median layer of cuticle; PMS, protractor muscle sarcoplasm. (Reprinted from B.Y. Endo, 1983).

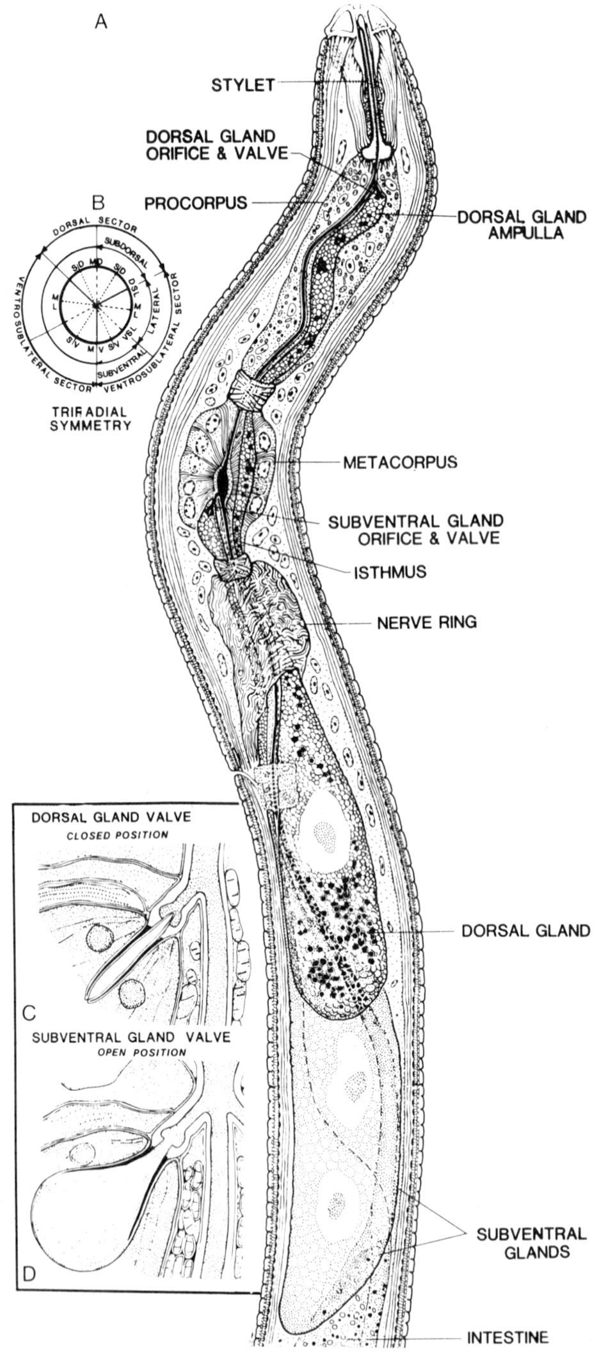
A
STYLET
DORSAL GLAND
ORIFICE & VALVE
B
PROCORPUS
DORSAL GLAND
AMPULLA
TRIRADIAL
SYMMETRY
METACORPUS
SUBVENTRAL GLAND
ORIFICE & VALVE
ISTHMUS
NERVE RING
DORSAL GLAND VALVE
CLOSED POSITION
DORSAL GLAND
C
SUBVENTRAL GLAND VALVE
OPEN POSITION
SUBVENTRAL
GLANDS
D
INTESTINE

cone material in *M. incognita*; however, an undivided band of shaft material extends into the cone of *H. glycines*. The shaft material is reduced anteriorly in the cone as a narrow dorsal region. Although the lumen of the stylet cone is central and triradiate near the stylet base of *M. incognita*, it becomes ventral and oblong near its terminus. In contrast, the lumen of the cone of *H. glycines* males and juveniles is circular in cross section throughout the length of the cone.

In the shaft region, membranes modified as junctional complexes extend posteriorly and fold inwardly to attach near the junction of the stylet cone and shaft. Apparently these junctional complexes together with the stomatal lining that includes the vestibule extension, vestibule, and stomatal opening and designated as a guiding apparatus when viewed with light microscopy (Baldwin and Hirschmann, 1976).

Recent ultrastructural observations of other Tylenchid species and juvenile *H. glycines* reveal the flexible nature of the stomatal wall (vestibule extension). Infolding of the wall during stylet projection clarifies the nature of the guiding apparatus (De Grisse, 1977; Endo, 1983; Fig. 4b, c).

OESOPHAGEAL REGION

In males of *H. glycines* the most anterior part of the procorpus comprises six large cells surrounding an inner complex of the oesophageal lumen with its lining, nerve processes and the ampulla of the dorsal oesophageal gland (Fig. 5). Three of the cells are sarcoplasmic components of the protractor muscles as described for the male and J2 of *H. glycines* (Baldwin et al., 1977; Endo, 1984). These protractor muscle cells are separated by three per-radial stylet shaft cells. Secondary stylet protractor muscle cells lie in the apices of the shaft cell and are bound externally by the protractor muscle cells (Endo, 1984).

The ampulla of the dorsal gland extension is considered to be a single unit whose contents have a continuum of dense spheres, microtubules, ribosomes, and mitochondria in a matrix of fine granules (Baldwin et al., 1977). Similarly, in juvenile *H. glycines* secretion granules and microtubules are observed through the cytoplasmic matrix of the dorsal gland extension that contains numerous ribosomes and polyribosomes (Endo, 1984). The mobility of the secretion granules in the procorpus of juvenile *H. glycines* can be observed with video enhanced light microscopy (unpublished data).

The dorsal gland ampulla and its contents is probably regulated by the two nerve processes that border the dorsal side of the ampulla, as seen in males of *M. glycines* and *M. incognita* (Baldwin et al., 1977). Similarly,

Fig. 5 ← A. Diagram of longitudinal section through the oesophageal region of *H. glycines*. B. Terminology of body regions of nematodes with emphasis on triradiate symmetry of the oesophagus (after Coomans, 1979). C. Longitudinal section of a closed valve or end apparatus within a dorsal gland ampulla; note the infolded valve membrane curves into the valve duct lumen. D. The open valve or end apparatus within one of a pair of subventral gland ampullae. Reprinted from B. Y. Endo, 1984).

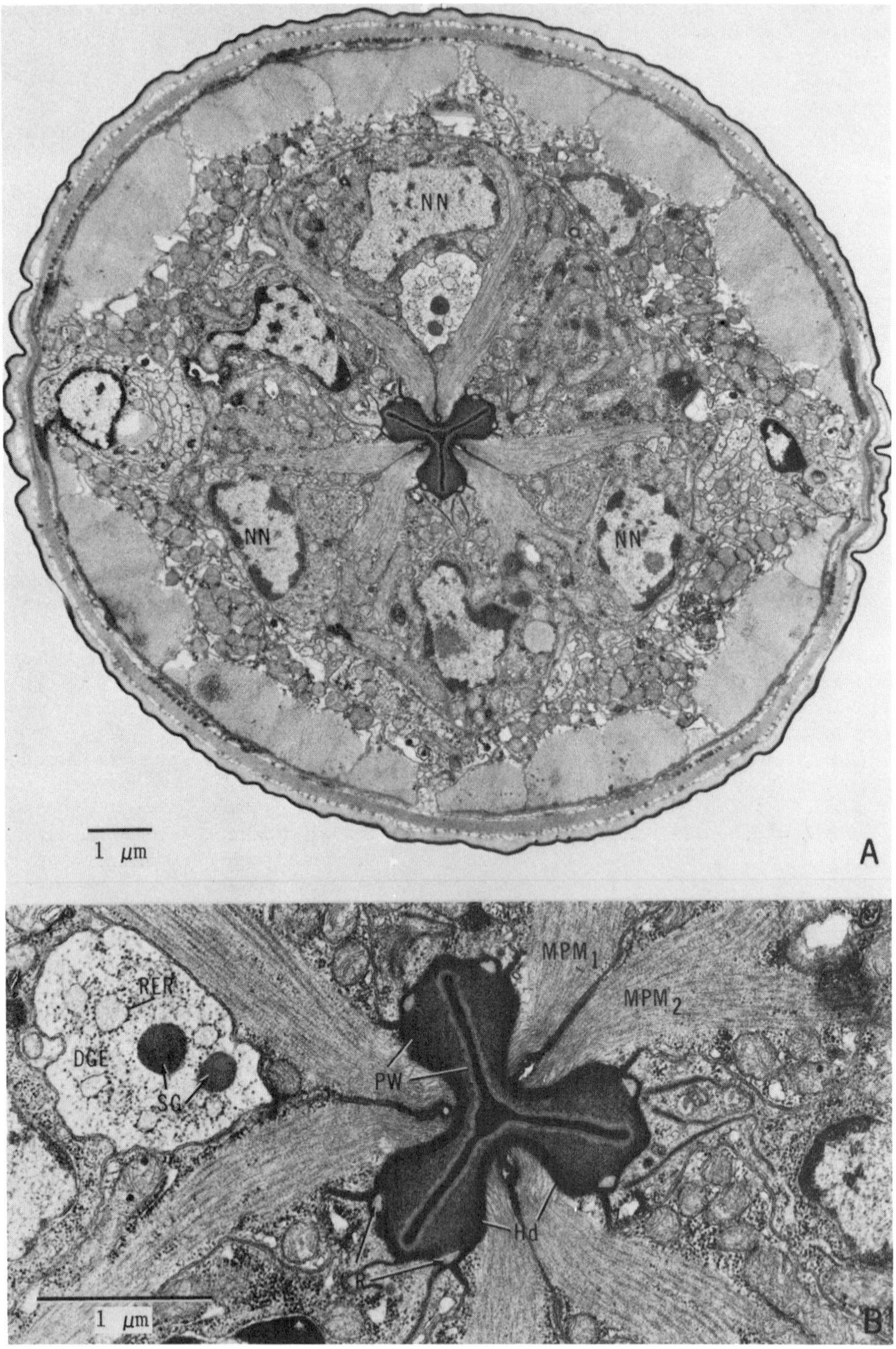

Fig. 6 A. TS through metacorpus of J2 *H. glycines* showing the triradiate pump chamber and muscles, neurons (NN) and somatic or sarcoplasmic

such a nerve process filled with neurosecretory granules lies adjacent to the ampulla of the dorsal gland in juvenile *H. glycines* (Endo, 1984).

The interaction between the host cell and the contents of the ampulla of the dorsal gland is mediated by a tetraradiate valve. In an open position, this valve allows movement of the ampulla contents into the lumen of the oesophagus during salivation. This event is apparently followed by a closure of the valve during ingestion of host contents. Although the dorsal oesophageal valves are observed in the closed positions in males or juvenile *H. glycines*, they have not been observed ultrastructurally in an open position. In contrast, closed and open functional modes have been observed in the subventral valves of J2 *H. glycines* (Endo, 1984; Fig. 5).

The oesophageal lumen of tylenchids is circular in cross section in the procorpus before becoming triradiate in the valve of the metacorpus (Yuen, 1968; Baldwin and Hirschmann, 1976; De Grisse, 1977; Natasasmita, 1979; Endo, 1984). The lumen remains triradiate throughout the isthmus, but as in juvenile *H. glycines*, becomes dilated and ovate before closing at the oesophago-intestinal valve.

At the base of the procorpus or the anterior of the metacorpus, a complex arrangement of muscles surrounds the central core of the procorpus. These muscles appear to function as constraining muscles in *Aphelenchoides blastophthorus* Franklin (Shepherd et al., 1980). In addition to a possible supportive function, this muscle complex in J2 *H. glycines* apparently restricts the movement of secretion granules and cytoplasm of the dorsal gland extension. Cross sections of this region reveal numerous microtubules with accumulations of secretion granules on either end of the restricted dorsal gland extension (Endo, 1984).

The general morphology of the metacorpus of cyst nematodes is similar to that described for *Ditylenchus dipsaci* (Kühn) Filipjev (Yuen, 1968). Myofilaments are attached to the interradiate walls of the triradiate valve of the metacorpus pump. Contraction of the three groups of paired muscles apparently opens the triradiate lumen to allow the ingestion of host cell contents during feeding. The radii of the metacorpus lumen wall appears not to be supported by myofilaments. However, membranes of adjacent cells make direct contact with the tubercles or ridges that border the termini of the triradiate valve (Baldwin et al., 1977; Endo, 1984; Fig. 6). Nuclei within the metacorpus support the muscle elements and supporting cells as well as nerve processes that occur throughout the pump body. The nerve processes are laden with neurosecretory granules and are especially prominent adjacent to the dorsal wall of the dorsal gland extension and the borders of the subventral gland extensions that form subventral ampullae in the metacorpus of J2 *H. glycines* (Endo, 1984).

Fig. 6 (Cont'd) cells. B. Enlargement of metacorpus pump showing the attachment of the pump muscles (MPM) to the pump chamber wall. The pair of electron-translucent "ridges" on each arm of the pump wall in longitudinal section appear as spine-like ridges (CR) that follow the contour of the regions of the pump walls (PW). The dorsal oesophageal gland extension (DGE) has cytolasmic components that include secretion granules (SG), dilated rough endoplasmic reticulum (RER), and microtubules. Hd, hemidesmosomes. (Reprinted from B.Y. Endo, 1984).

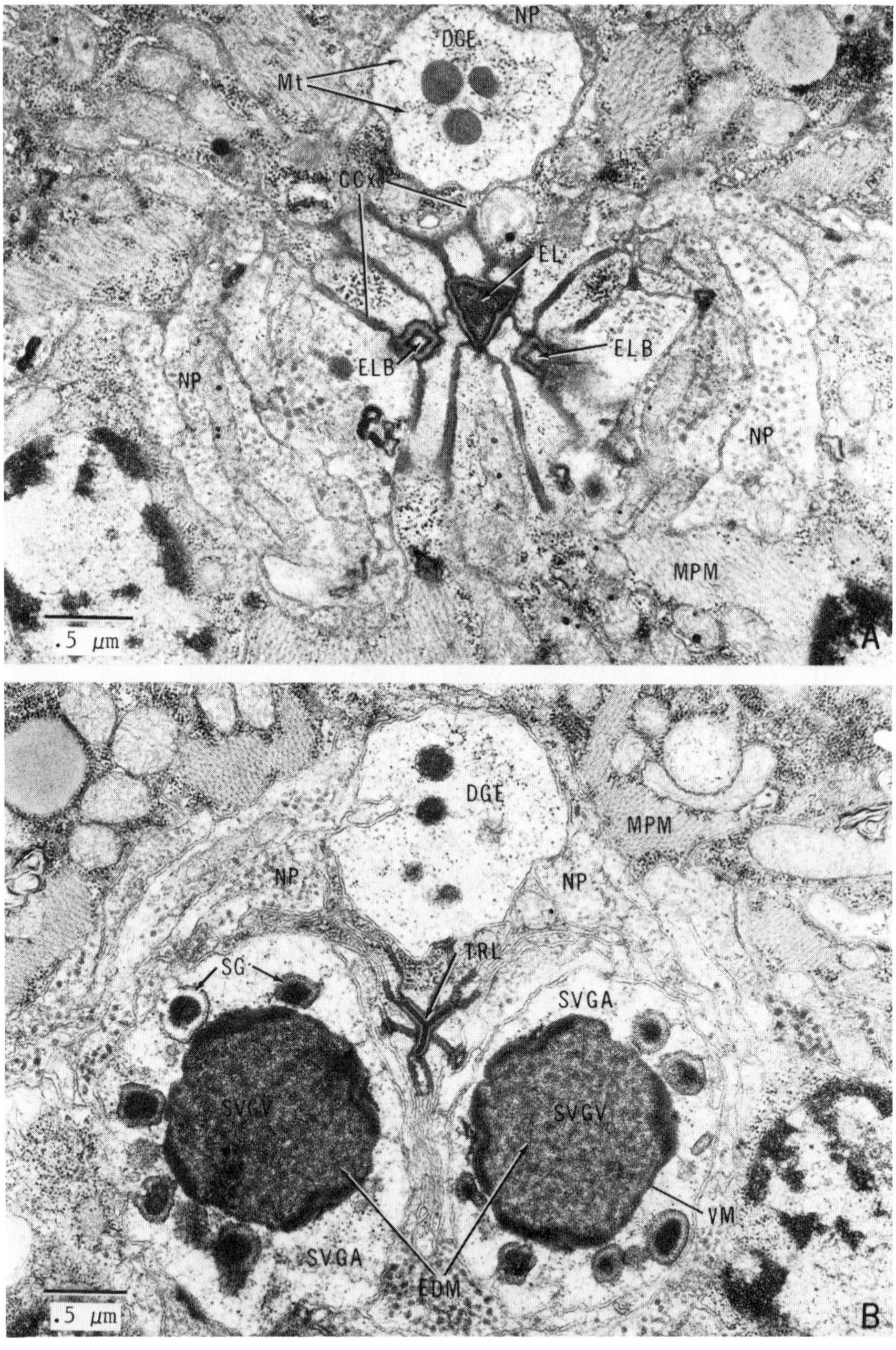

Fig. 7 A. TS through posterior metacorpus of J2 *H. glycines* showing the triangular shape of the central oesophageal lumen (EL) and the

Tetraradiate-shaped valves occur at the terminus of the subventral gland extensions. The lumen of the subventral valves are tetraradiate in cross section when closed, and circular when open (Fig. 7). Connection between the ampullae contents and the central lumen of the oesophagus is through two short bell-shaped channels with thick cuticular walls that are continuous with the wall of the central lumen and the metacorpus valve. Posteriorly, this cuticular branch supports and blends into the supporting arms of the subventral gland valves. Secretory granules in the ampullae of the subventral gland have irregular membrane surfaces bounded by an electron translucent region which in turn borders a central electron dense zone (Endo, 1984).

The manner in which these subventral valves open and close in relation to the metacorpus valve and the independent action of the dorsal oesophageal valves during salivation is not fully documented. The increased resolution available from video-enhanced light microscopy will provide opportunities to determine oesophageal functions that can be verified with current and future ultrastructural observations.

The lobe in *H. glycines* males consists of a dorsal gland with a process that terminates anteriorly as an ampulla in the procorpus and of two indistinctly separated subventral glands with processes that terminate in the posteriad, subventral sectors of the metacorpus (Baldwin et al., 1977). This arrangement of the gland lobe is the same as in J2 *H. glycines* except that membranes clearly define the individuality of the paired subventral gland cells (Endo, 1984). Dorsal and subventral glands of J2s, as well as males, have extensive accumulations of rough endoplasmic reticulum, mitochondria, golgi complexes, and secretion granules. In the living state, glandular secretion granules in juveniles may be observed as they move through the gland extensions using video-enhanced light microscopy (unpublished data).

ANTERIOR SENSORY ORGANS

In an excellent summary of the anterior sensilla of nematodes, Coomans (1979) uses the term sensillum, defined in Webster's Dictionary as a simple epithelial sense organ composed of one or more cells with a nerve connection

---

Fig. 7 (Cont'd) square cross-sectional shape of its branches (ELB) that end in each of two subventral gland ampullae. These cuticular wall structures are connected with a complex of cells (CCx) and membrane junctions. B. TS of open valves (SVGV) within the subventral gland ampullae slightly posterior to that in A. The valve walls have characteristic single and membrane boundaries (VM) that confine the electron-dense material (EDM) which apparently accumulates from the secretory granules (SG). The closed triradiate oesophageal lumen (TRL) extended from terminus of the metacorpus pump vestibule to the valve in the muscular anterior isthmus. The triradiate lumen is here not actuated by muscles but appears to be stabilised by membrane junctions and cells that are adjacent to and include the dorsal gland extension (DGE) and the subventral gland ampullae (SVGA). MPM, metacorpus pump muscle; Mt, microtubules; NP, nerve process (Reprinted from B. Y. Endo, 1984).

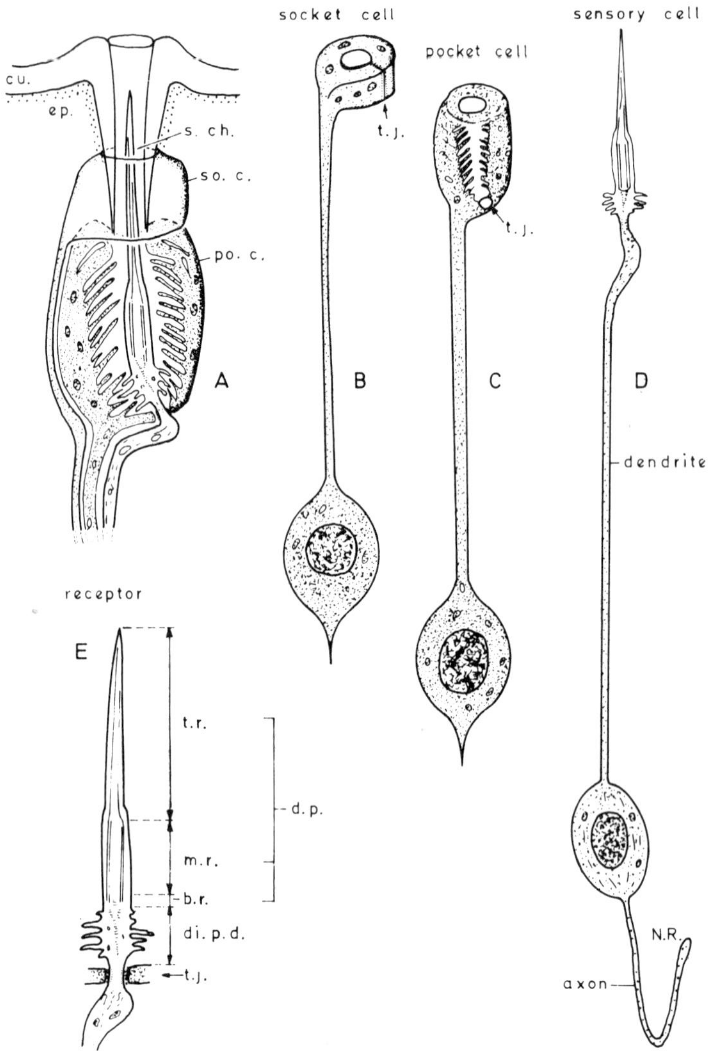

Fig. 8 Basic structure of nematode sensilla. A. Distal part with the three components. B. Supporting or socket cell. C, Gland or pocket cell. D. Sensory cell. E. Ciliary receptor. b.r., basal region; cu, cuticle; dp, dendritic process; di.p.d, distal part of dendrite; ep, epidermis; m.r, median region; n.r, nerve ring; po.c, pocket cell; s.ch, sensillar channel; so.c, socket cell; t.j, tight junction; t.r, terminal region. (Reprinted from A. Coomans, 1984).

that usually takes the form of a spine, plate, rod, cone, or peg. The basic structure of nematode sensillum consists of a neuronal and non-neuronal commponent (Fig. 8). The non-neuronal cell that surrounds the neuronal component of the sensillum can be considered as a specialised epidermal cell with a continuous supporting function and as an intermediate secretory function when forming the cuticular lining of the sensilla channels. Ward et al. (1975) named this supporting cell a socket cell. In *Caenorhabditis elegans* (Maupas) Dougherty, Coomans (1979) considers a second non-neuronal cell of a sensillum as a specialised epidermal cell with probably a continuous function, but with variable activity. He termed this a sensillar gland cell or sensillar gland as in an amphidial gland when referring to the amphid as a sensillum. This same sensillar gland cell is referred to as a pocket cell (Ward et al., 1975) and a goblet cell (De Grisse, 1977).

The neuronal part of the sensory organ consists of a nerve cell body. The anterior extensions of this cell body are referred to as dendrites (Coomans, 1979) and axons (McLaren, 1976; Wergin and Endo, 1976). According to Coomans (1979), the neuronal components of nematode sensilla would be classified as bipolar sensory neurons with the distal parts extending as dendrites with dentritic processes at their termini such as the cilia; cells near the nerve ring; and axons that lead to the nerve ring.

Consistent with the arrangement of the anterior sensory organs as depicted by DeConinck (1965) (Fig. 9) the inner labial, outer labial, cephalic, amphidial, and accessory sensilla are described in the J2 and males of *H. glycines* (Baldwin & Hirschmann, 1975a; Endo, 1980). In terms of function of the anterior sensory organs, Ward et al. (1975) classified these structures into two types, the papillary and amphidial, which are assumed to be responsible for mechano- and chemo-reception, respectively. In the Tylenchids, the inner labial receptors do not terminate as papillae but connect with the exterior through pores. Thus, amphids and inner labial receptors of cyst nematodes are considered to be chemoreceptors, while the outer labial and cephalic receptors are classified as tactoreceptors (Fig. 10). In *C. elegans*, papillary nerve fibres enter directly into the nerve ring, whereas the amphidial nerve fibres enter the ventral ganglion which is a posterior extension of the nerve ring.

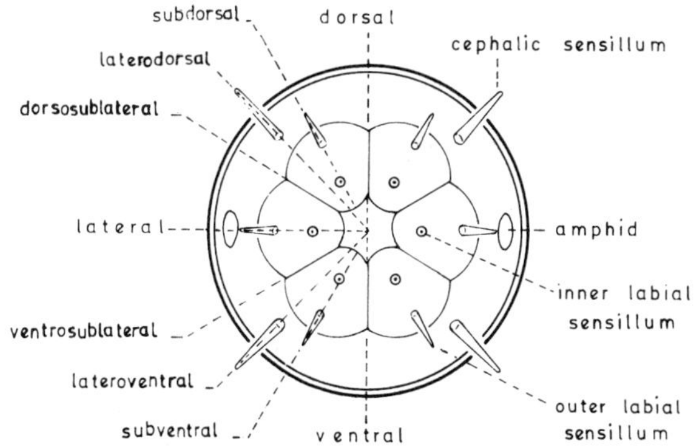

Fig. 9 Basic arrangement of the anterior sensilla in nematodes (modified after DeConinck, 1942). (Reprinted from A. Coomans, 1979).

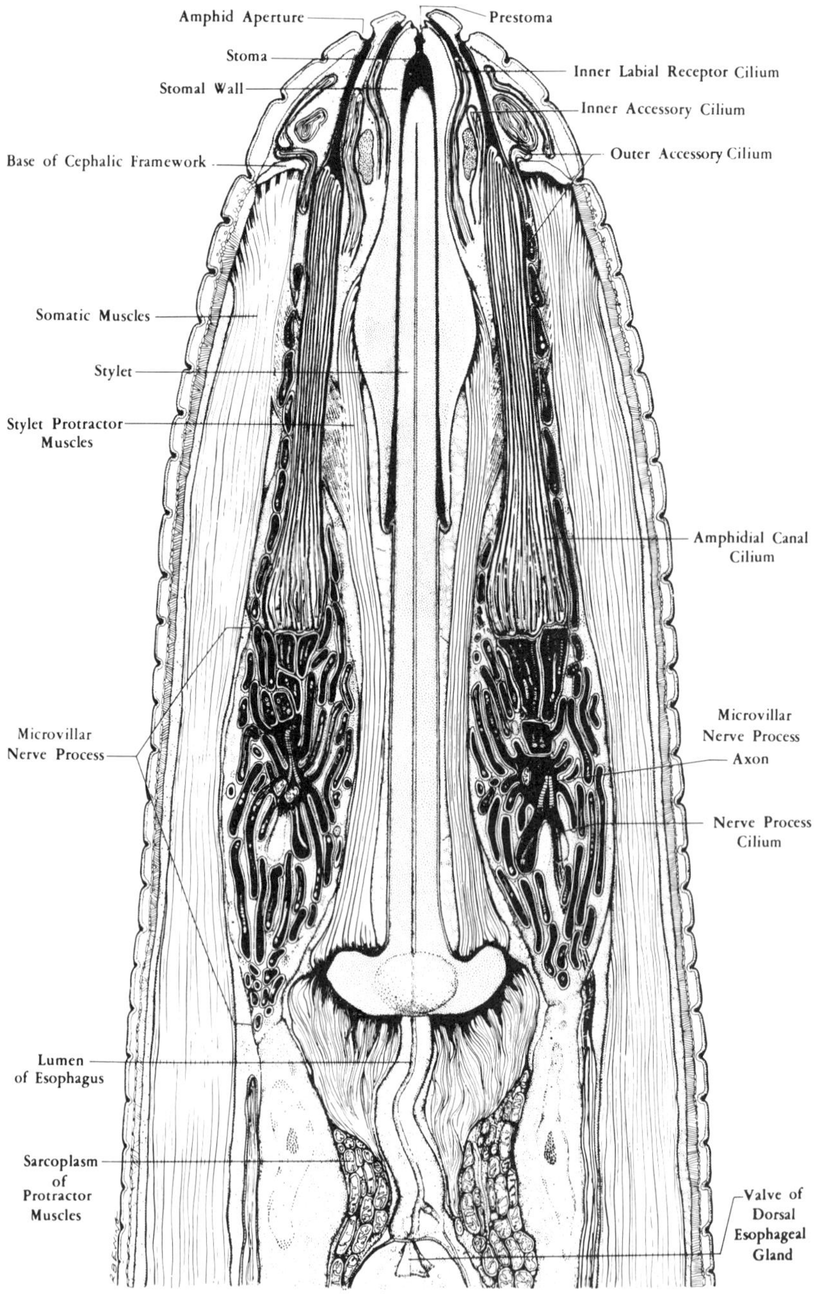

Fig. 10 Composite median dorsoventral view of a juvenile of the soybean cyst nematode, *H. glycines*. Three of seven amphidial cilia are shown. The axon that forms the microvillar nerve process is ventral to the nerve bundle that gives rise to the amphidial and accessory cilia. The cephalic framework includes the base shown here and the hexaradiate cuticular structures anterior to it. (Reprinted from B.Y. Endo, 1979).

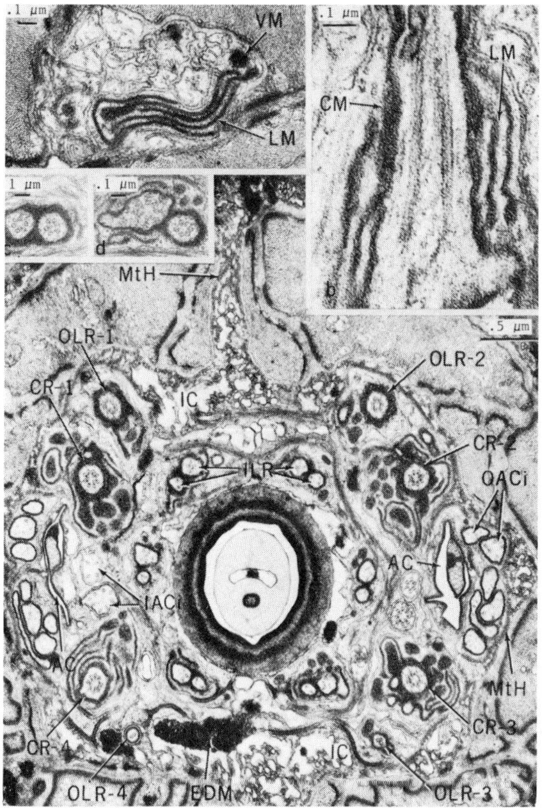

Fig. 11 A section below the cephalic framework of *H. glycines*. Convoluted membranes (CM) and their invaginations into sheath cells that form laminar membranes (LM) are shown in insets (a) and (b). Insets (c) and (d) show two sections through the basal regions of a single inner labial receptor (ILR). The cephalic receptors (CR-1, CR-2, CR-3, CR-4) have up to eight doublets and four singlet microtubules per cilium. Sections of the outer labial receptor cilia (OLR-1 and OLR-2) were made through the basal region of the cilia while the others (OLR-3 and OLR-4) were sectioned anteriad. Inner accessory receptor cilia (IACi) appear circular in outline and lack the membrane complexes of electron-dense deposits that characterise outer accessory receptor cilia. The high number of outer accessory cilia (OACi) in the anterior region is attributed to branching of those cilia present in the basal region. The sheath cells of the various receptor cilia are bordered by a dense network of microtubules (MtH) that extend from the hypodermal layer between the cuticle and somatic musculature to these receptor sites. The hypodermis bordered by these microtubules form irregular channels (IC) that may have electron-dense materials (EDM), especially in the anterior regions of the cephalic framework. AC, amphidial canal; LM, laminar membranes; VM, vesicular membranes. (Reprinted from B.Y. Endo, 1980).

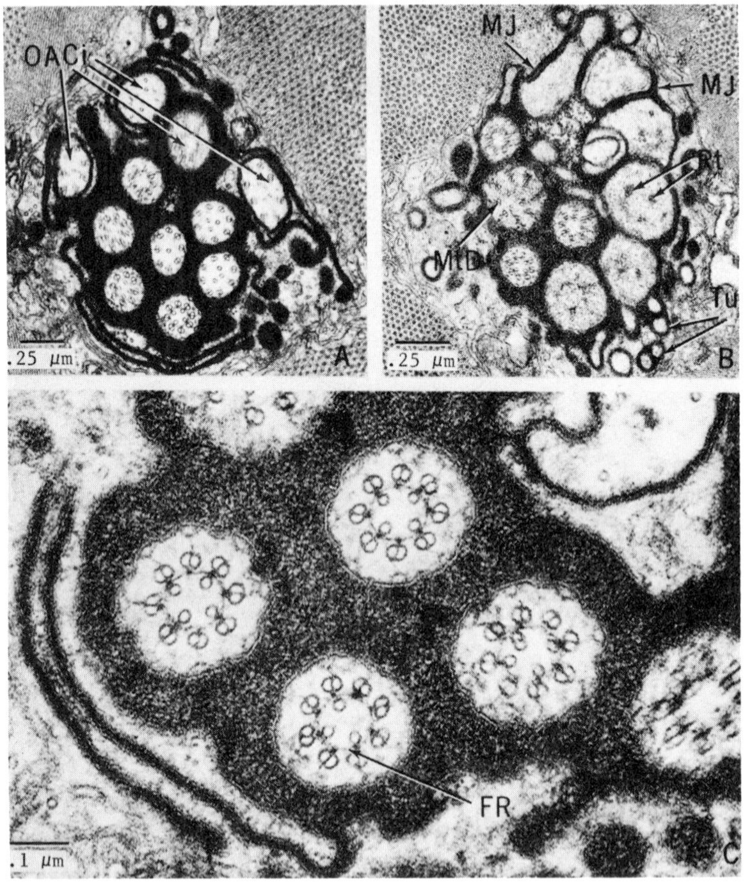

Fig. 12 Sections through the basal regions of the cilia of the amphidial canal and the outer accessory receptors, *H. glycines* A. Electron-dense material is dispersed throughout the interciliary spaces at a level where microtubule doublets and singlets have a symmetrical arrangement. B. A Transverse section near the junction of the cilia and supporting axons that shows a dispersed arrangement of microtubule doublets (MtD), ciliary rootlets (Rt), membrane junctions (MJ) between cilia and a sheath cell, and portions of the tubular (Tu) or microvillar outgrowth of the nerve process. OACi, outer accessory cilia. C. High magnification of a portion of the basal region of the amphidial canal and outer accessory cilia. Four of the cilia show clearly defined peripheral doublets of microtubules in each cilium. The eight evenly spaced doublets of microtubules are attached to a fibrillar ring (FR) centripetally. Each doublet also has a fibrillar strand that extends from the apposed boundaries of their membranes to the cilium membrane. The strand from each doublet traverses the axoplasm to form bifid attachments to the membrane of each cilium. The four singlets that occur within the central ring are attached to the ring opposite alternate microtubule doublets. (Reprinted from B.Y. Endo, 1980).

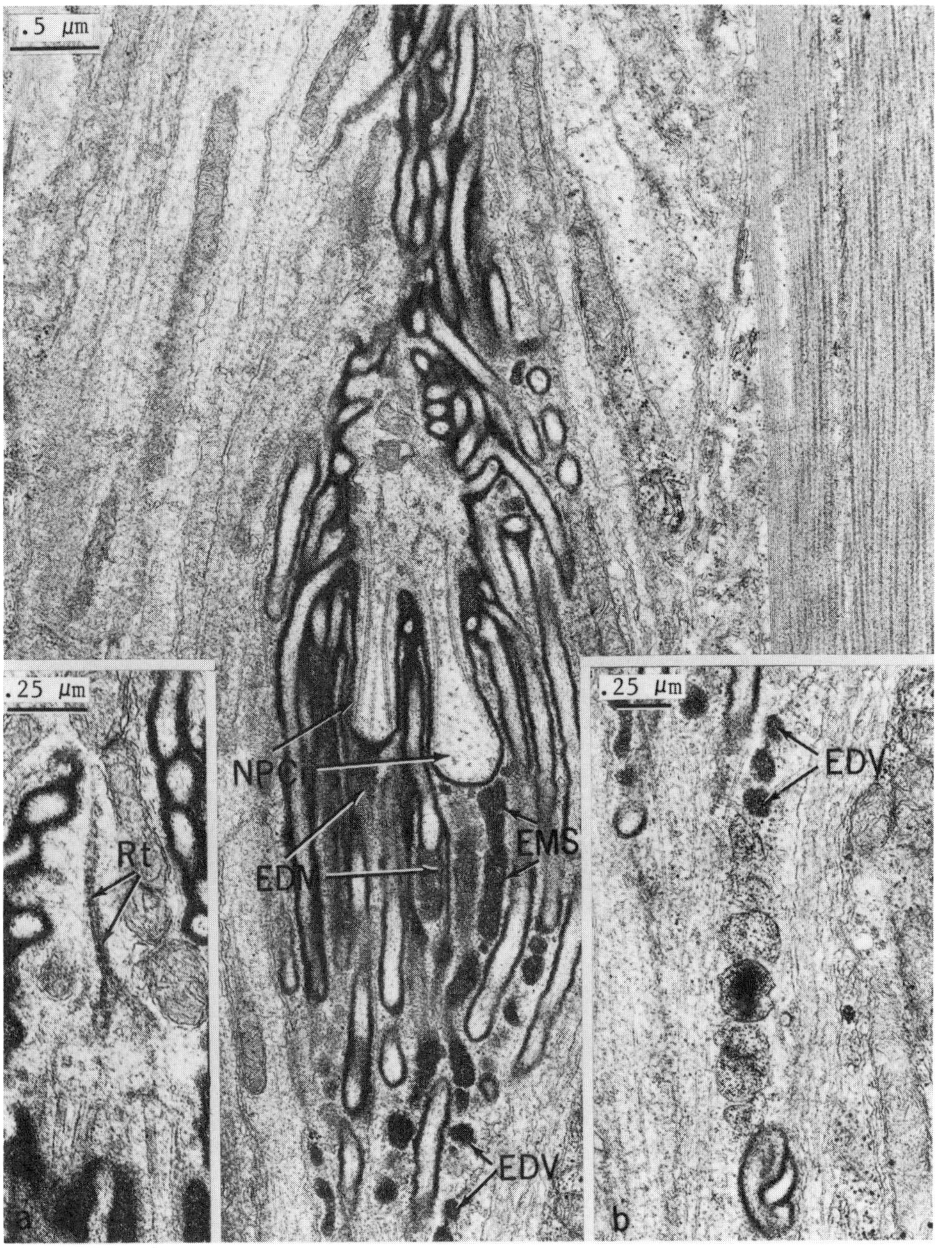

Fig. 13 L.S. of a microvillar nerve process with paired cilia (NPCi) in *H. glycines*. Inset (a) shows the rootlet (Rt) that supports the paired cilia. Inset (b) shows the terminals of some microvilli of the nerve process. Electron-dense materials (EDM) are retained in elongate membrane-like sacs (EMS) and vesicles (EDV). Some electron-dense vesicles (EDV) appear to distend into larger vesicles with granular contents. (Reprinted from B.Y. Endo, 1980).

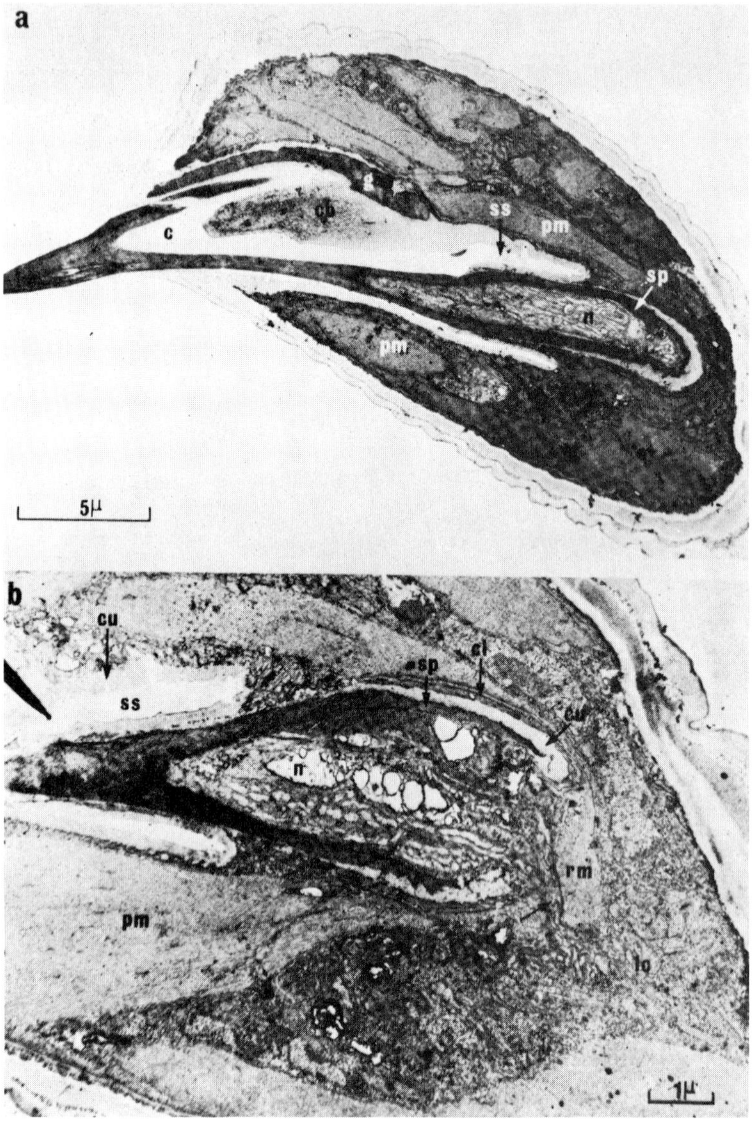

Fig. 14 A. L.S. through the tail tip of an *H. rostochiensis* male, showing the spicule base embedded in the body tissues, with a large nerve (n), and the cuticular guiding bar (cb) projecting into the cloaca (c). g, gubernaculum; pm, protractor muscle; sp, scleroprotein sheath; ss, spicular sac. B. L.S. through the spicule base, showing the connection between the spicular nerve (n) and the lateral chord (lc) (at arrow). The spicule protractor muscle (pm) is attached at one end to the spicule base and at the other to the hypodermis of the ventral body wall. cn, cuticle; cl, compound layer; rm, retractor muscle; sp, scleroprotein layer; ss, spicular sac. (Reprinted from S.A. Clark, A.M. Shepherd and A. Kempton, 1973).

Although the inner labial receptors of males of J2 *H. glycines* and other Tylenchids have cuticularised channels with access to the exterior, only one of the paired cilia appears to have general access to the exterior. The shorter cilium of this paired receptor may act as a tactoreceptor, as is the case for the outer labial and cephalic receptors (Baldwin and Hirschmann, 1975a; De Grisse, 1977; Endo, 1980). The basal body regions of the cilia of the inner labial, outer labial, and cephalic sensilla for *H. glycines* males and J2s are located just below the cephalic framework where they usually display easily recognisable microtubules as rings of eight doublets and four singlets with interconnecting fibrils (Endo, 1980; Fig. 11).

Among the basal body regions of sensilla that are located just posteriad to the base of the cephalic framework are supplementary or accessory sensilla. They are situated between the stomatal wall and the amphidial canal. Natasasmita (1979) established the structure of the accessory sensilla by relating the basal body components to the lamellae located in the lateral, subdorsal, and subventral head sectors. Similar lamellae found in the cephalic sectors of *M. incognita* juvenile are interpreted as nervous structures and reported as terminal accessory axons (Endo and Wergin, 1977).

The relationship between the accessory cilia and the lamellar membranes within the cephalic framework is further clarified in the J2 of *H. glycines* (Endo, 1980). Two types of lamellae occur as part of the accessory receptor system. One type is associated with cilia that have their basal body regions adjacent to those of the amphidial cilia, and another type of cilia with the basal body regions just below the cephalic framework and located just internal to the amphidial canal cuticle. The function of these lamellar membranes and their possible role as photoreceptor are speculative. Coomans (1979) suggests that these receptor surfaces are not unlike certain photoreceptors reported in other nematode species such as *Oncholaimus vesicarius* Weiser (Burr and Webster, 1971; Burr and Burr, 1975).

Tylenchids usually have seven canal receptors for each amphid, but this number can vary (Chen and Wen, 1972; Baldwin and Hirschmann, 1975a; Wergin and Endo, 1976; De Grisse, 1977; Coomans, 1979). Cross sections of *H. glycines* juveniles show the basal body regions of the amphidial cilia and the adjacent accessory receptors with the microtubular configurations of eight doublets and four singlets for each cilium (Fig. 12). The number of microtubules and presence of rootlets vary as serial sections are made through the basal body region. Furthermore, as sections are made through the basal body regions of the accessory sensilla related to the amphid and those just below the level of the cephalic framework, the intermembrane spaces of the lamellae differ. In juvenile *H. glycines*, the lamellae of the amphid-associated outer accessory sensilla have dense electron deposits, whereas intermembrane accumulations are minimal between lamellae associated with the inner accessory sensilla that have their basal body regions located just posteriad to the cephalic framework.

In *M. incognita* (Wergin and Endo, 1976) and *H. glycines* (Endo, 1980), a nerve process with microvilli has been observed with much the same pattern as described for *C. elegans* (Ward et al. 1975). The microvilli of *M. incognita* and of *H. glycines* are part of the lateral nerve bundles. The microvilli of the nerve process are evaginations of dendrites (axons) that make tubular invaginations into the amphidial gland cell. Each of these modified dendrites terminates in a pair of short cilia, with characteristic microtubules in their basal body regions (Endo, 1980; Fig. 13).

In nematode nerve processes dendritic elements are present in association with the pores at the terminals of spicules. Clark et al. (1973) made scanning and transmission EM observations of eleven species of *Heterodera* and showed that the terminals of spicules of *H. (Heterodera) rostochiensis* are basically bifid, whereas those of *H. (Globodera) rostochiensis* have single points. In sections of spicules, they also observed, using transmission electron microscopy, that each spicule contained a large nerve (Fig. 14) that tapered toward the spicule tip and terminated as two dendrites associated with the pair of pores located close to the tip of each spicule. Clark et al. (1973) concluded that the spicules are sensitive organs that are extensively innervated and are virtually cuticular-sheathed nerves. They further suggest that spicules with their terminal pores and dendrites may play a dual role in mating and in transferring sperm.

CONCLUSIONS

Ultrastructural studies provide a better understanding of the relationship of the hypodermis and the cuticle. Future studies should focus on the nature and source of moulting hormones which, if modified, could be used as a mechanism for control. Ultrastructural observations show that the stylet protractor muscle arrangements and attachment sites are similar among various nematode species within the Tylenchidae. Furthermore, fine structure studies show how the stomatal wall folding and attachment of related membranes are related to the stylet guiding ring as observed with light microscopy. The ultrastructure of the oesophagus of juveniles and males of root-knot and cyst nematodes provide a background to study the interaction of the oesophagus and its relation to host cells during salivation and food ingestion. Closely associated with nematode host interaction are the anterior sensory organs or sensilla that regulate nematode movement during host penetration or feeding. Sensilla, such as amphids, have neurosecretory functions that also include the secretion of plug material at the feeding site.

Further investigations on the ultrastructure of nematodes should emphasise anatomical changes that occur during moulting with a view that weak points in development could be sites for nematode control. Recent studies on the stylet initiating zone illustrate the complex nature and cell arrangements related to stylet replacement. Further investigations are necessary to document the sequence of events that lead to complete stylet development during the moulting process. Finally, the combination of ultrastructural information with video-enhanced light microscopy can provide a better understanding of the developmental process of nematodes from which new control strategies may be developed. Among other challenging areas of investigation are the sites and nature of pheromone synthesis related to reproductive physiology. The loss of many currently used nematicides will require future control practices that may depend on biological and alternative chemicals for the disruption of normal host-parasite interactions.

REFERENCES

Baldwin, J.G., 1983, Fine structure of body wall cuticle of females of *Meloidodera charis*, *Atalodera lonicerae*, and *Sarisodera hydrophila* (Heteroderidae), J. Nematol., 15:370.

Baldwin, J.G. and Hirschmann, H., 1975a, Fine structure of cephalic sense organs in *Heterodera glycines* males, J. Nematol., 7:40.

Baldwin, J.G. and Hirschmann, H., 1975b, Body wall fine structure of the anterior region of *Meloidogyne incognita* and *Heterodera glycines* males, J. Nematol., 7:175.

Baldwin, J.G. and Hirschmann, H., 1976, Comparative fine structure of the stomatal region of males of *Meloidogyne incognita* and *Heterodera glycines*, J. Nematol., 8:1.

Baldwin, J.G., Hirschmann, H. and Triantaphyllou, A.C., 1977, Comparative fine structure of the oesophagus of males of *Heterodera glycines* and *Meloidogyne incognita*, Nematologica, 23:239.

Bird, A.F., 1968, Changes associated with parasitism in nematodes. III. Ultrastructure of the egg shell, larval cuticle, and contents of the subventral oesophageal glands in *Meloidogyne javanica*, with some observations on hatching, J. Parasitol., 54:475.

Bird, A.F., 1971, "The Structure of Nematodes." Academic Press, New York and London.

Bird, A.F. and Rogers, G.E., 1965a, Ultrastructure of the cuticle and its formation in *Meloidogyne incognita*, Nematologica, 11:224.

Bird, A.F. and Rogers, G.E., 1965b, Ultrastructure and histochemical studies of the cells producing the gelatinous matrix in *Meloidogyne*, Nematologica, 11:231.

Burr, A.H. and Burr, C., 1975, The amphid of the nematode *Oncholaimus vesicarius* ultrastructural evidence for a dual function as chemoreceptor and photoreceptor, J. Ultrastructure Res., 51:1.

Burr, A.H. and Webster, J.M., 1971, Morphology of the eyespot and description of two pigment granules in the oesophageal muscle of a marine nematode, *Oncholaimus vesicarius*, J. Ultrastructure Res., 36:621.

Byers, J.R., and Anderson, R.V., 1972, Ultrastructural morphology of the body wall, stoma, and stomatostyle of the nematode, *Tylenchorhynchus dubius* (Bütschli, 1873) Filipjev, 1936, Can. J. Zool., 50:457.

Chen, T.A., and Wen, G.Y., 1972, Ultrastructure of the feeding apparatus of *Pratylenchus penetrans*, J. Nematol., 4:155.

Clark, S.A., Shepherd, A.M. and Kempton, A., 1973, Spicule structure in some *Heterodera* spp., Nematologica, 19:242.

Coomans, A., 1979, The anterior sensilla of nematodes, Revue Nematol., 2:259.

DeConinck, L.A.P., 1942, De symmetrie - verhoudingen aan het vooreinde der vrijlevende nematoden, Natuurw. Tijdschr., 24:29.

DeConinck, L.A.P., 1965, in "Traite de Zoologie: Nemathelminthes," P.P. Grasse, ed., Masson, Paris.

De Grisse, A.T., 1977, De ultrastruktuur van het zenuwstelsel in de kop van 22 soorten plantenparasitaire nematoden, behorende tot 19 genera (Nematoda: Tylenchida), D.Sc. Thesis, Rijksuniversiteit Gent.

De Grisse, A.T., Lippens, P.L. and Coomans, A., 1974, The cephalic sensory system of *Rotylenchus robustus* and a comparison with some other Tylenchids, Nematologica, 20:88.

Endo, B.Y., 1980, Ultrastructure of the anterior neurosensory organs of the larvae of the soybean cyst nematode, *Heterodera glycines*, J. Ultrastructure Res., 72:349.

Endo, B.Y., 1983, Ultrastructure of the stomatal region of the juvenile stage of the soybean cyst nematode, *Heterodera glycines*, Proc. Helminthol. Soc. Wash., 50:43.

Endo, B.Y., 1984, Ultrastructure of the esophagus of larvae of the soybean cyst nematode, *Heterodera glycines*, Proc. Helminthol. Soc. Wash., 51:1.

Endo, B.Y., 1985, Ultrastructure of the head region of molting second-stage juveniles of *Heterodera glycines* with emphasis on stylet formation, J. Nematol., 17:112.

Endo, B.Y. and Wergin, W.P., 1977, Ultrastructure of anterior sensory organs of the root-knot nematode, *Meloidogyne incognita*, J. Ultrastructure Res., 59:231.

Günther, B. and Kämpfe, L., 1966, Bau und veranderung des integumentes im entwicklungs-zyklus cystenbildender nematoden, Zool. Anz. Suppl., 30:152.

Lamberti, F. and Taylor, C.E., 1979, "Root-knot Nematodes (*Meloidogyne* species). Systematics, Biology and Control," Academic Press, London, New York, San Francisco.

Maggenti, A., 1981, in: "General Nematology," M.P. Starr, ed., Springer-Verlag, New York, Heidelberg and Berlin.

McLaren, D.J., 1976, Nematode sense organs, Adv. Parasitol., 14:195.

Natasasmita, S., 1979, Transmission and scanning electron microscope observations on *Tylenchulus semipenetrans* juveniles (J2), males and females, Ph.D. Thesis, Rijksuniversiteit Gent.

Shepherd, A.M., Clark, S.A. and Dart, P.J., 1972, Cuticle structure in the genus *Heterodera*, Nematologica, 26:313.

Shepherd, A.M., Clark, S.A. and Hooper, D.J., 1980, Structure of the anterior alimentary tract of *Aphelenchoides blastophthorus* (Nematoda: Tylenchida, Aphelenchina), Nematologica, 26:313.

Ward, S., Thomson, N., White, J.G. and Brenner, S., 1975, Electron microscopical reconstruction of the anterior sensory anatomy of the nematode *Caenorhabditis elegans*, J. Comparative Neurol., 160:313.

Wergin, W.P., 1982, Bibliography on the use of scanning electron microscope (SEM) in studies of plant parasitic nematodes, Scan. Electr. Micros., 2:831.

Wergin, W.P. and Endo, B.Y., 1976, Ultrastructure of a neurosensory organ in a root-knot nematode, J. Ultrastructure Res., 56:258.

Wisse, E. and Daems, W.T., 1968, Electron microscopic observations on second-stage larvae of the potato root eelworm *Heterodera rostochiensis*, J. Ultrastructure Res., 24:210.

Yuen, P.H., 1967, Electron microscopical studies on *Ditylenchus dipsaci* (Kuhn). I. Stomatal region, Can. J. Zool., 45:1019.

Yuen, P.H., 1968, Electron microscopical studies on *Ditylenchus dipsaci*. II. Oesophagus, Nematologica, 14:385.

DISCUSSION

Stone queried whether enough was known about the moulting process in nematodes to make comparisons with insects, particularly with regard to moulting hormones. Endo considered that whilst there was much research ongoing with nematodes (e.g. that on moulting hormones by D.J. Chitwood at Beltsville, Maryland), a lot could be learnt from the extensive investigations with insects. Asked if there was any evidence for the presence of photoreceptors, Endo replied that structure and morphology gave some support to the suggestion of Burr and Burr (J. Ultrastruct. Res., 1975) that amphidial sense organs of certain nematode species may have a dual sensitivity and that nematode photoreceptors may possibly have evolved from chemoreceptors. Wyss commented that during the course of filming the behaviour of *H. schachtii* juveniles (J2) he had noted their response to light, under conditions in which temperature was controlled by filters.

In a discussion on the evolutionary pathways of chemoreceptors in ecto- and endo-parasitic nematodes, Baldwin referred to the small size of the phasmids in male Heteroderids and suggested that there was an evolutionary inclination to a loss of phasmids in these forms. In response to queries relating to particular structures, Baldwin confirmed that the three openings of the dorsal gland in *Meloidodera* all opened into the same duct, so they were unlikely to be three different types. It was acknowledged that secretory granules would have to be liquified to form saliva and Baldwin suggested that this might be effected by a kind of reverse kinocystosis.

TESTING HYPOTHESES OF PHYLOGENY OF HETERODERIDAE

James G. Baldwin

Department of Nematology, University of California
Riverside, CA 92521

INTRODUCTION

Knowledge of phylogeny of Heteroderidae is essential to achieving stable meaningful systematics. Stable systematics has been elusive. New genera are frequently described and genera are variously grouped into subfamilies and other taxa by a number of authors. Luc et al. (1978) pointed out "heterogenity and inconsistencies" within subfamilies and proposed abolishing existing subfamilies. They suggested that arrangement of genera into meaningful (i.e. phylogenetic) groups must await additional information. Currently, characters useful for interpreting phylogenetic relationships are scarce. Nevertheless, systematists have relied on these limited characters to propose several competing hypotheses of phylogeny of Heteroderidae (Wouts, 1973; Stone, 1977; Krall and Krall, 1978; Ferris, 1979, 1985). Further advancement requires testing, refining and retesting of these hypotheses.

INITIAL HYPOTHESES AND APPROACH TO TESTING

Dr Virginia Ferris (1979)* proposed a phylogeny of Heteroderidae, using cladistic methods, and challenged us to test it (Table 1). Her approach is distinct from previous hypotheses for Heteroderidae because it infers common ancestry, not from overall similarity but on the basis of shared features which are derived (apomorphic) from previously existing features (primitive or plesiomorphic). The special value of shared derived features for interpreting phylogeny is apparent in some simplified examples. Such as relationships among lizards, mice and humans might be interpreted from the occurrence of mammary glands in mice and humans; this shared derived feature is the result of common ancestry. Conversely, the presence of a tail, a primitive feature, is not reliable for proposing common ancestry of lizards and mice! Similarly, an abridged phylogeny of Heteroderidae (Fig. 1) illustrates how a similar primitive en face pattern might be inherited in juveniles of distantly related *Meloidodera charis* Hopper and *Cactodera eremica* Baldwin & Bell; clearly this shared primitive feature is not a valid basis for proposing a unique common ancestor between

*More recently, Ferris (1985) has revised this phylogeny based, in part, on our tests of the original hypothesis, some of which are reviewed in this paper.

Table 1 Characters of Heteroderidae, their polarity and resulting cladogram (after Ferris, 1979).

Character	Name	Primitive State	Derived State
1	Vulva	subequatorial	subterminal
2	♀ Cuticle	annulated	not annulated
3	Spicules	short (<30 μm)	long (> 30 μm)
4	Anus-vulva	widely separated	close together (4); or located on prominence (4+)
5	♀ Body wall	no cyst	rugose (5) or cyst (5+)
6	Phasmids J2	lenslike	porelike
7	Perivulva	not fenestrated	fenestrated
8	♀ Shape	irregularly or lemon shaped	round or pear shaped

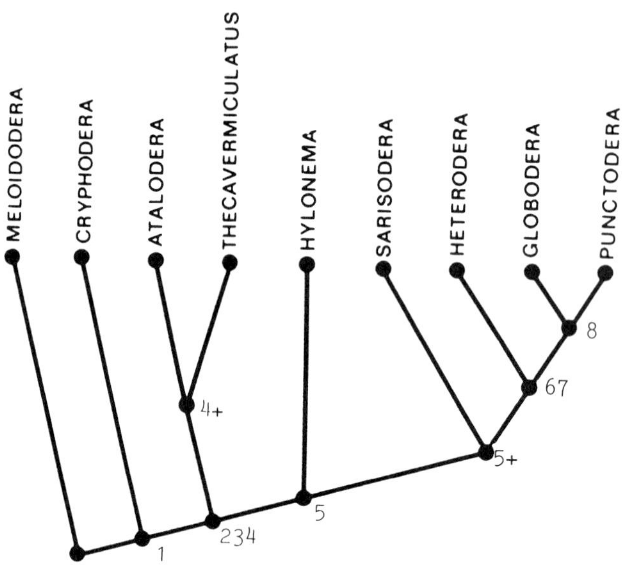

the two species. Conversely, juveniles of *Sarisodera hydrophila* Wouts & Sher and *Rhizonema sequoiae* Cid del Prado Vera, Lownsbery & Maggenti, share specialised derived lip patterns, which with several other shared derived features, support common ancestry between the two species (Othman and Baldwin, 1985b).

While the value of distinguishing between primitive and derived features (determining polarity) may be evident, the means by which the

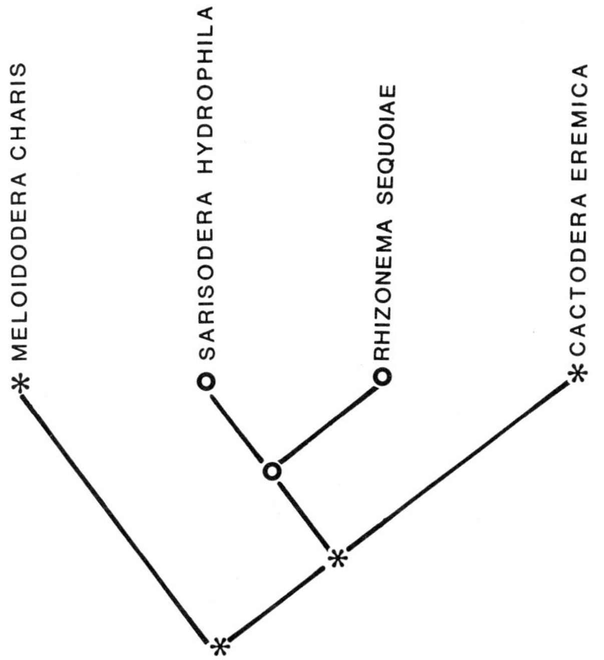

Fig. 1 Abridged phylogeny of Heteroderidae showing possible divergent distribution of a primitive character state. *, primitive lip pattern; 0, derived lip pattern.

distinction can be made may be less clear. Polarity is also a hypothesis that is subject to testing and revision (Eldredge and Cracraft, 1980). Hypotheses of polarity might be based on outgroup comparisons, i.e. if a character occurs in two or more states in Heteroderidae and one of these states if also present in a closely related group (outgroup) that character state is considered primitive within Heteroderidae. There is a body of evidence that Hoplolaimidae is a reliable outgroup and may share a unique common ancestry with Heteroderidae (Wouts and Sher, 1971, Baldwin, 1983). This hypothesis can be used, for example, to propose polarity of the surface cuticle pattern of female heteroderids which may be annulated or irregular. It could be argued that the irregular pattern evolved within Heteroderidae, and that the annulated pattern is ancestral, because annulation occurs throughout the outgroup, Hoplolaimidae.

Development might also be useful for hypothesising polarity because general (primitive) character states tend to appear earlier in development than specialised (derived) characters (K. Von Baer as discussed by Eldredge and Cracraft, 1980). For example, all heteroderid juveniles have annulated cuticles, but only in certain species does the more specialised irregular pattern develop at maturity.

Ferris (1979) hypothesised polarity of eight characters to propose a phylogeny of Heteroderidae (Table 1). This phylogeny was the most parsimonious explanation of the available data, with shared derived characters to support every proposed common ancestor. For example, *Sarisodera, Heterodera, Globodera* and *Punctodera* were believed to share cysts (Table 1). The most parsimonious explanation is that the cyst evolved once and was inherited by these four genera, i.e. they have a common ancestor. This simple explanation is acceptable unless and until new evidence, such as incongruence with the distribution of a set of other characters, indicates an alternate hypothesis, i.e. the cysts are not homologous, having evolved more than once by parallel or convergent evolution.

Ferris (1979) emphasised that the proposed phylogeny of Heteroderidae was not intended as conclusive but as a point of departure for testing, refining and retesting. Several general categories have been investigated at the University of California, Riverside, for testing the hypothesis of phylogeny of Heteroderidae:

(1) New taxa. New species may refute a proposed phylogeny or strengthen it by readily being accommodated into a similar scheme. Since Ferris' (1979) phylogeny several new species and the genera *Afenestrata, Dolichodera, Rhizonema,* and *Verutus* have been described. A phylogeny including these new taxa will test the original scheme.

(2) Re-evaluation of classical characters. A phylogeny may be falsified if characters used in developing the scheme are later shown to have been assigned a doubtful polarity or if the characters are incorrectly ascribed. In the phylogeny by Ferris (1979), for example, a cyst is attributed to *Sarisodera*. However, Baldwin and Bell (1985) reported that the type species, *S. hydrophila,* lacks a cyst, whereas a second species, *Afenestrata africana (= Sarisodera africana)* Luc, Germani & Netscher has a cyst. Similarly, *Meloidodera* and *Hylonema* were miscoded in the original scheme as having juveniles with lens-like phasmids (Table 1); most species of *Meloidodera* and the only species of *Hylonema* have pore-like phasmids. Thus, reassessment of classical characters may indicate a basis for revision of phylogeny (e.g. Ferris, 1985).

(3) New characters and character states. Hypotheses of phylogeny of Heteroderidae would be strengthened if they were based on a large number of reliable characters. While new characters may include non-morphological parameters such as ecology, biochemistry, and cytology, the focus of our ongoing investigations is on detailed morphology through combined transmission electron (TEM), scanning electron (SEM) and light microscopy (LM). New characters revealed through this approach include layers of the cuticle in females and lip patterns of juveniles, males, and females. New character states may also be revealed, e.g. with SEM the surface patterns of females coded by Ferris as annulated and not annulated (Table 1) may be concluded to be a transformation series with many intermediate expressions (Othman and Baldwin, unpublished observations).

(4) Detection of homoplasy. It is highly unlikely that all characters are equally correlated with the evolutionary history of Heteroderidae because of homoplasy (parallel and convergent evolution). The detection of homoplasy is essential for testing and refining hypotheses of phylogeny. A high degree of coincidence in the distribution of a large suite of characters throughout Heteroderidae would strengthen arguments of homology (absence of homoplasy) of each. Conversely, certain exceptional incongruent characters may be suspect as homoplasic, particularly if similar environmental selection pressures might favour repeated evolution of a character state. The stylet of Tylenchida and that of certain Dorylaimida is an example of homoplasy. A unique common ancestor is not proposed for these taxa because a large suite of reliable characters rejects such an interpretation. A common selection pressure for development of an adaptation to plant feeding is clear, and detailed observations

of morphology and development indicate fundamental differences between the two types of stylets. Similarly, homoplasy within Heteroderidae might be identified by introduction of additional reliable characters, knowledge of function and selection pressures, and comparative detailed morphology.

Ongoing studies which are designed to provide new insights for testing and refining hypothesis of phylogeny of Heteroderidae are discussed in four areas:

host responses;
layering of the female body wall cuticle;
cuticular surface patterns (SEM);
phasmids.

HOST RESPONSES

To test hypotheses of phylogeny of Heteroderidae "host response" was introduced as a new character (Mundo-Ocampo and Baldwin, 1983a, b, c, 1984). Initially, and supported by previous observations, it was hypothesised that (i) host responses vary throughout Heteroderidae, and (ii) the type of host response is a function of the nematode and is stable regardless of conditions such as host species. *Heterodera* spp. and *Globodera* spp. have been described on several hosts as inducing similar syncytia with wall ingrowth (e.g. Feldmesser, 1953; von Sembdner, 1963; Endo, 1964). In contrast, Taylor et al. (1978) described a "unique" single uninucleate giant cell induced by *Hylonema ivorense* Bernard on three hosts; wall ingrowths were not reported. To test the initial hypothesis, host responses were examined in several species of thirteen genera of Heteroderidae using a combination of LM, SEM and TEM (Table 2). The response was observed in more than one host, where feasible, and in several cases the response of different heteroderid genera, as well as *Meloidogyne* spp. were examined on the same host. Each nematode species consistantly induced a characteristic response, regardless of host.

Meloidodera spp., *Cryphodera utahensis* Baldwin, Mundo-Ocampo & Othman, *Hylonema ivorense, Sarisodera hydrophila,* and *Rhizonema sequoiae* each induce a single uninucleate giant cell (Cid del Prado Vera and Lownsbery, 1984; Mundo-Ocampo and Baldwin, 1983b, c, 1984; Taylor et al., 1978; Fig. 2A). These host cells are fundamentally similar with respect to the uninucleate condition; all lack wall ingrowths, but pit fields with numerous plasmodesmata occur between the giant cell and adjacent vascular tissue. Finer differences, such as morphology of nuclei and wall thickenings of the giant cell, were interpreted as specific to the nematode species or genus (Mundo-Ocampo and Baldwin, 1984).

In contrast to the single uninucleate giant cell, host responses of *Atalodera* spp. and *Thecavermiculatus* spp. induce similar multinucleate syncytia superficially similar to those of *Heterodera* and *Globodera* (Mundo-Ocampo and Baldwin, 1983a; Fig. 2B). Unlike syncytia of the cyst-forming species, however, wall ingrowth are lacking; instead pit fields and abundant plasmodesmata are present.

If it is accepted that the single uninucleate giant cell is the original or primitive state and the syncytium (which begins development as a single enlarged uninucleate cell) is the derived state, then the pattern of host responses throughout Heteroderidae is generally congruent with Ferris' (1979) original hypothesis of phylogeny (Fig. 3). The syncytium of *Atalodera* and *Thecavermiculatus* appears to be an exception. In this case

Table 2 Host responses induced by Heterodereridae (modified and expanded from Mundo-Ocampo and Baldwin, 1984).

Species	Host response	Cell wall ingrowths
Meloidodera floridensis	SUGC[a] (Ruehle, 1963; Mundo-Ocampo & Baldwin, 1983c)	Absent
M. charis	SUGC (Heald, 1978; Mundo-Ocampo & Baldwin, 1983c)	Absent
M. belli	SUGC (Mundo-Ocampo & Baldwin, 1983c)	Absent
Cryphodera utahensis	SUGC (Mundo-Ocampo & Baldwin, 1984)	Absent
Atalodera ucri	Syncytium (Mundo-Ocampo & Baldwin, 1983a)	Absent
A. lonicerae	Syncytium (Mundo-Ocampo & Baldwin, 1983a)	Absent
Thecavermiculatus gracililancea	Syncytium (Mundo-Ocampo & Baldwin, unpublished)	Absent
Hylonema ivorense	SUGC (Taylor et al., 1978)	Not reported
Sarisodera hydrophila	SUGC (Mundo-Ocampo & Baldwin, 1983b)	Absent
Rhizonema sequoiae	SUGC (Cid del Prado Vera & Lownsbery, 1984)	Absent
Afenestrata africana	Syncytium (Baldwin & Bell, 1985)	Absent
Heterodera spp.	Syncytium (e.g. Endo, 1964)	Present
Globodera rostochiensis	Syncytium (e.g. Feldmesser, 1953)	Present
Cactodera eremica	Syncytium (Baldwin & Bell, 1985)	Present
Punctodera chalcoensis	Syncytium (Suarez et al., 1985)	Not reported
Verutus volvingentis	Syncytium (Cohn et al., 1984)	Not reported

[a]Single uninucleate giant cell.

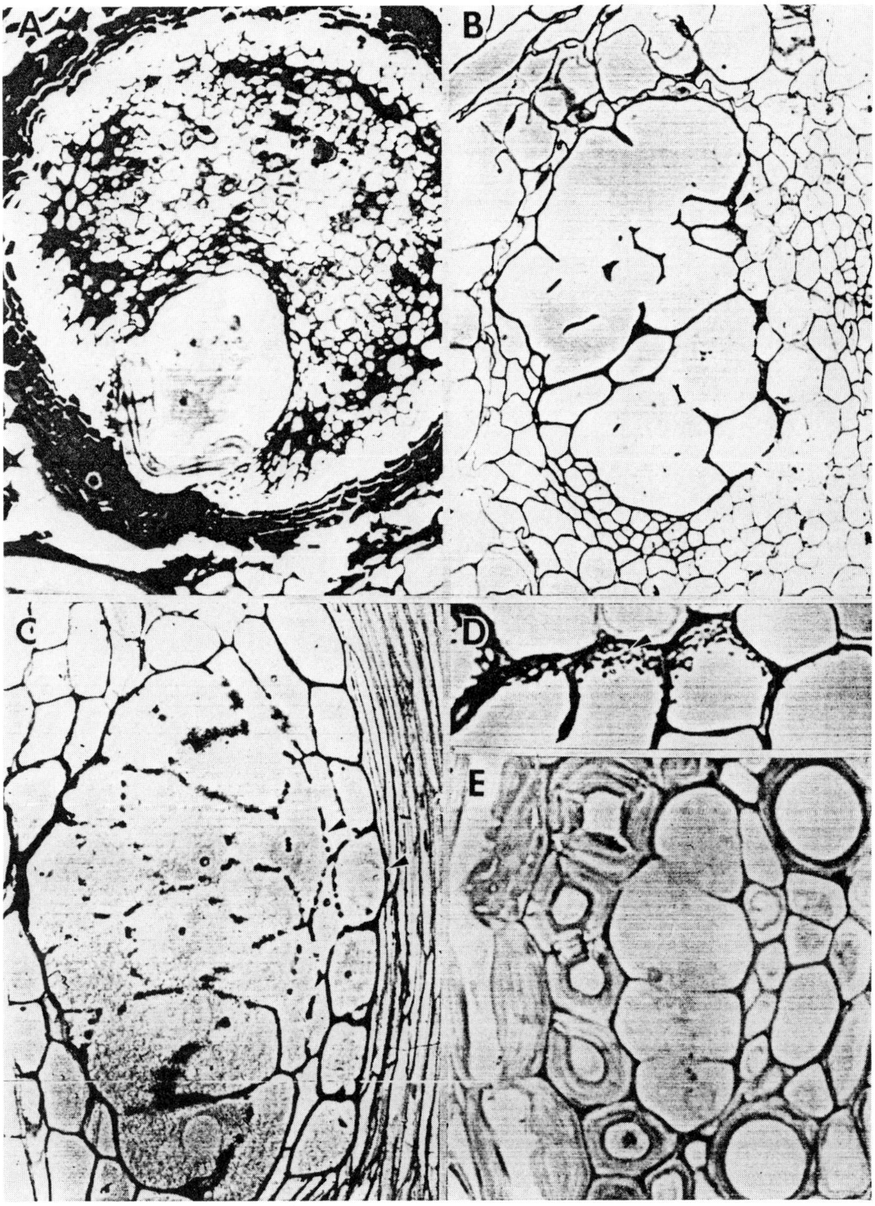

Fig. 2 Host responses of Heteroderidae. A) Single uninucleate giant cell induced by *Sarisodera hydrophila* on *Salix lasiolepis*. B) Multinucleate syncytium with wall ingrowth (arrow) induced by *Cactodera eremica* on *Atriplex confertifolia* (after Baldwin and Bell, 1985). C) Multinucleate syncytium with pit fields (arrows) induced by *Atalodera lonicerae* on *Lonicera involucrata*. D) Enlargement of wall of multinucleate syncytium induced by *C. eremica* (section adjacent to that shown in B) showing wall ingrowths (arrow). E) Syncytium induced by *Afenestrata africana* on *Panicum maximum* (after Baldwin and Bell, 1985).

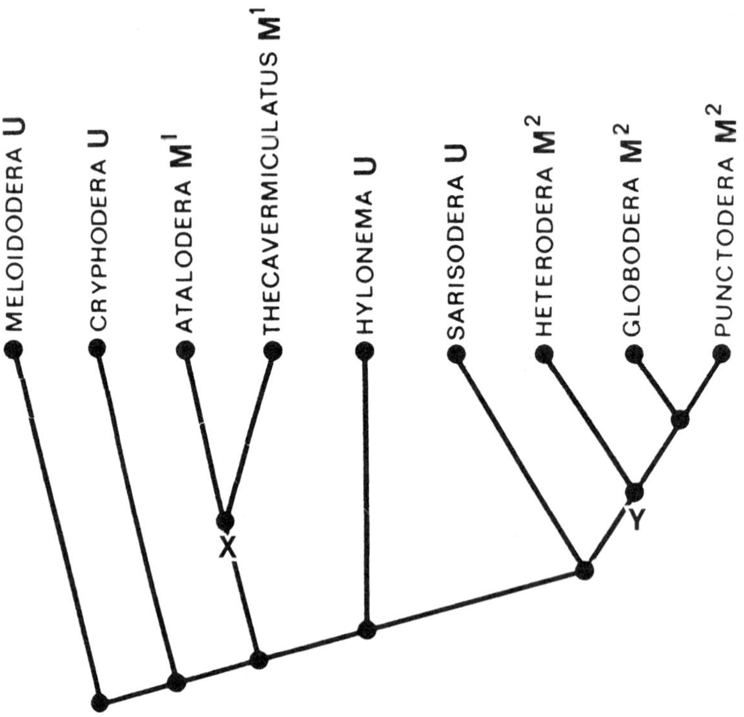

Fig. 3 Phylogeny of Heteroderidae proposed by Ferris (1979) in relation to distribution of single uninucleate giant cell (U), multinucleate syncytia believed to have wall ingrowth (M^2) with ancestor Y.

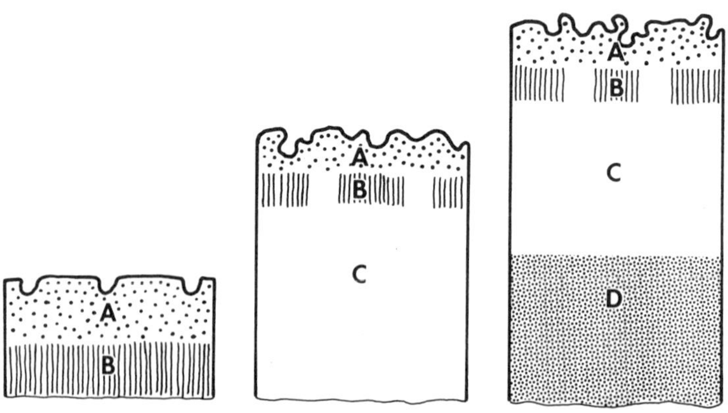

Fig. 4 Layering of body wall cuticle of Heteroderidae in which males and second-stage juveniles have layers A, B and females of species have A, B, C whereas other species have A, B, C, D. (Redrawn from Shepherd et al., 1972).

the phylogeny requires parallel evolution, i.e. the syncytium arose independently in the common ancestor (X) of *Atalodera* and *Thecavermiculatus*, and again in the ancestor of *Heterodera* sensu lato (*Heterodera*, *Globodera* and *Puntodera* with ancestor Y). The hypothesis of parallel evolution is plausible since the two types of syncytia are fundamentally different with respect to the mechanisms of transporting nutrients, i.e. wall ingrowths versus pit fields.

While not yet included in a phylogenetic analysis, recent observations of host responses in additional species (Table 2) seem to be phylogenetically informative. The single uninucleate giant cell of *Rhizonema* (Cid del Prado Vera and Lownsbery, 1984) was predicted on the basis of other characters which indicate a common ancestor with *Sarisodera*. Similarly, the multinucleate syncytium with wall ingrowths in *Cactodera* might be expected based on similar responses in other cyst forming Heteroderidae (Baldwin and Bell, 1985; Fig. 2C). On the other hand, *Punctodera chalcoensis* Stone, Sosa Moss & Mulvey, while not yet examined with TEM, appears, with LM to lack typical wall ingrowth, contrary to our predictions (Suarez et al., 1985). In addition, *Afenestrata* (Baldwin and Bell, 1985; Fig. 2E), an unusual cyst-forming heteroderid, and *Verutus* (Cohn et al., 1984) a noncyst-forming genus, induce unusually diminutive syncytia. The presence of a syncytium in *Verutus* (Cohn et al., 1984) may be incongruent with Ferris' (1985) more recently proposed phylogeny. However, the significance of these responses must be considered in combination with conventional and other new characters for a revised phylogenetic analysis of Heterodiderae. In addition, future investigations correlating particular host responses with specific biochemical agents of the nematode "saliva" will provide a tool for finer characterisation of host-response differences among species and for testing hypotheses of homology and homoplasy.

LAYERING OF THE FEMALE BODY WALL CUTICLE

LM and TEM observations of the body wall cuticle of mature females reveal new characters useful for testing hypotheses of phylogeny of Heteroderidae (Baldwin, 1983; Cliff and Baldwin, 1985). The basic or primitive pattern of cuticle layering which occurs throughout Tylenchida and in second-stage juveniles and males of Heteroderidae consists of the cortex (A) and striated (B) layers. Shepherd et al. (1972) found that an inner layer (C) is formed as females of *Heterodera* sensu lato enlarge; in some species an additional layer (D) also develops (Fig. 4). Observations of species of ten genera indicate diversity of layering throughout the family (Fig. 4). It is hypothesised that the presence of D with C is derived relative to C alone. This is supported by the presence of an inner layer in adults of the outgroup, Hoplolaimidae, which appears to be homologous with C. In addition, D is interpreted as derived because D forms subsequent to C as heteroderid females mature.

Distribution of D throughout Heteroderidae generally is consistent with Ferris' phylogeny (Fig. 5). A striking discrepancy, however, is that D is shared by *Atalodera*, *Sarisodera*, as well as *Globodera*, *Cactodera*, *Punctodera*, but is absent in *Heterodera* and *Afenestrata*. At least two hypotheses are indicated: (i) absence of D is due to a reversal in a unique ancestor of *Heterodera* - *Afenestrata* and (ii) *Atalodera*, *Sarisodera*, *Globodera*, *Cactodera* and *Punctodera* share a common ancestor not shared by *Heterodera* - *Afenestrata* which is consistent with the possibility that the cyst arose independently in the ancestor of *Heterodera* and again in a separate ancestor of *Globodera* and closely related genera.

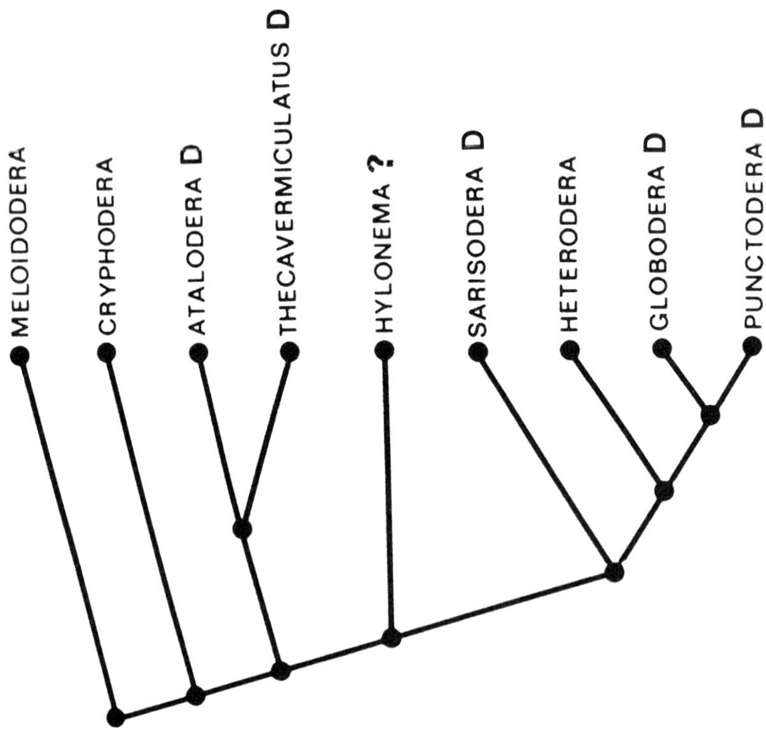

Fig. 5 Phylogeny of Heteroderidae proposed by Ferris (1979) in relation to distribution of D-layer in the female cuticle. Those unlabelled have a C-layer but lack D. The cuticle of females of *Hylonema* has not been examined.

These hypotheses can be tested by consideration of the congruence of the body wall character with a suite of other reliable characters, and by tests of homology of cysts through comparative morphological, developmental, and biochemical investigations.

Othman and Baldwin (1985a) predicted similar layering in *Rhizonema sequoiae* and *Sarisodera hydrophila* because many shared derived characters indicate a common ancestor between the two genera. Curiously, *R. sequoiae* has simple layering excluding D, whereas *S. hydrophila* has elaborately thick layers including D (Fig. 6). External morphology may be useful in interpreting these differences; *Rhizonema sequoiae* has an annulated surface, except in the cone area where the pattern is irregular, whereas *Sarisodera hydrophila* is annulated far anteriorly, but is otherwise irregular (Fig. 7). As the cross sections for our observations were from midbody, it is hypothesised that the cuticle in the irregular posterior region of *Rhizonema* may be layered with D similar to *Sarisodera*, whereas the annulated anterior portion of *Sarisodera* females may resemble the midbody of *Rhizonema*. If these prove to be correct, it indicates the need to consider layering throughout the body for the interpretation of relationships among additional Heteroderidae.

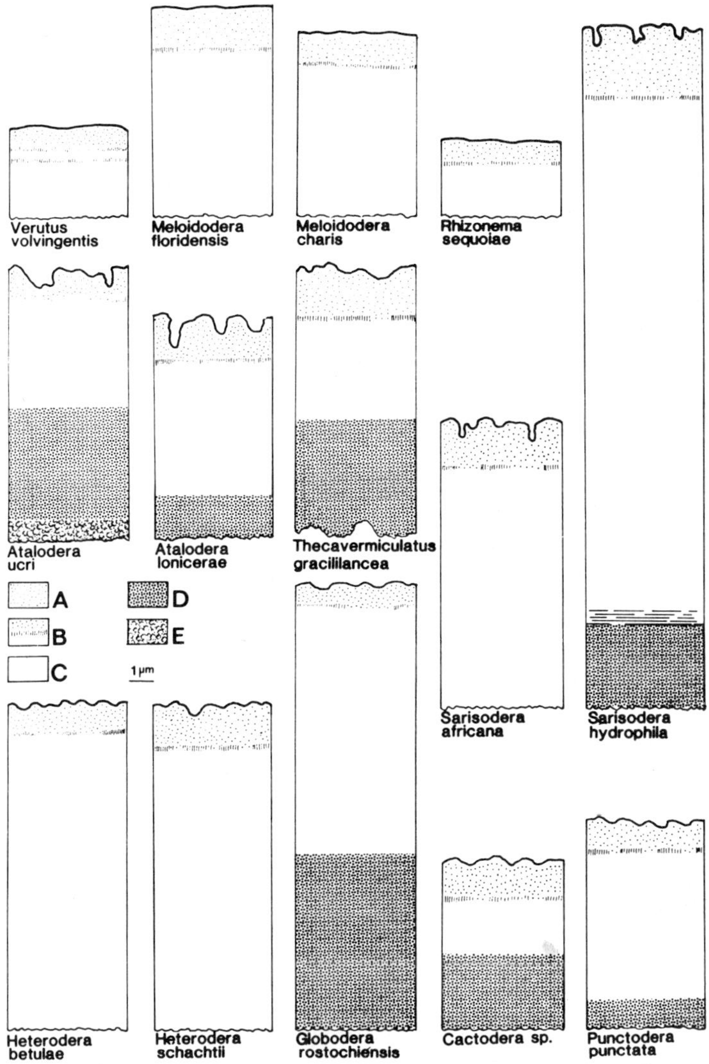

Fig. 6 Diagram illustrating layering of cuticle in midregion of mature females of 14 species of Heteroderidae (Cliff and Baldwin, 1985).

The difference in layering of *S. hydrophila* and *Afenestrata africana (= Sarisodera africana)*, both with irregular surface patterns, is enlightening (Fig. 6). This difference, in conjunction with additional contrasting characters, contributes to the justification for placement of *S. africana* in the separate genus *Afenestrata* (Baldwin and Bell, 1985). A number of shared derived characters may indicate that *Afenestrata* is a sister group of *Heterodera*, with which it also shares a similarly layered cuticle.

The thin cuticle of *Verutus* is unique by the presence of multiple B layers (Fig. 6). This feature is consistent with other characters, such as host response and lip patterns, which seem to distinguish it from

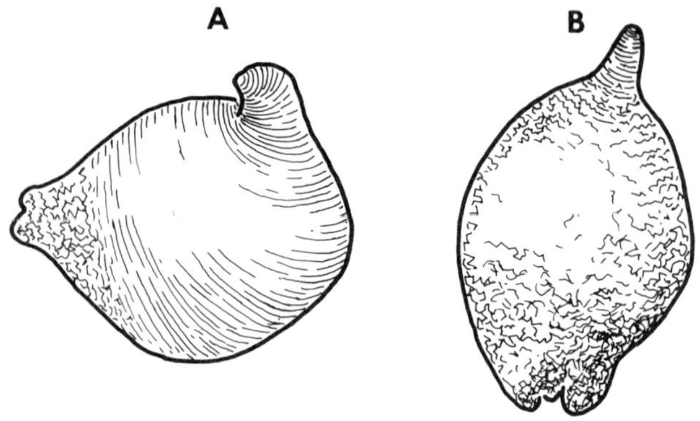

Fig. 7 Diagrammatic representation of distribution of annulated and zig-zag surface patterns of females of *Rhizonema sequoiae* (A) and *Sarisodera hydrophila* (B).

Meloidodera, a genus with which it shares a number of primitive characters including a subequatorial vulva (Esser, 1981). However, relative to derived characters, such shared primitive characters are not reliable for inferring common ancestry.

The phylogenetic position of *H. betulae* Hirschmann & Riggs, like *Verutus*, is controversial. Krall and Krall (1978) placed *H. betulae* in *Cactodera*, but Mulvey and Golden (1983) believed it to be distinct from other *Heterodera*. *H. betulae* is distinguished from *Cactodera* spp. by the absence of D in the former (Fig. 6). While *H. betulae* has A, B, C layers similar to *Heterodera* spp., the layers are infused with an electron dense material, and therefore the cuticle is distinct from any other Heteroderidae examined.

Clearly body wall layering provides useful new characters, which when considered in conjunction with additional characters, can contribute to refining hypotheses of phylogeny of Heteroderidae.

CUTICULAR SURFACE PATTERNS (SEM)

The potential of SEM in characterising this family has been demonstrated for cone patterns by Green (1975), and for lip patterns of second-stage juveniles by Stone (1975). Since these studies, improved preparation methods and instrumentation minimise artifacts and allow comparison of even finer details. Our ongoing observations at the University of California include body regions and life stages previously not considered for characters, previously unexamined species and genera, and comparison of samples sufficiently large for interpreting variability. Results of this study are incomplete and not yet fully compiled. Preliminary observations indicate, however, that new reliable characters may include cuticle patterns of the tail region of males, the origin and terminus of areolation of lateral lines, and morphology of phasmid openings. In addition, cuticle patterns of the female body are now shown to have many character states which comprise a transformation series throughout the family. Our observations confirm that lip patterns are particularly rich in reliable characters. Although previous investigations have primarily considered lip patterns of second-stage juveniles, our findings indicate that adults frequently have more specific derived characters which are

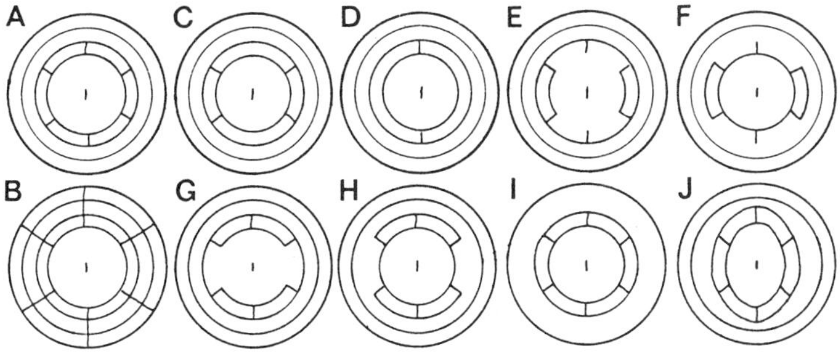

Fig. 8 Diagrammatic representation of characters and character states of lip patterns of second-stage juveniles (J2) and males of Heteroderidae. A) Primitive pattern of J2 in which round labial disc is surrounded by six lips followed by additional head annulations. B) Primitive pattern of male identical to J2 (A) except with longtidinal striae. C-J) Derived modifications of primitive patterns. C) Fusion of adjacent submedial lips. D) Fusion of submedial lips with lateral lips. E) Fusion of submedial lips with labial disc. F) Fusion of submedial lips with head annule. G) Fusion of lateral lips with labial disc. H) Fusion of lateral lips with head annule. I) Fusion of head annules. J) Elongation of labial disc.

particularly useful for phylogenetic analysis. This might be predicted by the generalisation that more specialised characters tend to occur later in development (Eldredge and Cracraft, 1980).

Our SEM work on lip patterns does not yet permit presentation of phylogenetic analyses or even complete matrices of character distribution for the family. Nevertheless, our approach to the recognisation of characters and the designation of polarities is discussed, as well as findings on intrageneric and interspecific variability of non-cyst forming groups.

Previously, lip patterns have been treated as a single character and only species with identical patterns were interpreted as sharing the character state. Our observations suggest, however, that a given lip pattern is a complex of several characters, which tend to be inherited independently from one another. Each character is assigned polarity based on the hypothesis that the primitive lip pattern which occurs throughout Tylenchida consists of a round labial disc surrounded by a first annule consisting of six lips: two subdorsal, two subventral, and two lateral (Baldwin et al., 1983; Fig. 8). A number of additional annules occur on the head region. In males of Heteroderidae it is tentatively hypothesised that longitudinal striae on the head region are primitive and the absence of such striae derived. This hypothesis is supported by the occurrence of similar longitudinal striae on many Hoplolaimidae. Effective use of lip patterns in phylogenetic analysis is dependent on understanding their variability at every taxonomic level. It has been found, for example, that even among closely related species of a given genus, patterns may vary from primitive to highly derived as seen in the genus *Meloidodera*, where juveniles of *M. charis* Hooper and *M. belli* Hooper have primitive hexaradiate lip patterns (Othman and Baldwin, 1985a; Fig. 9). Thus it was surprising to discover that the type species, *Meloidolera floridensis* Chitw., Hannon & Esser, has a strikingly different highly derived pattern in which the labial disc is

SPECIES J2 SEX	EN FACE PATTERN	FUSION OF ADJACENT SUBMEDIAL LIPS	FUSION OF SUBMEDIAL LIPS WITH LATERAL LIPS	FUSION OF SUBMEDIAL LIPS WITH LABIAL DISC	FUSION OF SUBMEDIAL LIPS WITH HEAD ANNULES	FUSION OF LATERAL LIPS WITH LABIAL DISC	FUSION OF LATERAL LIPS WITH HEAD ANNULES	FUSION OF HEAD ANNULES	ABSENCE OF LONGITUDINAL STRIATIONS (MALE)	ELONGATION OF LABIAL DISC
Meloidodera floridensis		+	–	–	–	–	+	+	N/A	+
M. charis *M. belli*	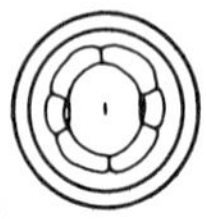	–	–	–	–	–	–	–	N/A	–
M. eurytyla		+	–	–	–	–	–	–	N/A	–
M. n.sp. (Tenn)	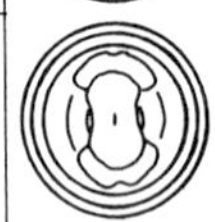	+	–	–	–	–	– (+)	–	N/A	+
M. n.sp. (Alaska)	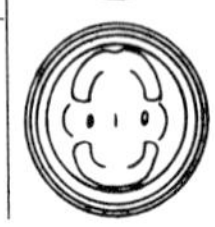	+	–	+	–	+	–	–	N/A	+

Fig. 9 Representation of lip pattern characters of second-stage juveniles of *Meloidodera* spp.

elongate, adjacent submedial lips are fused, lateral lips are fused with head annules, and head annules are almost completely fused with one another. Similarly, male lip patterns of *M. charis* and *M. belli* are primitive, including longitudinal lines on the head region, whereas the male lip pattern of *M. floridensis* is highly derived, with virtually all parts fused except for the labial disc (Fig. 10). These observations might initially suggest that *Meloidodera* is polyphyletic, i.e. species within the genus do not share a unique common ancestor (Othman and Baldwin, 1985a). However, subsequent investigations of additional species have revealed a number of intermediate patterns and a clear transformation series. For example, second-stage juveniles of *M. eurytyla*, as well as two undescribed new species from Alaska, resemble *M. charis* and *M. belli* except that adjacent submedial lips are fused as in *M. floridensis* (Fig. 9). The lip pattern of an additional new species from Tennessee is further derived by elongation of the labial disc. Similarly, lip patterns of males of *Meloidodera* indicate a transformation series; for example, males of *M. eurytyla* are intermediate between *M. charis* and *M. floridensis*. In *M. eurytyla* lateral lips are fused with head annules, and these head annulations and longitudinal lines are highly disorganised with considerable fusion (Fig. 10).

Lip patterns of other groups, such as the *Atalodera* – *Thecavermiculatus* complex are more consistent than *Meloidodera* (Othman et al., 1985). In the juveniles of these species the lateral lips tend to be fused with the

SPECIES MALE SEX	EN FACE PATTERN	FUSION OF ADJACENT SUBMEDIAL LIPS	FUSION OF SUBMEDIAL LIPS WITH LATERAL LIPS	FUSION OF SUBMEDIAL LIPS WITH LABIAL DISC	FUSION OF SUBMEDIAL LIPS WITH HEAD ANNULES	FUSION OF LATERAL LIPS WITH LABIAL DISC	FUSION OF LATERAL LIPS WITH HEAD ANNULES	FUSION OF HEAD ANNULES	ABSENCE OF LONGITUDINAL STRIATIONS (MALE)	ELONGATION OF LABIAL DISC
Meloidodera floridensis	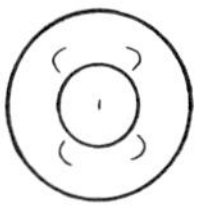	+	-	-	+	-	+	+	+	-
M. charis *M. belli*	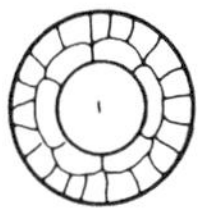	-	-	-	-	-	-	+	-	-
M. eurytyla	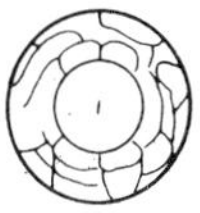	-	-	-	-	-	+ (-)	+ (+)	+ (-)	-
M. n.sp. (Tenn)	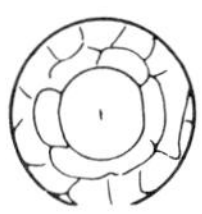	±	±	-	-	-	-	+ (-)	+	-

Fig. 10 Representation of lip pattern characters of males of *Meloidodera* spp.

labial disc, and may or may not have fusion of adjacent submedial lips (Fig. 11). For example, submedial lips are fused in *M. ucri* Wouts & Sher, but not in *T. andinus* Golden, Franco, Jatala & Astocaza. In *A. lonicerae* Wouts, adjacent submedial lips may be separate, fused, or in a given individual one pair may be separate and the other fused (Fig. 12). Examination of over 350 topotype juveniles indicates that no particular lip pattern is predominant in this polymorphic species. Similar polymorphism occurs, for example, in juveniles of certain populations of *Heterodera schachtii* Schmidt and *Globodera rostochiensis* (Woll.) Behr. (Othman and Baldwin, unpublished observations).

In many other species and populations one pattern is predominant and variants are less common. The *Sarisodera* - *Rhizonema* complex is of interest in this respect (Othman and Baldwin, 1985b). Juveniles of *S. hydrophila* have a consistant derived pattern in which submedial lips are enlarged and adjacent lips are fused. The pattern of most topotype juveniles of *R. sequoiae* are nearly identical except for a few individuals in which submedial lips are separate (Fig. 13). Variants of males may be more significant in interpreting phylogeny because their characters tend to be more specialised. The typical pattern of males of *S. hydrophila* includes fusion of all lips. This differs from males of *R. sequoiae* in which lateral lip remain separate; yet some topotype males of *S. hydrophila* occur which also have separate lateral lips and are nearly indistinguishable from *R. sequoiae* (Fig. 14).

Since lip patterns are variable, even within a given population, large samples are often needed to characterise a species. Furthermore, since lip patterns are highly variable among species of many genera, patterns cannot

SPECIES J2 SEX	EN FACE PATTERN	FUSION OF ADJACENT SUBMEDIAL LIPS	FUSION OF SUBMEDIAL LIPS WITH LATERAL LIPS	FUSION OF SUBMEDIAL LIPS WITH LABIAL DISC	FUSION OF SUBMEDIAL LIPS WITH HEAD ANNULES	FUSION OF LATERAL LIPS WITH LABIAL DISC	FUSION OF LATERAL LIPS WITH HEAD ANNULES	FUSION OF HEAD ANNULES	ABSENCE OF LONGITUDINAL STRIATIONS (MALE)	ELONGATION OF LABIAL DISC
Atalodera ucri		+	– (+)	–	–	–	–	–	N/A	–
A. lonicerae	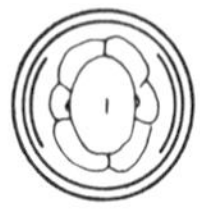	±	– (+)	–	–	–	–	–	N/A	–
Thecavermiculatus gracililancea	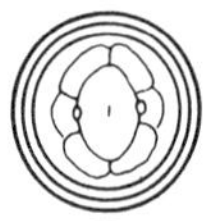	–	– (+)	–	–	–	–	–	N/A	–
T. andinus	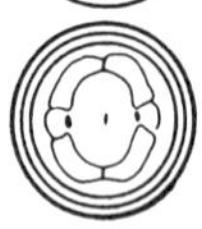	–	– (+)	–	–	–	–	–	N/A	–
T. crassicrustata		–	– (+)	–	–	–	–	–	N/A	–

Fig. 11 Representation of lip pattern characters of second-stage juveniles of *Atalodera* spp. and *Thecavermiculatus* spp.

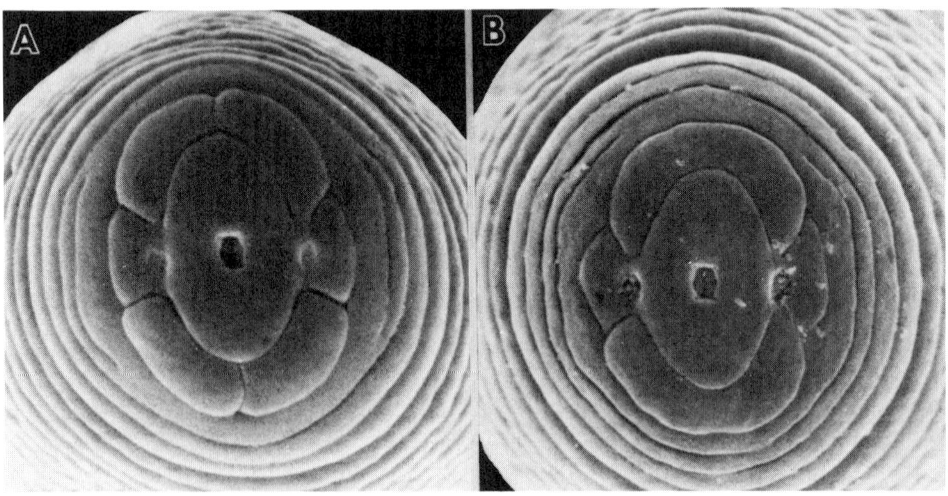

Fig. 12 Polymorphism of lip pattern characters of second-stage juveniles of *Atalodera lonicerae*. A) Separate submedial lips. B) Fused submedial lips. (After Othman and Baldwin, 1985).

SPECIES J2 SEX	EN FACE PATTERN	FUSION OF ADJACENT SUBMEDIAL LIPS	FUSION OF SUBMEDIAL LIPS WITH LATERAL LIPS	FUSION OF SUBMEDIAL LIPS WITH LABIAL DISC	FUSION OF SUBMEDIAL LIPS WITH HEAD ANNULES	FUSION OF LATERAL LIPS WITH LABIAL DISC	FUSION OF LATERAL LIPS WITH HEAD ANNULES	FUSION OF HEAD ANNULES	ABSENCE OF LONGITUDINAL STRIATIONS (MALE)	ELONGATION OF LABIAL DISC
Sarisodera hydrophila	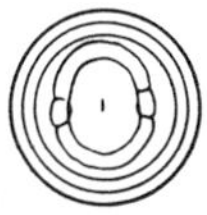	+	–	–	–	– (+)	–	–	N/A	–
Rhizonema sequoiae	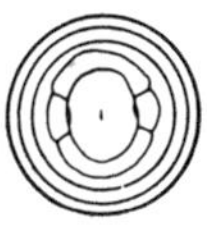	+	–	–	–	– (+)	–	–	N/A	–
R. sequoiae (variant)	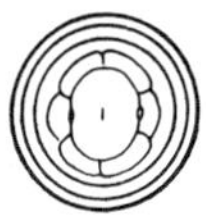	(-)	–	–	–	–	–	–	N/A	–

Fig. 13 Representation of lip pattern characters of second-stage juveniles of *Sarisodera hydrophila* and *Rhizonema sequoiae*.

arbitrarily be treated as diagnostic "generic characters". Rather, they are useful in defining genera when considered in conjunction with a suite of additional features.

Regardless of variability, lip patterns have been found to be useful in contributing evidence for testing hypotheses of monophyly. Perhaps the clearest example is *Afenestrata africana* (= *Sarisodera africana*). It has been noted that the type species *Sarisodera hydrophila* lacks a cyst, induces a single uninucleate giant cell, and has a thick complex cuticle; these characters are in striking contrast to *A. africana*. Just as the lip patterns indicate several shared derived characters between *Sarisodera* and *Rhizonema* species these also confirm that *A. africana* does not share a common ancestor with these species (Baldwin and Bell, 1985; Othman and Baldwin, 1985b). In juveniles and males of *A. africana* the labial disc is elongate and submedial lips are fused with the labial disc; this is in contrast to *Rhizonema* - *Sarisodera*, but resembles that of certain *Heterodera* (Figs. 13, 15A, B). These observations confirm that the group, *S. hydrophila* and *Afenestrata* is polypheletic, but suggest that *Afenestrata* and *Heterodera* share a common ancestor.

It has been noted that *Verutus* and *Meloidodera* share many primitive characters, but shared derived characters are needed to support hypotheses of monophyly between the two genera. However, lip patterns of *Verutus*, while highly derived, are apparently unique to the genus (Othman and Baldwin, 1985a). In juveniles, fused submedial lips are broadened in the four medio-lateral positions and amphid openings are usually large (Fig. 15C). In males, submedial lips are broadened as in juveniles and are also fused with lateral lips (Fig. 15D). The lip patterns of *Verutus* cannot be interpreted as part of the transformation series of *Meloidodera* (Figs. 9, 10), and it is hypothesised that the two genera are not monophyletic. In contrast to Ferris (1985) it is also hypothesised that *Verutus* does not form a monophyletic group with *M. charis* and *M. belli*, a

SPECIES / MALE SEX	EN FACE PATTERN	FUSION OF ADJACENT SUBMEDIAL LIPS	FUSION OF SUBMEDIAL LIPS WITH LATERAL LIPS	FUSION OF SUBMEDIAL LIPS WITH LABIAL DISC	FUSION OF SUBMEDIAL LIPS WITH HEAD ANNULES	FUSION OF LATERAL LIPS WITH LABIAL DISC	FUSION OF LATERAL LIPS WITH HEAD ANNULES	FUSION OF HEAD ANNULES	ABSENCE OF LONGITUDINAL STRIATIONS (MALE)	ELONGATION OF LABIAL DISC
Sarisodera hydrophila	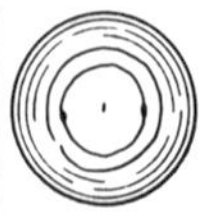	+	+	-	-	-	-	-	+	-
S. hydrophila (variant)	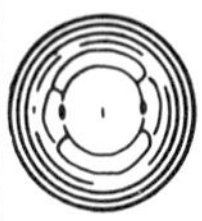	+	-	-	-	-	-	-	+	-
Rhizonema sequoiae	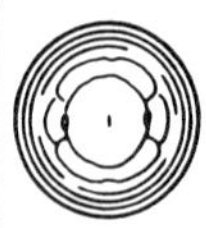	+	-	-	-	-	-	-	+	-

Fig. 14 Representation of lip pattern characters of males of *Sarisodera hydrophila* and *Rhizonema sequoiae*.

hypothesis that may be tested by investigation of the pore-like phasmids in these species (see below).

Characters clearly detailed by SEM, specifically lip patterns, evolve quickly; patterns may have mosaic distributions, high variability, and frequent homoplasy. This does not preclude their use in interpreting phylogeny, but does indicate the importance of thorough investigation, sophisticated treatment and evaluation in conjunction with a large suite of additional characters.

PHASMIDS

The phasmid in second-stage juveniles of Heteroderidae is a useful character whether in the form of a lens-like structure or a pore. Ferris (1979) considered the pore phasmid derived; the two character states fit her proposed phylogeny well and were congruent with other characters (Table 1). However, re-evaluation of the coding for these characters indicated the need for further consideration reflected in a revised phylogeny (Ferris, 1985). For example, *Meloidodera* was previously coded as having a lens-like phasmid but only *M. floridensis* has a lens; juveniles of other species of *Meloidodera* have pore phasmids (Othman and Baldwin, 1985a). In addition, *Hylonema* has a pore rather than a lens-like phasmid (Luc et al., 1978) and *Sarisodera hydrophila* with a lens-like phasmid contrasts with *Afenestrata africana (= Sarisodera africana)* which has a pore phasmid (Baldwin and bell, 1985). When, in addition, new species and genera are considered including *Verutus* (pore), *Rhizonema* (lens-like) and *Dolichodera* (pore), distribution of phasmid character states is a mosaic throughout Heteroderidae (Fig. 16). Proposing common ancestry based on the shared derived character, pore phasmid, would suggest unlikely groupings such as certain *Meloidodera* with *Heterodera*. The relationship is unlikely because the distribution of pore phasmid is incongruent with many other characters and yields an apparently nonparsimonious "phylogeny" (Table 1).

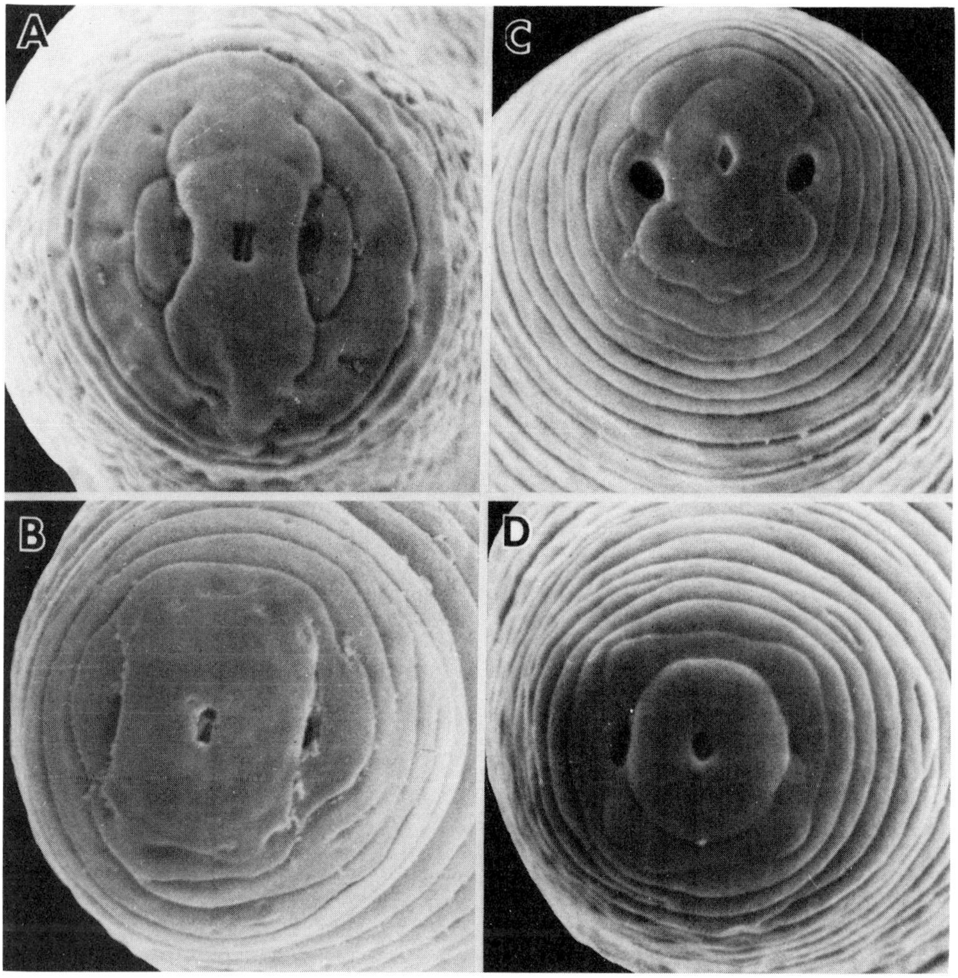

Fig. 15 Lip patterns of *Afenestrata africana* and *Verutus volvingentis*. A) Second-stage juveniles of *Afenestrata africana*. B) Male of *Afenestrata africana*. C) Second-stage juvenile of *Verutus volvingentis*. D) Male of *Verutus volvingentis*. (After Othman and Baldwin, 1985a, b).

A more reasonable hypothesis is that the pore phasmid evolved independently in *Meloidodera* and *Heterodera*. It is likely that two groups or pore phasmids resulting from parallel evolution would be morphologically and perhaps developmentally distinct. Such fundamental differences, although not apparent at the LM level, could be revealed by TEM; thus, TEM could be useful in testing the hypothesis of parallel evolution of pore phasmids. Initially, we selected for comparison phasmids of second-stage juveniles of *M. floridensis* (lens-like), *M. charis* (pore), and *H. schachtii* (pore). The rationale for these choices is to characterise the exact nature of the two types of phasmids in *Meloidodera*, and in addition, to determine if the pore phasmid in a species where primitive character states predominate

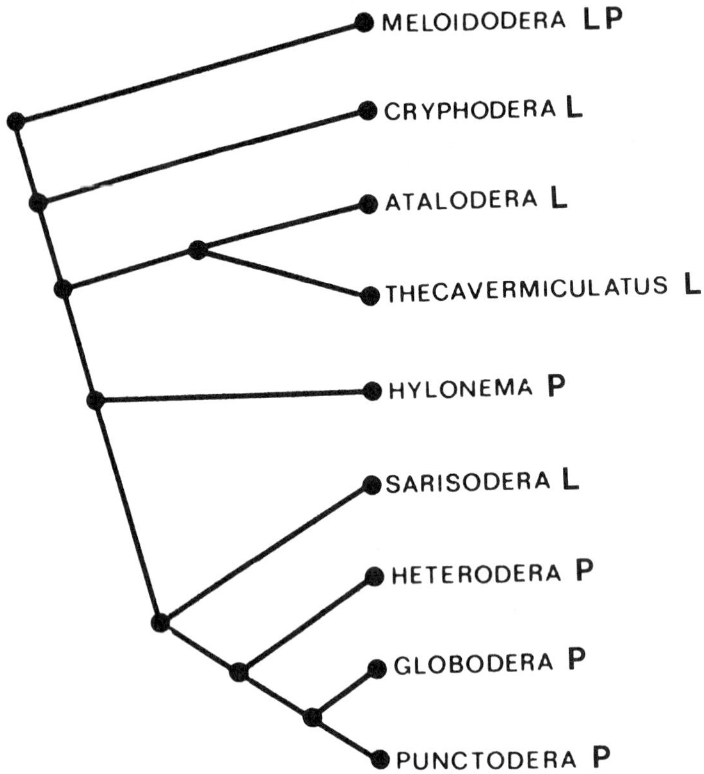

Fig. 16. Phylogeny of Heteroderidae proposed by Ferris (1979) in relation to distribution of lens (L) and pore (P) phasmid.

(*M. charis*) is equivalent (homologous) to the pore phasmid in species where derived character states predominate (*H. schachtii*).

In second-stage juveniles of *H. schachtii* the phasmid consists of an external pore, cuticle-lined duct, socket cell, sheath cell and single dendrite receptor (Baldwin, 1985; Fig. 17). The dendrite terminates in a receptor cavity enclosed by the sheath cell. The wall of the receptor cavity is deeply infolded resulting in a maze of channels which extend, with vesicles, throughout the posterior half of the sheath cell (Fig. 18).

Ongoing TEM investigations of phasmid of *M. floridensis* and *M. charis* indicate two main areas of contrast with *H. schachtii* (Baldwin, unpublished observations), namely the external opening of the sheath cell (Fig. 18). In *M. floridensis* the narrow external opening leads to the "lens" which is a large cuticle-lined oval ampulla within the lateral hypodermal chord (Fig. 18C). In *M. charis* a similar ampulla occurs, although it is smaller, round (versus oval) and more superficial, being partly embedded in the body wall cuticle of the lateral chord (Fig. 18B).

The sheath cells of phasmids of *M. floridensis* and *M. charis* are similar to one another and distinct from that of *H. schachtii* (Fig. 18D-F). In both species of *Meloidodera* the receptor cavity lining is continuous with closely associated pairs of membranes which extend throughout the sheath cell and contact with the channels and vesicles of *H. schachtii*. The sheath cell of *M. floridensis* is larger than that of *M. charis*, and

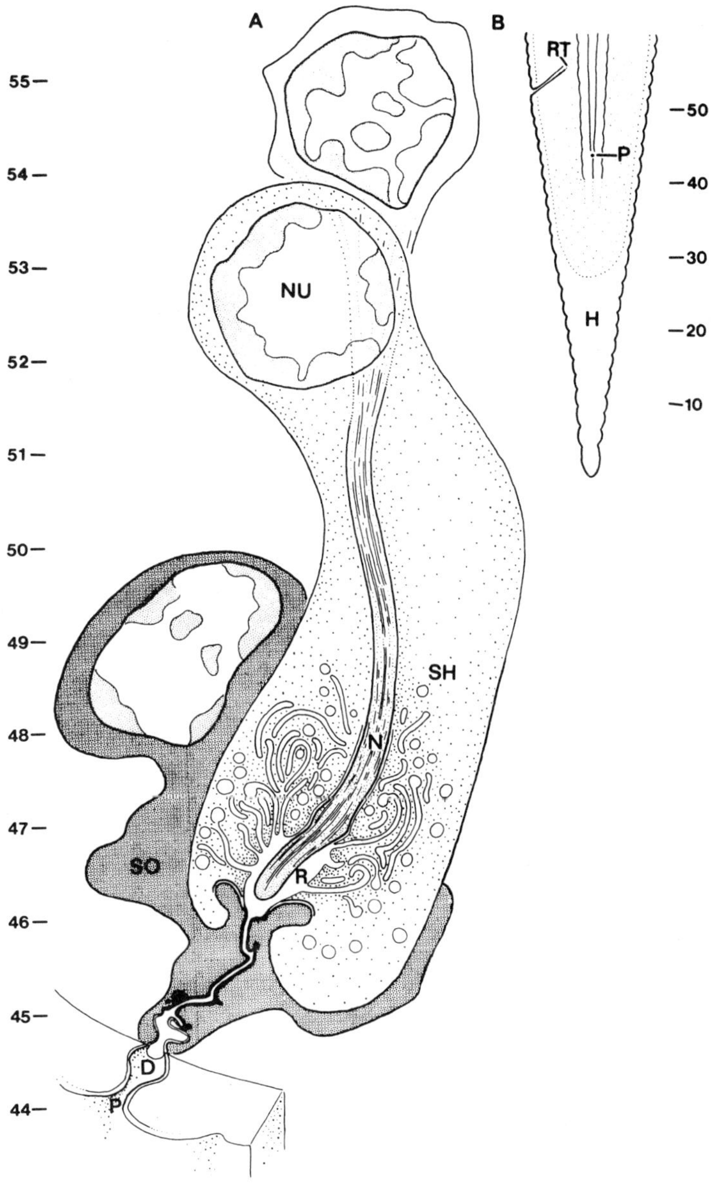

Fig. 17 Diagram of phasmid of second-stage juvenile of *Heterodera schachtii*. A) Detailed view indicating relationships among dendrite of neuron (N), sheath (SH) and socket (SO) cells. B) Tail showing position of phasmidial pore (P). D, duct, H, hyaline (cuticular) tail terminus; NU, nucleus, R, receptor cavity, RT, rectum. Scale units are in micrometers and indicate distance from the posterior end. (Baldwin, 1985; reproduced by permission).

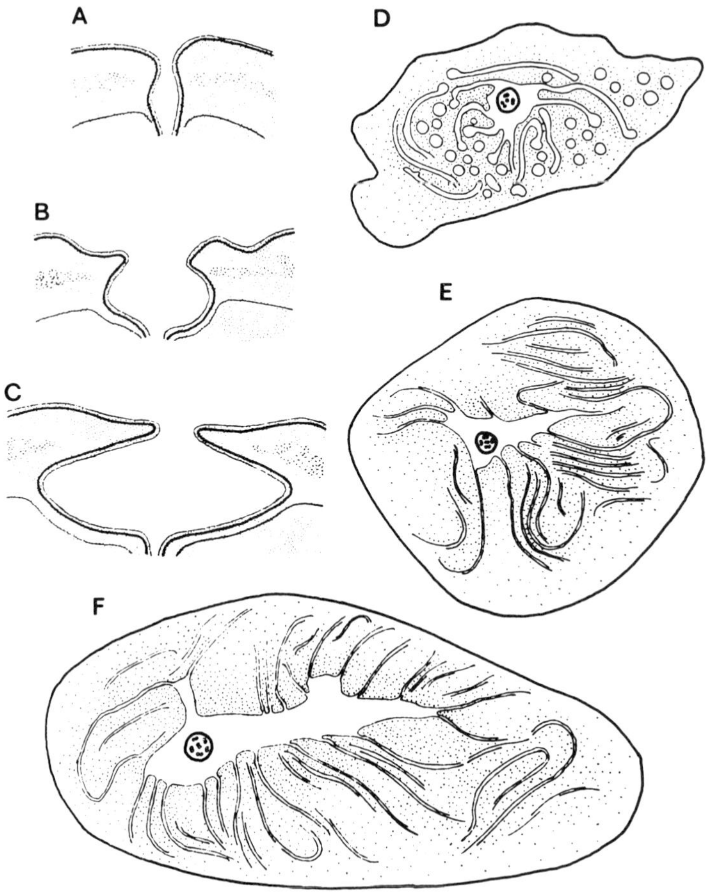

Fig. 18 Diagrammatic representation of phasmids of second-stage juveniles of Heteroderidae from tracings of transmission electron micrographs (A-C). Cross section of opening at the same relative scale. A) *Heterodera schachtii*. B) *Meloidodera charis*. C) *Meloidodera floridensis*. (D-F) Cross sections through portion of sheath cell including receptor cavity at same relative scale. D) *Heterodera schachtii*, E) *Meloidodera charis*, F) *Meloidodera floridensis*.

the internal membranes tend to be less regularly arranged in the former (Fig. 18D-F).

Although results of this study are incomplete, it is believed that preliminary findings are consistent with the hypothesis of parallel evolution, i.e. the pore phasmid of *M. charis* and *H. schachtii* are not homologous; apparent similarity with the light microscope is superficial. While there is strong evidence for this hypothesis apart from fine structural studies, homologies and homoplasies of other phasmids are less apparent. For example, we now raise the question as to whether the pore phasmids of *Verutus* and *Hylonema* are a similar character state to that of *M. charis*, or if they more nearly resemble that of *H. schachtii*; in these cases the type of phasmid may have a profound influence on interpreting phylogeny of Heteroderidae.

Our fine structural investigations also are useful in testing Ferris' (1979) hypothesis that the pore phasmid of Heteroderidae is derived. We have found that the lens-like structure of *M. floridensis* is a large ampulla similar to that shown through TEM investigation of "enlarged phasmids" in Hoplolaimidae (e.g. Wang and Chen, 1985). If Hoplolaimidae is accepted as an outgroup of Heteroderidae, the lens phasmid is supported as the primitive state. This interpretation if strengthened by our observations of development, since juveniles with lens-like phasmids such as *M. floridensis* mature to adults with pore-like phasmids. Ongoing TEM developmental studies of the phasmid indicate that adults of cyst-forming Heteroderidae typically do not have phasmids, suggesting a tendency toward progressively greater reduction of these organs in Heteroderidae (Carta and Baldwin, unpublished observations).

APPLICATIONS AND CONCLUSIONS

The purpose of these combined studies on host responses, body wall layering, SEM surface morphology, and phasmid fine structure is to test and refine hypotheses of phylogeny of Heteroderidae as a basis for developing a natural taxonomic system. The large pool of characters generated from these studies will be combined with data from previous investigations to develop a matrix of the distribution of each character throughout the family. Some characters will be rejected (tentatively) if insufficient data are available for coding, or knowledge of the character is otherwise insufficient for interpreting phylogeny. We will then attempt to assign polarity to each character, based on testing by outgroup comparisons and investigations of development. These data will be analysed using a variety of computer programs primarily designed to identify the most parsimonious phylogenetic system, i.e. the simplest one substantiated by the pool of shared derived characters. Our experience with similar analyses of Merliniinae (Powers, 1984) indicates that repeated analyses are required. Initial analyses may point to errors in coding, which can be resolved by re-examination, and unresolved relationships may be clarified as additional characters are introduced. Our analysis of Merliniinae resulted in several alternate phylogenies; however, certain broad phylogenetic patterns were repeatedly substantiated. Similarly, we anticipate a consensus of certain broad aspects of competing phylogenies of Heteroderidae. It is where such broad patterns are clearly demonstrated that alterations in the taxonomic system, needed to reflect these relationships, are justified. However, obscure relationships are also useful because they direct our work to areas most productive for contributing to the continuing process of testing and refining hypotheses of phylogeny of Heteroderidae.

ACKNOWLEDGEMENT

The investigations described in this paper were undertaken in collaboration with my former student Dr M. Mundo-Ocampo and my present student Mr A. A. Othman whose assistance is gratefully acknowledged.

REFERENCES

Baldwin, J.G., 1983, Fine structure of body wall cuticle of females of *Meloidodera charis*, *Atalodera lonicerae* and *Sarisodera hydrophila* (Heteroderidae), J. Nematol., 15:370.

Baldwin, J.G., 1985, Fine structure of the phasmid of second-stage juveniles of *Heterodera schachtii* (Tylenchida: Nematoda), Can. J. Zool., 63:534.

Baldwin, J.G. and Bell, A.H., 1985, *Cactodera eremica* n.sp., *Afenestrata africana* (Luc et al., 1973) n.gen., n.comb., and an emended diagnosis of *Sarisodera* Wouts and Sher, 1971 (Heteroderidae), J. Nematol., 17:187.

Baldwin, J.G., Luc, M.C. and Bell, A.H., 1983a, Contribution to the study of the genus *Pratylenchoides* Winslow (Nematoda: Tylenchida), Rev. Nematol. 6:11.

Cid del Prado Vera, I. and Lownsbery, B.F., 1984, Histopathology and host range studies of the redwood nematode, *Rhizonema sequoiae*, J. Nematol., 16:68.

Cliff, G.M. and Baldwin, J.G., 1985, Fine structure of the body wall cuticle of females of eight genera of Heteroderidae, J. Nematol., 17:286.

Cohn, E., Kaplan, D.T. and Esser, R.P., 1984, Observations on the mode of parasitism and histopathology of *Meloidodera floridensis* and *Verutus volvingentis* (Heteroderidae), J. Nematol., 16:256.

Eldredge, N. and Cracraft, J., 1980, Phylogenetic patterns and the evolutionary process. Method and theory in comparative biology. Colombia Univ. Press, New York.

Endo, B.Y., 1964, Penetration and development of *Heterodera glycines* in soybean roots and related anatomical changes, Phytopathology, 43:79.

Esser, R.P., 1981, *Verutus volvingentis* n.gen., n.sp. (Heteroderidae: Tylenchida) in Verutinae n.subf., a phytoparasitic nematode infesting buttonweed in Florida, Proc. Helminthol. Soc. Wash., 48:220.

Feldmesser, J., 1953, The life cycle of the golden nematode, *Heterodera rostochiensis* Wollenweber, 1923, on tomato and a cytological study of its effects on the host. Doctoral Dissertation, New York Univ.

Ferris, V.R., 1979, Cladistic approaches in study of soil and plant parasitic nematodes, Amer. Zool., 19:1195.

Ferris, G.R., 1985, Evolution and biogeography of cyst-forming nematodes, OEPP/EPPO Bull., 15:123.

Green, C.D., 1975, The vulval cone and associated structures of some cyst nematodes (genus (*Heterodera*), Nematologica, 21:134.

Heald, C.M., 1978, Histological studies of *Meloidodera charis* infecting ridgeseed,euphorbia, and okra and additional host range studies, Nematropica, 8:12 (Abstract).

Krall, E.L. and Krall, K.A., 1978, (Revision of plant nematodes of the family Heteroderidae on the basis of the trophic specialization of these parasites and their co-evolution with their host plants.) Fitogel'mintologicheskie issledovaniya Moscow, USSR; "Nanka.", 39, (in Russian).

Luc, M., Taylor, D.P. and Cadet, P., 1978, Description of a new tropical Heteroderidae, *Hylonema ivorense* n.gen., n.sp., and a new outlook for the family Heteroderidae (Nematoda: Tylenchida), Rev. Nematol., 1:73.

Mundo-Ocampo, M. and Baldwin, J.G., 1983a, Host-parasite relationships of *Atalodera* spp. (Heteroderidae), J. Nematol., 15:234.

Mundo-Ocampo, M. and Baldwin, J.G., 1983b, Host response to *Sarisodera hydrophila* Wouts and Sher, 1971, J. Nematol., 15:259.

Mundo-Ocampo, M. and Baldwin, J.G., 1983c, Host responses to *Meloidodera* spp. (Heteroderidae), J. Nematol., 15:544.

Mundo-Ocampo, M. and Baldwin, J.G., 1984, Comparison of host response of *Cryphodera utahensis* with other Heteroderidae and a discussion of phylogeny, Proc. Helminthol. Soc. Wash., 51:25.

Othman, A.A. and Baldwin, J.G., 1985a, Comparative morphology of *Meloidodera* spp. and *Verutus* sp. (Heteroderidae) with scanning electron microscopy, J. Nematol., 17:297.

Othman, A.A. and Baldwin, J.G., 1985b, Comparative morphology of *Sarisodera hydrophila*, *Rhizonema sequoiae*, and *Afenestrata africana* (Heteroderidae) with scanning electron microscopy, Proc. Helminthol. Soc. Wash., 53:69.

Othman, A., Baldwin, J.G. and Bell, A.H., 1985, Comparative morphology of *Atalodera* spp. and *Thecavermiculatus* spp. (Heteroderidae) with scanning electron microscopy, J. Nematol., (in press).

Powers, T.O., 1983, Systematic analysis of Merliniinae Siddiqi, 1971 (Nematoda: Dolichoderidae), Ph.D. Thesis, University of California, Riverside, CA.

Ruehle, J.L., 1962, Histopathological studies of pine roots infected with lance and pine cystoid nematodes, Phytopathology, 52:68.

Shepherd, A.M., Clark, S.A. and Dart, P.J., 1972, Cuticle structure in the genus *Heterodera*, Nematologica, 18:1.

Stone, A.R., 1975, Head morphology of second-stage juveniles of some Heteroderidae (Nematoda: Tylenchoidea), Nematologica, 21:81.

Stone, A.R., 1977, Recent developments and some problems in the taxonomy of cyst-nematodes, with a classification of the Heteroderoidea, Nematologica, 23:273.

Suarez, Z., Sosa-Moss, C. and Inserra, R.N., 1985, Anatomical changes induced by *Punctodera chalcoensis* in corn roots, J. Nematol., 17:242.

Taylor, D.P., Cadet, P. and Luc, M., 1978, A unique host-parasite relationship between *Hylonema ivorense* (Nematoda: Heteroderidae) and the roots of a tropical rain forest tree, Rev. Nematol., 1:99.

von Sembdner, G., 1963, Anatomie der *Heterodera-rostochiensis*-Gallen an tomatenwurzeln, Nematologica, 9:55.

Wang, K.C., and Chen, T.A., 1985, Ultrastructure of the phasmids of *Scutellonema brachyurum*, J. Nematol., 17:175.

Wouts, W.M., 1973, A revision of the family of Heteroderidae (Nematoda: Tylenchoidea) III. The subfamily Ataloderinae, Nematologica, 19:279.

Wouts, W.M. and Sher, S.A., 1971, The genera of the subfamily Heteroderinae (Nematoda: Tylenchoidea) with a description of two new genera, J. Nematol., 3:129.

DISCUSSION

Responding to a question on the similarity of host response in different families and genera of nematodes, (e.g. *Naccobus* and *Rotylenchulus* compared with the Heteroderidae) Baldwin said that this illustrated parallel evolution. Stone observed that the group of nematodes was very large and thus it is not surprising to see parallel evolution exhibited in a situation where there is similar selection pressure with shared genes. Roberts asked if lip patterns and other morphological features could be interpreted in terms of selection advantage and evolutionary significance. Baldwin considered it difficult to interpret features in those terms and Stone added that differences in genital patterns might be interpreted in terms of evolutionary significance. However, Baldwin and others concluded that there was insufficient information about selection pressure in relation to genetic drift and the linkage of characters and hence the reason for hypothesising as a way of concentrating attention on such problems and seeking solutions.

With reference to questions on the structure of the body wall, Balwin said that *H. schachtii* lacks a D layer regardless of the maturity of the individual or the body region.

THE DEVELOPMENT OF INDIVIDUALS AND POPULATIONS OF CYST NEMATODES ON PLANTS

J. Willem Seinhorst

formerly: Instituut voor Plantenziektenkundig Onderzoek
Wageningen
The Netherlands

INTRODUCTION

In the root infesting tylenchs there are the extremes of the truly migratory nematodes and the truly sedentary nematodes, and in between a relatively small group with somewhat intermediate characters. All species achieve the same biological objective: to survive with a minimum effort and low reproduction rates, in which they are helped by the virtual absence of interference from predators or parasites.

The truly migratory species (e.g. *Tylenchorhynchus*) live on the contents of normal root cells, mostly of the cortex, moving from one to another after relative short feeding times. They are polyphagous and above all have the capacity to leave "a sinking ship" without being harmed. When food ceases to be in sufficient supply, e.g. because of intraspecific competition, intervals between feedings simply become longer and when food ceases to be supplied growth processes and egg laying stop without much delay. Eggs still without a shell are decomposed and the materials are reincorporated in the body. At this stage of relative inactivity (the nematodes keep moving about) they can survive for many years. Mortalities are often of the order of 50 per cent per year or less. When food again becomes available growth and egg laying are resumed. As a result, populations consist of constant proportions of juveniles and adults, e.g. about 30 per cent adults in *Pratylenchus penetrans* Cobb (Seinhorst and Kuniyasu, 1971) and about 25 per cent adults in *Rotylenchus uniformis* Thorne.

Sedentary tylenchs (Heteroderidae and related groups) always feed on host cells, mostly of the stele, which are induced by the nematodes to become so-called transfer cells. These are cells with a much denser than normal protoplasm and are either enlarged single cells or formed by cell fusion (as in Heteroderidae). Contrary to normal cells these transfer cells are capable of supplying a sufficient quantity of food continuously for the nematodes not only to become adult females, but also to become much larger and produce some five to ten times as many eggs as migratory species, when there is little competition for space and food by other individuals. Because the nematodes do not need to move, they can afford to lose their body musculature and thus the females become first sausage shaped and, ultimately, pear, lemon shaped or spherical cysts. Their ultimate volume is forty to a hundred times that of the adult male, which becomes sausage shaped at the end of the second stage but becomes vermiform again inside the moulted skins of the third and fourth juvenile stages. Females are, therefore, large enough to produce and contain within their bodies five to ten times the number of eggs produced by the vermiform females of migratory root infesting tylenchs. Another major difference between migratory and cyst nematodes is that in the former egg production is closely associated with feeding, while cyst nematode females first accumulate sufficient reserves to produce all eggs. Adult females removed from roots still produce the normal number of eggs.

The passive life of cyst nematodes is restricted to the second stage juvenile in the egg, which also is the only stage capable of a long survival. Once hatched from the egg the death rate of the nematodes is of the order of 50 per cent per week unless they penetrate into a host root and start feeding. The reduction of passive survival to the egg stage also leads to the possibility of having the second stage juveniles in the eggs activated only under conditions suitable for survival, i.e. when host roots are present. However, not all cyst nematodes have adopted this mechanism, i.e. hatching only after stimulation by a host root diffusate, as in *Globodera* species, *Heterodera schachtii* Schmidt, *H. glycines* Ichinohe, *H. goettingiana* Liebscher and several others. Hatching of second stage juveniles from eggs of *H. avenae* Woll. and *H. trifolii* Goffart is stimulated by exposure to low temperature for a certain time. From an agricultural point of view this has the advantage that the latter species are much more easily controlled by crop rotation than the former.

The advantage of the larger egg production in sedentary nematodes feeding on large transfer cells, compared with migratory species, is the associated risk of total disaster by malfunctioning or death of the food source.

In the intermediate species only the adult stage is sedentary and swollen. In the citrus nematode (*Tylenchulus semipenetrans* Cobb) females feed in the juvenile stage and remain vermiform, but *Rotylenchulus* juveniles go through all the moults to the adult stage without feeding.

LIFE CYCLE OF CYST NEMATODES

The morphology, anatomy and biological activities (penetration in host roots, feeding and competition for food, and the reactions they cause in host roots) of the various cyst nematode species are very similar. The life cycle can be considered to begin with the hatching of the second stage juvenile from the egg in the mother cyst which is in the soil. The juvenile penetrates into a host root near the tip, starts to feed on a developing so-called transfer cell, after having become immobile, and develops through three moults to a vermiform male or a round, lemon or pear shaped female, which from the fourth juvenile stage onwards is visible on the roots as a small white or yellow sphere about 0.5 mm in diameter. The female fills with eggs, turns brown and becomes detached from the root. Development from penetration into a root to the mature cyst takes about a month at suitable temperatures.

Hatching of eggs

Cyst nematode eggs hatch when they have been stimulated by a substance diffusing from host roots in the case of *Heterodera schachtii*, *H. glycines*, *H. goettingiana*, *H. carotae* (Jones) *H. cruciferae* (Franklin) *Globodera rostochiensis* (Woll.), *G. pallida* (Stone), *G. tabacum* (Lownsberry &

Lownsbery) and probably several other species; and by exposure to low temperature during a certain time in *H. avenae* (hatching at <5°C) and *H. trifolii* (hatching at >15°C). Even after exposure to host root diffusate only a few eggs of *G. rostochiensis* and *G. pallida* hatch within the three month period after the first moult of the juvenile in the egg. These few nematodes occasionally give rise to a second generation. The delayed hatching avoids the severe population reduction of the nematodes caused by hatching at a time when a new life cycle cannot be completed before the host dies, or perhaps more importantly when too few new roots are produced to accommodate a sizeable population. Eggs of *H. schachtii* and probably several other *Heterodera* species can be stimulated to hatch immediately after the moult is completed in the egg.

Generally only about 5 per cent of the eggs of *G. rostochiensis* and *G. pallida* hatch in water, but up to 30 per cent of those of beet cyst nematodes. However, about 35 per cent of potato cyst nematode eggs hatch in the soil without the stimulus from host roots during a short period in spring as soon as the temperature is high enough (den Ouden, 1960). The mortality of these juveniles is about 50 per cent per week. Spontaneous hatching is practically the only cause of decrease of populations in the soil in the absence of a host (den Ouden, 1960). It is not known whether a similar phenomenon occurs in other species that require a stimulus from host roots for hatching.

Penetration of juveniles in roots and further development

The second stage juvenile penetrates into a host root a short distance from the root tip, then moves through the cortex towards the tip and penetrates the layer that is to become the pericycle. After coming to rest the nematode punctures a developing vascular cell with its stylet, introduces saliva and induces the cell contents to rapidly become more dense. Vacuoles disappear and neighbouring cells fuse as parts of the separating cell walls dissolve. The nematode feeds on the contents of this developing transfer cell, swells, loses its muscles and thus becomes immobile and moults three times. The fourth moult of females takes place inside the skin of the third stage juvenile. The fourth stage female juvenile is flask shaped. Because of its size it splits the cortex of the root and the nematode becomes visible externally; only the thin anterior part of the adult female remains within the root. After the last moult the female soon fills with eggs, spontaneously if the species is parthenogenic (e.g. *H. trifolii*) or after mating with one or more males if it is amphimictic. The males are attracted by a volatile substance produced by the female (Green, 1971).

After becoming sausage shaped in the second stage, the male goes through three moults. It becomes vermiform again in the fourth stage but remains in the third stage skin until after the fourth moult. The adult male emerges from the fourth stage cuticle about three weeks after penetration of the second stage juvenile into a root. The ultimate volume of a male is 1/40 - 1/100 that of a female; also, the transfer cell on which the male feeds remains much smaller than that of a female juvenile. More extensive descriptions and illustrations of the development of females and males of cyst nematodes are in Raski (1950), Chitwood and Buhrer (1950), Franklin (1951) and Williams (1978).

The eggs are retained within the female body in most species, but in some a proportion (e.g. 30 per cent in *H. carotae*) is contained in a gelatinous egg sac similar to that of *Meloidogyne* species. After all the

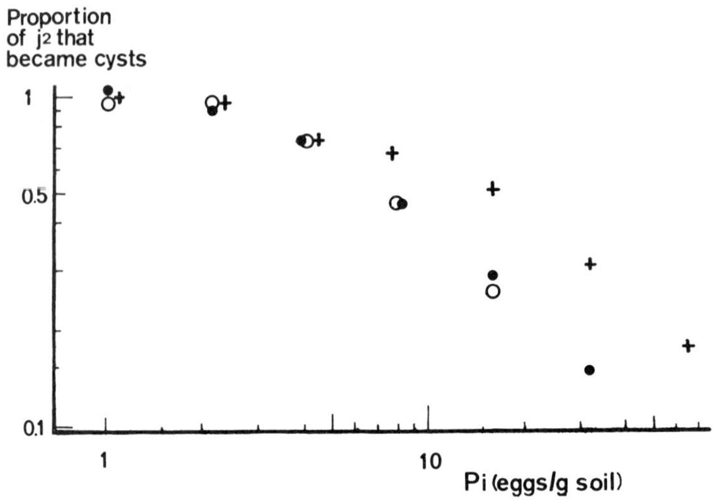

Fig. 1 Proportion of second stage juveniles of potato cyst nematodes in roots of two potato cultivars that became cysts at a range of initial egg densities (*Pi*, eggs per g soil). Scale of vertical axis gives ratios between these proportions and the proportion at *Pi* <2 eggs per g soil.

+ cv. Ehud with *G. pallida*, pathotype Pa3; O cv. Irene with *G. pallida*, pathotype Pa 3; O cv. Irene with *G. rostochiensis*, pathotype Ro 1.

	a	Pf_{max}
cv. Ehud Pa 3	40	150 eggs per g soil
cv. Irene Pa 3	47	85 eggs per g soil
cv. Irene Ro 1	70	150 eggs per g soil

a = maximum rate of multiplication as in equation (1). On cv. Ehud according to Seinhorst (1983), on cv. Irene according to Seinhorst and Oostrom (1984). See also Fig. 3.

eggs have formed the female dies and becomes a cyst filled with eggs. The white or yellow skin turns brown (cf. *G. rostochiensis* and *G. pallida*) and becomes leathery. The embryos in the eggs develop into first stage juveniles which moult to the second stage. In this stage they remain dormant as long as they do not receive the appropriate stimulus to hatch or conditions are unsuitable for activity (low temperature, drought). Some species are resistant against very dry conditions. *G. rostochiensis* and *G. pallida* populations survived dry storage in a refrigerator for 15 years. Others only survive the moderate drying that occurs in dry soil outdoors (*H. avenae*, *H. trifolii*) or storage in an atmosphere with 40 per cent moisture (*H. avenae*) (Meagher, 1974).

ADAPTATION OF POPULATION SIZE TO THE AVAILABLE QUANTITY OF FOOD (SPACE FOR TRANSFER CELLS)

To grow to the adult stage, a juvenile of a cyst nematode must feed on a single transfer cell during its entire life. This cell must be large enough and last long enough to supply sufficient food to allow the nematode to develop to a certain minimum size. At large densities the available space is limited by adjacent transfer cells. This results in the exclusion from competition for food of large proportions of the developing juvenile population at large initial population densities in the soil and the roots. Fig. 1 demonstrates the density dependency of the proportion of the initial population of second stage juveniles of *G. pallida* in roots of cv. Ehud and of *G. pallida* and of *G. rostochiensis* in roots of cv. Irene that became cysts in experiments by Seinhorst (1983) and Seinhorst and Oostrom (1984), respectively. To calculate survival (as a proportion of the numbers of juveniles per unit weight of roots, that became cysts at initial egg densities <2 eggs per g soil) it was assumed that numbers of juveniles that penetrated per unit weight of root were proportional to initial egg density (Jones and Kempton, 1982). Further, root weight, which was not determined, was assumed to be proportional to dry haulm weight. The difference between cvs Ehud and Irene with *G. pallida* is probably due to the larger space available in the roots of cv. Ehud than in those of cv. Irene, as indicated by the larger maximum density on cv. Ehud. This is a difference between experiments, not between the two cultivars. There are no indications that competition for food retards the development of cyst nematodes, as it does in migratory nematodes. According to Trudgill (1972) *Meloidogyne* juveniles separated from root tissue at a certain age proceeded to moulting only if they were larger than a certain minimum size. Probably this is only the mechanism by which juveniles of root knot and cyst nematodes are excluded from further development in root tissue. The exclusion of juveniles from competition for food by a density dependent mortality is a fundamental difference with the effect of competition for food on populations of migratory tylenchs, where it simply slows down or even stops growth and egg production until food becomes available again. It remains to be demonstrated what happens to transfer cells associated with dying nematodes.

POPULATION INCREASE ON HOST PLANTS

Some cyst nematode species produce only one new generation per season (*G. rostochiensis*, *G. pallida*, *H. avenae*, *H. trifolii*) and others two or more depending on the length of the season and favourable conditions (*H. schachtii*, *H. glycines* and several others). The relation between egg numbers of the parent generation and cyst and egg numbers of the next generation is readily observed in species with one generation per season; it can be explained in terms of competition for space in the host roots and reduction of the size of the host due to the nematode attack.

With more than one generation the explanation is more difficult. At the beginning of the season plant roots extend into soil with a moisture content at field capacity; the nematodes are then fully active. As the roots continue to grow they reduce the water content of the surrounding soil; the roots also collect all the activated nematodes which in turn damage them. In Western and Central Europe the first juveniles of beet cyst nematodes (*H. schachtii*) that penetrate into beet roots become adult about mid June, by which time the plant may have used up most of the water in the surrounding soil. Therefore, hatching of second generation eggs and penetration into roots by second stage juveniles may depend on summer rainfall and on the formation of new roots (which may be associated with this rainfall). The density of the new generation may be many times that of the parent and, therefore, capable of reducing the growth of invaded roots considerably more than the first generation. Thus the size of the food attack: per unit weight or volume of soil) is, according to equation (4) governed by two independent parameters, the maximum rate of multiplication a and the maximum density M. The presence of a in equation (3) does not mean that M depends on a, but only that when a is large (more eggs produced per egg initially present) fewer eggs of the initial population are needed to produce the individuals needed to exploit a certain proportion of the available space than when a is small (see also Fig. 1).

If the competition between transfer cells for space which leads to the elimination of developing nematodes also brings about the death of the associated transfer cells a more complicated relation between initial and final egg densities would exist. There is no clue yet how to estimate the relative volume of transfer cells associated with the eliminated individuals at the moment of the cessation of their feeding. All that can be said is that if most of the transfer cells that remained too small to support further growth of a nematode were not taken over by remaining individuals the numbers of eggs produced per unit weight or volume of root would have a strong negative regression upon initial egg density.

The volume of males of cyst nematodes is about 1/40 to 1/100 of that of females. Therefore, the volume of food they need is negligible in comparison to that needed by females at small to medium nematode densities. However, at large densities there may be eight to ten times as many males as females (Trudgill, 1967), needing about 20 per cent of the quantity of food that is needed by the females. The presence of males may, therefore, be expected to reduce the number of eggs produced per unit weight of root at large initial egg densities by up to 20 per cent, compared with smaller initial egg densities.

With more than one generation in a season population increase at small initial densities may become very large (e.g. *H. schachtii* on beet). First and second generations cannot be separated as the first juveniles of the latter may already have started to hatch and invade roots in the top layers of the soil, while deeper in the soil and further from the centre of the plant first generation juveniles may still be hatching. Therefore, it is difficult to determine a true maximum multiplication rate of a single generation, whereas in a second generation there may already be competition for food between individuals. Densities at the end of the season often are close to or equal to the maximum density (ceiling level of Jones, 1965) even at initial densities much smaller than those at which noticeable damage occurs. Only these maximum densities are of agronomic importance.

Effect of reduction of ultimate size of plant on reproduction of nematodes

Theories of the relation between initial and final egg densities of cyst nematodes on a host crop can only be tested if the effects of growth reduction and damage to the root tissue on the size of the food source are taken into account. This size will be closely related to the size of the plant. The relation between nematode population density Pi and relative plant size and root mass y can be assumed to be

$$y = m_y + (1 - m_y)z^{Pi-T} \text{ for } Pi > T \text{ and } y = 1 \text{ for } Pi < T \quad \ldots (6),$$

with m_y a constant <1, mostly between 0.1 and 0.4 in pot experiments with potato cyst nematodes (Seinhorst, elsewhere in this publication). A similar relation

$$f = m_f + (1 - m_f)z^{Pi-T} \quad \ldots . (7)$$

source is reduced and consequently the density of the adult second generation and the numbers of eggs it produces.

The relation between population density of the parent generation and of its offspring (initial and final population density on a crop in the case of one generation per season)

At very small population densities there is very little competition for space and food. The invading nematodes can, therefore, develop unhampered and produce the maximum number of eggs per individual initially present under the external conditions pertaining. Final population density is proportional to initial density:

$$Pf = aPi \quad .... \ (1)$$

and a is the maximum rate of multiplication (or reproduction) for the given combination of nematode and host plant under the given external conditions. It is by no means the same for different fields or seasons. For potato cyst nematodes it varied from 2.5 times to 150 times in 16 fields investigated in the Netherlands during eleven years, with a geometric mean of 24 (Seinhorst, elsewhere in this publication).

If initial nematode densities are not very small the rate of multiplication is less than the maximum because of intraspecific competition for space for the transfer cells and for food. If it is assumed that firstly, second stage juveniles arrive at and penetrate into the root tips at random time intervals and, therefore, that the transfer cells they induce are initially distributed randomly in the stele; and secondly that the total available space is exploited at large initial nematode densities, irrespective of elimination of individuals, then the numbers of eggs produced can be considered to be proportional to the volume of the transfer cells and, therefore, to the proportion of the available space occupied by them. With a random distribution of individuals this proportion approximates to the equation $1 - q^{Pi}$ if $1 - q$ is the proportion occupied at $Pi = 1$. The number of eggs produced at initial egg density Pi then is

$$Pf = c(1 - q^{Pi}) \quad .... \ (2)$$

assuming that the density of the second stage juveniles in the roots is proportional to the egg density in the surrounding soil (Jones and Kempton, 1978). If at very large densities this proportion decreases e.g. according to a factor p^{Pi} (competition resulting from random overlapping of territories) with $q<p<1$, the result is indistinguishable from that according to equation (2).

For $Pi \to 0, PfPi^{-1} \to a$ (equation (1)). According to the theorem of de l'Hôpital then

$$c = a(-{}^{e}\log q)^{-1} \quad .... \ (3)$$

and equation (2) becomes

$$Pf = a(1 - q^{Pi})(-{}^{e}\log q)^{-1} \quad .... \ (4)$$

For $Pi \to \infty$, $Pf \to a(-{}^{e}\log q)^{-1} = M$ .... (5) (Seinhorst, 1967).

The relation between initial and final egg densities per unit volume or weight of root (in the case of absence of growth reduction due to nematode

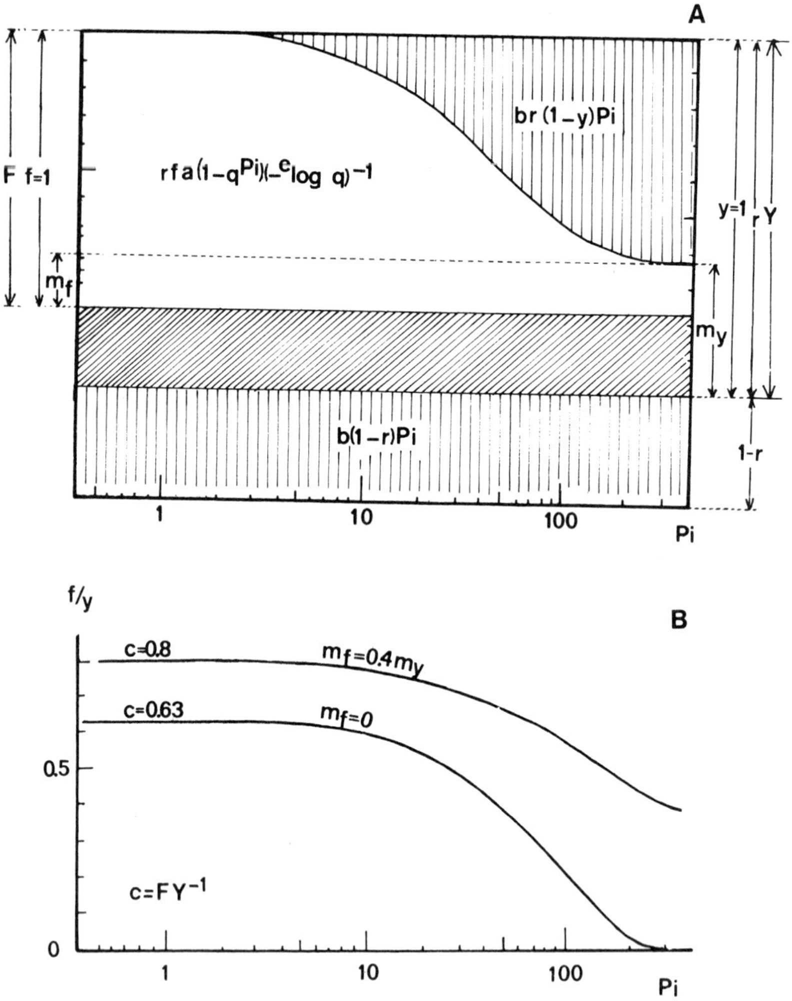

Fig. 2 A. Graphical presentation of the relative values of the different terms in equations (8), (9) and (10) at different values of *Pi*. B. Ratios at a range of *Pi* values between relative size of food source and that of root system at values of m_f according to A.

for $Pi > T$ and $f = 1$ for $Pi < T$ can be assumed to exist between *Pi* and the relative size of the food source. The relation between *Pi* and population density of a second generation *Pf* (= the measured final density if the soil was inoculated with eggs) then becomes:

$$Pf = f\,a(1 - q^{Pi})(-^e\log q)^{-1} \quad .... \ (8)$$

if equation (4) applies. If the inoculation was done with cysts the final population in the soil would also contain the eggs of the parent population that were not stimulated to hatch by the presence of host roots. If it is

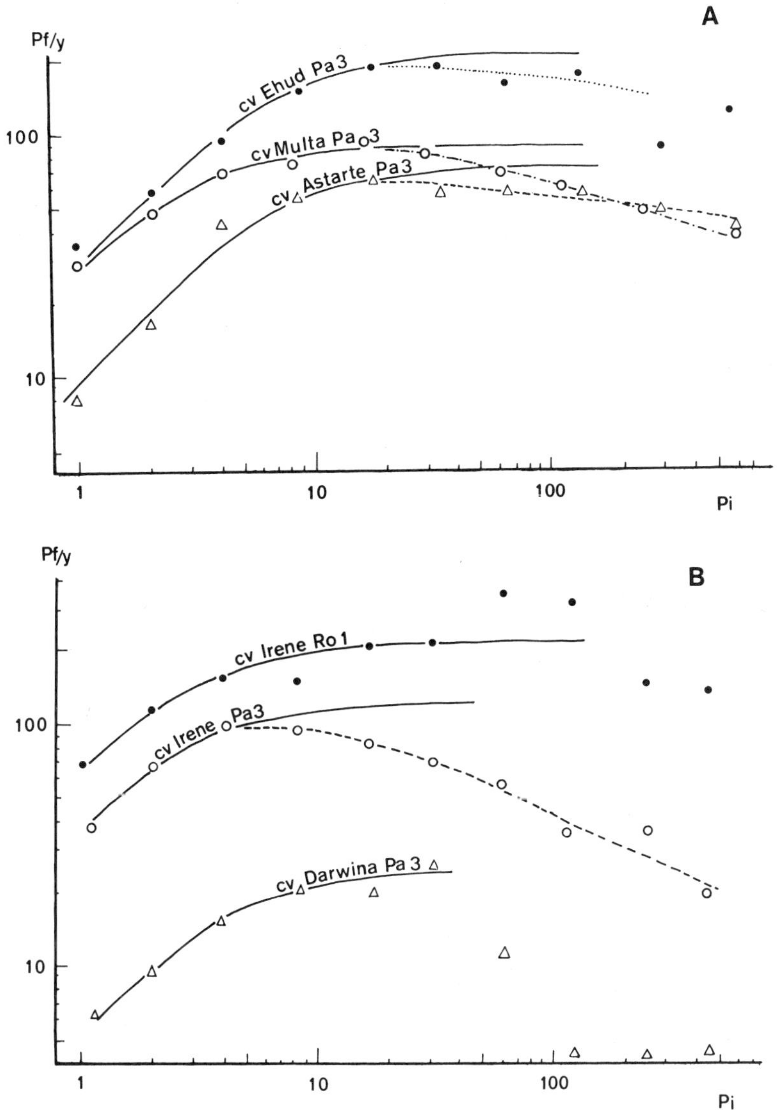

Fig. 3 Relation between initial egg density of *G. rostochiensis*, pathotype Ro 1 and *G. pallida*, pathotype Pa 3, (*Pi*, scale in eggs per g soil) and final densities of these nematodes (*Pf*) on various potato cultivars (eggs per g soil) divided by ratio *y* between haulm weight at nematode density *Pi* and that at densities < 2 eggs per g soil according to population data of Seinhorst (1983), Fig. 3, and Seinhorst and Oostrom (1984), and plant weight data according to Seinhorst et al. (in preparation). The scale of *Pi* for Pa 3 in Seinhorst (1983) Fig. 3 is not for eggs per g soil as indicated erroneously but for eggs per 2 g soil.

assumed that the proportion of the parent population that is stimulated to hatch is y according to equation (5) then the final population is:

$$Pf = fa(1 - q^{Pi})(-{}^{e}\log q)^{-1} + b(1 - y)Pi \quad .... \ (9),$$

in which $1 - y$ (with y according to equation (5)) is the proportion of the parent population that is not induced to hatch by root diffusate (proportion that does hatch is assumed to be proportional to plant size); b is the proportion of these eggs that also does not hatch spontaneously during the experiment; and f is the relative size of the food source according to equation (7). In Fig. 2B, c is the proportion of the root mass in the absence of nematodes (Y in Fig. 2A) that could have served as a food source for the nematodes (F in Fig. 2A). If $c = 1$, then $m_f = m_y$ and if $c < 1$, then $m_f < m_y$.

For field experiments equation (8) must be extended with the provision that a certain proportion of the population may stay out of reach of host roots (up to 30 per cent for potato cyst nematodes in potato fields according to Seinhorst and den Ouden, 1980). This results in:

$$Pf = r\{fa(1 - q^{Pi})(-{}^{e}\log q)^{-1} + b(1 - y)Pi\} + b(1 - r)Pi$$

$$= rfa(1 - q^{Pi})(-{}^{e}\log q)^{-1} + b(1 - ry)Pi \quad .... \ (10)$$

in which r is the proportion of the population that is induced to hatch from the eggs at $y = 1$ ($Pi < T$ in equation (6)). A graphical presentation of the different terms in equation (10) is given in Fig. 2A.

Test of the hypothesis that the food source is exploited efficiently irrespective of the death of developing nematodes due to competition for food (equations (4), (6), (8) and (9))

Final egg densities in two pot experiments with *G. rostochiensis* and *G. pallida* on various potato cultivars (Seinhorst, 1983, Seinhorst and Oostrom, 1984) divided into relative haulm weights, as an approximation of the relative sizes of the food sources in these experiments (Seinhorst et al., in preparation), Pfy^{-1}, were plotted against initial egg density Pi in Fig. 3. A curve according to equations (4) and (8) appeared to fit well to the observations up to at least $Pi = 20$ eggs per g soil, except for *G. pallida*, pathotype Pa 3, on cv. Irene. Here the deviation of observations from the curve according to equation (4) already begins at $Pi = 5$ eggs per g soil. It could be concluded that in many cases the food source is exploited efficiently despite the death of large proportions (up to 85 per cent) of the juveniles that penetrated into the roots. In most cases the smaller than expected numbers of eggs produced according to equation (4) might, up to $Pi = 100$ eggs per g soil, be largely due to an overestimation of the size of the food source by haulm weight (see Fig. 2B). Also, the space taken by juvenile males may have affected the numbers of eggs produced. However, for *G. pallida* on cv. Darwina at $Pi > 50$ eggs and cv. Ehud at more than 100 eggs per g soil and for *G. rostochiensis*, pathotype Ro 1, on cv. Irene at > 200 eggs per g soil there must have been another cause for egg production that was much smaller than proportional to root weight. There may have been a relation to "early senescence" (Seinhorst, elsewhere in this publication), which may have affected the functioning of the transfer cells before the females became adult.

Difference between relative size of the food source and relative root weight

Especially in the field, roots may grow under conditions (e.g. low temperature) in which nematodes are not active. This results in a higher minimum yield than when the nematodes are active. However, as cyst nematode juveniles penetrate only into root tips, those parts of the roots that grew when the nematodes were inactive cannot later provide a food source. Thus, the maximum size of the food source is smaller than the total root mass. This not only results in a barely measureable reduction of M (equation (4)), but also in a smaller minimum size m_f of the food source than the relative minimum root weight m_y (equations (6) and (7)), m_y being largely the roots grown while the nematodes were inactive (Fig. 2).

Size of the root system and the proportion of the population that hatches

The value of b in equations (9) and (10) can be determined by measuring population densities in fallow pots or plots at the beginning and the end of an experiment or the growing season. An estimate of r (equation (10)) is found by measuring the reduction caused by growing a highly resistant cultivar at a non-damaging initial nematode density (for potato cyst nematodes: $Pi < 10$ eggs per g soil) or by growing a susceptible cultivar treated with a large dosage of a non-killing nematostatic chemical e.g. aldicarb (Seinhorst and den Ouden, 1980). It may vary with plant density and size.

It may be assumed that the proportion of the population in a pot or plot that is induced by host roots to hatch is closely related to the extension and, therefore, to the weight of the root system present in the soil volume sampled, but exact observations are not available. In pot experiments it may be larger than y for root weight (according to equation (6)) if the root density in the absence of nematodes exceeds that required to obtain a complete hatch. According to Weischer (1959) root diffusates only induce potato cyst nematode juveniles to hatch from eggs if the distance between egg and root is not more than one cm. In pot experiments problems are avoided by inoculation of the soil with egg suspensions instead of with cysts.

Fitting curves according to equations (8) and (10) to observations on initial and final egg densities in pot and field experiments

When pots are inoculated with eggs and, therefore, the final population in cysts consists only of newly produced eggs, relative root weights and relative sizes of the food source can be compared as was done in Fig. 2. If the soil was inoculated with cysts, curves according to equation (10) can be fitted to the observation as was done by Greco et al. (1982). Discrepancies between y (observed) and f as estimated from the relation between Pi and Pf according to the best fitting curve can, at small to medium values of Pi, be dealt with easily, while at large Pi they are obscured by the large values of $b(1 - y)Pi$ (approaching $b(1 - m_y)$).

An effort was made in Fig. 4 to fit curves according to equation (10) to initial and final egg densities of *G. rostochiensis* on potato (cv. Bintje) in two groups of square metre plots in different parts of a field. The plots had been laid out in rows parallel to the long axes of two infested patches similar to the patch of Fig. 9 of Seinhorst (1986). Population densities in the plots ranged from 0 to 200 eggs per g soil. "True" initial densities (Pi) were estimated assuming linear gradients for log egg densities in successive plots in the rows, as described for log

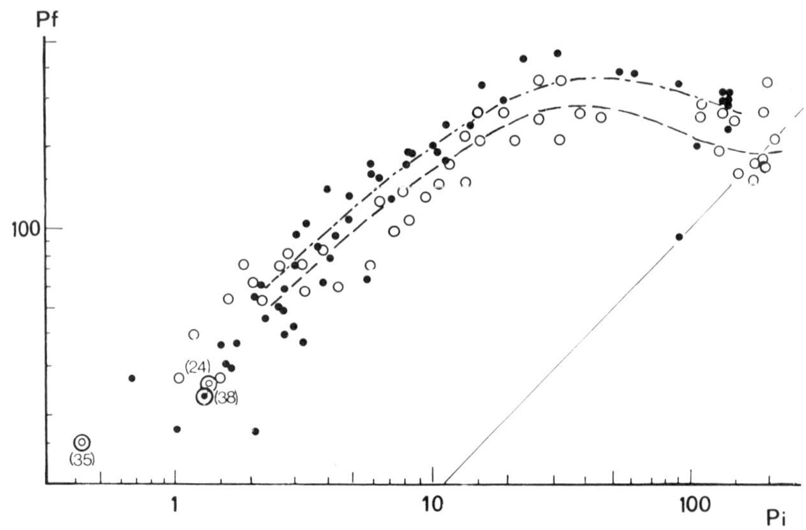

Fig 4. Relation in two infested patches in the same field between population densities of *G. rostochiensis* before planting and after growing potatoes (cv. Bintje) (*Pi* and *Pf*, scales in eggs per g soil). Graphs according to equation (10) with $r = 0.7$, $b = 0.65$, $m_y = 0.6$, $m_f = 0.1$ and for
O : $a = 32$ and $M = 1{,}580$ eggs per g soil
O : $a = 38$ and $M = 1{,}880$ eggs per g soil
o , o averages if egg density in the numbers of plots mentioned between brackets. All plots in uniformly infested areas. The points indicate maximum rates of multiplication and used in Table 1 of Seinhorst (elsewhere in this publication) $PfPi^{-1} = r(a - b) + b$.

cyst densities by Seinhorst (1982b). The values of *Pi* in Fig. 4 are those according to the lines drawn in Fig. 9 of Seinhorst (1986); those of *Pf* are the actual counts. Apparently both maximum rates of multiplication and maximum densities differ from the different parts of the field and the first also for different parts of the rows, both with maximum ratios of about 1 to 1.5 at Pi >5 eggs per g soil, but 1 to 2.5 at Pi < 2 eggs per g soil. Such differences will maintain large scale variations of potato cyst nematode densities in fields. However, the same values of maximum rates of multiplication seem to apply to large enough areas to contain density ranges in patches of 5 to 200 eggs per g soil. Although the nine parameters in equation (10) seem to give a considerable degree of freedom of manipulation, there appeared to be only very few combinations of values that gave a reasonable fit of curves to the observations, in this case curves for small m_f (0.1) combined with large m_y (m_y and m_f as in equations (6) and (7)). The value of m_y is the same as of m for tuber weights in this experiment (Fig. 1 of Seinhorst, 1982a).

The example given and the nature of the data used demonstrate that curves according to equation (10) can only be sensibly fitted to data from relatively small areas in a field e.g. to population densities in small plots to measure relative nematode multiplication on different host crop

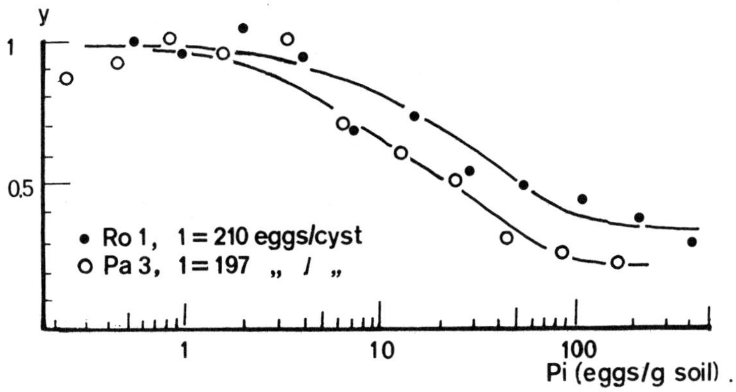

Fig. 5 Relation between initial egg density *Pi* (eggs per g soil) and relative numbers of eggs per cyst *y* of *G. rostochiensis*, pathotype Ro 1 and *G. pallida*, pathotype Pa 3, on cv. Irene.

cultivars. For planning long term rotations and control strategies, expected maximum rates of multiplication and/or maximum densities must be estimated. This can only be done on the basis of frequency distributions of these rates and densities determined in several fields and years. The probably very limited applicability of maximum rates of multiplications (due to their great variability) is discussed by Seinhorst (elsewhere in this publication). Long term rotation schemes can probably be based on the much less variable maximum densities of cyst nematodes with more than one generation in a season (*H. schachtii* on beets, *H. glycines* on soy bean).

Effect of intraspecific competition on cyst content

According to Jones and Perry (1978) the numbers of eggs produced by *H. goettingiana* females in a field experiment were independent of initial population density and, therefore, not affected by competition of these females for space or food. The same could be derived from the numbers of eggs per cyst of *G. rostochiensis* populations from field experiments (.e.g. that of Fig. 4). However, Rao and Peachey (1965) and Seinhorst (1967, 1983) found that competition for food in *G. rostochiensis* and *G. pallida* populations in pot experiments not only reduced the proportion of invading juveniles that became females (Fig. 1) but also the average number of eggs per female (cyst) (Fig. 5). At very large initial egg densities of potato cyst nematodes, cyst contents in the new generation were in some cases one fifth of those at small initial densities.

Relation between initial nematode density and rate of multiplication

The relation between initial egg density and rate of multiplication can be demonstrated by superimposing lines indicating equal rates (lines at 45° to the axes) to graphs like those of Figs. 1, 4 and 6 and as was done in Fig. 1 of Seinhorst (elsewhere in this publication). As the relation was thought to be of practical importance efforts were made to find linear expressions for it to facilitate statistical treatment of observations (Raeuber and Stelter, 1970; Mugniéry, 1976). However, of the two terms of equations (8) and (9) one is related linearly to *Pi* and the

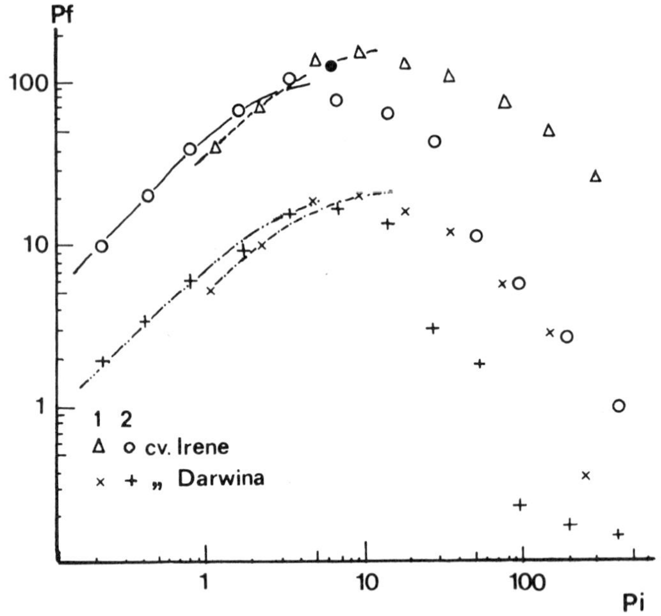

Fig. 6 Relation between initial and final egg densities *Pi* and *Pf* (in eggs per g soil) of *G. pallida*, pathotype Pa 3, on cvs Irene and Darwina. 1 : after Seinhorst (1983), 2 : after Seinhorst and Oostrom (1984).

other exponentially and separate graphs according to these terms (whether or not divided by *Pi*) can have very different position relative to each other, depending on the value of *a*; a linear transform covering a sufficient range of values of *a* does not exist. Application of arbitrary linear equations can only lead to large errors of estimations of $PfPi^{-1}$. Also a larger sampling error of *Pi* than of *Pf* is translated into a considerable overestimation of $PfPi^{-1}$ at small *Pi*.

HOST STATUS

The parameters *a* and $M = a(-^{e}\log q)^{-1}$ in equations (3), (5), (9) and (10) determine the status of a plant as a host for the nematode population under investigation. However, *a* and perhaps also to a lesser extent *M* may vary from experiment to experiment, due to differences in external conditions. Therefore, comparison with a standard is advisable e.g. when the susceptibility of potato cultivars to potato cyst nematodes is to be determined (Seinhorst, 1983; Seinhorst and Oostrom, 1984). Values of *a* can be determined directly if initial nematode densities are made small enough; one or fewer eggs per g soil for *G. rostochiensis* or *G. pallida* on potatoes, but about 0.2 or fewer eggs per g soil for *H. avenae*. The difficulties arising with species of cyst nematodes producing more than one generation in a season have already been mentioned. An arbitrary solution could be to restrict the time during which nematodes penetrate into roots by inoculating only a small portion of the soil in pots with a small number of nematodes (local nematode density should be kept small) and stop the experiment as soon as adult females start to produce eggs. As these females do not depend on the continuing intake of food for further egg production, the plant tops can be cut off and the soil allowed to dry, thus preventing the hatching of the new eggs.

Values of *M* (equation (5)) can be derived from the *Pf* values observed at a range of *Pi* values. If the latter were obtained by inoculation of the soil with egg suspensions, curves according to equation (8) can be fitted to the observations, estimating *T* from the plant weights at the end of the experiment or using known values e.g. two eggs per g soil for potato cyst nematodes on susceptible cultivars. If only values of *M* on different cultivars are to be compared, calculated values can be replaced by the maximum values of *Pf* in the experiments, provided *T* is the same for the ratios between multiplication rates on the cultivars to be compared are independent of *Pi* up to the maximum values of *Pf*. Seinhorst (1983) and Seinhorst and Oostrom (1984) found this to be the case for *G. pallida* pathotype Pa 3, on cvs Darwina and Irene (Fig. 6). This is not necessarily always the case for potato cyst nematodes on potatoes or for other nematode and plant species (cf. *G. pallida*, pathotype Pa 3, on cvs Ehud and Astarte in Fig. 3 of Seinhorst, 1983). Maximum rates of multiplication and maximum densities may vary independently of each other (Seinhorst, 1983). The ratio of maximum densities of *H. schachtii* on susceptible and resistant *Raphanus sativus* subsp. *oleifera* probably is much larger than that of the maximum rates of multiplication (Kaai, 1984). Therefore, growing it (as a green manure crop) may reduce large population densities of beet cyst nematode but not small ones.

REFERENCES

Chitwood, B.G. and Buhrer, Edna M., 1946, The life history of the golden nematode of potatoes, *Heterodera rostochiensis* Wollenweber, under Long Island, New York conditions, Phytopathology, 36:180.

Fisher, J.M., 1969, Investigations on fecundity of *Aphelenchus avenae*, Nematologica, 15:22.

Franklin, Mary T., 1951, The cyst forming species of *Heterodera*, Imp. Bur. Agr. Parasit., Farnham Royal, 147 pp.

Greco, N., di Vito, M., Brandonisio, A., Giordano, I. and Marinis, G. de., 1982, The effect of *Globodera pallida* and *G. rostochiensis* on potato yield, Nematologica, 28:379.

Green, C.D., 1971, Mating and host finding behaviour of plant nematodes, in "Plant Parasitic Nematodes" Vol. 2, B.M. Zuckerman, W.F. May and R.A. Rohde, eds., Academic Press, New York.

Jones, F.G.W., 1956, Soil populations of beet eelworm (*Heterodera schachtii* Schm.) in relation to cropping II, Microplot and field results, Ann. appl. Biol., 44:25.

Jones, F.G.W. and Kempton, R.A., 1978, Population dynamics, Population models and Integrated Control, in "Plant Nematology", J. Southey, ed., MAFFA AS Publ. GD/1, HMSO, London.

Jones, F.G.W. and Perry, J.N., 1978, Modelling populations of cyst nematodes (Nematoda: Heteroderidae), J. appl. Ecology, 15:349.

Kaai, C., 1948, Invloed van kruisbloemige groenbemesters op de populatie van *Heterodera schachtii*, in "Jaarverslag 1983, Instituut voor Plantenziektenkundig Onderzoek", p.128.

Meagher, J.W., 1974, Cryptobiosis of the cereal cyst nematode *Heterodera avenae* and effects of temperature and relative humidity on survival of eggs in storage, Nematologica, 20:323.

Mugniéry, D., 1976, Etablissement d'un modèle de dynamique de population d'*Heterodera pallida* Stone: Application à un cas pratique de lutte intégrée, Annales de Zoologie-Ecologie animale, 8:315.

Ouden, H. den., 1960, Periodicity in spontaneous hatching of *Heterodera rostochiensis* in the soil, Nematologica Suppl. II, 101.

Raeuber, A. and Stelter, H., 1970, Die Endverseuchung in Abhängigkeit von der Anfangspopulation bei *Heterodera rostochiensis* Woll., Biol. Zentralbl., 89:87.

Rao, G.N. and Peachey, J.E., 1965, The effect of adding larvae of potato cyst nematode to potatoes grown in pots, Plant Pathology, 14:15.

Raski, D.J., 1950, The life history and morphology of the sugar beet nematode, *Heterodera schachtii* Schmidt, Phytopathology, 40:135.

Seinhorst, J.W., 1967, The relationships between population increase and population density in plant parasitic nematodes. II Sedentary nematodes, Nematologica, 13:157.

Seinhorst, J.W., 1982a, The relationships in field experiments between population density of *Globodera rostochiensis* before planting potatoes and the yield of potato tubers, Nematologica, 28:277.

Seinhorst, J.W., 1982b, The distribution of cysts of *Globodera rostochiensis* in small plots and the resulting sampling errors, Nematologica, 28:285.

Seinhorst, J.W., 1983, Relation between population density of potato cyst nematodes and measured degrees of susceptibility (resistance) of resistant potato cultivars and between this density and cyst content in the new generation, Nematologica, 30:66.

Seinhorst, J.W., 1986, The estimation of densities of nematode populations in field and whole fields. II: Sampling and Sampling Error, (in preparation).

Seinhorst, J.W. and Kuniyasu, K., 1971, Interaction of *Pratylenchus penetrans* and *Fusarium oxysporum* forma *pisi* race 2 and of *Rotylenchus uniformis* and *F. oxysporum* f. *pisi* race 1 on peas, Nematologica, 17:444.

Seinhorst, J.W. and Oostrom, A., 1984, Comparison of multiplication rates of three pathotypes of potato cyst nematodes on various susceptible and resistant potato cultivars, Meded. Fac. Landbouww. Rijksuniv. Gent, 48:605.

Seinhorst, J.W. and Ouden, H. den., 1980, The effect of different dosages of aldicarb on the multiplication at small population densities of *Globodera rostochiensis* on potato in rotavated and non-rotavated plots, Neth. J. Pl. Path., 86:147.

Seinhorst, J.W., Schomaker, C. and Been, T., 1987, The relation between initial egg density of *Globodera rostochiensis* and *G. pallida* and the growth, final haulm and tuber weights and dry matter contents of various potato cultivars, (in preparation).

Trudgill, D.L., 1967, The effect of environment on sex determination in *Heterodera rostochiensis*, Nematologica, 13:263.

Trudgill, D.L., 1972, Influence of feeding duration on moulting and sex determination of *Meloidogyne incognita*, Nematologica, 18:476.

Williams, T.D., 1978, Cyst Nematodes: Biology of *Heterodera* and *Globodera*, in "Plant Nematology", J. Southey, ed, MAFFA AS Publ. CD/1, HMSO, London.

DISCUSSION

B'Chir asked if the three factors: time of infestation, host response and effect of micro-organisms are included in the model. Seinhorst replied that population dynamic models do not need to consider ecological parameters. He added that with reference to PCN in western Europe, the roots of the potato plant are in soil which is plentifully supplied with water and the first generation of the nematode is very regular, but the second generation is unpredictable. Andersen pointed out that 60 per cent of the juvenile nematodes die in the first week and queried whether this rate of reduction in population density would be affected by the tolerance of cultivars. Seinhorst said that he had no real experience with tolerant plants but observed that there was a large variation in invasion and multiplication. Sikora asked if penetration of the roots by second stage juveniles was correlated with root size and Seinhorst responded that there was no evidence in the literature to support or deny this supposition.

PHYSIOLOGY OF HATCHING

Roland N. Perry

Rothamsted Experimental Station
Harpenden, Herts, AL5 2JQ
England

INTRODUCTION

Cyst nematode hatching has interested research workers ever since Baunacke (1922) first demonstrated that eggs hatch in response to host root diffusates. A natural hatching stimulus, such as diffusate, points to potential control strategies. The gap between theory and field application has yet to be bridged, although several recent developments in research into the hatching mechanisms of cyst nematodes indicate that measures which influence nematode behaviour may eventually be feasible as part of an integrated control programme. Such an approach may avoid environmental problems, which are an increasingly important consideration.

Research into hatching mechanisms has two main objectives: either to prevent hatch in the presence of the host plant or to initiate hatch in the absence of the host. The former objective is less desirable as the cysts and contents would remain in the soil. Initiation of hatch has the additional advantage of being more attainable, judged by results from recent research.

Knowledge of the various stages in the hatching sequence from hatch stimulation to eclosion (Fig. 1) is important if control targets are to be identified accurately. Cinematographic techniques were used to study the behavioural responses involved when the juvenile of *Globodera rostochiensis* cuts a polar slit in the eggshell to facilitate eclosion (Doncaster and Shepherd, 1967; Doncaster and Seymour, 1973). These observations have been discussed in the context of active responses of other nematode species during hatching (Perry and Clarke, 1981). However, as Doncaster and Shepherd (1967) pointed out, in *G. rostochiensis* "usually at least 3 days elapsed between the application of root diffusate to the eggs and the onset of vigorous movement within them". It is the events during these 3 days on which recent research has concentrated and the present review reflects this emphasis. Most work on hatching mechanisms has been carried out on *G. rostochiensis, Heterodera schachtii* and *H. goettingiana*.

THE ROLE OF TREHALOSE

Clarke and Hennessy (1976) showed that the fluid within the egg of *G. rostochiensis* which surrounds the juvenile contains trehalose at a

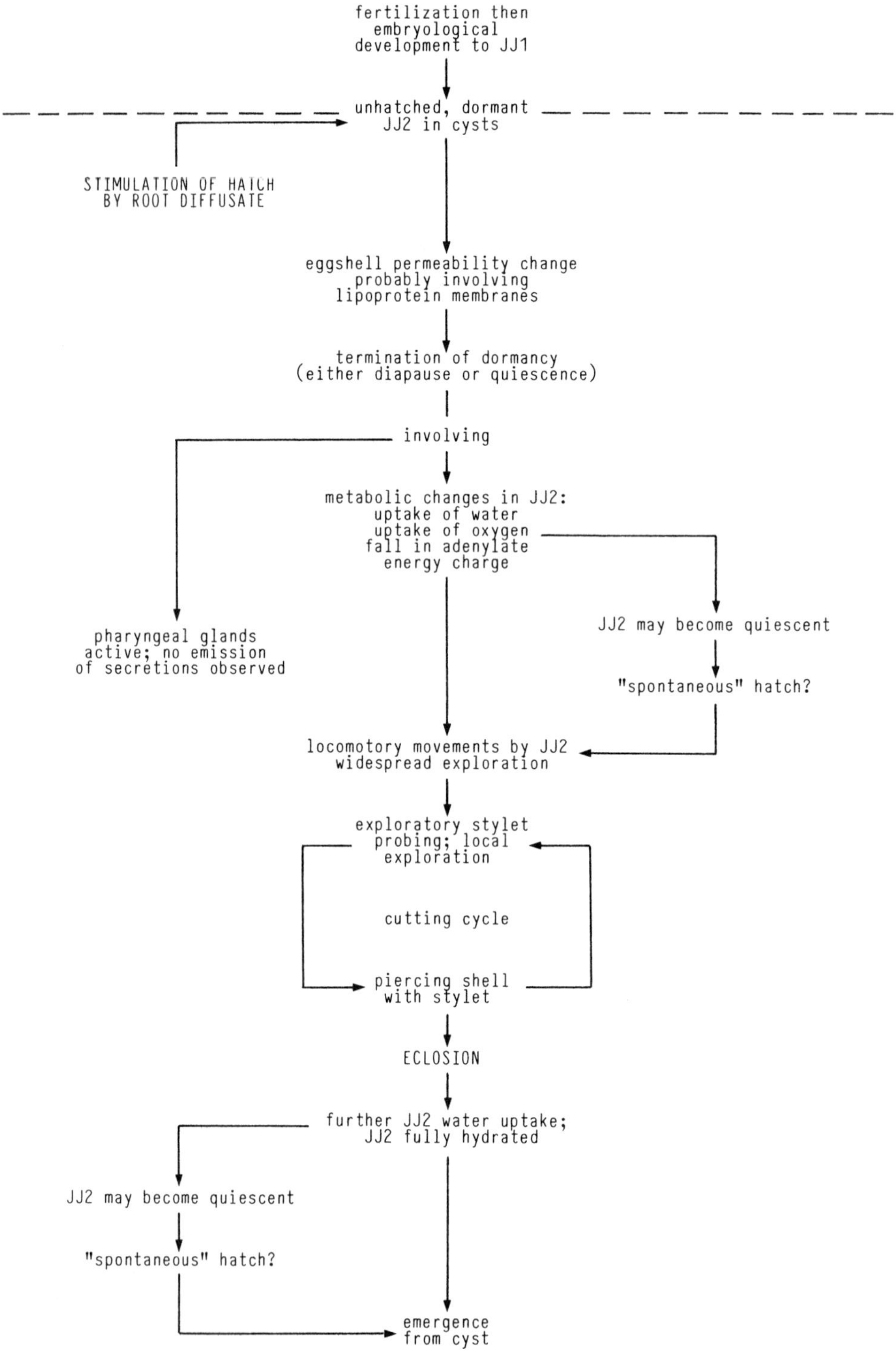

Fig.1. The sequence of events in the hatching of *Globodera rostochiensis* when stimulated by potato root diffusate (Diagram modified from Clarke and Perry, 1977).

concentration of 0.34M. The perivitelline fluid of *H. goettingiana* also contains trehalose but at a concentration of 0.5M (Perry et al., 1983). The osmotic pressure generated by the trehalose serves to keep the water content of juveniles of *G. rostochiensis* and *H. goettingiana* down to 67.3% and 66.5% respectively (Clarke et al., 1978; Perry et al., 1983). Although these values are only about 5% below the water content of active juveniles immediately before hatch, they are not commensurate with full mobility probably because the turgor pressure of the nematode is insufficient to antagonise the longitudinal musculature. In this state the juveniles are unable to start the cutting cycle leading to eclosion.

Thus, juveniles of *H. goettingiana* and *G. rostochiensis* are not in a suitable condition for hatching until the osmotic pressure of the egg fluid is reduced. In both species, hatching can be prevented even in the presence of root diffusates by increasing the osmotic pressure of the hatching solution (Clarke et al., 1978; Perry et al., 1983) but the levels of external osmotic pressure required preclude this as a possible control measure.

EFFECTS OF THE REDUCTION OF OSMOTIC PRESSURE

Evidence from investigations on *G. rostochiensis* and *H. goettingiana* indicates that root diffusates cause a loss of solutes, such as trehalose, from the egg fluid thus removing osmotic stress on the juvenile.

In *G. rostochiensis* the removal of osmotic stress results in the juvenile becoming metabolically active after stimulation by potato root diffusate (PRD). The oxygen consumption of cysts increases significantly and, as this occurs during the 24h immediately following addition of PRD, it is unlikely to be attributable solely to locomotor activity of the juvenile (Atkinson and Ballantyne, 1977a). Also, during this period there is a significant fall in the adenylate energy charge (Atkinson and Ballantyne, 1977b). These events accompany hydration of the juvenile, for Ellenby and Perry (1976) showed a significant increase in juvenile water content before hatching which reached a maximum for juveniles in free eggs within 24h after stimulation and for juveniles in encysted eggs within 48h. Leakage of trehalose has also been detected during this 48h period (Clarke, personal communication).

In vitro hatch of *H. goettingiana* can be obtained using pea root diffusate (Perry et al., 1980a) and field beans (Beane and Perry, 1983). There was an increase in water content of unhatched juveniles of *H. goettingiana* when exposed to pea root diffusate but the rate is much slower than that for *G. rostochiensis*; the maximum water content was not attained until 7d after stimulation (Perry et al., 1983). This may reflect variation between the species in the time taken to reduce the different initial trehalose concentrations of the perivitelline fluid; the response time of the juveniles may also differ and this point is discussed more fully below.

The unhatched *H. schachtii* juvenile already contains enough water (69%) to permit movement without exposure of the egg to root diffusate (Perry et al., 1980b) and no water uptake was detected before hatch (Perry, 1977). The greater water content of unhatched *H. schachtii* juveniles reflects a lower osmotic pressure of perivitelline fluid (0.3M) and this, plus the fact that *H. schachtii* second stage juveniles (JJ2) move in sugar solutions which would inhibit movement of *G. rostochiensis* JJ2 (Perry et al., 1980b), may be a partial explanation of the large water hatch of *H. schachtii*. However, there is an additional hatch induced by host root diffusates and it is possible that even a slight dilution of the egg fluid may be important.

THE IMPORTANCE OF THE EGGSHELL

The precursor of enhanced juvenile metabolism and subsequent activity leading to hatch appears to be a change in eggshell permeability which occurs within the 24h after exposure to PRD (Perry and Clarke, 1981). Ultrastructural and histochemical studies on the eggshell of *G. rostochiensis* showed that the shell consists of an outer vitelline layer, a chitinous layer and an inner lipid layer (Perry et al., 1982). As the lipid layer is the main permeability barrier of eggshells of many nematode species (Wharton, 1980), the discovery of this layer in *G. rostochiensis* suggests it may be of importance in the hatching response. There is no report of the existence of such a lipid layer in eggshells of *H. goettingiana* but, if present, it may differ in structure which could affect the rate of permeability change and thus the rate of JJ2 hydration.

The potential of the lipoprotein membrane as a novel target site for control agents can best be evaluated when the nature of the permeability change is understood. Clarke et al., (1967) showed that Ca^{2+} was a major inorganic constituent of the eggshell and it is known that Ca^{2+} stabilises biomembranes. As long ago as 1958, Ellenby and Gilbert suggested that cation transport was an essential part of the hatching mechanism of *G. rostochiensis* and that Ca^{2+} synergises the action of root diffusates. Atkinson and Ballantyne (1979) used La^{3+} and ruthenium red, which both inhibit many Ca^{2+}-mediated processes, and Ca^{2+} ionophores and suggested that hatching might be a calcium-mediated process. Atkinson and Taylor (1980a) suggested that Ca^{2+}, after its initial binding to the eggshell, is transported into the egg; there it may act as a secondary messenger (Atkinson and Taylor, 1980b). Recently Atkinson and Taylor (1983) reported that a sialoglycoprotein with a high affinity for Ca^{2+} participated in the process.

In contrast, Clarke and Hennessy (1981) found that La^{3+} was an effective hatching agent and that ruthenium red inhibited hatch by inactivating the hatching factor. Clarke and Hennessy (1983) also showed that *G. rostochiensis* eggs hatch in test solutions containing little or no free Ca^{2+} and suggested that the enhanced hatch that Ellenby and Gilbert (1958) found after the addition of Ca^{2+} to the test solutions may be attributed to the ion-exchange effects of dilute solutions in the presence of cyst protein. Thus, Clarke and Hennessy (1981, 1983) considered that free Ca^{2+} is not essential for hatching and that Ca^{2+} transport mechanism is probably not involved in the permeability change.

However, Ca^{2+} may have a structural role relating to eggshell permeability. Clarke and Perry (1985a) used X-ray microanalysis to examine the Ca^{2+} content of eggshells of *G. rostochiensis* and suggested that there are three types of Ca^{2+} binding sites. The more strongly bound Ca^{2+} is bound to sites in the outer layers and appears not to be involved in the hatching mechanism. Another site is the lipoprotein layer and exposing the eggshell to artificial hatching agents and decationised PRD caused the removal of Ca^{2+}; Ca^{2+} was not displaced by non-hatching agents. The third type of site, presumably also located in the lipoprotein layer, becomes available when eggshells are treated with PRD and results in the eggshells binding additional Ca^{2+} (Atkinson et al., 1980; Clarke and Perry, 1985a). Hatching agents, by binding to or replacing Ca^{2+}, may change lipoprotein membrane structure and allow the leakage of molecules such as trehalose. Atkinson and Taylor (1983) also used X-ray microanalysis to investigate eggshells of *G. rostochiensis* but their conclusion, that the relevant cation binding sites are located on the innermost layer of the eggshell, was based on an experimental protocol which has been questioned by Clarke and Perry (1985a).

In several respects, the changes in Ca^{2+} content of *H. schachtii* eggshells in response to various treatments parallel the pattern for *G. rostochiensis* (Clarke and Perry, 1985b). The eggshell contains Ca^{2+}, most of which is inaccessible to chelating agents and is therefore thought to be located in an inner layer; the presence of lipoprotein membranes in the eggshells of *H. schachtii* remains to be confirmed. The eggshell Ca^{2+} content is diminished by treatment with artificial hatching agents but not by non-active agents and exposure to host root diffusate causes a considerable increase in Ca^{2+} content (Clarke and Perry, 1985b). Thus, in *H. schachtii*, the eggshell permeability change may be Ca^{2+}-dependent and result in a slight but important dilution of the egg fluid.

Zinc, rather than Ca^{2+}, appears to be involved in the hatching mechanism of *H. glycines* (Tefft and Bone, 1984) and the biological activity of zinc as a hatching stimulus indicated that a zinc-dependent enzyme may mediate hatching. Recently, Tefft and Bone (1985a) have reported the occurrence of leucine aminopeptidase activity in the egg supernatant of *H. glycines* although root diffusate does not directly increase the activity of this enzyme (Tefft and Bone, 1985b). The possible involvement of enzymes in hatching of plant and animal parasitic nematodes was reviewed by Perry and Clarke (1981).

The inner lipoprotein layer of cyst nematodes presents an obvious target for control measures. A substance which alters lipoprotein membrane permeability may initiate the hatching sequence. This may be the explanation for the results of Kraus and Sikora (1983) who found that the herbicide diallate significantly increased the hatch of *H. schachtii*. Diallate is a thiocarbamate, a group known to cause membrane disintegration and altered permeability in plant cells, and it may have altered eggshell permeability characteristics causing increased hatch. However, it has not been possible to obtain similar results using *G. rostochiensis* (Perry, unpublished) possibly because of the bimodal action of PRD.

CONSEQUENCES OF THE BIMODAL ACTION OF PRD

As well as altering eggshell permeability, PRD has an additional role, which may be affected sometime after permeability change, which is to stimulate juvenile activity. Clarke and Hennessy (1984) showed that PRD stimulated movement and migration through sand columns of *G. rostochiensis* JJ2. The respiration and energy changes noted by Atkinson and Ballantyne (1977a, b) may be due to direct stimulatory effects of PRD on the juveniles; the termination of quiescence of unhatched JJ2 may also be due to this direct effect as well as to the leakage of trehalose. However, the effects of osmotic stress override any activation of the juvenile as *G. rostochiensis* eggs do not hatch in the presence of PRD if the solution has an osmotic pressure greater than 0.4M trehalose (Clarke et al., 1978).

The stimulatory effect of PRD may be important in migration of hatched juveniles if a concentration gradient is set up (Wallace, 1958, 1961) attracting juveniles to the root area. Papademetriou and Bone (1983) reported that *H. glycines* juveniles were attracted by soy-bean root diffusates. The enhanced susceptibility of *G. rostochiensis* to nematicides in the presence of PRD (Evans and Wright, 1982) may also be associated with their increased metabolic activity (Clarke and Hennessy, 1984).

Thus, the use of chemicals which affect membrane permeability may cause changes in the eggshell and egg trehalose content but may not result in large JJ2 hatch because for some JJ2 enhanced activity, derived from metabolic stimulation, is required for eclosion. However, such a control strategy for *G. rostochiensis* may be of use even if only about 30% of the cyst contents hatch; the treatment may also render the unhatched juveniles more susceptible to nematicides and to environmental extremes. Experiments on desiccation survival illustrate this latter point. Free *G. rostochiensis* JJ2 can survive drying for minutes only (Kämpfe, 1959) but the unhatched J2 is protected from desiccation effects by the eggshell (Ellenby, 1968) and probably by the perivitelline fluid (Clarke et al., 1978). Exposing cysts and eggs to PRD, for a period long enough for permeability change and trehalose leakage but too short for any hatching, followed by desiccation caused a decrease in hatch when the cysts and eggs were subsequently rehydrated and placed in PRD (Perry, 1983).

NATURAL HATCHING FACTORS

Host root diffusates which are needed to induce hatch of *G. rostochiensis*, *G. pallida*, *H. goettingiana* and *H. glycines*, for example, and to enhance the rate of hatch and total hatch of species like *H. avenae* and *H. schachtii* provide obvious control potential. Potato root diffusate contains between four to six components which stimulate hatch (Clarke, 1970; 1971). These hatching factors are moderately strong organic acids (Clarke, 1967), are rapidly inactivated at >pH8 and are fluorescent (Clarke, 1968). The hatching factors are present in only very small concentrations in the root diffusates and they can act at extreme dilutions. Masamune (1976) reported that hatching factors for *H. glycines* can act at dilutions as low as 10^{-14} g/ml.

The problems associated with obtaining pure samples of the hatching factors centre on the purification methods required to yield chromatographically homogeneous products. The introduction of procedures such as high performance liquid chromatography (HPLC) enables more effective purification of hatching factors. Masamune et al., (1982) used HPLC to isolate a natural hatching factor for *H. glycines* (as the p-bromophenacyl ester). The factor, glycinoeclepin A, was a dibasic acid with the molecular weight of 446 and a proposed formula of $C_{25}H_{34}O_7$. The hatch stimulating concentrates were prepared from root extracts of kidney bean rather than root diffusates and it is therefore not certain that the factor is present in diffusates or, if present, whether it is in this form. However, this work marks an important step in the eventual isolation and structural determination of the hatching factors of cyst nematodes. Only when this has been achieved can the reality of formulating analogues be assessed.

When details of the structure of hatching factors are known it may be possible to determine if and why diffusates from different potato cultivars vary in hatching activity. At present confusion exists. Evans (1983) considered that the log-activity values for various potato cultivars were probably a reflection of plant growth rather than intrinsic differences; diffusates from most cultivars hatched *G. rostochiensis* well, some hatched *G. pallida* well and some hatched it poorly. Turner and Stone (1981) concluded that evidence for diminished hatching of both species by *S. vernei* hybrids was not consistent. In contrast, Farrer and Phillips (1983) and Forrest and Phillips (1984) found that, in tests with *G. pallida*, some potato hybrids produce diffusate of reduced hatching activity. In part, different experimental protocols may be influential. Turner and Stone (1981) used the number of JJ2 hatching per gm of root to give an exact basis for comparison between cultivars, but Forrest and Phillips (1983) suggested that Turner and Stone included poorly grown root systems in their comparisons, reflecting one of the problems associated with pot tests. Unit root length may also be a more accurate measure of the physiological contribution of the root than root weight (Newman, 1966). However, the extent of the root system is clearly an important ecological consideration for root diffusate effects in the soil.

The use of root diffusate as a control measure to induce hatch of *G. rostochiensis* in the field in the absence of host plants was discounted some 30 years ago. Fenwick (1956) found that when PRD was added to soil it was quickly inactivated and he considered this approach impractical. Recent work has indicated that, while the use of root diffusate may not be a control agent alone, a re-examination of the feasibility of using PRD as part of an integrated control strategy is worthwhile; even a hatch of 30% may be of importance in areas, such as the Netherlands, where aggressive pathotypes mean that resistant or tolerant cultivars are less effective. In particular, research is needed to determine why PRD does not persist when added to soil and to discover whether these problems can be surmounted. If diffusates act to attract juveniles to roots by establishing a concentration gradient as suggested earlier, then persistence would seem feasible.

Tsutsumi (1976) suggested that PRD retained hatching activity for up to 100 d after the removal of the host plant and Perry et al., (1981) found that even 4 wk after plants were removed from pots the leachate induced >50% hatch in in vitro tests. Under field conditions PRD may induce a small hatch of juveniles after potato harvest and Whitehead (1977) considered that the increased hatch found when the artificial hatching agent picrolonic acid was incorporated in the soil after growing potatoes for 4 wk may have been due, not to the picrolonic acid, but solely to the redistribution of PRD when the soil was rotavated. A similar effect was noted with *H. goettingiana* where leachate from soil in which peas had been grown 12 wk earlier induced hatches of almost 25% in in vitro tests (Perry et al., 1980a).

Another aspect of particular relevance is that the period of exposure to PRD which is required to initiate the hatching sequence of *G. pallida* and *G. rostochiensis* is very brief. A weekly 5 min exposure for 3 wk gave a 43% hatch from free eggs of *G. pallida* (Forrest and Perry, 1980) and the same exposure for 5 wk induced >50% hatch from eggs of *G. rostochiensis* (Perry and Beane, 1982); soaking for 5 min in diffusate is also very effective in stimulating hatch from whole cysts of these species. Thus, continual exposure to PRD is not required to "trigger" hatch. The situation is slightly different with *H. goettingiana* where duration of stimulation by pea root diffusate is important; exposure periods of >18 h each week were necessary to achieve hatches commensurate with those obtained from continuous exposure to diffusate (Perry and Beane, 1983).

In the field there is a 35-40% "carry over" of viable cyst contents of *G. rostochiensis* from one year to the next (Jones and Parrott, 1969) which would seem unlikely if hatching is elicited by short exposures to PRD which may persist for some time. Obviously factors such as the degree of PRD persistence and whether diffusate reaches all the cysts are influential, but changes in the physiology of the juvenile (associated with dormancy, for example) may be of paramount importance as this could block the sequence of events leading to eclosion and thus override the effects of root diffusate. In any future use of diffusate for control purposes timing of application will clearly be a consideration.

PHYSIOLOGY OF UNHATCHED JUVENILES

The existence of a dispause in nematodes was reviewed by Evans and Perry (1976) who concluded that the best documentated example of diapause within the egg of a cyst nematode was in *H. avenae*, based primarily on the work by Banyer and Fisher (1971). More recently, Hominick et al., (1985) have demonstrated that dormancy in *G. rostochiensis* is a true diapause. Work by Rivoal (1983) has demonstrated a diapause in French populations of *H. avenae*. Perry and Clarke (1981) suggested that if the J2 was involved in altering eggshell permeability, perhaps by secreting enzymes in response to PRD stimulation, then a dormant J2 may not respond to the stimulation and be retained in an unaltered egg. However, the changes in eggshell Ca^{2+} induced by hatching agents indicate that the J2 is probably not involved in these initial stages. Thus, it may be possible for the hatching sequence to be initiated and eggshell permeability and trehalose characteristics to be altered but because the juvenile is diapausing or quiescent (Fig. 1) the sequence is delayed; eclosion may then occur some considerable time after, when dormancy requirements have been completed, without any additional hatching stimulus. Under these circumstances, hatch may have been classed as "spontaneous".

On a shorter timescale, emergence from *G. rostochiensis* cysts also varies. Hatching may continue for some months but the majority of JJ2 emerge from the cysts during the first 1-2 weeks after stimulation. After hatching, JJ2 remain infective for a limited period and Storey (1984) correlated infectivity with the extent of their initial lipid reserves. Robinson at al., (1985) demonstrated that exposure to PRD increased lipid utilization by the unhatched J2, a further example of enhanced metabolic response to PRD. These authors also showed that late hatching individuals had a significantly reduced lipid content at eclosion with subsequent impaired infectivity and delayed development compared with JJ2 emerging from the same cysts soon after stimulation. These effects did not result from initial differences in lipid content or differences in the time of stimulus reception. Thus, measures which delay hatch after stimulation by PRD may result in an important reduction in the numbers of JJ2 successfully invading the host plant.

CONCLUSION

Investigations into the hatching mechanisms are now at the stage where several approaches to possible future control strategies can be defined and explored. Such studies, linked with chemical characterisation of the hatching factors and possible use of slow-release granular formulations of compounds influencing eclosion, may provide, in the long term, the basis for novel cyst nematode control programmes which are environmentally acceptable.

REFERENCES

Atkinson, H. J. and Ballantyne, A. J., 1977a, Changes in the oxygen consumption of cysts of *Globodera rostochiensis* associated with the hatching of juveniles, Ann.appl.Biol., 87:159.

Atkinson, H. J. and Ballantyne, A. J., 1977b, Changes in the adenine nucleotide content of cysts of *Globodera rostochiensis* associated with the hatching of juveniles, Ann.appl.Biol., 87:167.

Atkinson, H. J. and Ballantyne, A. J., 1979, Evidence for the involvement of calcium in the hatching of *Globodera rostochiensis*, Ann.appl.Biol., 93:191.

Atkinson, H. J. and Taylor, J. D., 1980a, A calcium-binding site on the eggshell of *Globodera rostochiensis* with a role in hatching, XV Int. Symp.Nematol.Bari, 19.

Atkinson, H. J. and Taylor, J. D., 1980b, Evidence for a calcium-binding site on the eggshell of *Globodera rostochiensis* with a role in hatching, Ann.appl.Biol., 96:307.

Atkinson, H. J. and Taylor, J. D., 1983, A calcium-binding sialoglycoprotein associated with an apparent eggshell membrane of *Globodera rostochiensis*, Ann.appl.Biol., 102:345.

Atkinson, H. J., Taylor, J. D. and Ballantyne, A. J., 1980, The uptake of calcium prior to the hatching of the second-stage juvenile of *Globodera rostochiensis*, Ann.appl.Biol., 94:103.

Banyer, R. J. and Fisher, J. M., 1971, Effect of temperature on hatching of eggs of *Heterodera avenae*, Nematologica, 17:519.

Baunacke, W., 1922, Untersuchungen zur Biologie und Bekämpfung des Rübennematoden *Heterodera schachtii* Schmidt, Arb.biol.Abt.Berl., 11:185.

Beane, J. and Perry, R. N., 1983, Hatching of the cyst nematode *Heterodera goettingiana* in response to root diffusate from bean (Vicia faba), Nematologica, 29:361.

Clarke, A. J., 1967, The potato cyst-nematode hatching factor, Rep.Rothamsted Exp.Sta. 1966, 107.

Clarke, A. J., 1968, Potato cyst-nematode *Heterodera rostochiensis* hatching factor, Rep.Rothamsted Exp.Sta.1967, 118.

Clarke, A. J., 1970, Hatching factors and sex attractants of cyst-nematodes, Rep.Rothemsted Exp.Sta.1969, 179.

Clarke, A. J., 1971, Hatching factors and the sex attractants of potato cyst-nematode, Rep.Rothamsted Exp.Sta.1970, 147.

Clarke, A. J. and Hennessy, J., 1976, The distribution of carbohydrates in cysts of *Heterodera rostochiensis*, Nematologica, 22:190.

Clarke, A. J. and Hennessy, J., 1981, Calcium inhibitors and the hatching of *Globodera rostochiensis*, Nematologica, 27:190.

Clarke, A.J. and Hennessy, J., 1983, The role of calcium in the hatching of *Globodera rostochiensis*, Rev.Nematol., 6:247.

Clarke, A.J. and Hennessy, J., 1984, Movement of *Globodera rostochiensis* (Wollenweber) juveniles stimulated by potato-root exudate, Nematologica, 30:206.

Clarke, A.J. and Perry, R.N., 1977, Hatching of cyst-nematodes, Nematologica, 23:350.

Clarke, A.J. and Perry, R.N., 1985a, Egg-shell calcium and the hatching of *Globodera rostochiensis*, Int.J.Parasitol, 15:511.

Clarke, A.J. and Perry, R.N., 1985b, The role of egg-shell calcium in the hatching of *Heterodera schachtii*, Nematologica (in press).

Clarke, A.J., Cox, P.M. and Shepherd, A.M., 1967, The chemical composition of the egg-shells of the potato cyst nematode, *Heterodera rostochiensis* Woll., Biochem.J., 104:1056.

Clarke, A.J., Perry, R.N. and Hennessy, J., 1978, Osmotic stress and the hatching of *Globodera rostochiensis*, Nematologica, 24:384.

Doncaster, C.C. and Seymour, M.K., 1973, Exploration and selection of penetration site by Tylenchida, Nematologica, 19:137.

Doncaster, C.C. and Shepherd, A.M., 1967, The behaviour of second-stage *Heterodera rostochiensis* larvae leading to their emergence from the egg, Nematologica, 13:476.

Ellenby, C., 1968, Desiccation survival in the plant parasitic nematodes, *Heterodera rostochiensis* Wollenweber and *Ditylenchus dipsaci* (Kühn) Filipjev, Proc.Roy.Soc.B, 169:203.

Ellenby, C. and Gilbert, A.B., 1958, Influence of certain inorganic ions on the hatching of the potato root eelworm *Heterodera rostochiensis* Wollenweber, Nature,Lond., 182:925.

Ellenby, C. and Perry, R.N., 1976, The influence of the hatching factor on the water uptake of the second stage larva of the potato cyst nematode *Heterodera rostochiensis*, J.Exp.Biol., 64:141.

Evans, A.A.F. and Perry, R.N., 1976, Survival Strategies in Nematodes, in "The Organisation of Nematodes", N.A. Croll, ed., Academic Press, New York and London.

Evans, K., 1983, Hatching of potato cyst nematodes in root diffusates collected from twenty-five potato cultivars, Crop Protection, 2:97.

Evans, S.G. and Wright, D.J., 1982, Effects of the nematicide oxamyl on the life cycle stages of *Globodera rostochiensis*, Annal.appl.Biol., 100:511.

Farrer, L.A. and Phillips, M.S., 1983, In vitro hatching of *Globodera pallida* in response to *Solanum vernei* and *Solanum tuberosum* x *Solanum vernei* hybrids, Rev.Nematol., 6:165.

Fenwick, D.W., 1956, The breakdown of potato-root diffusate in soil, Nematologica, 1:290.

Forrest, J.M.S. and Perry, R.N., 1980, Hatching of *Globodera pallida* eggs after brief exposures to potato root diffusate, Nematologica, 26:130.

Forrest, J.M.S. and Phillips, M.S., 1984, The effect of *Solanum tuberosum* x *S. vernei* hybrids on hatching of the potato cyst nematode *Globodera pallida*, Ann.appl.Biol., 104:521.

Hominick, W.M., Forrest, J.M.S. and Evans, A.A.F., 1985, Simultaneous experiements documenting diapause in *Globodera rostochiensis* and variability in hatching trials, Nematologica (in press).

Jones, F.G.W. and Parrot, D.M., 1969, Population fluctuations of *Heterodera rostochiensis* Woll. when susceptible potato varieties are grown continuously, Ann.appl.Biol., 63:175.

Kämpfe, L. 1959, Uber Moglichkeiten der "physiologischokologischen" Arbeitweise in der Nematodenforschung, Verh.dt.zool.Ges., 25:378.

Kraus, R. and Sikora, R.A., 1983, Effects of the herbicide diallate, alone and in combination with aldicarb, on *Heterodera schachtii* population levels in sugar beet, Z.PflKrankh.PflSchutz, 90:132.

Masamune, T., 1976, Purification of the hatching substance of the soybean cyst nematode, 34th Annual Meeting Chem.Soc.Japan (abst.), 91.

Masamune, T., Anetai, M., Takasugi, M. and Katsui, N., 1982, Isolation of a natural hatching stimulus, glycinoeclepin A, for the soybean cyst nematode, Nature, Lond., 297:495.

Newman, E.I., 1966, A method of estimating the total length of root in a sample, J.appl.Ecol., 3:139.

Papademetriou, M.K. and Bone, L.W., 1983, Chemotaxis of larval soybean cyst nematodes *Heterodera glycines* Race 3, to root leachates and ions, J.Chem.Ecol., 9:387.

Perry, R.N., 1977, Water content of the second-stage larva of *Heterodera schachtii* during the hatching process, Nematologica, 23:431.

Perry, R.N., 1983, The effect of potato root diffusate on the desiccation survival of unhatched juveniles of *Globodera rostochiensis*, Rev. Nematol., 6:99.

Perry, R.N. and Beane, J., 1982, The effect of brief exposures to potato root diffusate on the hatching of *Globodera rostochiensis*, Rev. Nematol., 5:221.

Perry, R.N. and Beane, J., 1983, The hatching of *Heterodera goettingiana* in response to brief exposures to pea root diffusate., Nematologica, 29:34.

Perry, R.N. and Clarke, A.J., 1981, Hatching mechanisms of nematodes, Parasitology, 83:435.

Perry, R.N., Clarke, A.J. and Beane, J., 1980a, Hatching of *H. goettingiana* in vitro, Nematologica, 26:493.

Perry, R.N., Clarke, A.J. and Hennessy, J., 1980b, The influence of osmotic pressure on the hatching of *Heterodera schachtii*, Rev.Nematol., 3:3.

Perry, R.N., Hodges, Julia, A. and Beane J., 1981, Hatching of *Globodera rostochiensis* in response to potato root diffusate persisting in soil, Nematologica, 27:349.

Perry, R.N., Wharton, D.A. and Clarke, A.J., 1982, The structure of the egg-shell of *Globodera rostochiensis* (Nematoda:Tylenchida), Int.J. Parasitol, 12:481.

Perry, R.N., Clarke, A.J., Hennessy, J. and Beane, J., 1983, The role of trehalose in the hatching mechanism of *Heterodera goettingiana*, Nematologica, 29:324.

Rivoal, R., 1983, Biology of *Heterodera avenae* in France, 3, Evolution of diapauses of FRI and FR4 races over several consecutive years: influence of temperature, Rev.Nematol., 6:157.

Robinson, M.P., Atkinson, H.J. and Perry, R.N., 1985, The effect of delayed emergence on the subsequent infectivity of second stage juveniles of the potato cyst nematode *Globodera rostochiensis*, Nematologica (in press).

Storey, R.M.J., 1984, The relationship between neutral lipid reserves and infectivity for hatched and dormant juveniles of *Globodera* spp., Ann.appl.Biol., 104:511.

Tefft, P. and Bone, L.W., 1984, Zinc-mediated hatching of eggs of soybean cyst nematode, *Heterodera glycines*, J.Chem.Ecol., 10:361.

Tefft, P.M. and Bone, L.W., 1985a, Leucine aminopeptidase in eggs of the soybean cyst nematode *Heterodera glycines*, J.Nematol., 17:270.

Tefft, P.M. and Bone, L.W., 1985b, Plant-induced hatching of the soybean cyst nematode *Heterodera glycines*, J.Nematol., 17:275.

Tsutsumi, M., 1976, Conditions for collecting the potato root-diffusate and the influence on the hatching of the potato cyst nematode, Jap.J.Nematol., 6:10.

Turner, S.J. and Stone, A.R., 1981, Hatching of potato cyst-nematodes *(Globodera rostochiensis, G. pallida)* in root exudates of *Solanum vernei* hybrids, Nematologica, 27:315.

Wallace, H.R., 1958, Observations on the emergence from cysts and the orientation of larvae of three species of the genus *Heterodera* in the presence of host plant roots, Nematologica, 3:236.

Wallace, H.R., 1961, The relation of parasitic nematodes to plants, Recent Adv.Bot., 1:407.

Wharton, D.A., 1980, Nematode egg-shells, Parasitology, 81:447.

Whitehead, A.G., 1977, Control of potato cyst-nematode, *Globodera rostochiensis* Rol, by picrolonic acid and potato trap crops, Ann. appl.Biol., 87:225.

DISCUSSION

Discussing the correlation between lipid content of juveniles and their successful invasion of roots, Perry referred to his investigations on lipid content in relation to the ability of the J2s to locate, penetrate and move through the root. He considered that the initial lipid reserves affect the ability of the juvenile to form a feeding site and to locate and penetrate the root. Only those J2s that emerged early from the cyst were successful in completing the invasion process; J2s hatching 4 to 5 weeks after stimulation had low lipid reserves and invariably failed to invade the roots. Sikora observed that J2s hatching from cysts were usually more successful than those hatching from free eggs in their ability to invade roots. Perry warned that the results of hatching experiments with free eggs should be interpreted carefully as they ignore hatching inhibitors in the cysts, which may also include trihalose. Carswell asked if the presence of trehalose affected the ability of unhatched J2s to survive dessication. Perry referred to work on trehalose in the USA and explained that both egg fluid trehalose and the egg membrane play a role in protecting the unhatched J2s from dessication, and alterations of membrane permeability and loss of trehalose renders the J2s more susceptible to dessication. Asked if diapause occurs in both PCN species and its affect on the use of hatching agents, Perry replied that the recent work of J. M. S. Forrest has shown that diapause occurs in *G. rostochiensis* as well as in *G. pallida* and that this would affect the timing of application of any substances to induce hatching.

Responding to a question, Perry said that root diffusate can be stored for several months at $5^{o}C$, but activity declines with time and he had not used it experimentally after 6 months storage.

HISTOLOGY AND ULTRASTRUCTURAL MODIFICATION INDUCED BY CYST NEMATODES

Burton Y. Endo

US Department of Agriculture, Agricultural Research Service
Plant Protection Institute, Beltsville, Maryland 2075

INTRODUCTION

Investigations on the histology, cytochemistry, and ultrastructure of syncytia and surrounding root tissue have emphasised the dynamic interaction that occurs between the host and parasite. These interactions can be better understood as more information is gained on the ultrastructure of the infective juveniles and the subsequent developmental stages of the nematode. The emphasis on cyst nematodes in this review should not diminish the importance of the root knot nematodes and other endoparasitic species that have previously been reviewed (Dropkin, 1969; Bird, 1974; Endo, 1975).

HOST PENETRATION

Initial penetration of Ladino clover roots by *Heterodera trifolii* Goffart was completed within 15 minutes (Mankau and Linford, 1960). Subsequent intracellular migration by second-stage juveniles of *Heterodera* has been described in host-parasite combinations such as *H. trifolii* in clover, *H. glycines* Ichinohe in soybeans, and *H. avenae* Woll. in oat roots (Endo, 1964; Johnson and Fushtey, 1966; Mankau and Linford, 1960). Host penetration by these juveniles resulted in the formation of galleries in the cortex. During the feeding process, the nematode stylet was thrust through the inner wall of a broken endodermal cell or cortical cell near the endodermis. Although cells of the endodermis and cortex may have been necrotic, the stylet tip was inserted into a living cell (Mankau and Linford, 1960; Johnson and Fushtey, 1966). The syncytia induced by *H. trifolii* and *H. avenae* in clover and oats, respectively, were confined primarily to the stele. Feeding sites were also located in the cortex, but the developing syncytia extended into the vascular system (Cole and Howard, 1958; Endo, 1964). Ross (1958) who reported that cortical cell damage was caused by invading juveniles, also noted the presence of necrotic cells adjacent to the nematode, especially in resistant plants. Whereas syncytia were formed and extensive disruption of vascular tissue occurred in the susceptible reaction, cells around the nematode head region in resistant plants were necrotic and disorganised (Ross, 1958). Many cells that appeared to be syncytia became necrotic after initial stimulation by the nematode (Endo, 1965). This observation was especially evident at the feeding site of the nematode.

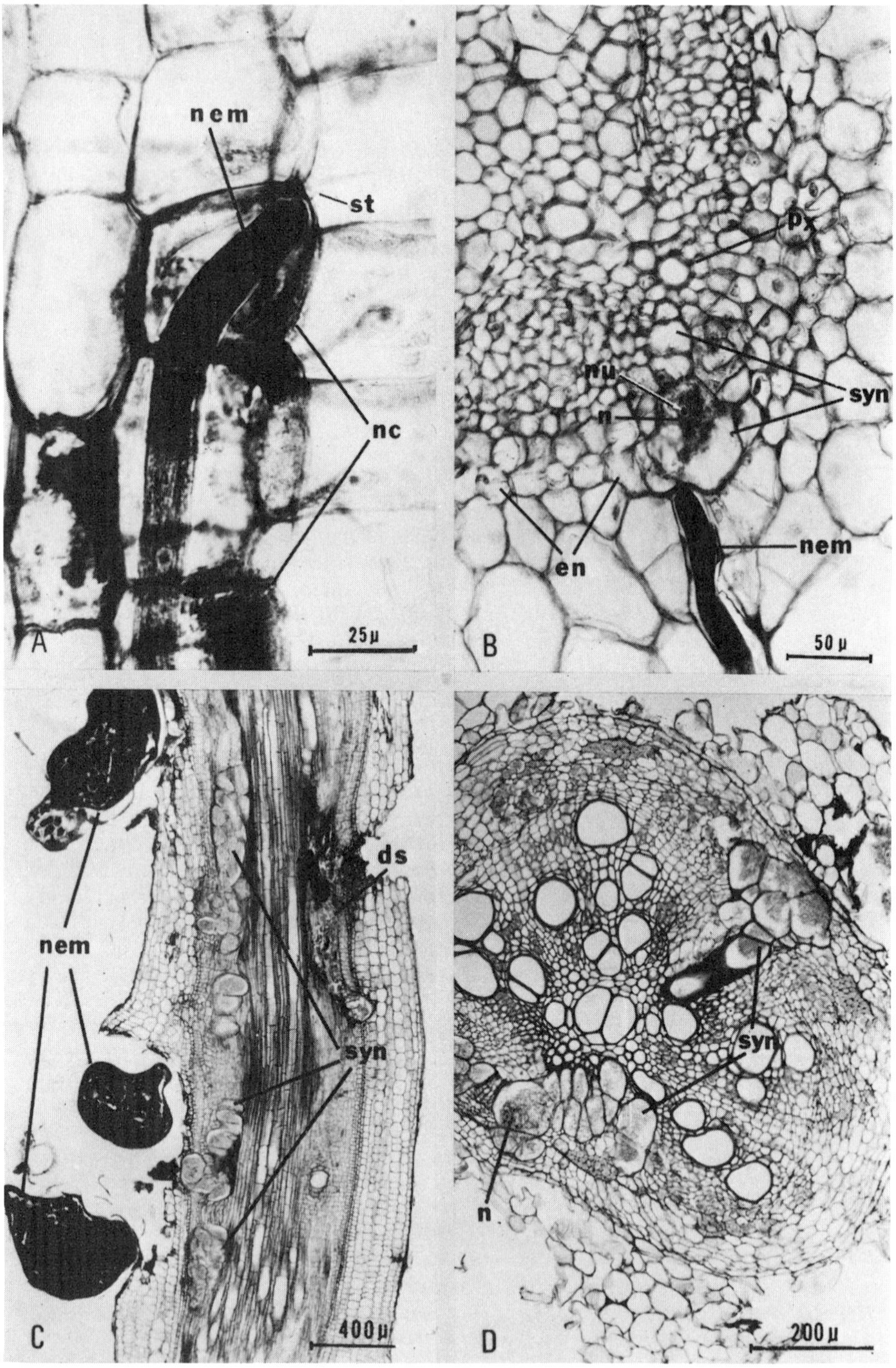
nem
st
nc
A
25µ
px
nu
n
syn
en
nem
B
50µ
ds
nem
syn
C
400µ
syn
n
D
200µ

INDUCTION OF SYNCYTIA

In the development of syncytia induced by *H. glycines* on soybeans, cell-to-cell continuity of host cytoplasm and the migration of nuclei among the cells were common features. The physical and chemical stress to the cell walls occurred during the early stages of nematode feeding and development. The progressive changes whereby the vacuolated vascular cells develop into the dense cytoplasmic cells of the syncytia that support nematode development, reflected the intensive host-parasite interaction. During rapid growth and development of syncytia, primary and secondary vascular tissues were inhibited and normal vascular tissue was lost. These changes were especially evident in heavy infections where syncytia occupied extensive regions of root tissue. In contrast, syncytia induced by males were usually restricted in size and tended to be at the margins of the vascular tissues primarily involving the cortex, endodermis, and pericycle. After maturation of the adult female cyst nematode, feeding terminated and the body cuticle was transformed into a cyst which provided long term survival for the enclosed eggs. After the active syncytium became necrotic and reduced in size, parenchymatous tissues invaded the region vacated by the deteriorating syncytium and its receding wall. In spite of regenerative parenchymatous tissue, the vascular system remained restricted. This deficiency resulted in limited nutrients and impeded plant growth (Fig. 1).

In root knot nematode infections autoradiographic and histological techniques on syncytial sites verified that polyploidy was induced through chromasomal proliferation in the absence of cytokinesis (Rubinstein and Owens, 1964; Huang and Maggenti, 1969). Microautoradiographs of sections of soybean roots showed that nuclei in syncytia induced by the soybean cyst nematode were quiescent. However, the study indicated that some tritiated thymidine was incorporated into nuclei of cells adjacent to the syncytium. These results suggest that DNA synthesis may account for the hypertrophy of the nuclei prior to their incorporation into the syncytium (Endo, 1971).

An ultrastructural study of carrot roots infected with *H. carotae* Jones emphasised the role of cell wall dissolution in syncytial growth and

Fig. 1 ⟵ Penetration and infection of the soybean cyst nematode, *Heterodera glycines* in cv. Lee soybean roots, a susceptible reaction: A. Intracellular migration of juvenile (nem) through cortex causing necrosis of cells (nc); note stylet (st) penetration of cortical cell wall (1 day after inoculation). B. Nematode (nem) in contact with early stage of syncytium (syn) extending from the endodermis (en) into the protoxylem region (px). Enlarged nucleus (n) surrounds hypertrophied nucleolus (nu). Other nuclei are present but not readily distinguishable (2 days after inoculation). C. Adult female nematodes (nem) adjacent to syncytia (syn) in vascular region. After the nematode matures and stops feeding, the syncytium (ds) deteriorates (24 days after inoculation). D. Vascular damage from syncytial development; syncytia inhibited wide regions of secondary xylem and phloem tissue. Note the cluster of nuclei in a syncytium (27 days after inoculation). (Reprinted from B. Y. Endo, 1971).

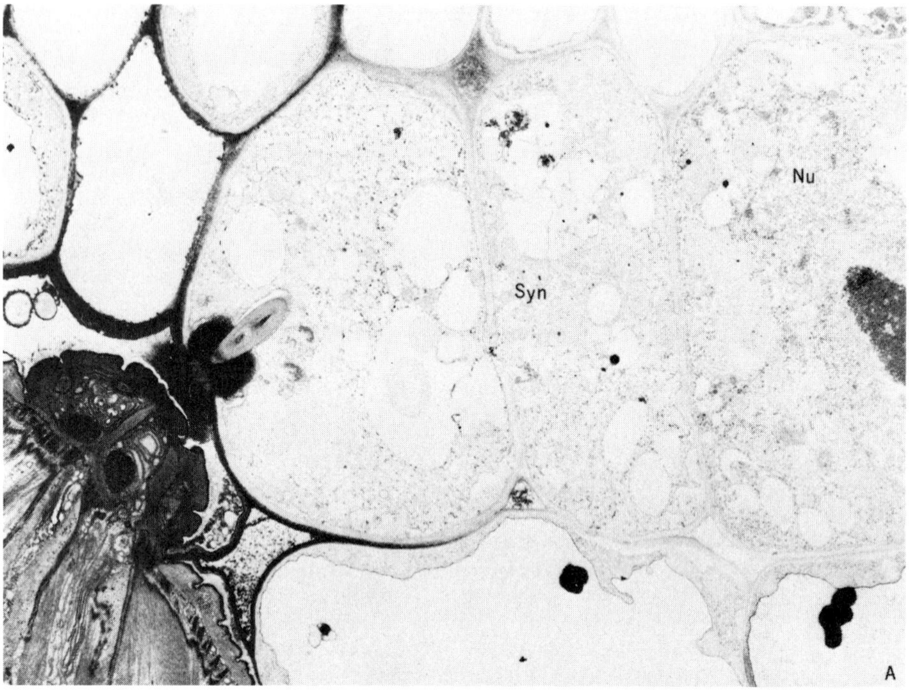

Fig. 2 Feeding of *H. glycines* on soybean roots. An infective second-stage juvenile with stylet extended into a syncytium (Syn) of resistant soybean cv. Pickett 3 days after feeding commenced. The plug and stylet are illustrated. The enlarged nematode indicates that feeding has occurred and the source of nutrition is the syncytium shown with its dense cytoplasmic content. The enlarged malformed nucleus (Nu) is characteristic of syncytial nuclei. (Reprinted from B. Y. Endo, 1978).

the thickening of remnant cell walls (Ambrogioni and Porcinai, 1972). Similarly, electron microscope studies of *H. glycines* in cv. Lee (susceptible) soybeans showed that cell wall perforations resulted in syncytium formation. Perforations occurred within 42 hours after inoculation (Gipson et al., 1971). The outer syncytial walls of clover and soybean roots were slightly thickened at the feeding site (Mankau and Linford, 1960; Endo, 1964, 1965, 1978).

Unique electron-dense plug-like deposits have been directly associated with the feeding sites of soybean cyst nematodes (Figs. 2, 3). Initially, the plug-like deposit, which appeared to emanate primarily from the amphidial pore, was fluid as it filled the space between the anterior portion of the nematode and the adjacent cell wall. In addition to having a possible role in guiding stylet movement, the deposit appeared to plug the cell wall during stylet retraction and nematode moult (Endo, 1978). Electron-dense feeding plugs were also associated with feeding sites of *H. schachtii* Schmidt in susceptible and resistant roots of *Raphanus sativus* (Wyss et al., 1984).

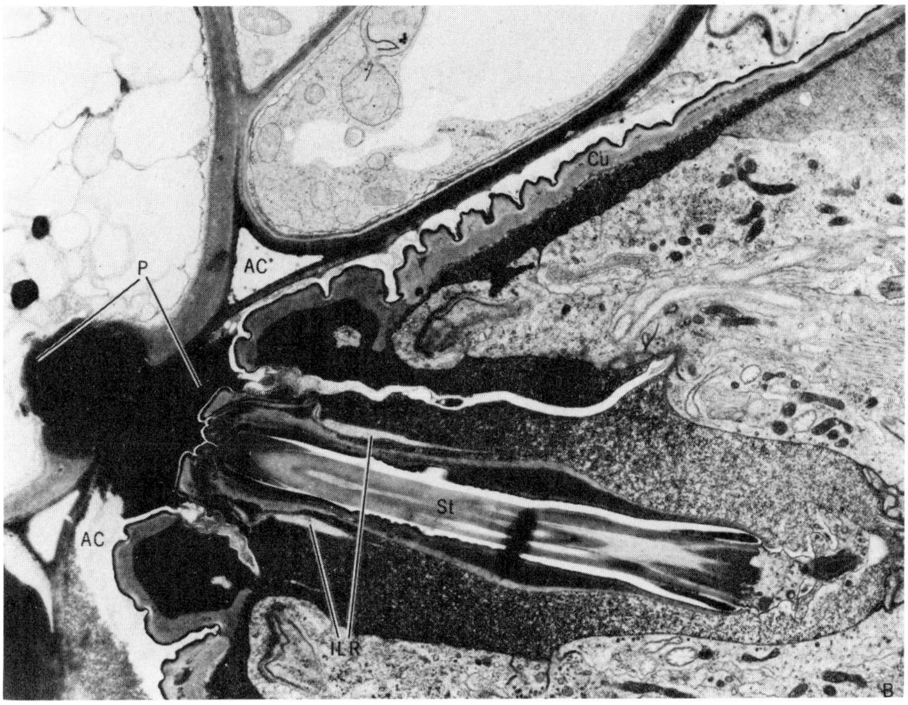

Fig. 3 Feeding of *H. glycines* on soybean roots 4 days after inoculation of the susceptible soybean cv. Lee. A juvenile is shown in a moulting stage. The dark granular material beneath the cuticle (Cu) and surrounding the stylet tooth (St) will be sloughed off during the moulting. Note the plug (P) in the syncytium wall. Inner labial receptors (ILR), amphidial canal (AC). (Reprinted from B. Y. Endo, 1978).

The presence of a feeding tube that extended from the stylet aperture of the cyst nematode, *H. schachtii,* in *R. sativus* (Wyss, et al., 1984) was also observed with light microscopy with the reniform nematode, *Rotylenchulus reniformis* Linford & Oliveira feeding on cowpea roots (Razak and Evans, 1976; Fig. 4) and with electron microscopy of soybean and cotton (Rebois et al., 1975; Rebois, 1980; Fig. 5). Razak and Evans (1976) proposed that feeding tubes secreted by *R. reniformis* might act filters to prevent the blockage of the stylet lumen during food ingestion. Ultrastructural observations of the feeding tube formed by *R. reniformis* (Rebois et al., 1975; Rebois, 1980) provided support for the suggestion by Wyss et al. (1984) that the tube was formed by rapidly hardening nematode saliva which was considered to be secreted from the dorsal oesophageal gland.

The infective juvenile of *H. schachtii* that induced the initial syncytial nurse cell near or within the vascular cylinder, ingested food for up to 1 hour. During this time the metacorpus pulsated almost continuously. At regular intervals, short rest periods occurred during which the nematode apparently defaecated (Müller et al., 1981).

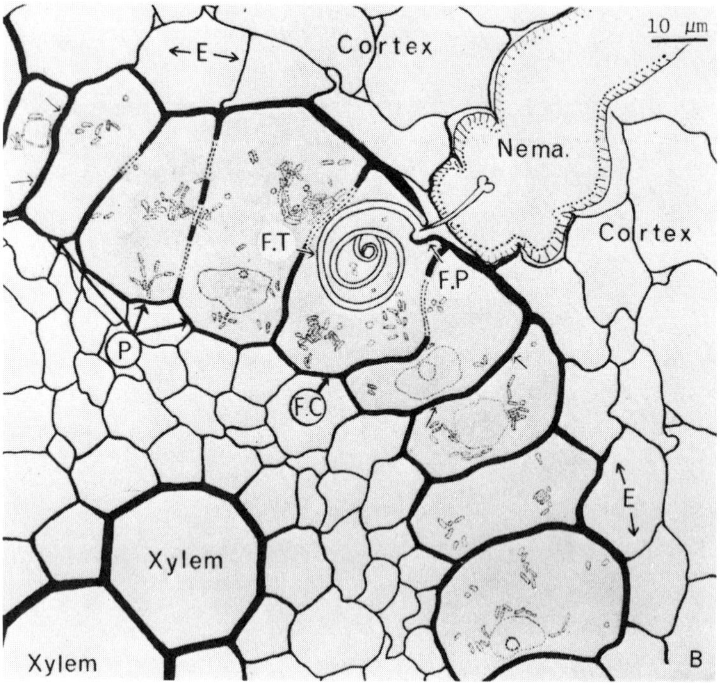

Fig. 4 Composite camera lucida drawing, from serial sections, of feeding tube and feeding zone of cowpea root with feeding *Rotylenchulus reniformis* female. Feeding cell (FC); feeding peg (FP); feeding tube (FT); endodermis (E); pericycle (P); limiting radial walls of the feeding zone (arrows). (Reprinted from A. R. Razek and A. A. F. Evans, 1976).

ULTRASTRUCTURE OF SYNCYTIAL CYTOPLASM

Wyss et al. (1984) proposed that the feeding tube served not only as a cytoplasmic filter, but also resulted in increased surface area of the plant plasmalemma, which extended into the cytoplasm of the syncytium around the feeding tube. Feeding activity was considered to assist in the "predigestion" of the surrounding cytoplasm (Wyss et al., 1984). Histological studies of *H. glycines* and *H. trifolii*, showed increased cytoplasmic density in host syncytia as the nematode fed and developed (Mankau and Linford, 1960; Endo, 1964). Within 42 hours after inoculation of susceptible soybeans with *H. glycines*, cells fed upon by the nematode exhibited numerous plastids and increased amounts of endoplasmic reticulum. The size of the large central vacuole of the initial syncytium diminished as the quantity of cytoplasm and the number of organelles increased. The amount of endoplasmic reticulum increased with the age of the syncytium and the cisternae formed as parallel arrays (Gipson et al., 1971). Similar changes were observed on carrot roots infested with *H. carotae* by Ambrogioni and Porcinai (1972) who reported that dense cytoplasm, which gradually invaded the central vacuole of the syncytium, had an abundance of plastids, mitochondria, dictyosomes, rough endoplasmic reticulum (RER),

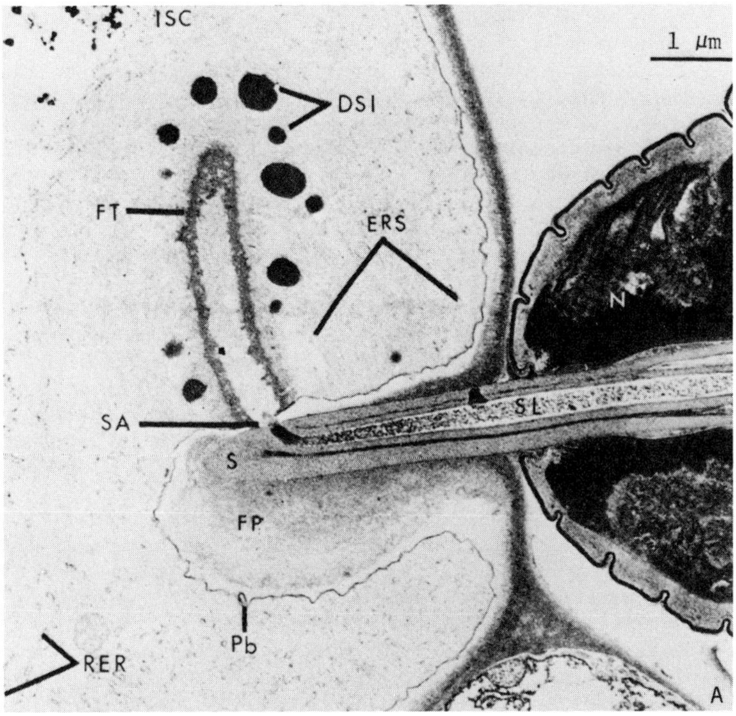

Fig. 5 Cross section through cotton root cv. 'Auburn 56' 3 days after inoculation with *R. reniformis*. Swirls of endoplasmic reticulum (ERS) are associated with the feeding tube (FT) near the stylet aperture (SA) side of the peg (FP). The central vacuole has been replaced by accumulations of endoplasmic reticulum, ribosomes, many small vacuoles, dense spherical inclusions (DSI), and other organelles. Portion of rough endoplasmic reticulum (RER) is scattered through the cytoplasm of the host. Initial syncytial cell (ISC); paramural bodies (Pb); stylet (S); stylet lumen (SL). (Reprinted from R. V. Rebois, 1980).

free ribosomes, and small vesicles that originated from the RER. These ultrastructural features were consistent with cytochemical studies that showed the presence of oxidoreductive and hydrolytic enzymes in syncytia induced by the soybean cyst and the root knot nematodes in resistant and susceptible soybean roots (Endo and Veech, 1970; Veech and Endo, 1970).

The initial response of resistant and susceptible soybean cultivars to soybean cyst nematodes appeared similar; syncytia in both cultivars were initiated by the invading juveniles (Endo, 1965; Gipson et al., 1971; Riggs et al., 1973). However, within 5 days after juvenile penetration, light microscope observations showed that syncytia in resistant cultivars had staining reactions that indicated cytoplasmic deterioration and necrosis (Endo, 1965). Ultrastructural examination of feeding sites in the resistant host 'Peking' revealed electron dense, vacuolate cytoplasm within a degenerating syncytium 4 days after inoculation. Associated with these cytoplasmic alterations within the syncytia were irregular cell wall thickenings considered to represent a mechanism of host resistance to nematode development. The cell wall thickenings associated with boundary formations and paramural bodies were thought to close or wall off plasmodesmata. In syncytia induced in resistant cultivars, walling off plasmodesmata is thought to prevent cell to cell movement of solutes. Such restrictions might cause a food supply deficit or a build up of toxic by-products (Riggs et al., 1973).

Wyss et al. (1984) studied the cytological changes that occurred in susceptible and resistant cultivars of *R. sativus* var. *oleiformis* infected by *H. schachtii* and found that host cell changes were identical for both cultivars within the first few days after the initial nematode invasion. Several studies of syncytia have shown common features of high metabolic activity e.g. an increase in cytoplasmic density, reduction of the central vacuole into numerous small vacuoles, abundance of ribosomes, polyribosomes, mitochondria, dictyosomes, and enlargement of nuclei (Gipson et al., 1971); Ambrogioni and Porcinai, 1972). Furthermore, within 3 days after inoculation, an extensive proliferation of smooth-surfaced endoplasmic reticulum had occurred; whereas, the resistant cultivars had rough-surfaced, flattened cisternae and increased vacuolisation. Ribosomes and polyribosomes were present in the cytoplasm of the susceptible cultivar, but very few adhered to the endoplasmic reticulum. In contrast, in the resistant *R. sativus* var. *oleiformis* cv. Pegletta, some ribosomes were part of the rough endoplasmic reticulum (RER) as well as being present as free ribosomes and polyribosomes within the cytoplasm. A similar relation apparently occurs in soybean roots infected by the *H. glycines*. Smooth surface ER predominated in the susceptible cultivar (Gipson et al., 1971) whereas RER was predominant in the resistant cultivar (Riggs et al., 1973). Wyss et al. (1984) discussed the implication of a resistance mechanism that would alter the nature of the endoplasmic reticulum and its genetically programmed effect on syncytial degeneration.

In comparing resistant and susceptible potato cultivars infected by *G. rostochiensis* (Woll.) Behr., Piegat and Wilski (1965) noted an increase in cytoplasmic density of cells adjacent to the feeding nematode. They noted the numerous vacuoles, the small starch grains in leucoplasts, and gradual decomposition of mitochondria in susceptible cultivars. In contrast, resistant plant roots showed no major changes in cytoplasmic density or number of vacuoles. However, they noted more numerous starch grains in plastids and a reduced tendency for mitochondrial degeneration. They proposed that mitochondria and cytoplasm of resistant plants may be able to neutralise the action of a substance secreted by juveniles whereas this capability was absent in susceptible plants. The biochemical nature of plant resistance to *H. rostochiensis* and other nematodes has been previously reviewed (Giebel, 1982).

CELL WALL BREAKDOWN

In contrast to root knot infections, where a lack or limited incorporation of new cells occurred during formation of giant cells, new cell incorporation and cell wall breakdown was the foundation for cyst nematode induced syncytia. In general, cell wall dissolution occurred in dicotyledonous and monocotyledonous hosts from early infections and into the extended feeding period of a developing nematode (Cole and Howard, 1958; Mankau and Linford, 1960; Endo, 1964; Johnson and Fushtey, 1966; Steele,

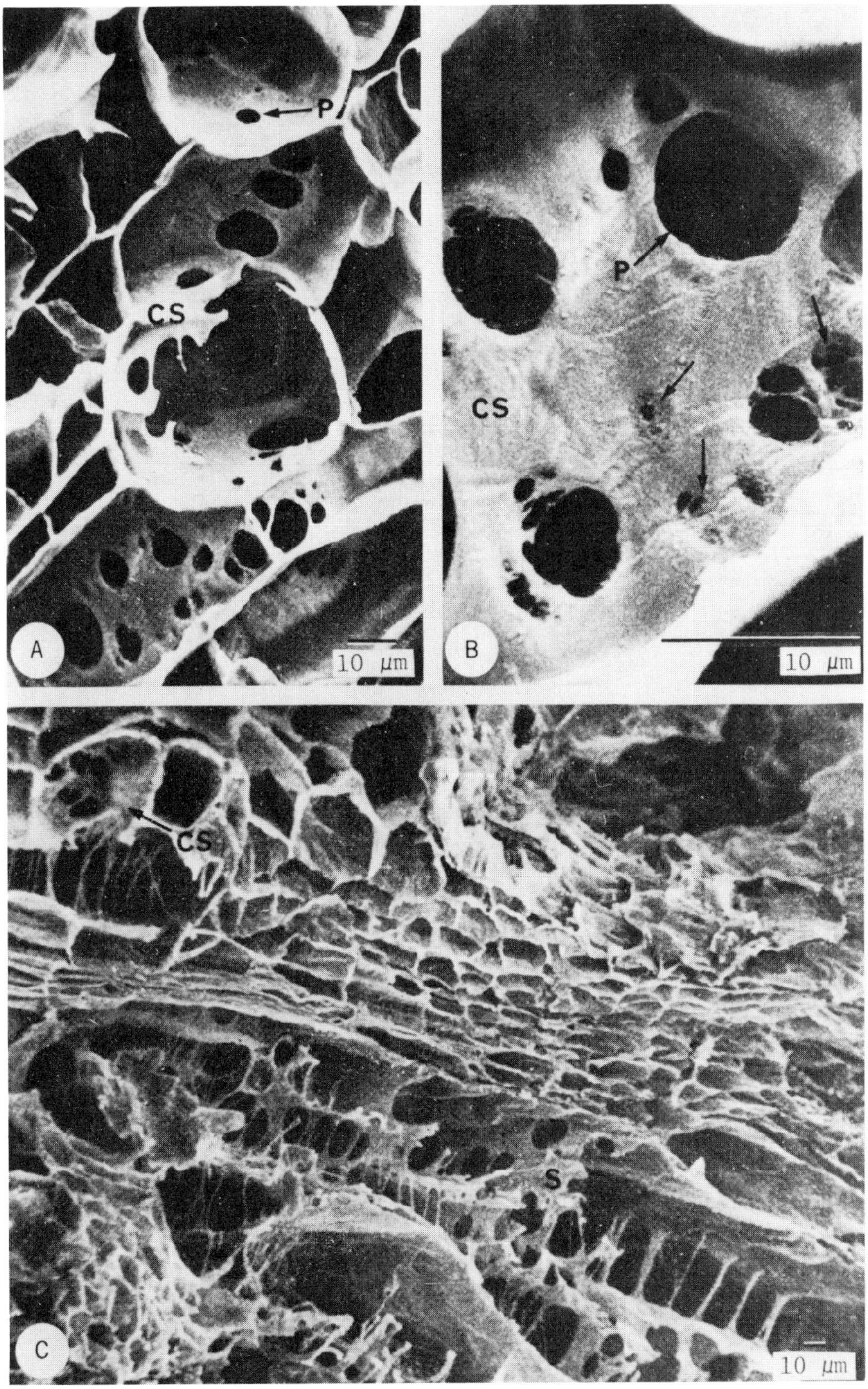

(caption overleaf)

1984). Scanning electron microscope observations of soybean roots infected with *H. glycines* (Fig. 6) and tobacco roots infected with *H. tabacum* Lownsbery & Lownsbery showed that cell wall dissolutions appeared to form at sites formerly occupied by pit fields. In the cortex, perforations were almost round and increased in diameter in the tiers of cells approaching the vascular system. The higher pit frequency in syncytia-incorporated parenchyma cells of the vascular region appeared consistent with the more extensive perforations found in the stelar cells. In citing examples for the role of enzymes that degrade cell wall components, authors have concluded that enzymes of plant origin are the primary cause of cell wall perforations. Cell wall dissolutions apparently occurred over a short time, possibly less than 24 hours; when a cell was incorporated into a syncytium, the wall digestion ceased (Jones and Dropkin, 1975).

Gommers and Dropkin (1977) lyophilised frozen sections and analysed giant cells and syncytia induced by *Meloidogyne incognita* (Kofoid & White) Chitw. in garden balsam and syncytia induced by *H. glycines* in soybeans. They found that giant cells and syncytia had concentrations of adenosine triphosphate (ATP), glucose-6-phosphate (G-6-P), and protein that were comparable to those in actively growing root tip tissue but contain four times more glucose. Free amino acids were more concentrated in giant cells than in syncytia. They concluded that pathogen and host influenced the altered metabolite concentrations in giant cells and syncytia.

Wall protuberances reported on hosts fed upon by cyst nematodes and root knot nematodes were apparently similar since they appeared to be associated with the transfer of nutrients of host plant conductive tissue to the giant cells or syncytia (Huang and Maggenti, 1969; Jones and Northcote, 1972). The transfer cell concept reported by Pate and Gunning (1972) was used to study the roots of *Coleus* sp. that were infected with root knot nematode (Jones and Northcote, 1972); (Fig. 7). They noted extensive protuberances on cell walls of giant cells that were adjacent to xylem vessels. The protuberances were less developed on walls adjacent ato sieve elements than on those next to xylem vessels, and were nearly absent from walls adjacent to parenchyma cells. Because solutes for syncytia were obtained from xylem and phloem elements, the presence of wall protuberances at the junctures between giant cells and the vascular elements indicated that they play a role in the movement of solutes across these cell walls. Wall protuberances were previously proposed to form as a result of selective nutrient flow across the plasmalemma (Jones and Northcote, 1972). Although syncytia induced by cyst nematodes differed

Fig. 6 Syncytia induced by *H. glycines* in soybean roots. A. Syncytial cells in the cortex (CS) have expanded and are more rounded than normal cortical cells. Perforations through the walls (P) are discrete; they increase in number towards the stele. B. Wall perforations within cortical syncytial cells (P) vary from limited removal of wall materials and the presence of strands across the perforations (unlabelled arrows) to complete removal of wall material. The perforations appear to be at sites previously occupied by pit fields. C. Within the stele the syncytium spreads longitudinally up and down the root, and the sizes of the wall perforations are larger. Syncytial cells in the stele (S). (Reprinted from M. G. K. Jones and V. H. Dropkin, 1975).

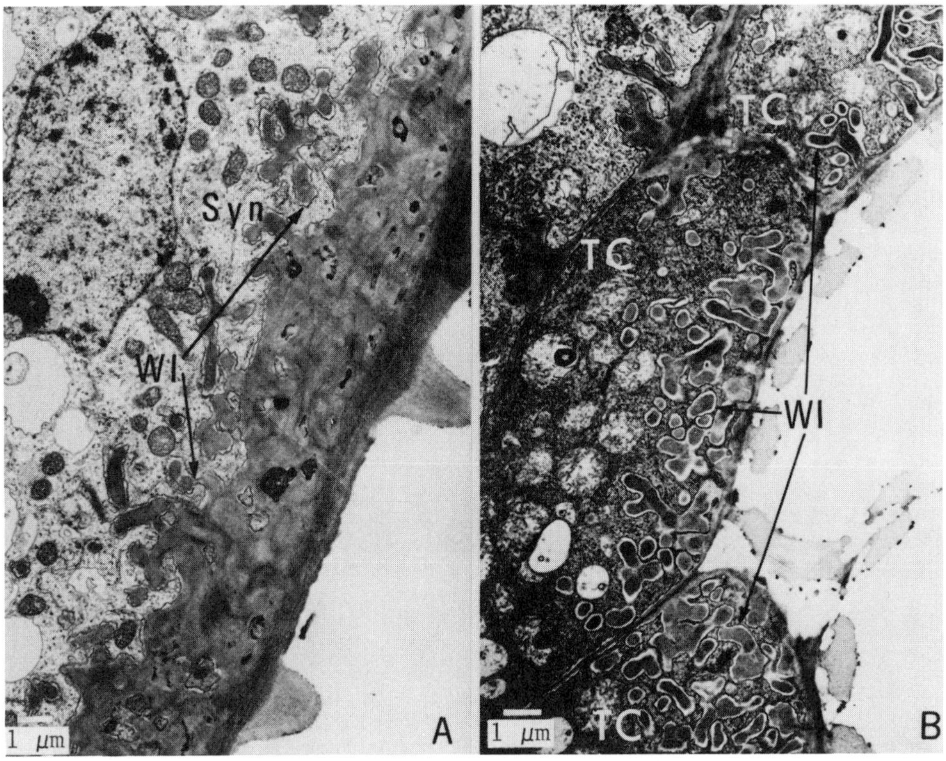

Fig. 7 Comparison of cell wall modifications. A. Cell wall ingrowths (WI) or protruberances within a syncytium (Syn) induced in *Impatiens balsamina*, 10 days after inoculation with *Meloidogyne incognita*. B. Cell wall ingrowths of transfer cells (TC) located in the cotyledonary node of *Galium aparine*. (Courtesy of B.E.S. Gunning and M.G.K. Jones).

from the multinucleated giant cells of root knot infections, the stimulation and accumulation of wall protuberances along walls of both types of infections were quite similar. Jones and Northcote (1972) further proposed that syncytia induced by *G. rostochiensis* in potato roots were multinucleate transfer cells and by the ingestion of syncytial contents, the juvenile was considered a nutrient sink. During syncytial expansion, sieve elements were usually crushed so that protuberance development continued primarily on syncytial walls adjacent to xylem elements.

CONCLUSIONS

Recent interest in isolating secretory granules from nematodes and determining their relationship to giant cell formation may provide additional information on the mechanism of syncytial induction and its relationship to host susceptibility and resistance (Hussey and Sundermann, 1984). The host and nematode both are apparently active in the stimulation of syncytia formation. The precise biochemical stimulant or inhibitor to these host-parasite interactions has not been described.

Recent advances in video-enhanced light microscopy together with specific biochemical tracers should help to provide clues to the sites of action of cell wall dissolution and subsequent syncytial development. Knowledge of nematode secretions as host cell stimulants or inhibitors should provide ways to modify their activity and possibly protect the host cells. Conversely, specific host cell components that inhibit nematode feeding and development could provide the much needed technology for nematode management.

Considerable progress has been made in determining the histological, biochemical and ultrastructural changes that occur at the feeding site of cyst nematodes. However, more observations are needed to show the consistency or differences of host-parasite interactions among cyst nematodes and their hosts. Future progress in host-parasite interactions should be enhanced by a multi-discipline approach to understand the changes that occur during the nematode development and the responses of the host. With current technology it should be possible to determine these host-parasite interactions by using certain biochemical markers specific for nematode morphology or plant structure. These markers could then be monitored by immunocytochemistry and viewed with video-enhanced light microscopy and verified with ultrastructural studies.

Although this review has emphasises the host-plant interactions related to the feeding aspects of the nematode, other avenues to develop nematode control could include the location and nature of the nematode moulting hormone whose disruption would inhibit nematode development. Finally, nematode control could be accomplished indirectly by the disruption of mating activity and therefore nematode reproduction through the development of mimics or inhibitors of sex pheromones.

REFERENCES

Ambrogioni, L. and Porcinai, G.M., 1972, Studio ultrastrutturistico delle cellule giganti prodotte da *Heterodera carotae* Jones, 1950 (Nematoda, Heteroderidae) in radici di carota, Redia, 53:437.

Bird, A.F., 1974, Plant response to root-knot nematode. Ann. Rev. Phytopathology, 12:69.

Cole, C.S. and Howard, H.W., 1958, Observations on giant cells in potato roots infected with *Heterodera rostochiensis*, J. Helminthol., 32:135.

Dropkin, V.H., 1969, Cellular responses of plants to nematode infections, Ann. Rev. Phytopath., 7:101.

Endo, B.Y., 1964, Penetration and development of *Heterodera glycines* in soybean roots and related anatomical changes, Phytopathology, 54:79.

Endo, B.Y., 1965, Histological responses of resistant and susceptible soybean varieties, and backcross progeny to entry and development of *Heterodera glycines*, Phytopathology, 55:375.

Endo, B.Y., 1971, Nematode-induced syncytia (Giant cells). Host-parasite relationships of Heteroderidae, in "Plant Parasitic Nematodes. II. Cytogenetics, Host-parasite Interactions, and Physiology", eds., B.M. Zuckerman, W.F. Mai and R.A. Rohde, Academic Press, New York and London.

Endo, B.Y., 1975, Pathogenesis of nematode-infected plants, Ann. Rev. Phytopath., 13:213.

B. Y. Endo, 1978, Feeding plug formation in soybean roots infected with the soybean cyst nematode, Phytopathology, 68:1022.

Endo, B.Y. and Veech, J.A., 1970, Morphology and histochemistry of soybean roots infected with *Heterodera glycines*, Phytopathology, 60:1493.

Giebel, J., 1982, Mechanism of resistance to plant nematodes, Ann. Rev. Phytopath., 20:257.

Gipson, I., Kim, K.S. and Riggs, R.D., 1971, An ultrastructural study of syncytium development in soybean roots infected with *Heterodera glycines*, Phytopathology, 61:347.

Gommers, F.J. and Dropkin, V.H., 1977, Quantitative histochemistry of nematode-induced transfer cells, Phytopathology, 67:869.

Huang, C.S. and Maggenti, A.R., 1969, Wall modifications in developing giant cells of *Vicia faba* and *Cucumis sativus* induced by root knot nematode, *Meloidogyne javanica*, Phytopathology, 59:931.

Hussey, R.S. and Sundermann, C.A., 1984, Electron microscopic cytochemistry of secretory granules of *Meloidogyne incognita*, Proc. First Intern. Congr. Nematol., 38.

Johnson, P.W. and Fushtey, S.G., 1966, The biology of the oat cyst nematode *Heterodera avenae* in Canada. II. Nematode development and related anatomical changes in roots of oats and corn, Nematologica, 12:630.

Jones, M.G.K. and Dropkin, V.H., 1975, Scanning electron microscopy of syncytial transfer cells induced in roots by cyst-nematodes, Physiol. Plant Pathol., 7:259.

Jones, M.G.K. and Northcote, D.H., 1972, Nematode-induced syncytium - A multinucleate transfer cell, J. Cell Sci., 10:789.

Mankau, R. and Linford, M.B., 1960, Host-parasite relationships of the clover cyst nematode, *Heterodera trifolii* Goffart, Univ. Ill. Agri. Exp. Sta. Bull., 667:3.

Müller, J., Rehbock, K. and Wyss, U., 1981, Growth of *Heterodera schachtii* with remarks on amounts of food consumed, Revue Nematol., 4:227.

Pate, J.S. and Gunning, B.E.S., 1972, Transfer cells, Ann. Rev. Plant Physiol., 23:173.

Piegat, M. and Wilski, A., 1965, Cytological differences in root cells of susceptible and resistant potato varieties invaded by potato root eelworm (*Heterodera rostochiensis* Woll.) larvae, Nematologica, 11:109.

Razak, A.R. and Evans, A.A.F., 1976, An intracellular tube associated with feeding by *Rotylenchulus reniformis* on cowpea root, Nematologica, 22:182.

Rebois, R.V., 1980, Ultrastructure of a feeding peg and tube associated with *Rotylenchulus reniformis* in cotton, Nematologica, 26:396.

Rebois, R.V., Madden, P.A. and Eldridge, B.J., 1975, Some ultrastructural changes induced in resistant and susceptible soybean roots following infection by *Rotylenchulus reniformis*, J. Nematol., 7:122.

Riggs, R.D., Kim, K.S. and Gipson, I., 1973, Ultrastructural changes in Peking soybeans infected with *Heterodera glycines*, Phytopathology, 63:76.

Ross, J.P., 1958, Host-parasite relationship of the soybean cyst nematode in resistant soybean roots, Phytopathology, 48:578.

Rubinstein, J.H. and Owens, R.G., 1964, Thymidine and uridine incorporation in relation to the ontogeny of root-knot syncytia, Contr. Boyce Thompson Inst., 22:491.

Steele, A.E., 1984, in "Plant and Insect Nematodes", ed. W.R. Nickle, Marcel Decker, New York and Basel.

Veech, J.A. and Endo, B.Y., 1970, Comparative morphology and enzyme histochemistry in root knot resistant and susceptible soybeans, Phytopathology, 60:896.

Wyss, U., Stender, C. and Lehmann, H., 1984, Ultrastructure of feeding sites of the cyst nematode *Heterodera schachtii* Schmidt in roots of susceptible and resistant *Raphanus sativus* L. var. *oleiformis* Pers. cultivars, Physiol. Plant Pathol., 25:21.

DISCUSSION

Trudgill referred to research by J.M.S. Forrest on carbohydrates on the heads of potato cyst nematodes and referred to the observation that juveniles leaving roots of host plants have very depleted amphidial carbohydrate production. Wyss said that carbohydrates from amphids appear to play a part in host identification. He thought that the depleted carbohydrate referred to by Trudgill is soon replaced.

Ibrahim asked if the feeding tube formed by trichodorids is similar to that formed by *Heterodera glycines*. Wyss replied that in the trichodorids it is definitely formed by salivation, whereas in *H. glycines* it is from a combination of amphidial secretions and saliva. Seinhorst referred to the large amount of material secreted by the amphids in trichodorids but Wyss said this is not related to the formation of the feeding tube. Baldwin suggested that amphid secretions may not be involved in anything but sensory activity.

GNOTOBIOLOGY OF CYST NEMATODES - ITS POTENTIAL IN BASIC RESEARCH

Urs Wyss and Ulrich Zunke

Institut fur Phytopathologie, Universität Kiel
D-2300 Kiel,
Federal Republic of Germany

INTRODUCTION

The significance of gnotobiology, that is the study of a single species in the absence of others or in the presence of only known species, was reviewed for phytonematology by Zuckerman (1971). As early as 72 years ago Berliner and Busch (1914) tried to cultivate a cyst nematode (*Heterodera schachtii* Schmidt) on seedling roots in agar, but Widdowson et al. (1958) were probably the first to observe the development of such a nematode (*Globodera rostochiensis* (Woll. Behr.)) in sterile root culture. Now it has become a common practice to cultivate members of the Heteroderinae (which comprise the cyst-forming genera) under gnotobiotic conditions on roots in agar medium. Lauritis et al. (1983b) list the species of cyst nematodes so far successfully maintained in monoxenic culture.

Apart from providing aseptic nematode inocula for well defined interaction experiments with plants, alone or in combination with other pathogens, the gnotobiotic culture of cyst nematodes introduces many possibilities in basic research. It allows a direct observation on the life cycle and, more importantly, enables specific investigations at any developmental stage. Fig. 1 illustrates a few of the possible fields in basic research, some of which will be considered in this chapter.

METHODOLOGY

Any species of the Heteroderinae may be reared 'in vitro', provided the agar medium sustains good growth of the roots to support nematode development and reproduction. Fig. 2a shows how well *Heterodera schachtii* develops on roots in aseptic agar culture with the appropriate nutrient medium (here modified White's medium according to Dropkin and Boone, 1966). Under such conditions egg production is comparable to that of females kept on roots in pot culture (Lauritis et al., 1983b). As with root knot nematodes (Orion et al., 1980), the composition of the growing medium should not impede the normal development of the nematode's nurse cell system. Actively growing root systems are obtained from axenic root explants or simply from single seedlings left in the agar medium in 9 or 14 cm Petri dishes. In most cases the cultures are kept in the dark, so that sucrose has to be added to the medium. The surface sterilisation of seeds and nematodes is now relatively easy. Different types of apparatus have been developed to sterilise infective J2-juveniles. We use Müller's (1970)

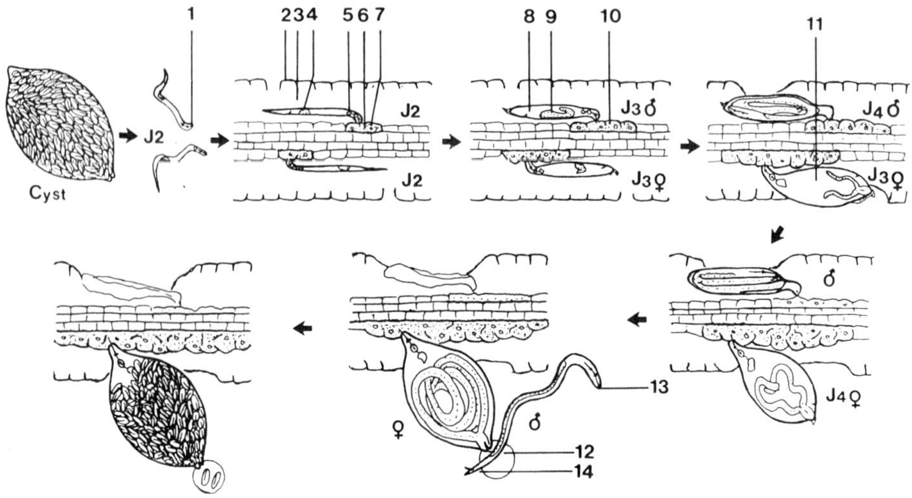

Fig. 1 Schematic representation of the life cycle of a cyst nematode (here *Heterodera schachtii*) and indications (1-14) of some possible fields in basic research to be performed under gnotobiotic conditions (drawing by F. Grundler). 1. Chemotactic behaviour of infective J2-juveniles; 2. Root selection and penetration; 3-6: In vivo observations* of: 3. Migration behaviour inside root; 4. Gonad development, cont. **; 5. Feeding mechanisms; 6. Dynamics in initial syncytial cells; 7. Ultrastructure of developing syncytia, cont.; 8. Growth and moulting processes after J3-juveniles have emerged from the root, cont.; 9. Sex expression, screening for resistance; 10. Analysis of biochemical events at feeding sites, cont.; 11. Infection processes by endoparasitic nematophagous fungi; 12. Chemical nature of sex attractants; 13. Chemotactic behaviour of males; 14. Function of spicular sense receptors.
* with the aid of special observation chambers and video-enhanced contrast microscopy.
** cont. = from this stage onwards.

method slightly modified, exposing the nematodes for 4 minutes to 0.1 per cent $HgCl_2$ before washing them thoroughly with sterile water. Once the sterile culture has been established, it can easily be maintained by transferring cysts with many eggs to new root systems. In one Petri dish hundreds of well developed cysts can be reared on a single root system which, for *H. schachtii*, allows at least two generations to develop. Stocks of sterile cysts can be kept for many months at low temperatures until they are needed for future experiments. The simple method described by Müller (1978) can be used for inoculations with definite numbers of juveniles obtained from sterile cysts.

IN VIVO OBSERVATIONS OF NEMATODE BEHAVIOUR INSIDE ROOTS, INCLUDING ROOT CELL RESPONSES

The study of the behaviour of cyst nematodes within roots has recently been made possible by the application of high resolution video-enhanced

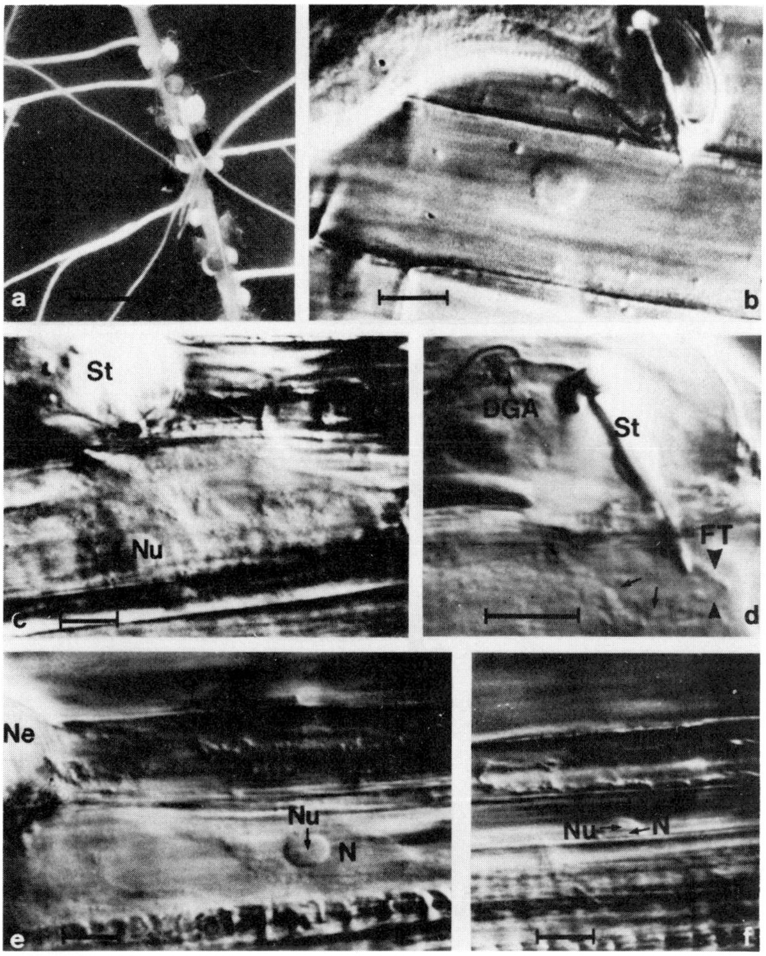

Fig. 2 Gnotobiotic culture of *Heterodera schachtii* and observations on the behaviour of J2-juveniles inside roots, including cell responses. (a) Females and cysts reared on roots of *Raphanus sativus* var. *oleiformis* in monoxenic agar culture. Bar = 1 mm. (b-f) In vivo observations with the aid of video-enhanced contrast microscopy on the behaviour of J2-juveniles inside roots of *Brassica napus* and on cell responses to feeding. All bars = 10 µm. (b) Intracellular migration through cortical cells. (c) Initial syncytial cell, about one day after induction. Nematode stylet (St) retracted during second phase of feeding cycle (see text). (d) Stylet tip reinserted during third phase. Note accumulation of secretory granules in dorsal gland ampulla (DGA), feeding tube (FT) emanating through stylet tip aperture and modified zone of cytoplasm (arrows) around the feeding tube. (e-f) Changes in the size of a nucleus (N) and its nucleolus (Nu) within the initial cell (e), about one day after induction and in a neighbouring, still unaffected cell (f). Ne = head of nematode. c-f from Film E 2904 (Wyss et al., 1985).

contrast microscopy (Wyss and Zunke, 1986). With simple observation chambers that cover a large root area (Wyss and Zunke, 1985) details of the migration of infective J2-juveniles inside roots (Fig. 2b) and of syncytium formation and maintenance close to or within the vascular cylinder can be traced and analysed. A 16 mm research film that was taken from video tapes and just published (Wyss et al., 1985) shows events such as those presented in Fig 2, b-f.

The film shows that the feeding behaviour of *Heterodera schachtii* consists of three phases. In the first, and by far longest phase, food is withdrawn from the syncytium by a continuous rapid pumping action of the median oesophageal bulb, which in J2-juveniles lasts about one hour and considerably more in each of the following developmental stages. At the end of the first phase, when the median bulb stops pumping, the stylet tip is retracted from the feeding site (Fig. 2,c) then gently moved forwards and backwards for a few minutes, before being reinserted and immediately followed by sporadic pumpings of the median bulb that last about one minute in the J2-stage. This second phase is, however, quite variable as the stylet may still be retracted when several sporadic pumpings of the median bulb occur. The film clearly shows that the contents of the valve of the two subventral oesophageal glands are depleted during the sporadic pumpings of the median bulb, so that in the second phase secretory membranes fluids from the subventral glands are passed into the intestine.

The third phase, lasting about 20 minutes in all developmental stages, is characterised by the total immobility of the median bulb. Secretory granules from the dorsal oesophageal gland then flow continuously forward through the gland's duct and accumulate in the ampulla behind the stylet knobs. Within the valve of the ampulla the granules are liquified and fluids can be seen to flow through the aperture of the continually protracted stylet tip, forming a tube. Around this tube the cytoplasm becomes modified (Fig. 2,d) and stays free of cell organelles. The possible function of feeding tubes formed by cyst nematodes has been discussed by Wyss et al. (1984) and Rumpenhorst (1984).

Metabolic activity within initial syncytial cells is soon much enhanced as revealed, for example, by the rapid increase in cytoplasmic density and nuclear volumes (Fig. 2,e-f) as well as in a pronounced acceleration of cytoplasmic streaming. In addition to the studies outlined, in vivo observations at high magnification help to support and extend ultra-structural examinations of the feeding apparatus with the transmission electron microscope (e.g. Endo, 1984).

IN VIVO OBSERVATIONS OF GROWTH AND DEVELOPMENT

Cyst nematodes are good subjects for studies on the growth and development of sedentary endoparasitic root nematodes, as in most cases the site of parasitism does not become obscured by proliferating root tissue. In the relatively thin roots that develop in aseptic agar culture, early developmental stages commonly break through the epidermis of the root and become exposed so that the dynamics of the moulting processes can be analysed at high magnification (Fig. 3,a,b). At a constant temperature of 25°C moulting of the individual developmental stages of *Heterodera glycines* Ichinohe lasts about one day (Lauritis et al., 1983b); in *H. schachtii* it is slightly longer (Müller et al., 1981a).

With the aid of the simple observation chambers mentioned before, gonad development of male and female juveniles can also be studied in detail.

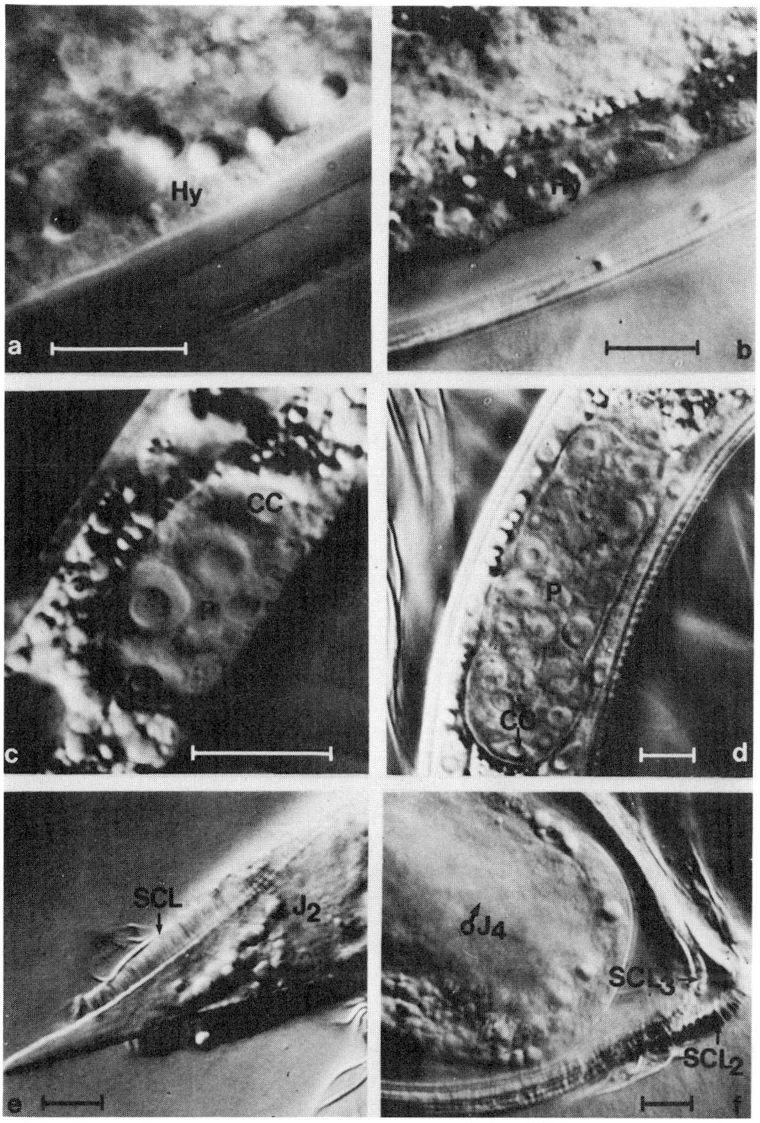

Fig. 3 Growth and development of *Heterodera schachtii* on roots in gnotobiotic agar culture, observed with video-enhanced contract microscopy. (a-b) Changes in the hypodermis (Hy) of a male juvenile that moults to the J4-stage. (b) 13 h after (a) Bars = 10 µm. (c-d) Growth of gonad primordium (P) of a male juvenile. (c) Primordium of an infective J2-juvenile, note two cap cells at both ends. (d) Primordium just after termination of second moult. Note cap cell only at posterior end (arrow) preserved. Bars = 10 µm. (e-f) Subcrystalline layers (SCL) on different developmental stages. (e) Tail region of a J2-juvenile, about one day before second moult. (f) Tail region of a male J4-juvenile. Note absence of SCL in this non-feeding stage. SCL2 and SCL3 = SCL of second and third developmental stages. Bars = 10 µm.

As an example, the gonad primordium of infective J2-juveniles of *H. schachtii* consists of six cells (Fig. 3,c) of which two are the so-called 'cap cells' (Papadopoulou and Triantaphyllou, 1982). Cell division takes place soon after the juveniles have started to feed at their permanent feeding site. At the end of the active (feeding) J2-stage the gonads are multicellular, but the future sex cannot yet be distinguished. Cell division continues during the second moult and at its termination sex differentiation just becomes evident (F. Grundler and U. Zunke, unpublished information). In the male juvenile primordium there is a cap cell situated only at the posterior tip (Fig. 3,d), and the anterior tip continues to grow forward by repeated cell divisions. In female juveniles there are cap cells at both tips; the gonad primordium moves towards the posterior end of the body and finally becomes typically V-shaped during the active J3-stage.

Another feature linked with nutrient withdrawal from the permanent feeding site is the formation of a subcrystalline layer (SCL) around the cuticle. A very thin layer is already visible soon after the infective *H. schachtii* juveniles have started to feed. Towards the end of the active J2-stage it is quite conspicuous and thickest around the tail region (Fig. 3,e). Evidence for the close association of SCL formation with nutrient uptake has been presented by Zunke (1985). In male juveniles only the J2 and J3 cuticles are covered by this layer, whereas in the non-feeding J4-juveniles it is never produced (Fig. 3,f). It is generally assumed that the SCL originates from waste products that exude through the cuticle, but the presumption that a colony of fungi immediately outside the cuticle may be involved in its formation (Brown et al., 1971) obviously can no longer be maintained.

The gnotobiotic culture of cyst nematodes on host roots grown in agar at a constant temperature allows continuous observation of the life cycle which has been recorded in this way for *H. schachtii* (Johnson and Viglierchio, 1969; Müller, 1978; Müller et al., 1981b; *H. zeae* (Lauritis et al., 1983a) and *H. glycines* (Lauritis et al., 1983b). A cinefilm of the life cycle of *H. schachtii* (Müller et al., 1981a) shows that for each feeding stage (J2 and J3 for males and all stages for females) there is a characteristic growth pattern consisting of regular alternations between gradual increases in body volume, followed by a sudden decrease, the extent of which is shown in Fig. 4,a,b. These rhythmical alterations in body size come to a temporary rest only during moulting.

From direct observations and the analysis of cinefilm material (Müller et al., 1981b) it became evident that the gradual increase in body volume is invariably accompanied by a continuous pumping action of the median oesophageal bulb (Fig. 5) during which food is withdrawn from the permanent feeding site. The sudden collapse in body volume was attributed to defaecation, which view was confirmed in subsequent studies at high magnification (Wyss et al., 1985). Fig. 4,c,d shows the defaecation process of a young female. The assumption that salivary fluids may be released during the rest periods between phases of continuous ingestion was also confirmed (Fig. 5). Thus, each developmental stage passes through a series of cycles composed of ingestion, salivation and defaecation. From the data of continuous food ingestion and repeated changes in body volume the approximate amounts of food withdrawn by the different developmental stages could be calculated (Müller et al., 1981b).

After the final moult, females of *H. schachtii* (and most likely all other amphimictic species of the Heteroderinae) secrete a sex pheromone that strongly attracts males ready to mate (Fig. 4,e). The propagation of cyst

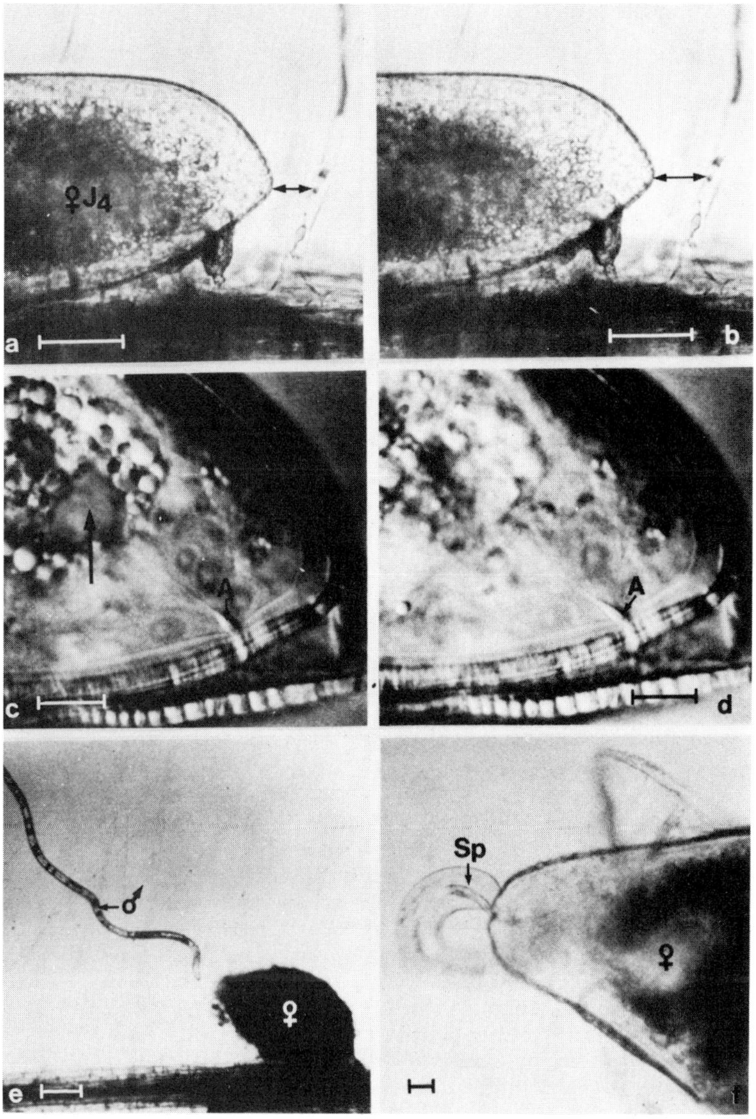

Fig. 4 Growth and development of *Heterodera schachtii* on roots in gnotobiotic agar culture. (a-b) Sudden decrease in body size of a young female J4-stage during defaecation. (b) Posterior end a few seconds after (a) Bars = 100 µm. (c-d) Posterior end of a young female at defaecation. (c) Arrow indicates dilated prerectum just before defaecation. (d) Just after defaecation. A = Anus. Bars = 10 µm. (e) A male on migration to a female secreting sex attractant. Bar = 100 µm. (f) Spicules (Sp) of a male having located vulva of a female. Bar = 10 µm. Figs. c-d from Film E 2904 (Wyss et al., 1985); e-f from Film C 1387 (Müller and Wyss, 1981).

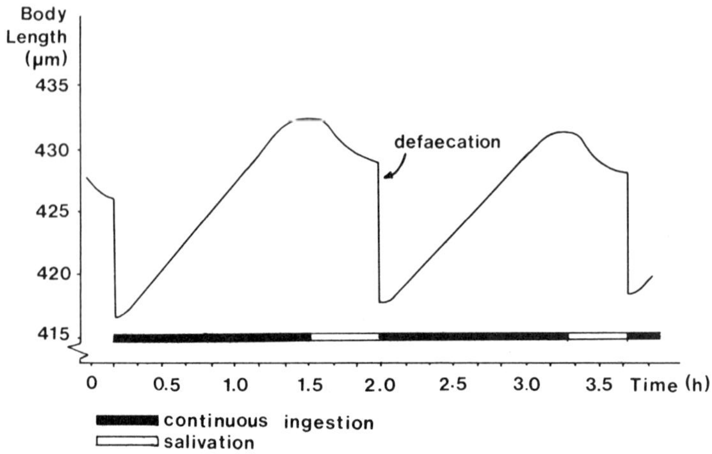

Fig. 5 Characteristic growth pattern of a female J3-juvenile of *Heterodera schachtii*. Continuous food ingestion is always accompanied by an increase in body size.

nematodes in aseptic agar culture offers excellent possibilities for well defined bioassays of sex attractants and for the characterisation of their chemical nature of which very little is yet known (Green, 1980). Basic research on pheromone communication will include sensory recognition of the chemotactic factors by the males. Fascinating new approaches in this field are reviewed by Zuckerman and Jansson (1984) and are worth pursuing. Characterisation of glycans at the binding sites for chemoattractant molecules and attempts to cap or remove these sites require the strict aseptic conditions provided by gnotobiotic cultures. In males of *Heterodera* spp. (and probably all other tylenchid nematodes) the chemotactic signals are perceived not only by receptors within amphids and labial sensillae but most likely also within the spicules that contain dendritic elements (Clark et al., 1973) and which are considered to have a sensory function during mating. This is confirmed in a film by Müller et al., (1981a) which shows that the spicules remain partially protracted during the long lasting exploration of the vulvar region before they are finally inserted into the vulva (Fig. 4,f).

SEX EXPRESSION AND SCREENING FOR RESISTANCE

The excellent review by Triantaphyllou (1973) on environmental sex differentiation of nematodes stressed the fact that obligatory amphimictic cyst nematodes are expected to have a strong genetic mechanism of sex determination and that an influence of the environment on sex differentiation cannot yet be supported by the bulk of experimental evidence. The many conflicting results and interpretations on unbalanced sex ratios presented at that time may primarily have been due to variations in experimental design and inherent errors when the fate of the inoculated nematodes could not be observed directly. Den Ouden (1960) was probably the first to examine sex differentiation of a cyst nematode (*G. rostochiensis*) by single inoculation of infective juveniles to potato seedlings grown in aerated agar plates in plastic bags. His observations support

Table 1 Sex ratio (♂♂:♀♀) of *Heterodera schachtii* in root systems of susceptible and resistant *Raphanus sativus* var. *oleiformis* cultivars grown in aseptic nutrient medium*.

	susceptible cv. 'Siletina'						resistant cv. 'Pegletta'					
	Intraspecific competition											
	No			Yes			No			Yes		
Root diameter (µm)	A	B	SR	A	B	SR	A	B	SR	A	B	SR
>510	1	24	(0.4)	9	122	2.0	5	34	7.5	6	72	17.0
510 - 435	2	23	(1.3)	17	135	1.7	6	41	9.3	18	56	13.0
420 - 345	6	44	(1.1)	12	258	1.4	21	92	10.5	52	245	23.5
330 - 255	3	82	(0.9)	34	348	2.1	11	115	13.4	53	388	42.1
240 - 165	1	31	(1.8)	17	134	5.4	5	18	∞	23	114	∞
<165	5	68	10.3	16	81	∞	20	44	43.0	34	82	∞

*Well developed root systems were inoculated in agar medium (25°) and then transferred 2-3 days later into liquid nutrient medium (25°).

Susceptible cv. n = 13; resistant cv. n = 16. A = total number of nematodes that did not develop beyond the J3-stage. B = total number of male or female J4 or adults. SR = sex ratio (♂♂:♀♀), determined 14 days after J2 inoculation. ( ) = sex ratio not significantly different from 1 (P = 0.05). Intraspecific competition = distance between two nematodes < their double body length.

the likelihood that sex differentiation is influenced by external conditions. Mugniéry and Fayet (1981; 1984), who also worked with agar culture techniques (although not aseptic) and inoculated host roots of *G. pallida* (Stone) Behr. and, respectively, *G. rostochiensis* with precise numbers of J2-juveniles, came to the conclusion that sex expression in both species must be governed by environmental factors.

Müller (1985) inoculated radicles of susceptible and resistant oil radish (*Raphanus sativus* var. *oleiformis*) and mustard (*Sinapis alba*) with 1, 4, 40 and 60 surface sterilised J2-juveniles of *H. schachtii* and transferred them two days later to nutrient agar in Petri dishes. Using many replicates and with careful evaluation he obtained clear evidence that sex of *H. schachtii* is environmentally controlled. Grundler et al. (1984) obtained similar results. Data from Grundler's work (some unpublished) are presented in Table 1 which clearly shows that differentiation towards the male sex is enhanced by adverse nutritional conditions imposed by resistance (degeneration of syncytia a few days after their induction), space for syncytium development (root diameter) and crowding (intraspecific competition). Great care was taken to check the root system for every invaded nematode. A large proportion of those that had not developed beyond the J3-stage had become moribund in the second moult or degenerated in the J3-stage. In those where an examination of gonad development was still possible, it became evident that both sexes were affected. It may be concluded that the gnotobiotic culture of cyst nematodes on roots in agar medium provides a reliable approach for the study of the influence of the environment on sex differentiation.

Another great advantage of the gnotobiotic culture of cyst nematodes is the screening of breeding lines for resistance to these nematodes. As already emphasised by Müller (1978) and Lauritis et al. (1982) it may be both economic and reliable, as many problems associated with screening in pot cultures are eliminated. Independent of seasonal variations it allows a direct evaluation of resistance responses and nematode development under well defined environmental conditions with precise inoculum levels from homogeneous nematode populations. For cyst nematodes that develop around 25°C the evaluations can be terminated after about two weeks. The disadvantage of being unable to regenerate whole plants from root explants as mentioned by Lauritis et al. (1982), can easily be overcome by keeping the nematodes on roots of intact seedlings or, in the case of potatoes, on shoots which are transferred back to light after evaluation.

CYTOLOGICAL AND BIOCHEMICAL CHANGES AT FEEDING SITES DURING NEMATODE DEVELOPMENT

The monoxenic culture of cyst nematodes on roots of susceptible and resistant cultivars grown in nutrient agar medium undoubtedly provides a safe assessment of the cytological and biochemical changes that occur at feeding sites during nematode development. To our knowledge few attempts have so far been made in this direction. Stender et al. (1982) and Wyss et al. (1984) were probably the first to apply this method in studying the ultrastructure of feeding sites at precise stages of nematode development. Rice (1983) used a similar approach, but roots of her test plants did not grow in agar under strict aseptic conditions. Fig. 6,a,b) shows ultrastructural changes at feeding sites of *Heterodera schachtii* in roots of a susceptible oil radish cultivar, 3 and 16 days after nematode invasion. The in vivo observations of feeding tube formation and associated modification of the surrounding cytoplasm mentioned previously are clearly shown in Fig. 6,a. In the resistant oil radish cultivar cell responses to nematode attack were initially the same as in the susceptible one. Distinct ultrastructural differences between the feeding sites of the two cultivars were first noted three days after nematode invasion and manifested themselves in the nature of the endoplasmic reticulum and in the degree of vacuolisation. Two days later cytoplasm and cell organelles in syncytia of the resistant cultivar were in a state of disintegration (Wyss et al., 1984). It appears to be a common feature that cyst nematodes first induce a compatible response in roots of resistant plants, but a few days later the syncytium disintegrates by mechanisms not yet fully understood.

For the elucidation of the biochemical changes at feeding sites during growth of cyst nematodes the rapid progress in the development of highly sensitive micro-analytical techniques should now be of great help. Jones (1981) lists some of the possible approaches. Krauthausen and Wyss (1982,

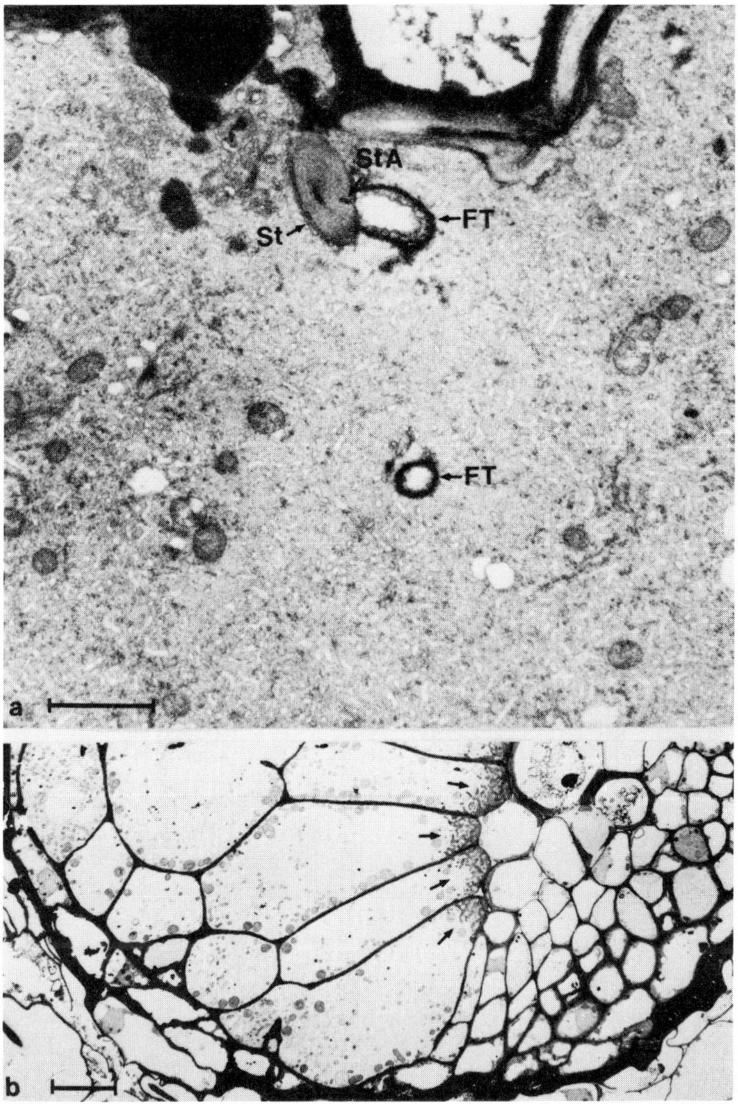

Fig. 6 Cross sections through roots of a susceptible cultivar of *Raphanus sativus* var. *oleiformis* grown in gnotobiotic agar culture and invaded by *Heterodera schachtii*. (a) Initial syncytial cell, three days after nematode invasion. Note feeding tube (FT) emanating through stylet aperture (StA) and cross section through a feeding tube within the zone of modified cytoplasm free of organelles. Bar = 1 µm. (b) Part of established syncytium within vascular cylinder, 16 days after nematode invasion. Arrows = wall ingrowth. Bar = 10 µm. a,b in parts from Wyss et al., 1984.

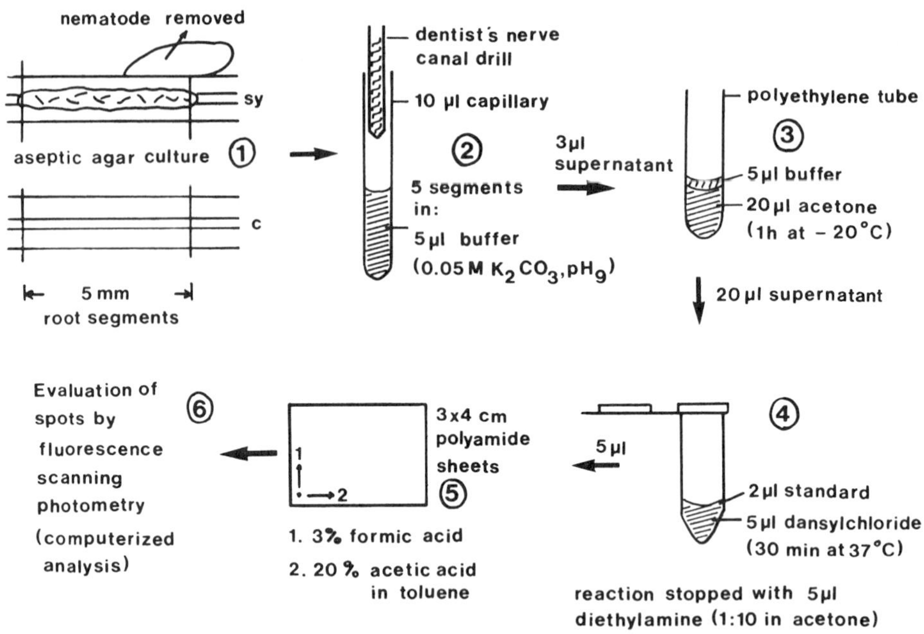

Fig. 7 Determination of relative changes in the pattern of free amino acids at *Heterodera schachtii* feeding sites. Schematic representation of main steps in microanalysis*. 1. Removal of nematodes from feeding sites without injury (from J3-stage onwards), 5 mm segments cut from roots containing syncytia (sy) and from non-inoculated roots (c). 2. Homogenisation of samples (5 root segments per sample). 3. Precipitation of proteins. 4. Dansylation process (couples free amino acids to fluorescent dansylchloride). Standard solution: α-phenylglycine and sarcosine. 5. Micro-chromatography of dansylated amino acids. 6. Evaluation of dansylated substances in relation to internal standards with computerised digital picture analysis.
*For details see Krauthausen and Wyss, 1982.

1984) analysed qualitative changes in free amino acids and water soluble proteins at *H. schachtii* feeding sites during nematode development under gnotobiotic conditions. Fig. 7 shows the main steps in the evaluation of relative changes in the pattern of free amino acids, based on the method developed by Neuhoff (1973).

For such studies it is essential to remove all nematodes uninjured from their feeding sites otherwise meaningless results will be obtained; and the appropriate controls should be determined in preliminary tests (Krauthausen and Wyss, 1982). As it is impossible to remove J2-stages uninjured from within the roots, first attempts can be made from the J3-stage onwards. In the relatively thin roots grown in agar medium these stages are frequently exposed and with care can be separated from their feeding sites. Table 2 shows the significant changes of free amino acids in the total pool during *H. schachtii* development in two Cruciferae hosts.

Table 2 Significant (P = 0.05) relative changes of free amino acids within the total pool that occurred during *Heterodera schachtii* development at feeding sites of *Raphanus sativus* var. *oleiformis* (R) and *Brassica napus* (B) root systems grown in aseptic agar culture*.

Host	R	B	R	B	R	B	R	B	R	B	R	B
Developmental stages	A		B		C		D		E		F	
Valine					+	+	+		+			
Leucine					+			+				
Isoleucine					+		+					
Proline	+	+										
Phenylalanine					+							
Tryptophane						+						
Tyrosine					+	+		+		+		+
Glutamine + Asparagine[1]	-	-	-	-	-	-	-	-				
Cystine/Cysteine									+		+	+
Ornithine											-	-
GABA[2]		+										
Ethanolamine[3]	+	+	+	+								

+ = relative increase; - = relative decrease compared to segments of control roots. Developmental stages: A = J3 (♀ or ♂), 7 days after inoculation (DAI); B = J4 (♀), 12 DAI; C = ♀, 19 DAI; D = ♀, 26 DAI; E = ♀, 33 DAI; F = dead ♀, 40 DAI. 1. Could not be separated by computerised digital picture analysis. 2. γ-amino butyric acid. 3. An amine. n = 6; Amino acids not evaluated: Serine; Threonine. *After Krauthausen (1983).

The similar changes in two different host species indicate that some of the amino acids may indeed play an important role during the nematode's development. They should be considered in future metabolic studies, including autoradiography, to elucidate their function in a successful host-parasite association. The example presented is just one of many possibilities for the study of biochemical events at the site of interaction under well defined conditions. The gnotobiotic culture of cyst nematodes is ideal for such studies, and without being over optimistic, we are confident that it will become an indispensible tool in basic research.

REFERENCES

Berliner, E. and Busch, K., 1914, Über die Züchtung des Rübennematoden (*Heterodera schachtii* Schmidt) auf Agar., Biol. Centralbl., 34:349.

Brown, G., Callow, R.K., Green, C.D., Jones, F.G.W., Rayner, J.H., Shepherd, A.M. and Williams, T.D., 1971, The structure, composition and origin of the subcrystalline layer in some species of the genus *Heterodera*, Nematologica, 17:591.

Clark, S.A., Shepherd, A.M. and Kempton, A., 1973, Spicule structure in some *Heterodera* spp., Nematologica, 19:242.

Dropkin, V.H. and Boone, W.R., 1966, Analysis of host-parasite relationships of root-knot nematodes by single-larva inoculations of excised tomato roots, Nematologica, 12:225.

Endo, B.Y., 1984, Ultrastructure of the esophagus of larvae of the soybean cyst nematode, *Heterodera glycines*, Proc. Helminthol. Soc. Wash., 51:1.

Green, C.D., 1980, Nematode sex attractants, Helminthol. Abstracts, Series A, 49:327.

Grundler, F., Zunke, U. and Wyss, U., 1984, Untersuchungen zur Geschlechtsdetermination bei *Heterodera schachtii*, Mitt. Biol. Bundesanst., 223:207.

Johnson, R.N. and Viglierchio, D.R., 1969, Sugar beet nematode (*Heterodera schachtii*) reared on axenic *Beta vulgaris* root explants. I. Selected environmental factors affecting penetration, Nematologica, 15:129.

Jones, M.G.K., 1981, Host cell responses to endoparasitic nematode attack: structure and function of giant cells and syncytia, Ann. appl. Biol., 97:353.

Krauthausen, H-J., 1983, Einfluß des sich entwickelnden Zystennematoden *Heterodera schachtii* Schmidt, 1871 auf freie Aminosäuren und Proteine im Parasitierungsbereich verschiedener Wirtspflanzen, Dissertation, Universität Hannover.

Krauthausen, H-J. and Wyss, U., 1982, Influence of the cyst nematode *Heterodera schachtii* on relative changes in the pattern of free amino acids at feeding sites, Physiol. Pl. Pathol., 21:425.

Krauthausen, H-J. and Wyss, U., 1984, Mikroanalyse von Proteinen im Parasitierungsbereich des sich entwickelnden Zystennematoden *Heterodera schachtii*, Z. Pflkrankh. PflSchutz., 91:41.

Lauritis, J.A., Rebois, R.V. and Graney, L.S., 1982, Screening soybean for resistance to *Heterodera glycines* Ichinohe using monoxenic cultures, J. Nematol., 14:593.

Lauritis, J.A., Rebois, R.V. and Graney, L.S., 1983a, Life cycle of *Heterodera zeae* Koshy, Swarup, and Sethi on *Zea mays* L. axenic root explants, J. Nematol., 15:115.

Lauritis, J.A., Rebois, R.V. and Graney, L.S., 1983b, Development of *Heterodera glycines* Ichinohe on soybean *Glycine max* (L). Merr., under gnotobiotic conditions, J. Nématol., 15:272.

Mugniéry, D. and Fayet, G., 1981, Détermination du sexe chez *Globodera pallida* Stone, Rev. Nématol., 4:41.

Mugniéry, D. and Fayet, G., 1984, Détermination du sexe de *Globodera rostochiensis* Woll. et influence des niveaux d'infestation sur la penetration, le développement et le sexe de ce nématode, Rev. Nématol., 7:233.

Müller, J., 1970, Ein einfaches Verfahren zur Oberflächensterilisierung von Nematoden, Nematologica, 16:154.

Müller, J., 1978, L'élevage monoxénique d'*Heterodera schachtii* sur crucifères et son application pour la sélection des plantes résistantes, Rev. Nématol., 1:47.

Müller, J., 1985, Der Einfluß der Wirtspflanze auf die Geschlechtsdeterminierung bei *Heterodera schachtii*, Mitt. Biol. Bundesanst., 226:46.

Müller, J. and Wyss, U., und Inst. Wiss. Film, 1981a, Entwicklung des Zystennematoden *Heterodera schachtii*, Publ. Wiss. Film., Sekt. Biol., Ser. 14, Nr.10/C 1387, 20 S.

Müller, J., Rehbock, K. and Wyss, U., 1981b, Growth of *Heterodera schachtii* with remarks on amounts of food consumed, Rev. Nématol., 4:227.

Neuhoff, V., 1973, Micromethods in Molecular Biology, Springer-Verlag, Berlin, Heidelberg, New York.

Orion, D., Wergin, W.P. and Endo, B.Y., 1980, Inhibition of syncytia formation and root-knot nematode development on cultures on excised tomato roots, J. Nematol., 12:196.

Ouden, H. den., 1960, A note on parthenogenesis and sex determination in *Heterodera rostochiensis* Woll., Nematologica, 5:215.

Papadopoulou, J. and Triantaphyllou, A.C., 1982, Sex differentiation in *Meloidogyne incognita* and anatomical evidence of sex reversal, J. Nematol., 14:549.

Rice, S.L., 1983, The anatomical response to cyst nematode invasion in potatoes with resistance derived from *Solanum vernei* and *S. tuberosum* sp. *andigena*, Dissertation, Univ. Birmingham.

Rumpenhorst, H.J., 1984, Intracellular feeding tubes associated with sedentary plant parasitic nematodes, Nematologica, 30:77.

Stender, C., Lehmann, H. and Wyss, U., 1982, Feinstrukturelle Untersuchungen zur Entwicklung von Wurzel-Riesenzellen (Syncytien) induziert durch den Rübenzystennematoden *Heterodera schachtii*, Flora, 172:223.

Triantaphyllou, A.C., 1973, Environmental sex differentiation of nematodes in relation to pest management, Ann. Rev. Phytopathol., 11:441.

Widdowson, E., Doncaster, C.C. and Fenwick, D.W., 1958, Observations on the development of *Heterodera rostochiensis* Woll. in sterile root cultures, Nematologica, 3:308.

Wyss, U. and Zunke, U., 1985a, In vitro feeding of root parasitic nematodes, in "Plant Nematology Laboratory Manual", B. M. Zuckerman, W. F. Mai and M. B. Harrison, eds, Amherst, Univ. Massachusetts Agric. Exp. Sta., 91.

Wyss, U. and Zunke, U., 1985b, und Inst. Wiss. Film, *Heterodera schachtii* (Nematoda). Behaviour inside roots (rape). Film E 2904 der Enc. Cin., Göttingen.

Wyss, U. and Zunke, U., 1986, The potential of high resolution video-enhanced contrast microscopy in nematological research, Rev. Nematol., 9:91.

Wyss, U., Stender, C. and Lehmann, H., 1984, Ultrastructure of feeding sites of the cyst nematode *Heterodera schachtii* Schmidt in roots of susceptible and resistant *Raphanus sativus* L. var. *oleiformis* Pers. cultivars, Physiol. Pl. Pathol., 25:21.

Zuckerman, B.M., 1971, Gnotobiology, in "Plant Parasitic Nematodes," B. M. Zuckerman, W. F. Mai and R. A. Rohde, eds, Vol. II, pp. 159-184, Academic Press, New York.

Zuckerman, B.M. and Jansson, H.B., 1984, Nematode chemotaxis and possible mechanisms of host/prey recognition, Ann. Rev. Phytopathol., 22:95.

Zunke, U., 1985, Zur Bildung der subkristallinen Schicht bei *Heterodera schachtii* unter aseptischen Bedingungen, Nematologica, 31:117.

DISCUSSION

Wyss answered several questions which arose from his lecture and the films on "the life cycle of *H. schachtii*" and on "behaviour of *H. schachtii* in rape roots". He explained that the population of *H. schachtii* used did not produce egg masses in a gelatinous sac in gnotobiotic culture. Endo added that no difficulties had been encountered in observing *H. glycines* eggs being deposited in tissue culture systems. Asked to expand on the secretion and excretion of metabolites through the body cuticle, Wyss said that further information would be derived from X-ray microanalysis. If juveniles (J2) are introduced into agar plates they may remain active outside the roots for several days without any observable changes in the cuticle; but several hours after the nematodes start to feed on the roots, the sub-crystalline layer (SCL) is formed. Thus the SCL is associated with nematode feeding and as it is formed in gnotobiotic conditions it cannot be considered to result from association with fungi or other microorganisms.

In the film it was noted that the juveniles developing near the stele were pointing in different directions, whereas previous published information suggests that they always point to the root tip. Wyss said that from several observations of root infestations, it can be concluded that juveniles may be orientated randomly within the roots.

BIOCHEMICAL REACTIONS AND INTERACTIONS

Giuseppe Zacheo

Istituto di Nematologia Agraria, C.N.R.
70126 Bari
Italy

INTRODUCTION

During their evolution plants not only acquired the ability to adapt to hostile environments but also versatile defense mechanisms against their enemies. Pathogens, on the other hand, have evolved to overcome the resistance mechanisms of the plants. Very little is known about this plant defense mechanism, but it appears to depend on events that are initiated soon after the plant cell and the pathogen come into contact. It is evident that there must be a general as well as a specific system in the host-parasite interaction, as a result of combined biochemical components, characteristic of normal host tissue and of the pathogens themselves. An appreciation of the two separate metabolic developments is necessary to interpret and distinguish between the metabolic changes due to host or to pathogen activity. In plants attacked by nematodes selective changes occur in the metabolism either as a consequence of the establishment of a compatible (susceptibility) host-pathogen interaction or as a result of incompatibility (resistance) between host and parasite.

Susceptibility and resistance are relative terms describing the state of the host plant confronting nematodes. Both terms have been used with wide connotations, at the level of cells, tissues, and individuals as well as in reference to field populations. Each term is generally used in the connotation of the two opposite extremes of continuous disease ratings (Ouchi, 1983).

Hosts and parasites coevolved in nature to accumulate genes for balanced coexistence and these genes are presumably functioning during both offensive and defensive interactions. Different strategies have evolved among the various genera of phytoparasitic nematodes, but all serve in one way or another to render the strategies of the hosts non-functional. The resistance in host plants is usually controlled by single dominant genes (Day, 1974; Webster, 1975; Price et al., 1978; Wood, 1982). This type of resistance, which is usually race-specific, is referred to as vertical resistance (Van der Planck, 1963), as distinct from horizontal resistance where several genes with small effects, control a race non-specific resistant response in the plant. Horizontal resistance is generally effective against all races of a parasite but the plants are only partially resistant. Rohde (1972) considers that poor hosts of *Meloidogyne* spp and *Heterodera* spp are probably of this type. Thus from the view point of evolution it seems that plant resistance to pathogens is initially non-specific; susceptibility is built up later, is specific and formed during the process of host selection by the pathogen and its adaptation to the host.

Whilst there is some knowledge of the biochemical changes that occur in plants following nematode attack, this has not yet been extended to an overall interpretation of the defense mechanisms. In the process of investigating the events of pathogenesis, it is often assumed that differences between susceptible and resistant hosts are indicative of host susceptibility or resistance. This notion, if valid, is useful with regard to secondary metabolism but is more tenuous with regard to primary metabolism (Veech, 1981).

Some authors have focused their attention on biochemical changes in plant tissues infected with nematodes, comparing infected plants or parts with comparable non-infected tissue. Their objective was to characterise the metabolic changes in infected cells and in healthy cells at various distances from them. The investigations indicated an acceleration of metabolic activity in cells adjacent to infected tissue which decreased exponentially in proportion to the distance from the site of infection (Tomiyama, 1983).

The purpose of this chapter is to examine the changes of the biochemical components, in healthy and infected plants, following *Heterodera* spp attack, as a basis for the elucidation and understanding of the host-parasite interaction.

JUVENILE PENETRATION

The frequency and path of penetration by juveniles of *Heterodera* species is generally similar in resistant and susceptible plants (Müller, 1978). In the susceptible roots the juvenile nematodes cause necrosis during their migration. The head of the juvenile may become embedded in the endodermis or pericycle but the rest of the body remains in the cortex (Mankau and Linford, 1960). Cells surrounding the head of the nematode are induced to modify their metabolism because of products introduced by the nematode. As an adaptation to the parasitism, root cells transform into "syncytia" that are rich in organic nutrients essential for nematode development. The host-parasite relationship is controlled by the genetic coding of both host and parasite and the close proximity of the syncytia to the vascular tissue seems to be a prerequisite for optimum nematode development. In contrast, it is interesting that the response of soybeans resistant to *H. glycines* Ichinohe commences with the development of an apparently normal syncytium before the hypersensitive reaction is manifest (Riggs et al., 1973).

Three to four days after juvenile penetration the syncytium begins to degenerate and is surrounded increasingly by necrotic cells (Huijsmann et al., 1969; Yu and Steele, 1981). Failure to establish a feeding relationship with the host tissue and subsequent death of the nematode is referred to as a necrotic hypersensitive reaction. This leads to the question as to whether necrosis inhibits nematodes or whether necrosis represents only a macroscopic expression of successive events of the mechanism used by the resistant plant to inhibit or prevent nematode attack. Studies have shown that nematodes freely penetrate the roots of hosts and non-hosts alike and that whilst incompatibility may occur at this stage, it is rare.

AMINO ACID CONTENT

The concentration of free amino acids in the fibrous roots of the resistant cultivar *Beta patellaris* were unchanged by infection with *H. schachtii*, Schmidt, but the concentration of total amino acids, aspartic acid, glutamic acid and glutamine were significantly increased in the fibrous roots of the susceptible *Beta vulgaris* by *H. schachtii* infection (Doney et al., 1970). Gommers and Dropkin (1977) found a 1.7 fold increase in free amino acid concentration in *Heterodera*-induced syncytia in soybean; they report also that concentrations of ATP, glucose-6-phosphate, fructose-1.6-diphosphate and 6-phosphogluconate were comparable to those of actively growing root tip tissue.

Kranthausen and Wyss (1982) using a microanalytical method, reported that glutamine and asparagine decreased significantly in the roots of oil radish, *Raphanus sativus* var. *oleiformis*, and *Brassica napus* from the early stages of parasitism until the females of *H. schachtii* had produced eggs. Isoleucine, leucine, valine, phenilalanine and triptophan increased during egg production of females. Changes in the pattern of free amino acids were the same in roots of susceptible and resistant oil radish. The significance of differences in amino acid patterns occurring in susceptible and resistant cultivars during *H. schachtii* infestation are not yet clear in relation to the host-parasite interaction. However, Giebel and Stobieka (1974) attribute an important role to proline and hydroxyproline in plant bio-resistance during *Heterodera* infestation. They report that 50 per cent more hydroxproline was present in roots of resistant solanaceous plants than in those of susceptible plants. The proline/hydroxyproline ratio decreased in resistant and increased in susceptible potato roots invaded by nematodes. Giebel (1982) speculated that enzymes produced by *Globodera rostochiensis* (Woll.) Behr. can decompose glycoproteins to free hydroxyproline. In this way this amino acid, in resistant plants, would inhibit the activity of indole-acetic acid (IAA); on the contrary, in susceptible plants, it would allow IAA to function.

PHENOLS AND LIGNIFICATION

The phenols are the most commonly produced compounds in the metabolic response of plants to viruses, bacteria, fungi and nematode infection (Kosuge, 1969). Phenols accumulate in infected tissues (Epstein, 1972) either by direct synthesis through the altered respiratory pathway of the host or by the release from precursors such as glycosides (Rohde, 1972). Giebel (1970) reported that phenols occur in bound form as glycosides in several solonaceae infested by *G. rostochiensis* and in peas susceptible or resistant to *H. goettingiana* Leibscher (Giebel and Lamberti, 1977). Their activation requires decomposition to free phenols and this is done by nematodes when they secrete β-glycosidases into the host tissue (de Maeseneer, 1964; Mountain and Patrick, 1959; Giebel, 1970).

Phenol accumulation and products of phenolic oxidation have been associated with cellular browning in plants infected by several phyto-parasitic nematodes (Chang, 1969; Giebel, 1970; Giebel et al., 1970; Hung and Rohde, 1973; Giebel and Lamberti, 1977). On the evidence that β-glycosidases hydrolyse phenols Wilski and Giebel (1966) present a detailed hypothesis to explain resistance of potatoes to *G. rostochiensis*. The secretion of β-glycosidases by the biotype B of the nematode was so little that necrosis failed to occur, whereas enzyme activity was much greater in the biotype A, and the necrotic response was evident. Wilski and Giebel (1966) assume that plants are susceptible to pathotype B because the unidentified phenolic glycosides are not hydrolysed to the necrosis-inducing aglycon. On the contrary, plants are resistant to pathotype A presumably because they contain phenolic glycosides. The release of the phenolic aglycon is then involved in the host necrosis, which leads to resistance. As a consequence, in plants susceptible to pathotype A, the phenolic compounds released from glycosides by β-glycosidases influence the activity of IAA-oxidase, favouring IAA accumulation; in resistant plants a system for the destruction of IAA is present (Tomasweski and Thimann, 1966). Root extracts from plants susceptible to *G. rostochiensis* were characterised by a low ratio of mono- to polyphenols and they strongly inhibited the enzymatic destruction of IAA. Root extracts of plants resistant to the nematode had a high ratio of mono- to polyphenols and did not inhibit IAA destruction.

Phenylalanine ammonia-lyase (PAL) and tyrosine ammonia-lyase (TAL) activity increases in plants resistant to viruses, fungi and nematodes (Farkas and Szirmai, 1969; Kuc, 1982). The activity of PAL was higher in potato roots resistant to *G. rostochiensis* than in susceptible roots and it increased in both kinds of plants after nematode invasion (Giebel, 1973). The activity of TAL after infection increased in both susceptible and resistant roots, but slightly more in the latter. Cinnamic acid and p-coumaric acid, the products of PAL and TAL activity respectively, are lignin precursors and cofactors of IAA-oxidase.

Histological tests of necrosis induced in resistant potatoes by *G. rostochiensis* showed a positive reaction for lignin-like substances and particularly in cells close to the head and excretory pore of nematodes. Giebel et al. (1970) suggest that the processes of lignin formation are stimulated by the enzyme glycosidase secreted by the nematode. Thus, the activated system of IAA-oxidase occurring in resistant plant, favours the production of free phenols, which are precursors of lignin. In the necrotic cells of resistant roots an active peroxidase was found. Differences in the activity of peroxidase in susceptible and resistant potatoes indicate that there are phenolic compounds in the roots that act as cofactors of peroxidase (Giebel et al., 1971).

GIEBEL'S HYPOTHESIS OF RESISTANCE MECHANISM

On the basis of investigations on the host-*Heterodera* biochemical interaction, Giebel (1974) proposed a complex theory to explain the biochemical mechanism of plant resistance to nematodes. IAA, mono- and polyphenols, aglycones, β-glucosidase, IAA-oxidase, tyrosinase, peroxidase, are thought to interact and, depending on their relative concentrations or activities, to result in a susceptible or resistant reaction. All these cofactors and enzymes seem to act in such a way as to favour IAA activity. Some authors (Jones, 1980; Veech, 1981) disagree with Giebel's hypothesis which correlates phenols, IAA, β-glycosidases with resistance. They consider the hypothesis unproven "because there is a conceptual problem in the design of several of the experiments presented in its support" (Kaplan and Keen, 1980). Giebel's hypothesis also suffers in the light of recent reports on pathogen-induced lignin (Grisebach, 1977; Asada et al., 1979; Vance et al., 1980).

Induced lignification around penetration sites or infection centres is accompanied by increased activity of several enzymes such as PAL, TAL, peroxidases. This concerted increase in enzyme activity may reflect activation or induction of multienzyme complexes, such as postulated to regulate cinnamic acid and lignin biosynthesis in different plants (Stafford, 1974;

Czichi and Kindl, 1975). Lignin formation in a disease situation is a localised response and some authors (Maule and Ride, 1976; Vance and Sherwood, 1976) suggest that sufficient lignin precursors can be formed at localised sites for lignification to occur. Whether these responses involve de novo synthesis of phenolic compounds or conversion of preformed constituents is still open to question.

PAL, cinnamic acid-4-hydroxylase, cinnamic acid-CoA-lyase and wall-bound peroxidase are enzymes involved in lignin biosynthesis and their activity is enhanced by ethylene. The effect of ethylene on lignification may be mediated through the hydroxyproline rich proteins in cell walls. These proteins are thought to act as cell wall structural components and they are strictly related to the ethylene production; it has been experimentally demonstrated that L-canaline and other inhibitors efficiently inhibit the production of ethylene and the synthesis of hydroxy-proline-rich proteins (Toppan et al., 1982). Whitmore (1978) showed that condensation products of coniferyl alcohol bound to hydroxyproline-rich proteins in a reaction catalysed by peroxidase. From these studies it would seem that lignification in the host-parasite interaction is a very complex system which cannot be explained simply by IAA-IAA oxidase changes (Giebel, 1974).

ASCORBATE FREE RADICAL-REDUCTASE, PEROXIDASES AND SUPEROXIDE-DISMUTASES

There is some evidence which correlates ascorbic acid (AA) with resistance to *H. goettingiana* in peas. Pea lines resistant or susceptible to *H. goettingiana* showed similar biochemical behaviour to tomato plants resistant and susceptible to *Meloidogyne incognita* (Arrigoni et al., 1979). In the pea lines, as in the tomato plants, nematode attack caused an increase of AA in resistant and a decrease in susceptible plants (Table 1).

Oxidation of AA results in the formation of two compounds through the successive loss of two electrons: ascorbate free radical (AFR), an intermediate compound after the loss of the first electron, and dehydro-ascorbic acid (DHA) the final product of the reaction. Of the two, only AFR can efficiently be reconverted into AA because of the action of the AFR reductase present in the cells, which catalyses the transfer of electrons from NADH (-nicotinamide adenine nucleotide) to AFR.

There was an increase in AFR reductase activity in resistant pea lines infested with *H. goettingiana* but no change in infested susceptible lines (Zacheo et al., 1981). An increase in AFR reductase activity occurred in different cellular components examined - mitochondria, microsomes and soluble fraction - from resistant pea lines, with the largest increase in the microsomes (Zacheo et al., 1981). It is evident that a correlation exists between AFR reductase activity, AA content and nematode infestation (Fig. 1). There is an increase in AA synthesis and turnover during nematode infestation and it is reasonable to question if AA increase is required to convert the proline residues present in polypeptide chains to hydroxyproline residue (Arrigoni et al., 1977a). Potato cultivars attacked by *G. rostochiensis* (Giebel and Stobiecka, 1974) and tomato cultivars infested by *M. incognita*, showed an increase of protein-hydroxyproline in the cell walls in both resistant and susceptible cultivars. Conversely, the protein-hydroxyproline content of the mitochondrial fraction tended to decrease in infested roots of susceptible tomato plants and to increase in the resistant ones. On the basis of these observations it is assumed that the synthesis of mitochondrial and cell wall hypoproteins is directly

Table 1 Density of population of *Heterodera goettingiana* and ascorbic acid content in resistant (R) and susceptible (S) pea lines

Lines	No. nematodes/ g roots	Ascorbic acid µg/gdw
MG 101877 c (R)	-	1.47
" + Nematodes	1,407	4.10
MG 101956 a (R)	-	1.55
" + Nematodes	1,619	4.26
MG 101877 b (R)	-	1.59
" + Nematodes	2,407	2.86
MG 101956 b (S)	-	2.85
" + Nematodes	4,506	1.75
MG 101877 a (S)	-	4.49
" + Nematodes	4,820	1.93
MG 101793 (S)	-	7.16
" + Nematodes	5,893	3.05

correlated (Zacheo et al., 1977) and leads to the suggestion that AA content plays an important role in the host-parasite relationship.

Artificial supply of AA to susceptible tomatoes inhibits *M. incognita* invasion, while a decrease in its endogenous content in resistant tomatoes, leads to an increase of nematode penetration (Arrigoni et al., 1979). Experimental results indicate that it is not the endogenous level of AA that differentiates between susceptible and resistant plants, but the capability of biosynthesising and reconverting the main product of its oxidation, AFR, to the AA of the plant after nematode attack. It is interesting that the major activity increases particularly in the microsome fraction localised in the endoplasmic reticulum, a cellular constituent actively engaged in hydroxylation processes. In fact, as AA increases in resistant plants, there is also an increase in proteins containing hydroxyproline and, particularly, in mitochondrial proteins (Zacheo et al., 1977).

Mitochondrial hydroxyproline containing proteins seem to be linked to the development of cyanide-resistant respiration, which metabolic process should produce H_2O_2. Peroxidases should utilise H_2O_2 in generating the anionic radical of oxygen ($O_2^{\cdot-}$), the superoxide. The H_2O_2 reacting with superoxide gives rise to the hydroxyl radical, the highly oxidant agent in nature (Fig. 2). Peroxidases have been shown to be involved in the infection process by different pathogens (Wheeler, 1975; Arrigoni et al., 1981; Zacheo et al., 1982, 1983) but their role and action in plant resistance remains undefined. The difficulty in attributing a role to these enzymwa is partly due to the extremely

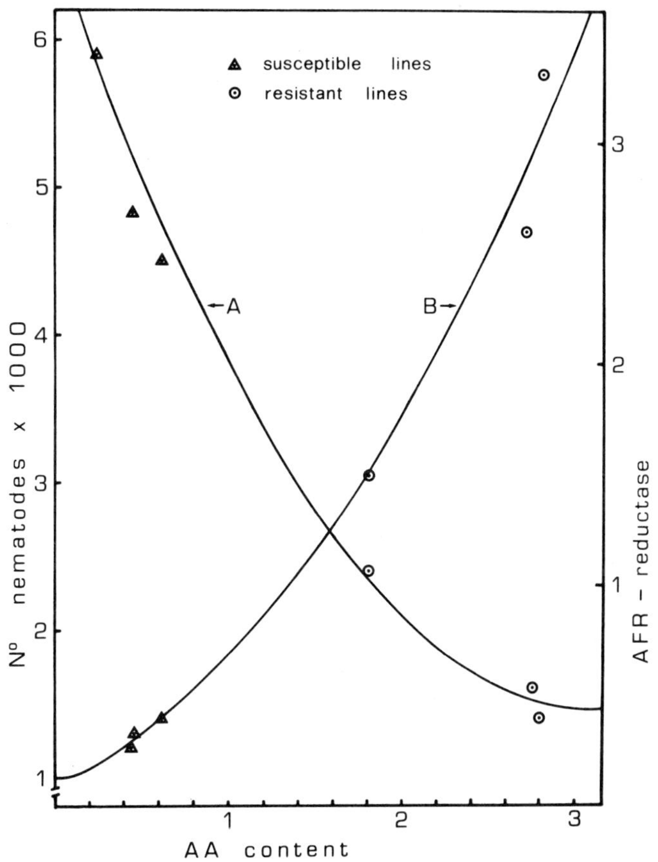

Fig. 1 Relationship between changes in ascorbic acid content and nematode populations in roots (A) and microsomal activity of AFR-reductage (B); ascorbic acid and AFR-reductase activity expressed as ratio infested/healthy.
$A : y = 4.81 - 1.59x + 0.14x^2$. $B : y = 0.10 + 0.32x + 0.26x^2$.

diversified enzyme populations and the considerable biochemical versatility that they exhibit (Hepler et al., 1972; Srivastava and Van Huystee, 1973).

It has recently been reported that neutrophils, when stimulated, release myeloperoxidase and H_2O_2 into the surrounding medium, thus being capable of destroying adjacent cells. This results in the liberation of a large quantity of $O_2^{\cdot-}$ which could result in the formation of hydroxyl radical and single oxygen; radical production in the cells should counteract the pathogens. The deleterious consequences of these radicals could be prevented by a protective action of some enzymes, such as superoxide dismutase (SOD). These enzymes, which are extremely rapid in action, must exercise strict control of the concentration of cellular $O_2^{\cdot-}$ radicals to avoid extensive damage to the various cytological structures (Lavell et al., 1973). Thus, peroxidases and SOD enzymes contribute to the determination of plant susceptibility to, or resistance against the pathogens.

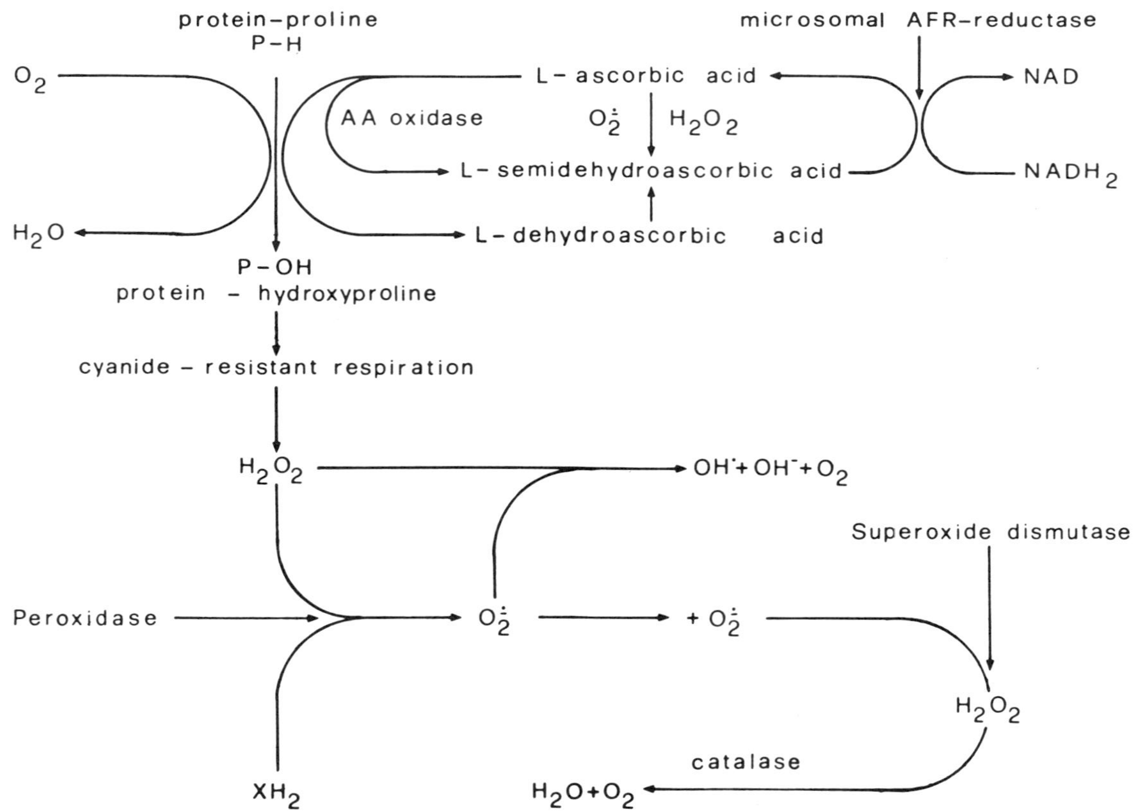

Fig. 2 Molecular events occurring in the hypothetical defence mechanism.

Table 2 Peroxidase activity in resistant (R) and susceptible (S) pea lines inoculated with *Heterodera goettingiana*

Lines	Peroxidase as ΔOD/min/mg proteins		
	Mitochondria	Microsomes	Soluble Fraction
MG 101877 c (R)	3.74	6.70	7.69
" + Nematodes	7.45	18.00	21.38
MG 101956 a (R)	4.06	11.27	15.39
" + Nematodes	6.43	24.60	24.08
MG 101877 b (R)	0.98	7.94	15.40
" + Nematodes	1.62	11.69	19.61
MG 101956 b (S)	3.48	12.66	12.87
" + Nematodes	2.55	11.12	13.29
MG 101877 a (S)	4.83	16.15	21.41
" + Nematodes	1.63	9.88	16.10
MG 101793 (S)	0.96	15.00	16.79
" + Nematodes	0.88	11.09	10.20

Susceptible and resistant pea lines, infested by *H. goettingiana*, showed remarkable changes in peroxidase activities during infection (Table 2). In susceptible lines nematode invasion induced a decrease in peroxidase activity, and an increase in resistant lines. SOD activity tested on the same pea lines during nematode invasion increased in the susceptible and decreased sharply in resistant lines. In resistant lines during the infestation process mitochondrial and microsomal SOD activity decreased sharply after inoculation. On the contrary, mitochondrial and microsomal SOD activity increased in susceptible lines (Table 3) (Arrigoni et al., 1981).

In conclusion, nematode attack leads to necrosis of the invaded tissues in resistant plants and the formation of syncytia in susceptible plants. Changes in certain biochemical components can induce either of the host responses, among them the oxy-redoxy system of AA being considered of primary importance. In resistant cultivars, increasing the synthesis of AA de novo, or reconverting AFR to AA by means of AFR-reductase, induces the increase of hypoproteins. Wienecke et al. (1982) present evidence that two kinds of hydroyproline rich proteins exist, one of which remains attached to the plasma membrane rather than being released into the cell wall. These proteins are synthesised at the endoplasmic reticulum and transported via the Golgi apparatus.

Susceptible cultivars do not seem to have the facility to initiate this kind of response to counteract nematode invasion. At present it is difficult to ascertain whether or not some of the proteins are involved

Table 3 Superoxidedismutase activity in resistant (R) and susceptible (S) pea lines inoculated with *Heterodera goettingiana*

	SOD as U/mg of proteins		
	Mitochondria	Microsomes	Soluble Fraction
MG 101877 c (R)	4.92	8.01	3.61
" + Nematodes	4.37	8.45	3.46
MG 101956 a (R)	4.70	10.83	5.14
" + Nematodes	3.20	8.87	4.90
MG 101877 b (R)	7.52	10.22	4.80
" + Nematodes	4.54	7.44	4.00
MG 101956 b (S)	4.27	5.53	4.18
" + Nematodes	6.26	7.92	4.34
MG 101877 a (S)	6.62	3.31	4.85
" + Nematodes	7.20	6.60	4.95
MG 101793 (S)	4.33	8.58	4.92
" + Nematodes	7.02	11.37	5.10

in the development of cyanide-resistant respiration and therefore in the resistance; however, others (Arrigoni et al., 1977b) correlated the increase of hypoproteins with development of this terminal respiratory chain. Also, resistant tomato roots infested with *M. incognita* showed both increase in cyanide-resistant respiration (Zacheo and Bleve Zacheo, 1985a; 1986) and synthesis of hydroxyproline proteins during nematode attack (Zacheo et al., 1977). Recently, production and localisation of oxygen free radicals have been reported in tomato roots resistant to *M. incognita* (Zacheo and Bleve Zacheo, 1985b) and in potatoes resistant to *G. rostochiensis* (unpublished data).

Thus, resistant plants could counteract nematode attack by increasing peroxidase activity and consequently oxygen radicals. Moreover, this process is facilitated by the low activity of SOD, whose function is the dismutation of radicals into H_2O_2, transformed then into $H_2O + O_2$ by catalase.

If this interpretation of the defence mechanism in resistant plants is sound, it also explains the lower peroxidase activity and the increase of SOD activity in susceptible plants. Thus, the lower peroxidase activity should generate radicals, which could be utilised by the plant to counteract nematode attack, and the increased SOD activity would neutralise the low amount of radicals present in the tissue.

REFERENCES

Arrigoni, O., Arrigoni-Liso, R. and Calabrese, G., 1977a, Ascorbic acid requirement for biosynthesis of hydroxyproline-containing proteins in plants, FEBS Letters, 82:135.

Arrigoni, O., De Santis, A., Arrigoni-Liso, R. and Calabrese, G., 1977b, Mitochondrial protein-hydroxyproline during the development of cyanide-insensitive respiration, Biochem. Biophys. Res. Comm., 74:1637.

Arrigoni, O., Zacheo, G., Arrigoni-Liso, R., Bleve-Zacheo, T. and Lamberti, F., 1979, Relationship between ascorbic acid and resistance in tomato plants to *Meloidogyne incognita*, Phytopathology, 69:579.

Arrigoni, O., Zacheo, G., Bleve-Zacheo, T., Arrigoni-Liso, R. and Lamberti, F., 1981, Changes of superoxide dismutase and peroxidase activities in pea roots infested by *Heterodera goettingiana*, Nematol. medit., 9:189.

Asada, Y., Ohguchi, T. and Matsumoto, I., 1979, Induction of lignification in response to fungal infection, in "Recognition and Specificity in Plant Host-Parasite Interactions", J. M. Daly and I. Uritani, eds, University Press, Tokyo.

Chang, L.M., 1969, The repellent effect of necrotic tissue on the nematode, *Pratylenchus penetrans* (Cobb, 1917), Filipjev et Schuurmans Stekhoven, 1941, University of Massachusetts, M.S. Thesis.

Czichi, U. and Kindl, H., 1975, A model of closely assembled consecutive enzymes of membranes: Formation of hydroxy-cinnamic acid from L-phenylalanine on thylakoids of *Dunaliella marina*, J. Physiol. Chemist, 356:475.

Day, P.R., 1974, Genetics of Host-Parasite Interactions, Freeman Press, San Francisco.

Doney, D.L., Fife, J.M. and Whitney, E.D., 1970, The effect of the sugar-beet nematode *Heterodera schachtii* on the free amino acids in resistant and susceptible *Beta* species, Phytopathology, 60:1727.

Epstein, E., 1972, Biochemical changes in technical root galls caused by an ectoparasitic nematode, *Longidorus africanus*: phenols, carbohydrates and cytokinins, J. Nematol., 4:246.

Farkas, G.L. and Szirmai, J., 1969, Increase in phenylalaline ammonial-yase activity in bean leaves infected with tobacco necrosis virus, Neth. J. Pl. Pathol., 75:82.

Giebel, J., 1970, Phenolic content in roots of some solanaceae and its influence on IAA-oxidase activity as an indicator of resistance to *Heterodera rostochiensis*, Nematologica, 16:22.

Giebel, J., 1973, Phenylalanine and tyrosine ammonia-lyase activities in potato roots and their significance in potato resistance to *Heterodera rostochiensis*, Nematologica, 19:1.

Giebel, J., 1974, Biochemical mechanisms of plant resistant to nematodes: a review, J. Nematol., 6:175.

Giebel, J., 1982, Mechanism of resistance to plant nematodes, Ann. Rev. Phytopathol., 20:257.

Giebel, J. and Stobiecka, M., 1974, The role of amino acids in plant tissue response to *Heterodera rostochiensis*. I. Proteinproline and hydroxyproline content in roots of susceptible and resistant solanaceous plants, Nematologica, 20:407.

Giebel, J. and Lamberti, F., 1977, Some histological and biochemical aspects of pea resistance and susceptibility to *Heterodera goettingiana*, Nematol. medit., 5:173.

Giebel, J., Krenz, J. and Wilski, A., 1970, The formation of lignin-like substances in roots of resistant potatoes under the influence of *Heterodera rostochiensis* larvae, Nematologica, 16:601.

Giebel, J., Krenz, J. and Wilski, A., 1971, Localization of some enzymes in roots of susceptible and resistant potatoes infected with *Heterodera rostochiensis*, Nematologica, 1:29.

Gommers, F.J. and Dropkin, V.H., 1977, Quantitative histochemistry of nematode induced transfer cells, Phytopathology, 67:869.

Grisebach, H., 1977, Biochemistry of lignification, Naturwissenschaftern, 64:619.

Hepler, P.K., Rice, R.M. and Terranova, W.A., 1972, Cytochemical localisation of peroxidase activity in wound vessel members of *Coleus*, Can. J. Bot., 50:977.

Huijsman, C.A., Klinkenberg, C.H. and Den Ouden, H., 1969, Tolerance to *Heterodera rostochiensis*, Woll. among potato varieties and its relation to certain characteristic of root anatomy, Europ. Potato J., 12:134.

Hung, C.L. and Rohde, R.A., 1973, Phenol accumulation related to resistance in tomato to infection by root-knot and lesion nematodes, J. Nematol., 5:253.

Jones, M.G.K., 1980, Micro-gel electrophoretic examination of soluble proteins in giant transfer cells and associated root-knot nematodes (*Meloidogyne javanica*) in balsam roots, Physiol. Pl. Pathol., 16:359.

Kaplan, D.T. and Keen, N.T., 1980, Mechanisms conferring plant incompatibility to nematodes, Rev. Nematol., 3:124.

Kosuge, T., 1969, The role of phenolics in host response to infection, Ann. Rev. Phytopathol., 7:195.

Kranthausen, H.J. and Wyss, U., 1982, Influence of the cyst nematode *Heterodera schachtii* on relative changes in the pattern of free amino acids at feeding sites, Physiol. Pl. Pathol., 21:425.

Kuc, J., 1982, The immunization of cucurbits against fungal, bacterial and viral diseases, in "Plant Infection. The Physiological and Biochemical Basis", Y. Asada, W.R. Bushnell, S. Ouchi, C.P. Vance, eds, Japanese Science Society Press, Tokyo.

Lavelle, F., Michelson, A.M. and Dimitrijevic, L., 1973, Biological protection by superoxide dismutase, Biool Biophys., Res. Comm., 55:350.

Mankau, R. and Linford, M.B., 1960, Host-parasite relationships of the clover cyst nematode, *Heterodera trifolii* Goffart, Ill. Agric. Exper. Sta. Bull., 667:1.

Masseneer, J. de, 1964, Leaf browing of *Ficus* spp. new host plant of *Aphelenchoides fragariae* (Ritzemabosi), Nematologica, 10:403.

Maule, A.J. and Ride, J.P., 1976, Ammonia-lyase and O-methyltransferase activities related to lignification in wheat leaves infected with *Botrytis*, Phytochemistry, 15:1661.

Mountain, W.b. and Patrick, Z.A., 1959, the peach replant problem in Ontario. VII. The Pathogenicity of *Pratylenchus penetrans* (Cobb, 1917) Filip. & Stek., 1941, Can. J. Bot., 37:459.

Müller, J., 1978, L'élevage monoxenique d'*Heterodera schachtii* sur crucifères et son application pour la selection des plantes résistantes, Rev. Nematol., 1:47.

Ouchi, S., 1983, Induction of resistance or susceptibility, Ann. Rev. Phytopathol., 21:289.

Price, M., Caviness, C.E. and Riggs, R.D., 1978, Hybridization of races of *Heterodera glycines*, J. Nematol., 10:114.

Riggs, R.D., Kimm, K.S. and Gipson, I., 1973, Ultrastructural changes in Pekin soybeans infected with *Heterodera glycines*, Phytopathology., 63:76.

Rohde, R.A., 1972, Expression of resistance in plants to nematodes, Ann. Rev. Phytopathol., 10:233.

Srivastava, D.P. and Van Huystee, R.B., 1973, Evidence for close association of peroxidase, polyphenoloxidase in IAA oxidase isozymes in peanut suspension culture medium, Can. J. Bot., 51:2207.

Stafford, H.A., 1974, The metabolism of aromatic compounds, Ann. Rev. Pl. Physiol., 25:459.

Tomaszweski, M. and Thimann, K.K., 1966, Interactions of phenolic acids, Metallic ions and chelating agents on auxin-induced growth, Pl. Physiol., 41:1443.

Tomiyama, K., 1983, Research on the hypersensitive response, Ann. Rev. Phytypathol., 21:1.

Toppan, A., Roby, D. and Esquerré-Tugayé, M.T., 1982, Cell surfaces in plant-microorganism interaction. III. In vivo effect of ethylene on hydroxyproline-rich glycoprotein accumulation in the cell wall of diseased plants, Pl. Physiol., 70:82.

Vance, C.P. and Sherwood, R.T., 1976, Regular lignin formation in red canarygrass in relation to disease resistance, Pl. Physiol., 57:915.

Vance, C.P., Kirk, T.K. and Sherwood, R.T., 1980, Lignification as a mechanism of disease resistance, Ann. Rev. Phytopathol., 18:259.

Van der Plank, J.E., 1963, "Plant Disease: Epidemics and Control", Academic Press, New York, London, 349 pp.

Veech, J.A., 1981, Plant resistance to nematode, in "Plant Parasitic Nematodes", B.M. Zuckerman, ed, Academic Press, New York.

Webster, J.M., 1975, Aspects of the host-parasite relationship of plant parasitic nematodes, in "Advances in Parasitology", B. Dawes, ed, Academic Press, London and New York.

Wheeler, H., 1975, Plant Pathogenesis, Springer-Verlag, Berlin, Heidelberg, New York.

Whitmore, F.W., 1978a, Lignin-carbohydrate complex formed in isolated cell walls of callus, Phytochemistry, 17:241.

Whitmore, F.W., 1978b, Lignin-protein complex catalyzed by peroxidase, Pl. Sci. Lett., 13:241.

Wienecke, K., Glas, R. and Robinson, D.G., 1982, Organelles involved in the synthesis and transport of hydroxyproline-containing glycoproteins in carrot root discs, Planta, 155:58.

Wilski, A. and Giebel, J., 1966, B-glucosidase in *Heterodera rostochiensis* and its significance in resistance to this nematode, Nematologica, 12:219.

Wood, R.K.S., 1982, Active Defense Mechanisms in Plants, Plenum Press, New York.

Yu, M.H. and Steele, A.E., 1981, Host-parasite interaction of resistant sugarbeet and *Heterodera schachtii*, J. Nematol., 13:206.

Zacheo, G., Lamberti, F., Arrigoni-Liso, R. and Arrigoni, O., 1977, Mitochondrial protein-hydroxyproline content of susceptible and resistant tomatoes infested by *Meloidogyne incognita*, Nematologica, 23:471.

Zacheo, G., Arrigoni-Liso, R., Bleve-Zacheo, T., Lamberti, F., Perrino, P. and Arrigoni, O., 1981, Changes of ascorbate free radical reductase in pea roots infested by *Heterodera goettingiana*, Nematol. medit., 9:181.

Zacheo, G., Bleve-Zacheo, T. and Lamberti, F., 1982, Role of peroxidase and superoxide dismutase activity in resistant and susceptible tomato cultivars infested by *Meloidogyne incognita*, Nematol. medit., 10:75.

Zacheo, G., Arrigoni-Liso, R., Bleve-Zacheo, T., Lamberti, F. and Arrigoni, O., 1983, Mitochondrial peroxidase and superoxide dismutase activities during the infection by *Meloidogyne incognita* on susceptible and resistant tomato plants, Nematol. medit., 11:107.

Zacheo, G. and Bleve-Zacheo, T., 1985a, Superoxide anion generation during the hypersensitive response in plants infested by nematodes, Phytopathol. medit., 23 (in press).

Zacheo, G. and Bleve-Zacheo, T., 1985b, Generation of superoxide anion and superoxide dismutase activity in tomato plants infected by nematodes, Elsevier Science Publishers R.V., Amsterdam.

Zacheo, G. and Bleve-Zacheo, T., 1986, Cyanide-resistant respiration in tomato roots infested by *Meloidogyne incognita*. (in press).

DISCUSSION

Kerry asked if the effect of *Heterodera goettingiana* on rhizobium was similar in susceptible and resistant pea lines, but Zacheo explained that he had not worked on this aspect. De Haan asked if the defence mechanism was specific for particular host/nematode relationships. Zacheo replied there is a similar general mechanism, which he had been studying. The biochemical changes depend on different host/nematode combinations and such changes are likely at the feeding sites. Wyss observed that as the changes are localised in the feeding site it is necessary to remove the nematodes from the sites before the localised chemical changes in the plant tissues can be analysed. Zacheo said that it is possible to take cells from around the necrotic areas for comparison. Trudgill asked whether Zacheo was satisfied that the explanation of resistance can be centred on the increase in phenylglycoside in the plant and differences in the phenylglycosidase activity in the nematode. Zacheo replied that glycosides cannot be central to the entire mechanism (which is not fully understood). The defense mechanism is certainly more complicated than Giebel envisaged.

CONCEPTS OF RESISTANCE, TOLERANCE AND SUSCEPTIBILITY IN RELATION TO CYST NEMATODES

David L. Trudgill

Scottish Crop Research Institute
Invergowrie, Dundee, DD2 5DA
Scotland

INTRODUCTION

Where one organism lives on or within and is dependent upon another there may be a range of relationships from symbiosis (mutualism) to predator/prey. Between are the parasite/host and, if the parasite damages its host, the pathogen/host relationship. Cyst nematodes cause damage and are therefore pathogens, but Koch's postulates cannot readily be applied to them; in part this is because they multiply and spread much less rapidly than many other plant pathogens and, in part, because the damage they cause is usually proportional to their population density at planting.

Definitions

Nematologists assess the resistance of a cultivar by its effect on nematode multiplication. In most other branches of plant pathology the numbers of the pathogens cannot readily be determined and disease symptoms are used instead. Because of these differences Nematologists use the terms relating to resistance and tolerance with meanings which often differ from those of other pathologists. However, all branches of plant pathology use resistance to describe the ability of the host to restrict or hinder pathogen invasion, development, or multiplication. Tolerance describes the extent to which the host is able to withstand infection without undue damage (Robinson, 1969). In nematology, resistance does not imply non-invasion so tolerance may be independent of resistance (Schaffer, 1971). The National Academy of Science (Anon, 1968) defines tolerance as "the ability of a host plant to survive and give satisfactory yields at a level of infection that causes economic loss to other varieties of the same host species". Resistance and tolerance of cyst nematodes is usually less than absolute and therefore the degree to which a cultivar possesses these attributes is often expressed in relative terms. The extremes of resistance and tolerance of cyst nematodes in the host may be represented in a four-celled chart:

		Host growth: Good	Host growth: Poor
Pathogen reproduction	Good	Tolerant/ non-resistant	Non-tolerant/ Non-resistant
	Poor:	Tolerant/ resistant	Non-tolerant/ resistant

The term "susceptible" is often used instead of non-resistant, but the two terms have subtly different meanings. Susceptibility is defined as the sum total of the qualities which make a plant a fit host for the pathogen (Robinson, 1969) and therefore relates to the positive effects e.g. larger root system, of the host on nematode multiplication whereas resistance relates to the negative effects.

Resistance and tolerance are terms which relate to the host; complementary terms describe the pathogen. Again, Nematologists use these terms with meanings which differ from those of other pathologists. Nematologists generally define differences in the ability of nematodes to multiply on a resistant genotype as differences in virulence. However, most plant mycologists and bacteriologists use virulence when discussing the ability of the pathogen to overcome vertical resistance alone (Vanderplank, 1963; Trudgill et al., 1985), and use aggressiveness to describe its ability to overcome horizontal resistance.

Differences in the capacity of nematodes to damage their host (i.e. complementary to tolerance/non-tolerance) are referred to as differences in pathogenicity. Often, virulence and pathogenicity go together but, just as tolerance can be independent of resistance, so pathogenicity can be independent of virulence or aggressiveness.

A four-celled chart for the terms relating to the nematode pathogen is:

		Pathogen reproduction	
		Good	Poor
Host growth	Good:	Non-pathogenic/ virulent	Non-pathogenic/ non-virulent
	Poor:	Pathogenic/ virulent	Pathogenic/ non-virulent

The terms relating to the plant and the nematode are relative and may be used with different meanings by different people. It is therefore essential that their meaning should always be defined when they are first used.

To summarise, the yield loss caused over several generations by cyst nematodes is the product of several factors. These include the virulence or aggressiveness and pathogenicity of the nematode population or species involved and the degree of resistance, susceptibility and tolerance of the host.

Pathotypes

Resistance and virulence are genetically determined and Flor (Flor, 1942; 1971) proposed that in virulent pathogens there are genes which are complementary to those for resistance in the host: the gene-for-gene hypothesis. A pathotype (host-specific race - see Cook and Evans, 1986) of a nematode is a biological entity which is distinguished by its inherent ability (or inability) to multiply on a given host genotype which embodies one or more genes for resistance. Andersen and Andersen (1982) and Trudgill (1986a) proposed that the genetic basis for resistance in the host should be known and that all individuals of the complementary pathotype should have an identical genetic base for virulence which differs from that found in other pathotypes.

Pathotype schemes are used within four species of cyst nematodes. That for *Heterodera avenae* Woll. is relatively soundly based, but those for *H. glycines* Ichinohe and the two species of potato cyst nematodes (PCN) are not. The pathotype scheme of Kort et al. (1977) for PCN does not fulfil Andersen and Andersen's criteria (Trudgill, 1986a); some of the clones used to differentiate pathotypes have resistance based on unknown genes derived from *Solanum vernei*. Moreover, their pathotypes are defined in such a way that individual nematodes may differ in their virulence (Trudgill, 1986a). However, we do not yet know whether the resistance in clones derived from *S. vernei* is horizontal, vertical, or part both, (for a recent review see Vanderplank, 1982) and whether it will be possible to select for the major changes in virulence (and hence pure pathotypes) often associated with vertical (major gene) resistance.

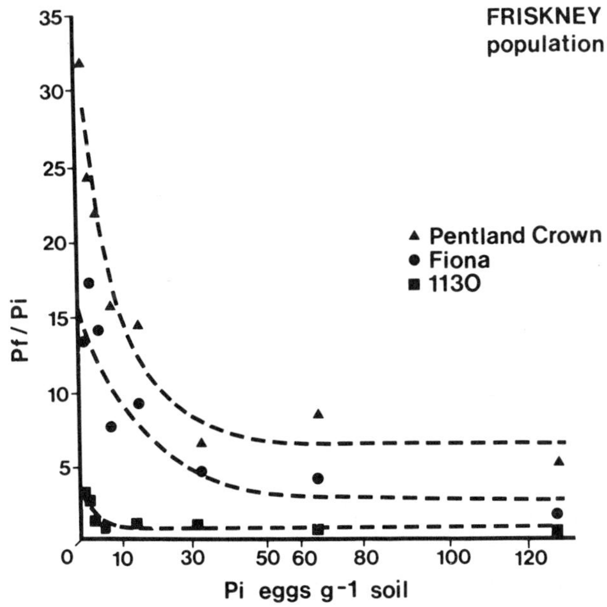

Fig. 1. The relationship between initial population density (P_i) of *Globodera pallida* from Friskney (England) and its multiplication rate (P_fP_i) on a non-resistant (Pentland Crown) and two partially resistant (Fiona, 11305 potatoes. Data from Phillips (1984).

RESISTANCE

Measuring resistance

In a breeding programme major gene (vertical) resistance can be readily identified in a pot test; resistant genotypes have no cysts on their roots whereas non-resistant should have many. Identifying and measuring the resistance of genotypes with only partial resistance is less easy, and defining such resistance is fraught with problems. The Dutch, German and, until recently, the British all defined categories of resistance to PCN in terms of absolute multiplication rates (P_fP_i). However, such an approach is unsound because the absolute multiplication rates of cyst nematodes on non-resistant and partially resistant plants are affected by many factors. Huijsman (1974), Chew (1979), Franco (1982), Phillips (1984) and Seinhorst (1984) all showed that the multiplication rate of cyst nematodes decreases as their population density increases (Fig. 1). In addition, Kort et al. (1977), Phillips et al. (1979) and Seinhorst (1984) reported that even in experiments using similar initial population densities and procedures and

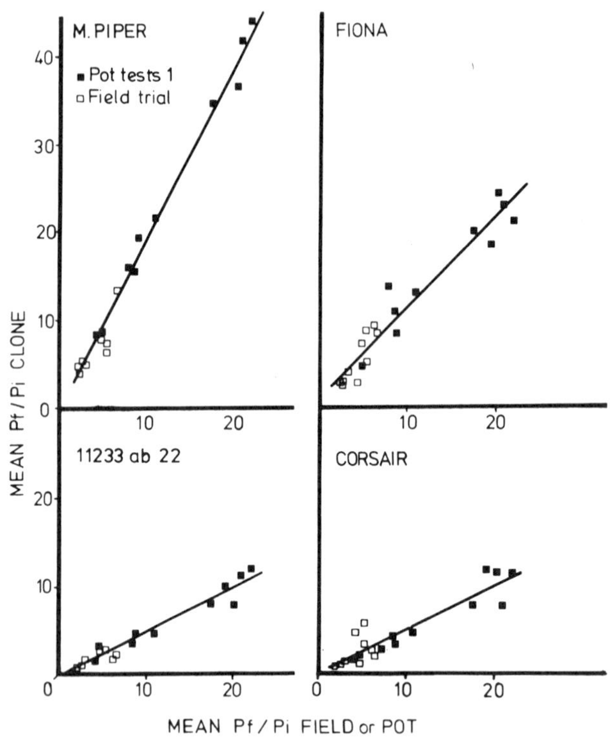

Fig. 2. The multiplication rate (P_fP_i) of *Globodera pallida* in a field test and a pot test on four potato cultivars/clones with different degrees of resistance in different environments as determined by the mean (P_fP_i) for all four cultivars/clones. The open and closed squares are for the field and pot results respectively. Data from Phillips and Trudgill (1986).

identical host/pathogen genotype combinations the over-all rates of multiplication varied greatly.

Another approach, which takes account of these variations, has been suggested by Phillips (1984, 1985) and Phillips and Trudgill (1986); they used genotype by environment interaction analyses to show that the ranking of a series of partly resistant cultivars remained constant even when experiments were done in different environments (e.g. pots and the field) (Fig. 2) or where there were large differences in over-all multiplication rates (Fig. 3). They therefore argued that, when testing unknown genotypes, cultivars of known resistance can and must be used as internal standards to delineate categories of resistance.

Expressing the numbers of cysts formed on partially resistant genotypes as a percentage of that on a non-resistant control can also be used to quantify resistance. However, non-resistant cultivars can differ in their susceptibility (Dale and Phillips, 1985) and therefore several should be included as controls. In practice, it would probably be best to use both

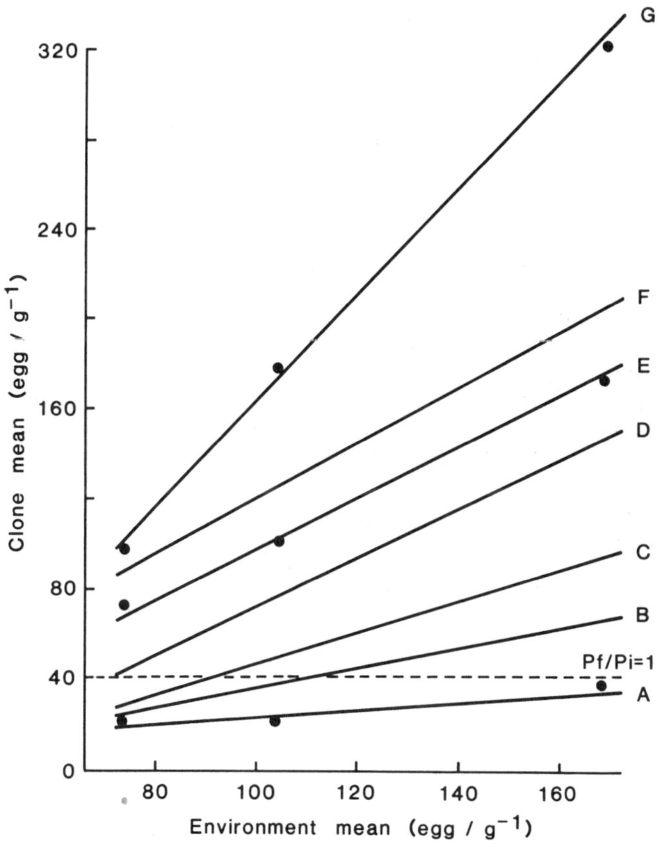

Fig. 3. Final population densities (P_f) of *Globodera pallida* in three experiments on seven increasingly resistant potato clones regressed against the mean P_i in each experiment. The individual values for clones A, E and G are shown. Data from Phillips (1985).

internal standards and percentages when attempting to measure partial resistance.

Measuring virulence

Virulence in the nematode is inversely proportional to resistance in the plant and, when both are incomplete, similar principles and problems apply to their measurement. One approach to estimating the proportion of virulent individuals in a heterogeneous population is to compare the numbers of cysts produced by the unknown with other, standard populations. However, a possible disadvantage of using standard populations is that their virulence may change during their multiplication and maintenance. An alternative is to express the numbers of cysts on the resistant genotypes as a percentage of that on the non-resistant controls. This enables the virulences of several populations to be compared independently of differences in their viability/fecundity (Phillips and Trudgill, 1983).

In tests for resistance or virulence environmental interactions must be minimised (Fisher, 1982; Trudgill et al., 1986a). The initial population density of the inoculum should be such that its potential multiplication rate is close to the maximum and is not affected by differences in host tolerance; Mugniery (1985) suggested that in pot tests with PCN an initial population density as low as 0.25 eggs/g soil should be used.

TOLERANCE

Measuring tolerance

The relationship between initial population density of root-feeding nematodes and the yield of plant dry matter is described by the equation $y = m + (1-m)z^{P-T}$ in which m is the minimum yield which is not affected by further increases in nematode density, T is the initial nematode population density at which yield is first reduced, P is the nematode population density at planting and z is a constant which probably reflects the pathogenicity of the nematode (see Seinhorst (1975) for a recent review). The yield Y at a given population density is expressed as a proportion of the nematode-free yield. Hence, unless m, z or T change, for any value of P the absolute yield losses should increase as the yield potential of the crop increases.

Theoretically, differences between cultivars in their tolerance of nematode damage could be due to differences in the values of T, z or m. For cyst nematodes the value of T is generally small and changes in T do not affect the slope of the regression. In contrast, changes in m or z do affect the slope of the yield-loss curve (Fig. 4). Huijsman et al. (1969) and Seinhorst and den Ouden (1971) used a range of initial population densities of PCN in pot tests to show that potato cv. Multa was more tolerant than cv. Libertas. Both groups of authors concluded that the difference was due mainly to differences in the value of m (Fig. 5). Chew (1979) used a similar approach to compare the tolerances of several oat genotypes to cereal cyst nematodes (CCN) and he also concluded that differences in tolerance were associated with differences in the value of m (the minimum yield).

Numerous authors have reported substantial differences between potato cultivars in their field tolerance of PCN, but only Trudgill and Cotes (1983a) attempted with limited success, to prepare a field site with a range of population densities. Others have used adjacent sites infested with few and many nematodes (Evans and Franco, 1979; Whitehead et al., 1980;

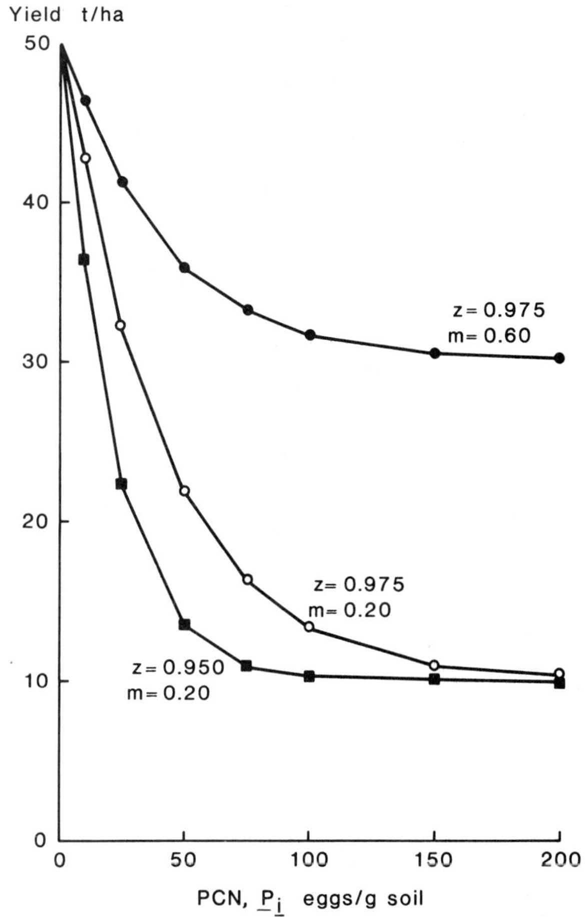

Fig. 4. Effect of different values of z and m in the equation $y = m + (1-m)z^{P-T}$ (see Seinhorst, 1975) on the relation between initial population density of potato cyst nematode and yield of a potato crops with a 50 t/ha potential.

Trudgill et al., 1983; Evans, 1984) or have applied various levels of oxime-carbamate nematicides at heavily infested sites and compared yield increases (Whitehead et al., 1980, 1984; Trudgill and Cotes, 1983a, 1983b; Trudgill et al., 1983; Trudgill et al., 1985). This last approach has obvious dangers, but Trudgill (1986b) has argued that, provided other pests are absent or are controlled by other means, oximecarbamate nematicides applied at heavily infested sites can be used to screen routinely for differences in tolerance. More recent studies (Phillips and Trudgill, unpublished) indicate that routine screening of breeders' clones for tolerance can be done more simply by comparing the untreated yields of single, widely-spaced plants with those of known standards in randomised block trials with ten or more replicates.

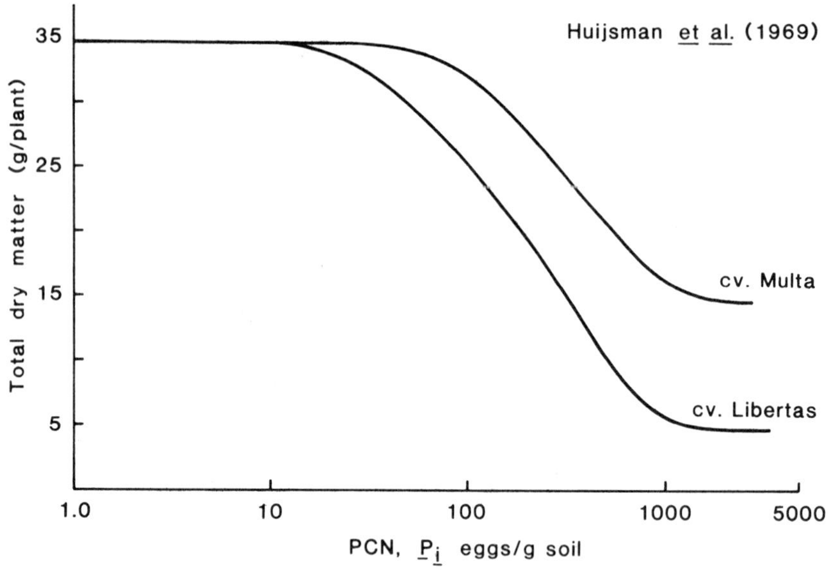

Fig. 5. The relationship between initial population density (P_i) of potato cyst nematode and total dry matter of two potato cultivars in pots. Data from Huijsman et al. (1969).

Relationship between resistance and tolerance

Whitehead et al. (1980, 1984) suggested that resistant potato plants could be expected to be more tolerant than susceptible because fewer juvenile nematodes invade their roots. However, detailed studies on oats (Chew, 1979) and potatoes (Forest et al., 1986) have shown that the roots of resistant and non-resistant cultivars are invaded equally. The reason why fewer cysts are found on the resistant cultivars is because more juveniles leave the roots and return to the soil, more died in situ, or more became male (Mugniery, 1985; Forrest et al., 1986).

The tolerance of resistant cultivars may also be greater than that of susceptible cultivars because the parasitic effect due to the withdrawal of food by the developing females is less. Trudgill et al. (1983, 1985) reported some evidence for this effect; the potato cv. Maris Piper was more tolerant of *Globodera rostochiensis* (Woll.) Behr., to which it possesses gene H resistance, than of *G. pallida* (Stone) Behr. to which it is not resistant. However, resistance does not always confer increased tolerance. The potato cv. Cara which, like cv. Maris Piper, also has H resistance to *G. rostochiensis* but is not resistant to *G. pallida* was equally tolerant of both species (Trudgill et al., 1983) whereas clone 11234, which has partial resistance to both species of PCN, was tolerant of neither (Trudgill et al., 1985). Similarly, Chew (1979) found no relationship in oats between resistance and tolerance of CCN. It should, however, be recognised that when two partially resistant cultivars which differ in tolerance are grown in heavily infested soil, the more tolerant will grow better and consequently appear to be less resistant.

Environmental effects on tolerance

The effect of cyst nematode damage on harvested yields is often modified by various environmental factors. Evans (1982) showed that some, but not all potato cultivars were markedly less tolerant of PCN in the presence of the fungus *Verticillium dahliae*. Tolerance may be increased by low temperatures at planting which decrease nematode activity proportionately more than crop growth. Increased fertiliser also increases the tolerance of potato to PCN (Trudgill, 1986b) and Seinhorst and Kozlowska (1977) showed that the value of m increased and plants become more tolerant the longer infection was delayed.

CONCLUSIONS

Resistance and tolerance are both relative concepts. They are independent; resistance measures plant effects which restrict nematode multiplication and tolerance measures the plant's ability to yield well when attacked by nematodes. Virulence and pathogenicity respectively define the nematode's ability to overcome resistance in the plant and its capacity to cause damage. Agressiveness is a term which is not widely used by nematologists but it describes the ability of a pathogen to overcome "horizontal" resistance. Usually these factors are often quantified by reference to known standards.

REFERENCES

Anderson, S., and Anderson, K., 1982, Suggestions for determination and terminology of pathotypes and genes for resistance in cyst-forming nematodes, especially *Heterodera avenae*, EPPO Bull., 12:379.

Anon., 1968, Plant-disease development and control, National Academy of Sciences Publication, 1956.

Chew, B., 1979, The study of genetic control of resistance to nematodes in oats (*Avena* spp): with observations on the nature of resistance and plant growth response to cereal cyst nematode, *Heterodera avenae*, Ph.D. Thesis, Aberystwyth, University College of Wales.

Cook, R., and Evans, K., 1986, Resistance and tolerance, in "Principles and Practice of Nematode Control", R.H. Brown and B.R. Kerry, eds., Academic Press, London.

Dale, M.F.B., and Phillips, M.S., 1985, Variation for the degree of susceptibility to the potato cyst nematode (*Globodera pallida*) with *tuberosum* ssp. *tuberosum*, Potato Research, 28:55.

Evans, K., 1982, Interactions between *Globodera pallida, G. rostochiensis* and *Verticillium dahliae* on four potato cultivars, Nematologica, 28:143.

Evans, K., 1984, Effects of infestation with *Globodera rostochiensis* (Wollen weber) Behrens Ro_1 in the growth of four potato cultivars, Crop Protection, 1:169.

Evans, R., and Franco, J., 1979, Tolerance to cyst-nematode attack in commercial potato cultivars and some possible mechanisms for its operation, Nematologica, 25:153.

Fisher, J.M., 1982, Towards a consistent laboratory assay for resistance to *Heteroders avenae*, EPPO Bull., 12:445.

Flor, H.H., 1942, Inheritance of pathogenicity in a cross between physiological races 22 and 24 of *Melampsora lini* , Phytopathology, 35:5.
Flor, H.H., 1971, Current status of the gene-for-gene concept, Ann. Rev. Phytopath., 9:275.
Forrest, J.M.S., Trudgill, D.L., and Cotes, L.M., 1986, The fate of juveniles of *Globodera rostochiensis* Ro_1 in roots of susceptible and resistant potato cultivars with gene H_1, Nematologica, (in press).
Franco, J., 1982, Potato cyst nematode, Annual Report for 1982, International Potato Centre, Péru, 56 pp.
Huijsman, C.A., 1974, Host plants of *Heterodera rostochiensis* Woll. and the breeding for resistance, EPPO Bull., 4:501.
Huijsman, C.A., Klinkenburg, C.H., and den Ouden, H., 1969, Tolerance of *Heterodera rostochiensis* Woll. among potato varieties and its relation to certain characteristics of root anatomy, Eur. Potato J., 12:134.
Kort, J., Ross, H., Rumpenhorst, H.J., and Stone, A.R., 1977, An international scheme for identifying and classifying pathotypes of potato cyst-nematodes *Globodera rostochiensis* and *G. pallida*, Nematologica, 23:333.
Mugniery, D., 1985, Resistance des solanacees *Globodera pallida* et influence de certains facteurs sur les resultats des tests de resistance, EPPO Bull., 15:199.
Phillips, M.S., 1984, The effect of initial population density on the reproduction of *Globodera pallida* on partially resistance potato clones derived from *Solanum vernei*, Nematologica, 30:57.
Phillips, M.S., 1985, Environmental differences and their effect on the assessment of quantitative resistance to potato cyst nematodes, EPPO Bull., 15:179.
Phillips, M.S., and Trudgill, D.L., 1983, Variations in the ability of *Globodera pallida* to produce females on potato clones bred from *Solanum vernei* or *S. tuberosum* ssp. *andigena* CPC 2802. Nematologica, 29:217.
Phillips, M.S., and Trudgill, D.L., 1986, A comparison of the assessment of partial resistance of potato clones to different populations and densities of *Globodera pallida* in a pot test and field trial, Nematologica, (in press).
Phillips, M.S., Forrest, J.M.S., and Hayter, A.M., 1979, Genotype x environment interaction for resistance to the white potato cyst nematode (*Globodera pallida*, pathotype E) in *Solanum vernei* x *S. tuberosum* hybrids, Euphytica, 28:515.
Robinson, R.A., 1969, Disease resistance terminology, Rev. appl. Mycol., 48:593.
Schaffer, J.F., 1971, Tolerance to plant disease, Ann. Rev. Phytopath. 9:235.
Seinhorst, J.W., 1975, Dynamics of the plant nematode system, in "Nematode Vectors of Plant Viruses", F. Lamberti, C.E. Taylor and J.W. Seinhorst, eds, Plenum Press, London.
Seinhorst, J.W., 1984, Relationship between population density of potato cyst nematodes and measured degrees of susceptibility (resistance) of resistant potato cultivars and between this density and cyst content in the new generation, Nematologica, 30:66.
Seinhorst, J.W., and den Ouden, H., 1971, The relation between density of *Heterodera rostochiensis* and growth and yield of two potato varieties, Nematologica, 17:347.

Seinhorst, J.W., and Kozlowska, J., 1977, Damage to carrot by *Rotylenchus uniformis*, with a discussion on the cause of increase of tolerance during the development of the plant, Nematologica, 23:1.

Trudgill, D.L., 1986a, Potato cyst nematodes: a critical review of the current pathotype scheme, EPPO Bull. 15:273.

Trudgill, D.L., 1986b, Prospects for decreasing yield losses caused by potato cyst nematodes, Ann. appl. Biol., 108:181.

Trudgill, D.L., and Cotes, L.M., 1983a, Differences in the tolerance of potato cultivars to potato cyst nematodes (*Globodera rostochiensis* and *G. pallida*) in field trials with and without nematicides, Ann. appl. Biol., 102:373.

Trudgill, D.L., and Cotes, L.M., 1983b, Tolerance of potato cyst nematodes (*Globodera rostochiensis* and *G. pallida*) in relation to the growth and efficiency of the root system, Ann. appl. Biol., 102:385.

Trudgill, D.L., Holliday, J.M., Mathias, P.L., French, N., Mackintosh, G.M., and Tones, S.J., 1983, Effect of the nematicide oxamyl on the multiplication of *Globodera rostochiensis* and *G. pallida* and on the haulm growth and yield of six potato cultivars with different levels of resistance and tolerance, Ann. appl. Biol., 103:477.

Trudgill, D.L., Mathias, P.L., and Tones, S.J., 1985, Effects of three rates of aldicarb and of different degrees of resistance and tolerance in potato cultivars on yield and post harvest population densities of the potato cyst nematodes *Globodera rostochiensis* and *G. pallida*, Ann. appl. Biol., 107:219.

Vanderplank, J.E., 1963, Plant Diseases: epidemics and control, Academic Press, New York.

Vanderplank, J.E., 1982, Host-pathogen interactions in plant disease, Academic Press, New York.

Whitehead, A.G., Tite, D.J., Fraser, J.E., French, E.M., and Short, L., 1980, Effects of aldicarb and oxamyl in peaty loam soils on potato cyst nematode, *Globodera rostochiensis* and on resistant and susceptible potatoes, J. Agric. Sci., 95:213.

Whitehead, A.G., Tite, D.J., Fraser, J.E., and Nichols, A.J.F., 1984, Differential control of potato cyst-nematodes, *Globodera rostochiensis* and *G. pallida* by oxamyl and the yields of resistant and susceptible potatoes in treated and untreated soils, Ann. appl. Biol., 105:231.

EFFECTS OF NEMATODE ATTACK ON THE GROWTH AND YIELD OF CROP PLANTS

J. Willem Seinhorst

formerly Instituut voor Plantenziektenkundig Onderzoek
Wageningen
The Netherlands

INTRODUCTION

Although it has been obvious from the first investigations on nematodes reducing the yield of crop plants (notably beet cyst nematode) that the extent of the damage was related to the degree of infestation of the soil, the exact nature of this relation remained vague in the minds of most investigators. Even so efforts to advise farmers on crop rotation against cyst nematode attack had already started in the first half of this century, most probably because the possibility of counting cysts in soil samples prompted investigators to attach a meaning to the numbers found. Hellinga (1942) even tried to determine a relation between numbers of *Heterodera schachtii* Schmidt cysts in samples taken in farmers' fields before sowing beets and the decrease in the quantity of sugar produced. A highly ambiguous approach to the problem prevailed. However, little research was done on quantitative relations and the uncertainty of these relations, due to emphasis on effects of varying external conditions but without proper demonstration. Nevertheless, an increasing number of plant parasitic nematode species were claimed, in part rightly, to be of economic importance; but again, often without convincing evidence that the nematodes reduced the yields of crop plants at the population densities occurring in the field, or even that they could reduce yields at all. Even for such extensively investigated species as potato and beet cyst nematodes and *Meloidogyne* species it remained obscure how much growth reduction could be expected at particular population densities, especially the smaller ones. Nevertheless, facilities were created to give advice to farmers on the basis of numbers of nematodes found in soil samples. As sampling errors were also unknown, classifications of degrees of infestion were sometimes used with class widths considerably smaller than the variability of the nematode counts in the samples.

Concepts of the mechanism which give rise to yield losses remained myths and fairy tales - extraction of food by the nematodes, blocking of vessels by transfer cells as a cause of the wilting of beets or mechanical damage to or killing of roots by the nematodes. None of these was ever demonstrated to be of importance. On the contrary, plant weights may be reduced by as much as 50 per cent without any visible signs of damage to the root system. Often in textbooks visible symptoms are related to the damaging characteristics of nematodes but very few of them contain even a bare minimum of discussion about the relation between nematode density and crop yield.

PATTERN OF RELATIONS BETWEEN NEMATODE POPULATION DENSITY AND PLANT WEIGHT

Assuming that a better insight into the relation between nematode population density and the growth and weight of plants could only be obtained if a general pattern applying to all cases could be found, Seinhorst (1960, 1965) collected all relevant data in the literature. These suggested an exponential relationship between nematode density at the time of sowing or planting and plant weight or length, resulting in similar S-shape curves in graphs of the relation between log nematode density and plant weight or length. In some cases the data fitted more or less to a relation according to the equation $y = z^P$, in which y is the ratio between plant weight and nematode density P and that in the absence of nematodes z is a constant < 1. However, a general relation fitting well to (almost) all observed relations emerged only after the introduction of two new concepts, a nematode density T, the tolerance limit, and the relative plant weight m, the minimum yield. The first serves to comply with the upward deviations of plant weights at small nematode densities from those according to a purely exponential relation between nematode density and plant weight. It can be interpreted as the density below which no reduction of growth and weight of the plant occurs, and also as the largest density at which the growth reducing effect(s) of the nematode are neutralised or eliminated completely by the plant, leaving the effect of a density $P - T$ reducing growth and yield at nematode densities > T. The minimum yield m was introduced to allow for the observation that in most experiments the ratio between plant weight at nematode density P and that in the absence of nematodes at large P (> 100T) seemed to approach to a minimum value, which often exceeds 0 considerably or at least that such a value must be supposed to apply where no observations at such large population densities were available, to obtain a good fit of the data to the general relation. Modifying the exponential relation with the two concepts led to the general relation

$$\left.\begin{array}{ll} y + m + (1 - m)z^{P-T} & \text{for } P > T\text{, and} \\ y = 1 & \text{for } P \quad T \end{array}\right) \quad \ldots. \ (1)$$

in which y = the ratio between plant weight at nematode density P at the time of sowing or planting and that in the absence of nematodes, z^{-T} = 1.05 (z^T = 0.95) in almost all cases (Seinhorst, 1965, still assumed z^{-T} = 1.1) so that

$$z^{P-T} = 0.95^{PT^{-1}-1},$$

and m = the minimum yield. According to Fig. 1, which brings the observations in thirteen experiments with different plant and nematode species together, there is a good fit of these observations to a relation between nematode density and plant weight according to equation (1).

The tolerance limit T has little to do with economic thresholds and the concept should be used only in relation to equation (1) and determined by adapting a curve according to this equation to the observations. Therefore T is a measure of the tolerance of the plant to the growth reducing effect of attack by the nematode species under investigation and under the conditions during the growth of the plant. The value of z^T can vary from 1 to 0.9. However, to get comparable estimates of T, it should be

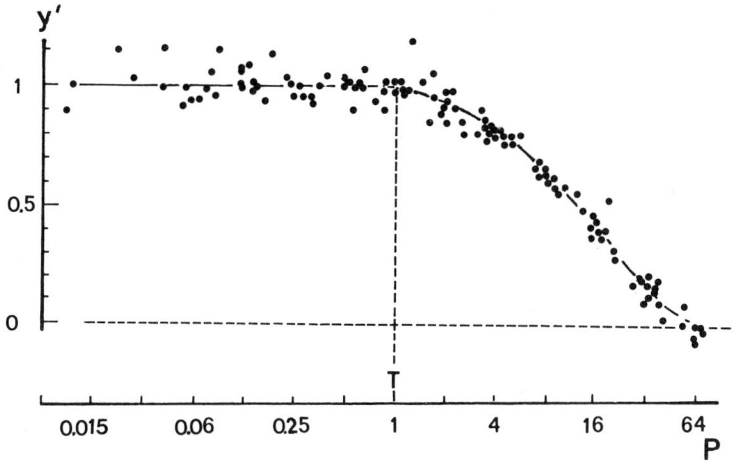

Fig. 1 The relation between nematode density at sowing or planting, P, and the weight of plants after a certain growing time (mostly several months) in thirteen experiments with several tylenchid nematode species and several plant species. Plant weights were plotted separately against log P for each experiment. Then curves were fitted according to equation (1) $y = m + (1 - m)0.95^{PT^{-1}-1}$ to these data (see Fig. 9), thus estimating values of T and m for each experiment separately. Then all values y were

$$y' = (y - m_x)(1 - m_x)^{-1}$$

m_x being the value of m in the experiment relating to y. Values of y' were again plotted separately against log P for each experiment, keeping the scale of log P and of y' the same. Fig. 1 was then obtained by superposition of all graphs with all T_x coinciding. The transformation of y to y' also transforms equation (1) to

$$y' = 0.95^{PT^{-1}-1}$$

which with all T_x coinciding, results in a single curve in Fig. 1.

standardised as the value for $z^T = 0.95$, this being the most common value in the xperiments analysed so far.

GROWTH AND GROWTH REDUCTION MODEL TO EXPLAIN GENERAL PATTERN

Seinhorst (1979a, 1981c) based a model to explain the pattern expressed by the arbitrary equation (1) on the following assumptions.

First, the ratio c_P of the rates of growth of plants of the same size (weight) at nematode density P and in the absence of nematodes, r_P and r_0 respectively, is independent of the size (weight) of these plants (although growth rates depend on plant size). In other words the growth reducing

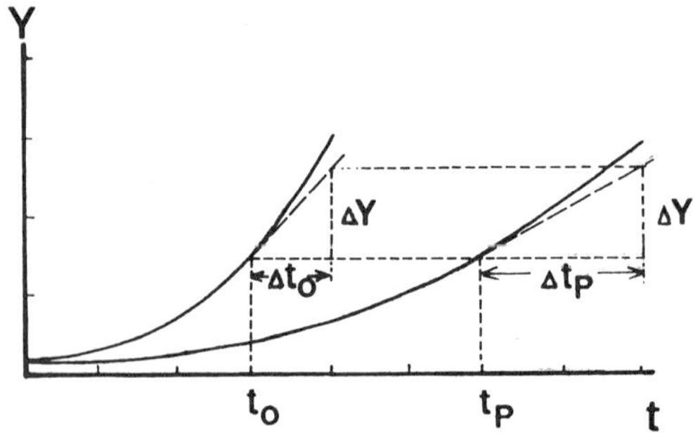

Fig. 2 Growth curves of a plant at two nematode densities with constant ratio

$$r_P r_0^{-1}$$

Y = plant weight, t = time after sowing or planting, t_0 and t_P = times plant must grow to obtain a certain weight at two different nematode densities, r_0 and r_P are the growth rates of plants of the same weight at nematode densities 0 and P.

effect of a nematode population remains the same during the growth of the plant, provided of course that growth conditions remain the same.

Second, $r_P r_0^{-1} = k + (1 - k)z^P$ .... (2)

with k and z constants <1 (adaptation of Nicholson's competition model).

The first assumption leads to a ratio between the time, as measured from the beginning of the attack by the nematodes, that the plant needs to reach a certain weight in the absence of nematodes and at nematode density

P, $t_0 t_P^{-1} = r_P r_0^{-1}$ .... (3)

(Fig. 2): if the same very small increase ΔY of the weight Y of a plant is obtained in a time Δt_0 in the absence of nematodes and in time Δt_P at a nematode density P then $\Delta Y = r_0 \Delta Y = r_0 \Delta t_0 = r_P \Delta t_P$. If $r_P = c_P r_0$ then $\Delta Y = c_P r_0 \Delta t_P = r_0 \Delta t_0$ from which $\Delta t_0 = c_P \Delta t_P$. This relation applies to all small time intervals for which $r_0 \Delta t_0 = r_P \Delta t_P$. Therefore

$$c_P \Sigma \Delta t_P = c t_P = \Sigma \Delta t_0 = t_0 \text{ and } t_0 t_P^{-1} = c_P = r_P r_0^{-1}$$

for all plant weights and irrespective of the growth curve of the plants without nematodes. The difference between the two is only one of the length of the unit for the time scale, that for nematode density P being $k + (1 - k)z^{P-1}$ times that in the absence of nematodes, if the second assumption also applies.

According to the two assumptions mentioned above P is constant at a given initial nematode density. This implies that growth and multiplication of the nematodes in the roots do not affect the magnitude of the growth reduction. This would be the case if a nematode affects growth only during a short period after having penetrated into a root, e.g. for cyst nematodes until the vascular tissue near the head of the juvenile has started to become a transfer cell. This could also apply to migratory and ectoparasitic nematodes as these will penetrate mostly in new roots and, therefore, not far from root tips. The first supposition also implies that at a given value of P the number of nematodes reducing the growth rate of a plant is at all times t proportional to the volume of root in which the growth reducing activity is being produced. This could certainly not be expected to apply to cyst nematodes with more than one generation in a season, e.g. *H. schachtii* and *H. glycines* Ichinohe and most *Meloidogyne* species. Little thought has been given so far to the effects of second and later generations of nematode populations on growth and development of plants (Seinhorst, 1978).

The constancy of the ratio $t_0 t_P^{-1}$ could be tested and was confirmed in two experiments, one with *H. avenae* Woll. on oats, comparing growth at one nematode density with that in the absence of nematodes from sowing until the emergence of the first panicles (Seinhorst, 1979b), and the other with two densities of *Globodera pallida* (Stone) Behr. causing growth reduction and four not doing so on cultivar Irene (Been and Schomaker, 1986).

Equation (3) suggests that exposing plants to different nematode densities desynchronises their development. Characteristics of plants of the same age (i.e. mineral contents, shoot to root weight ratios) should not be compared uncritically. comparison of characteristics of plants of the same weight may be more appropriate (Seinhorst, 1979b; Been and Schomaker, 1986).

The second assumption is formally identical with Nicholson's competition theorem (proportion of discrete units of space not occupied by P randomly distributed individuals $y' = z^P$, ($z < 1$) for the proportion $1 - k$ of the space within actual reach of the population). It is also used as an expression of the effect of intraspecific competition on the proportion of the space in the root system occupied and that of the food source exploited by a cyst nematode population (Seinhorst, see elsewhere in this publication). However, the mechanism by which intraspecific competition leads to a relation between nematode density and the reduction of the growth rate expressed by the second assumption is not obvious. It could be obtained in different ways. With cyst nematodes it could not be due to a decrease with increase in nematode density of the proportion of the second stage juveniles present in the soil near root tips and that penetrate into these root tips. According to Jones and Kempton (1978) this proportion is constant up to large population densities. Possibly it does not make much difference in terms of the growth reduction caused, whether one or more nematodes entered a single root tip simultaneously or almost so. Then only the proportion of invaded root tips at a given moment would determine the rate of growth at that moment. However, this probably requires far more root tips per unit volume of soil than actually occur. Another possibility is that the presence of a nematode triggers a certain number of cells surrounding it to produce a growth reducing substance or to stop that of a growth promoting one but that a second nematode cannot further affect the production of these cells. Growth is not directly related to the volume of transfer cells present in a root system. Also the growth of resistant plants (no transfer cells) can be reduced strongly by cyst nematode attack. The relation of the volume of transfer cells to initial nematode population density only becomes similar to that between

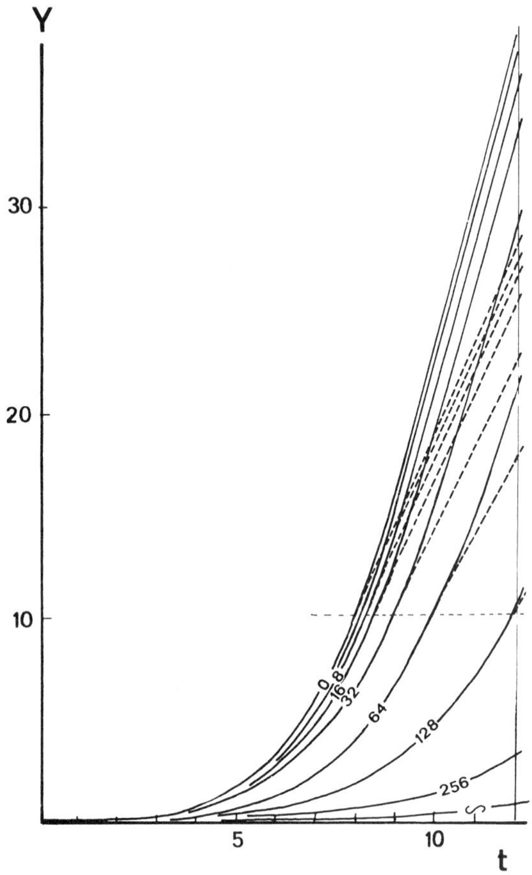

Fig. 3 Growth curves, related as in Fig. 2, of a plant at a range of relative nematode densities (figures with curves). Y = plant weight, t = time for sowing or planting. The curves for growth in the absence of nematodes (solid line: fast growth, broken line: slower growth from Y = 10 on) are arbitrary but very similar to those representing actual growth of potatoes and oats in pot experiments. Growth rates of plants of the same weight in the absence of nematodes (r_0) and at nematode densities P (r_P) are related according to equation (2):

$$r_P = r_0\ k + (1 - k)z^P \text{ with } k = 0.25$$

growth reduction and the latter some time after growth reduction became apparent.

Values of $k > 0$ indicate that there is a minimum growth rate which cannot be reduced further by the mechanism of growth reduction described by equation (2), either because external conditions prevent the invasion of roots for some time after sowing or planting or even the occupation of all available sites is insufficient to stop growth completely.

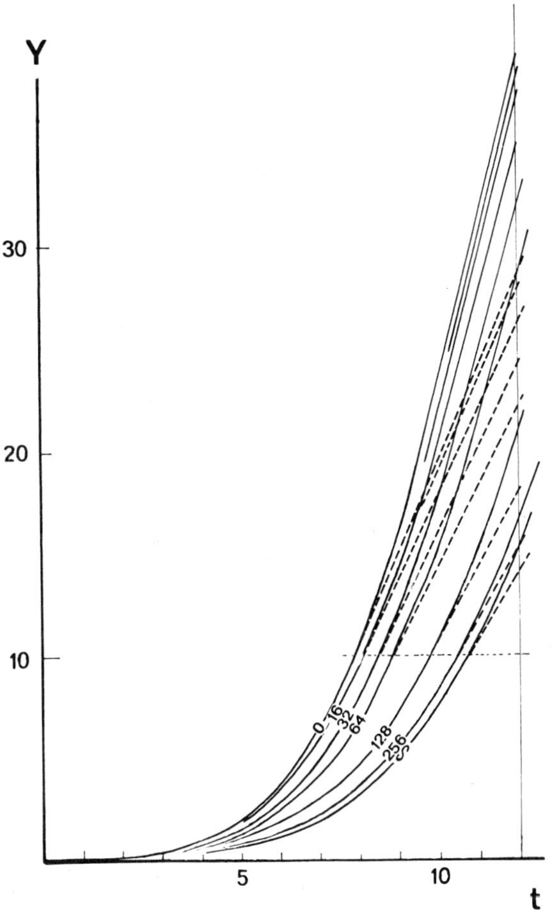

Fig. 4 Growth curves, related as in Fig. 2, of a plant at a range of relative nematode densities (figures with curves). y = plant weight, t = time for sowing or planting. The curves for growth in the absence of nematodes (solid line: fast growth, broken line: slower growth from y = 10 on) are arbitrary but very similar to those representing actual growth of potatoes and oats in pot experiments. Growth rates of plants of the same weight in the absence of nematodes (r_0) and at nematode densities P (r_P) are related according to equation (2):

$$r_P = r_0\ k + (1 - k)z^P \text{ with } k = 0.7$$

The second assumption was not tested directly as this would require experiments with several hundred pots for each combination of nematode/plant species to be investigated. Instead, the relation between nematode density and plant weight at various times after sowing or planting according to a growth model (Figs. 3 and 4) based on the two assumptions mentioned above was determined and compared with actual relations between nematode density and plant weight. Two combinations of initial exponential growth decreasing with an increase of plant weight by a constant amount per unit

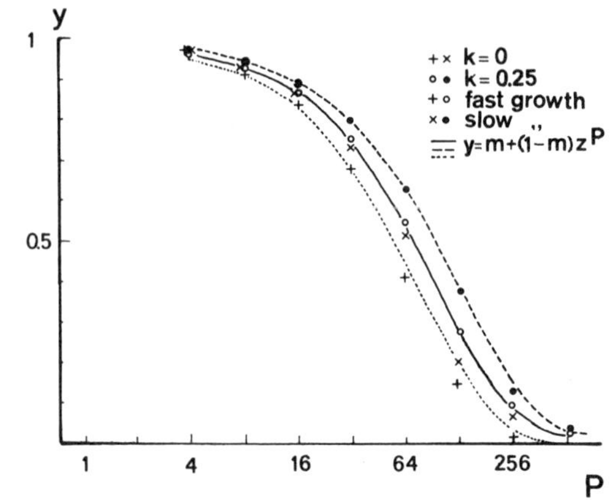

Fig. 5 Relation according to Fig. 3 (k = 0.25) and also for k = 0 derived from Seinhorst, 1979a, Fig. 5) between nematode density P and the ratio y between plant weight at nematode density P and that in the absence of nematodes at t = 12.

duration of time were investigated, with k = 0.25 and k = 0.7. Seinhorst (1979a) investigated the relations for k = 0. The results (Figs. 5 and 6; Fig. 5 in Seinhorst, 1979a) indicate a complete similarity of relative weights according to Figs. 3 and 4 at time t = 12 to those according to the equation

$$y = m + (1 - m)z^{P} \quad .... \ (4).$$

Further measurements indicate that this also applies to smaller values of t. However, for k = 0 (data from Seinhorst, 1979a Figs. 4 and 5) relative plant weights at nematode densities causing weight reductions larger than 50 per cent were a little smaller than according to equation (4). However, such a small value of k apparently is very unlikely as values of m smaller than 0.05 (k < 0.03) are at least very rare.

A much more important discrepancy is that between theoretical and observed relative plant weights at small nematode densities. The latter weights require curves according to an equation containing a factor z^{P-T} (with in most cases z^{T} = 0.95 and, therefore,

$$z^{P-T} = 0.95^{PT^{-1}-1} \text{ instead of } z^{P}$$

(possibly except for the first weeks after sowing). A satisfactory explanation has not yet been found. Seinhorst and Kozłowska (1977) and Seinhorst (1979a) suppose that it is due to a further increase in root density, where all nematodes have already entered older roots. This would occur later at larger than at smaller nematode densities. However, this could reduce weight reduction but not compensate entirely for the effect of an initial delay of growth. Another, but unlikely, explanation would be a slight growth stimulation by the nematodes. Unless this would lead to an exact compensation of the weight reduction at nematode densities $<T$, Fig. 1 contradicts this supposition.

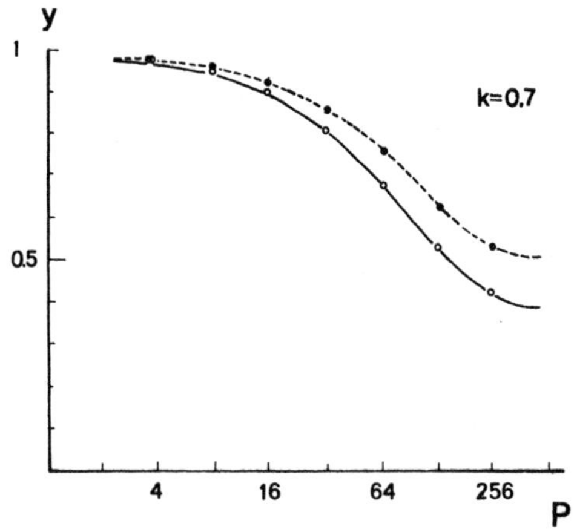

Fig. 6 Relation according to Fig. 3 (k = 0.7) and also for k = 0 derived from Seinhorst, 1979a, Fig. 4) between nematode density P and the ratio y between plant weight at nematode density P and that in the absence of nematodes at t = 12.

Growth retardation ("the same happens later") as suggested by the two suppositions upon which the growth model is based, most probably is a very important, if not the only, component of the effect of nematodes in the early stages of the growth of plants at nematode densities mostly up to 16T or 32T. However, this pure retardation is most probably restricted to the vegetative stages. The delay of the emergence of the panicles in oats attacked by *H. avenae* Woll. was much less than suggested by plant weight at time of emergence, interpreted as the result of growth retardation (Seinhorst, 1981b). Tuber formation of potatoes at various densities of *G. pallida* also began at smaller haulm weights although later than at smaller nematode densities even in a growth room with constant day length and constant day and night temperatures (Been and Schomaker, 1986). The resulting larger tuber to total weight ratio at given total weights than in the presence of non-damaging nematode densities did not affect the applicability of the first assumption of the growth model ($t_P t_0^{-1}$ constant for a given nematode density) to the data obtained in the growth room.

EFFECT OF CHANGE OF CONDITIONS DURING AN EXPERIMENT

In glasshouse and field experiments periodic changes of light intensity and temperature and, outside the tropics, systematic change of daylength can hardly be avoided. Increase of temperature and daylength results in increase of the growth rate of the plants with and without nematodes and therefore reduces the ratio $t_0 t_P^{-1}$ (imitated by making vertical pleats in Figs. 3 and 4). The ratio can be kept constant although by the proper reduction of the unit of time where growth is increased (putting fewer real days in it). Therefore, these changes do not affect the applicability of equation (1). They do, however, when as in an experiment by Seinhorst and den Ouden (1971), plants at small nematode densities cease growing, while those at larger densities continue to become larger. Moreover the shift of haulm growth to longer days by the nematode attack increased their length

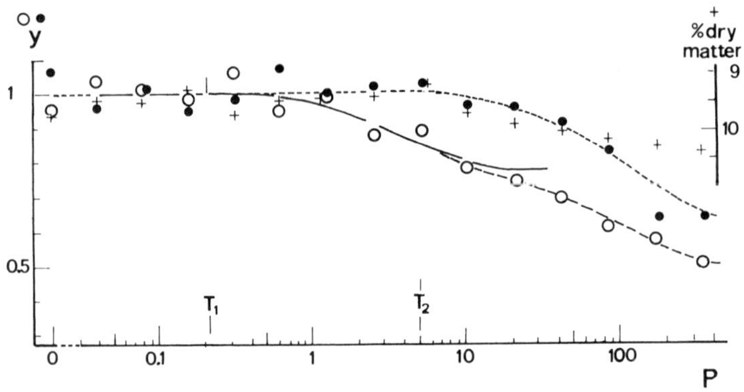

Fig. 7 Relation between population density of *Tylenchorhynchus dubius* (*P* in nematodes per g soil) and O: the relative weight of oat plants 35 days after sowing (*y* = 1 for 0.52 g dry matter per plant) • : relative water consumption per g plant per day (*y* = 1 for 25.8 g per g plant per day, 34 and 35 days after sowing) and + : dry matter content 35 days after sowing. After Seinhorst (1981a).

to weight ratio. In such cases the model must be adapted by introduction of the pattern of growth change into it, as was done by Seinhorst and den Ouden (1971).

A SECOND MECHANISM OF GROWTH REDUCTION

According to equation (1) the range of densities between y = 1 and y = 0.05(1 - *m*) is *T* to 58*T*. No material reduction of plant size would be possible beyond about 70*T*, if there was only one mechanism of growth reduction with the characters described above. However, several experiments with different plant and nematode species indicated a longer range. It was especially clear in an experiment with *Tylenchorhynchus dubius* (Bütschli) on oats in which the value of *m* associated with the nematode density range *T* to about 30*T* was 0.78 (Seinhorst, 1981a), (Fig. 7). With small values of *m* weight reductions over a greater range of nematode densities than could occur according to equation (1) are obscured by experimental error. Contrary to the weight reductions at small nematode densities, those at larger ones are associated with several diagnostic characters - decrease of water consumption per unit weight of plant, increase of dry matter content, necrotic edges of leaves (in certain cases) and, probably, reduction of potassium content. Seinhorst (1981a) concludes that tylenchids cause a reduction of the growth rate of plants by at least two mechanisms. The first mechanism operates at relatively small nematode densities (according to equation (1) over a range *T* to about 64*T*) with no diagnostic characters known other than decrease of relative weight in the vegetative state by retardation of development. The second mechanism generally becomes noticeable at nematode densities 16*T* to 32*T*, with the diagnostic characters described above. Observations on successive cuts of white clover attacked by *H. trifolii* Goffart suggest that sensitivity to the "second mechanism" of growth reduction increases and, therefore, shifts to smaller nematode densities with increase of plant age (Seinhorst, 1981a). Also for the second mechanism the relation between nematode density and plant weight could be described in several cases by equation (1) or equation (4). For the two mechanisms together the relation then becomes

$$y = \{m_1 + (1 - m_1)z_1^{P-T}\}\{m_2 + (1 - m_2)z_2^{P}\}$$
$$= \{m_1 + (1 - m_1)0.95^{PT^{-1}-1}\}\{m_2 + (1 - m_2)0.95^{cPT^{-1}}\} \quad \ldots . \text{ (5)}$$

with c = 16 - 32 in the cases mentioned by Seinhorst (1981a), but c = 8 for potato tuber weights in experiments by Seinhorst et al (1987). The second part of equation (5) is a purely arbitrary and, probably, not always a very accurate description of the effect of the second mechanism of growth reduction.

In potatoes attacked by potato cyst nematodes (*G. rostochiensis* (Woll.) Behr. and *G. pallida*) the relations between nematode density and growth, plant size and tuber weight are even more complicated, mainly because of a phenomenon called "early senescence" by Trudgill et al. (1975). While small nematode densities delay the dates at which the rate of increase of haulm length decreases, larger densities shift it to earlier dates (arrows in Fig. 8). At larger nematode densities it may lead to a much earlier cessation of haulm growth than at smaller ones (Fig. 8, Seinhorst et al., 1987). It is associated with a larger dry matter content of the haulms and, probably, also of the immature tubers than at smaller nematode densities although the former starts to be higher than at smaller densities at a density about a third of that at which "early senescence" becomes apparent. Water consumption per unit weight of senescent but not yet dying haulms was much smaller than that of those still growing (Seinhorst et al., 1986).

In pot experiments there were considerable differences between potato cultivars in sensitivity to "early senescence", and also in the density ranges at which the same cultivars with the same nematode population exhibited "early senescence" in different years. Both susceptible and resistant potato cultivars are sensitive to it.

PRACTICAL IMPORTANCE OF "SECOND MECHANISM OF GROWTH REDUCTION" AND OF "EARLY SENESCENCE"

Apparently the tolerance of plants to nematode attack over the full range of possible population densities of nematode species with one generation per year (*H. avenae, H. trifolii, G. rostochiensis, G. pallida*) and of species multiplying continuously but fairly slowly is expressed by a function with at least four parameters. However, for practical purposes only the relation between nematode density and plant yield in the range of nematode densities up to $20T$ is of importance. Generally it is uneconomic to grow a crop at nematode densities $>4T$. Expected yield losses of more than 10 per cent justify control (by chemical treatment and/or crop rotation) when a field is infested with potato or beet cyst nematodes (Seinhorst, 1982b). The extent of control needed does not relate to the loss expected at the existing nematode density but depends only on what is needed to reduce this density to that required for a maximum return. The latter relation is always one causing growth reduction only by the "first mechanism". Therefore, for practical purposes only the parameters T and m_1 in equation (5) need to be known (Seinhorst, 1982b). Fortunately T so far seems to be independent of external conditions (except extreme ones) and, therefore, a function only of nematode species and plant species or cultivar. It can be determined in pot experiments provided pots of sufficient size are used (10 litres for potatoes with a single haulm to be grown for a full season, 40 litres for beets). There are indications that

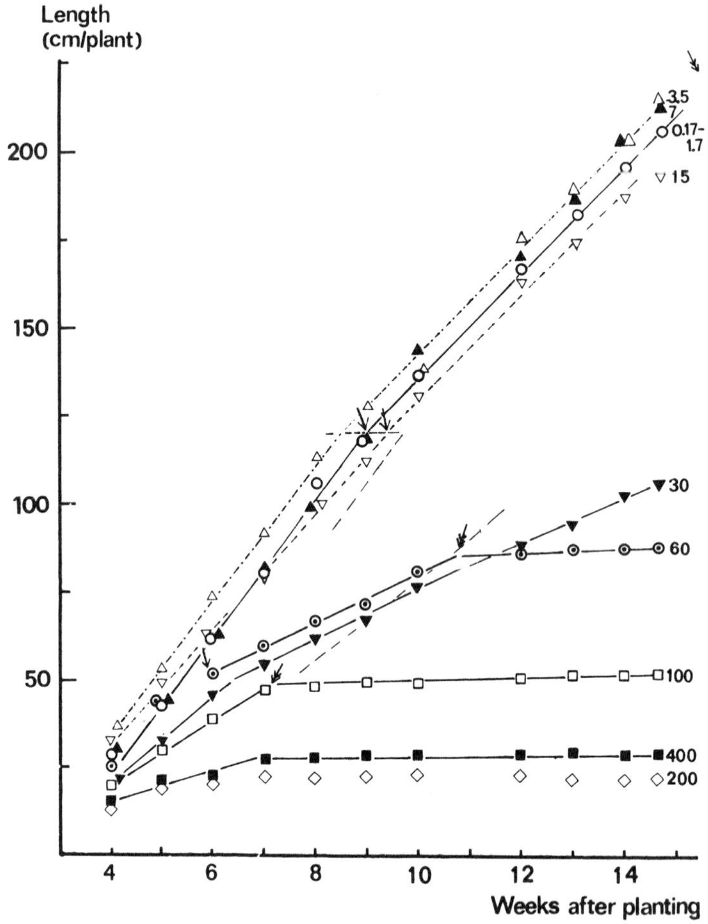

Fig. 8 Haulm lengths at a range of dates of the potato cv. Darwina grown at a range of initial egg densities of *Globodera pallida*, pathotype Pa 3 (numbers in graph are eggs per g soil) demonstrating early senescence at 60 to 400 eggs per g soil. Each point represents the average haulm length of two haulms each in a 10 litre pot, except those at 0.7 to 1.7 eggs per g soil which are averages of eight haulms.

the tolerance of (resistant) potato cultivars for potato cyst nematodes can be determined in experiments of short duration and therefore with pots much smaller than 10 litres.

MORE THAN ONE GENERATION OF CYST NEMATODES PER SEASON

Seinhorst (1978) imitated a second generation by reinoculating pots with oat plants, two months after sowing, with forty times the numbers of *H. avenae* eggs and juveniles (maximum 35 per g soil) present in the soil at sowing. This changed the relation between initial density and relative plant weight with $T = 0.8$ eggs and juveniles per g soil and $m = 0.05$ by superimposing a relation with T = 0.02 eggs and juveniles per g soil (when plant weight was plotted against the densities at sowing) and m = 0.85. However, real second generations will not be present completely at a single date but appear in a wave moving outwards and downwards from the centre of the plant. Seinhorst's (1978) experiment suggests that the growth and plant weight reductions are too small to justify control measures (which have to reduce initial nematode density to less than a fifth of the value of T for the first generation). However, the second generation might cause reactions associated with the "second mechanism of growth reduction" of which the disturbed water management of the plant especially might affect flowering and fruit setting, as is suspected to occur in tomato at initially small root knot nematode densities.

The second generation is produced in soil with already many roots. These may have depleted it of water. As a result hatching, movement of juveniles and penetration into roots may depend much more on external conditions, especially rainfall, than in the first generation, as the first roots of the host grow out into moist soil in spring. More research is needed to elucidate the role of second generations of cyst nematodes in causing (additional) crop losses, especially in beets and soybeans.

ESTIMATING SYSTEM PARAMETERS FOR SPECIFIC NEMATODE-PLANT COMBINATIONS

Values of the parameters T and m in equation (1) or T, m_1, m_2 and c in equation (5) can only be estimated by measuring plant weights at a sufficiently wide range of nematode densities. It can be done in pot and in field experiments. As the values of T are largely independent of external conditions, they can be determined in pot experiments. On the other hand m seems to vary considerably under field conditions (Seinhorst, 1982a). Ideally, the frequency distribution of its values should, therefore, be determined by doing a large number of field experiments, but because of the difficulty of arranging such experiments estimates of the possible variation will, most probably, have to be based on few observations.

Pot experiments

To measure growth reduction of plants caused by nematode attack much larger experiments are required than to determine relations between initial and final nematode densities over a certain period of growth of a plant. If density ranges 0,2^x x c with x = 1, 2, 3,....., are used the variation coefficient of average plant weight per density should not exceed 4 per cent. For potatoes this requires at least five plants (with single haulms) per density. The required variation coefficient for multiplication rates of about 8 per cent can be obtained with two single haulm potato plants in 10 litre pots. With oats five single plants per density of *H. avenae* in two litre pots gave the required accuracy (Seinhorst, 1981a).

The density range should preferably enable an estimate of the weight for y = 1 with a variation coefficient of not more than 2 per cent and, therefore, there should be at least four densities $<T$. With two densities with expected plant weights $<m + 0.05(1 - m)$ this makes twelve densities 0 to 2^xT with x = about -2, -1, 0, 1, 2,.....11, but a longer range is better.

Soil is easily inoculated with cyst nematodes either as cysts with eggs or as egg suspensions. Other species (especially trichodorids and *Longidorus* spp. but also tylenchids) may be difficult to handle. Cyst nematodes do well in almost all soils. Other species only survive in a limited number of soils and even in these their numbers may decrease markedly after inoculation. Mixing soil with and without nematodes in different proportions may be a solution but the soil should be the same and all portions treated similarly whether nematode-free or inoculated with nematodes. Mixing sterilised and unsterilised soil may lead to differences in growth and plant weight in different mixtures even without nematodes. Another drawback of mixing soil with and without nematodes is that only a limited range of densities can be obtained with an acceptable degree of accuracy. According to the published data, weights of plants grown in soil with different nematode densities obtained by manipulations which could also affect plant growth in the absence of nematodes, e.g. partial sterilisation and mixing of soils, generally did not fit to a relation expressed in equation (1), whereas plants grown in properly inoculated soil generally did (Fig. 1).

Pots should be large enough to allow a normal expansion of the root system during the experiment. If they are too small T tends to shift to larger values as soon as the roots have occupied the whole soil volume and all of the nematodes have entered or attached themselves to roots (Seinhorst and Kozłowska, 1977). Probably values of T in many combinations of a plant and nematode species do not change during the development of a plant growing in a sufficiently large pot or in the field. Then T can be determined in experiments of short duration using smaller pots than required for fully developed plants.

Field experiments

The greatest difficulty to surmount when arranging field experiments on the relation between nematode density and crop yield is to find or prepare areas of sufficient size with different nematode densities, avoiding an association of these densities with other effects on yield than by nematode attack. This again rules out chemical soil treatment as a means of manipulating nematode density. Jones (1957) used beet cyst nematode densities in microplots obtained by growing good and poor or non hosts. However, this resulted in hardly more than two different densities. Additional variation in numbers of nematodes counted in samples from the plots was largely sampling error. Thus, Seinhorst's (1965) fitting of curves to these data was not justified. The same applies to the curve fitting by Whitehead et al. (1980) to probably only three potato cyst nematode densities, obtained by chemical treatment followed by growing susceptible and resistant potatoes (Seinhorst, 1982a). These two relative failures do not mean that the approach was fundamentally wrong. The essential requirements are a sufficiently wide range of densities and sampling errors that are equally small at all densities. Growing different good hosts treated with different dosages of a systemic nematicide (aldicarb, oxamyl) in addition to growing untreated good and poor or non hosts could fill up the gap between the large and small densities created by the latter (Kaai, 1979). Systemic nematicides have an effect on crop yield other than by affecting nematode populations (Seinhorst and den Ouden, 1980), but it can be expected that this will have disappeared a year after the chemical was applied.

Seinhorst (1982a) made use of ranges of potato cyst nematode densities originating from the spreading of potato cyst nematode in fields from centres of (recent) infestation. An exception to the requirement for a random distribution of the treatments (different nematode densities) had to be accepted. This would certainly not have been justified if the

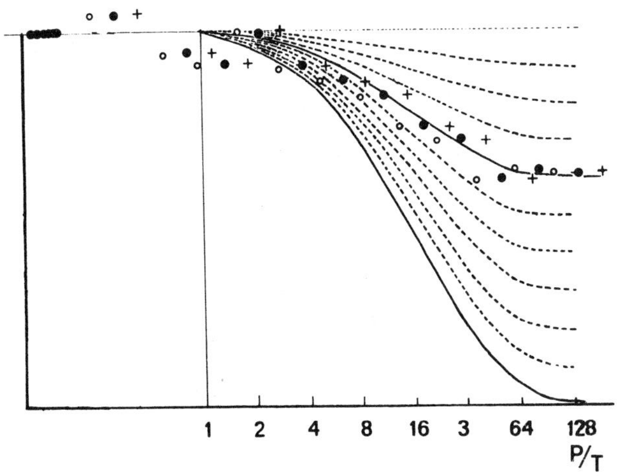

Fig. 9 Fitting a curve according to equation (1) to observed plant weights plotted against log P, in order to estimate T and m. The estimate of the yield at $y = 1$ is determined unambiguously here by the observations at P smaller than the smallest acceptable estimate of T, so that only a horizontal shift of the set of graphs relative to the plotted observations is necessary to choose the best fitting curve. When few observations at densities $<T$ are available, positions obtained by vertical shifts must also be compared. The yield data (from Seinhorst, 1982a) are plotted in three positions relative to the graphs, o : T overestimated, + : T underestimated, ● : best estimate of T.

patchiness had been due to differences in soil conditions, as in stem nematodes (*Ditylenchus dipsaci* (Kühn)) and probably also in certain cyst and other root infesting nematodes.

FITTING CURVES ACCORDING TO EQUATION (1) TO OBSERVED PLANT WEIGHTS

If plant weights are plotted against log nematode density and the scale of the latter is kept the same for different cases, the shape of the curve to be fitted depends only on the scale of $1 - m$. If a series of curves $(y - m)(1 - m)^{-1} = 0.95^{x-1}$ ($x = PT^{-1}$ = 1 to 200), drawn on a transparent sheet on a scale for log P equal to that in the graphs of the observed plant weights (maximum weight about twice scale for log 10), with the scale for $1 - m$ varying from 0.2 to two times that for log 10 in ten equal steps (Fig. 9) the proper curve can be found and the values of T and m estimated by placing the sheet with the curves on the plotted data, trying out what is the best fitting curve and what is its best position (Fig. 9). If necessary a curve can be interpolated between two given ones. If there is a substantial growth reduction by the so called "second mechanism" a curve can be fitted well only to the plant weights at small nematode densities with the points for those at the large densities below the curve, or to the plant weights at the large nematode densities with the points for those at the small densities below the curve.

Ferris et al. (1981) developed a computer algorithm to compute the best fitting curve. However, before any such fitting is tried it should be checked whether plant weights at the larger nematode densities were affected by the so called "second mechanism" of growth reduction. Assuming that plant weights were due only to the action of the "first mechanism" might lead to overestimation of *T* and of experimental error. Even then it would not produce better results than curve fitting by eye. Unless the latter can be done without much difficulty the estimates of *T* and *m* would be much too uncertain to be of much practical value, whatever method of estimating them was used.

VALUES OF THE PARAMETERS *T* AND *m* FOR SOME COMBINATIONS OF PLANT AND NEMATODE SPECIES

Application of equation (1) for prediction of crop losses to be expected at small nematode densities and to determine the amount of control needed to reduce nematode densities to levels at which maximum returns may be expected, requires the values of *T* and *m* and their degrees of uncertainty due to unpredictable variation to be known for the nematodes and crops involved. Seinhorst (1982a), Greco et al. (1982) and Seinhorst et al. (1986) found a value of *T* for *G. rostochiensis* and *G. pallida* of about 2 eggs per g soil and Kaai (1979c, 1980, 1981, 1982) a value of 3 eggs of *H. schachtii* per g soil for sugar beet, Brussel sprouts and cabbage. Greco et al. (1982) found 1.8 and 2.4 eggs of *H. schachtii* for sugar beet. In Australia the tolerance limit of wheat to *H. avenae* was 0.4 eggs per g soil (Meagher and Brown, 1974). In the Netherlands the tolerance limits of oats in pot experiments were 0.4 eggs per g soil (Seinhorst, 1981b) and 0.8 and 1.2 eggs per g soil (Seinhorst, 1981a) and in Sweden in field observations < 1 egg per g soil (Andersson, 1982).

Contrary to *T*, *m* appeared to be very variable in field experiments. Values found for *H. schachtii* on red beet were 0.3 and 0.6 (Kaai, 1950, 1979) and on Brussel sprouts 0.3 and 0.65 (Kaai, 1981, 1982). Seinhorst (1982a) derived values between 0.25 and 0.6 for potatoes attacked by *G. rostochiensis* from data from five field experiments in five years and on five different fields. Large values especially lead to a considerable overestimation of *T*, unless many observations are available and variability of average yields in classes of nematode densities is small (Seinhorst, 1982a). Probably all apparent differences between fields or years in tolerance of a crop plant to attack by a certain nematode species are due entirely to variation of *m*.

The causes of variability of *m* in fields are largely unknown. Delay of the beginning of attack is one such cause (Seinhorst, 1979c). This delay can be caused by low temperatures, at which the plants grow slowly but the nematodes are inactive, when nematodes are inactivated temporarily and/or only in part of the soil by a systemic nematicide or when they are killed only in the soil around the young plants as in row treatment. No further information on factors affecting the value of *m* is available.

REFERENCES

Andersson, S., 1982, Population dynamics and control of *Heterodera avenae* - A review with some original results, EPPO Bulletin, 12:463.

Been, T. and Schomaker, C., 1986, Quantitative analysis of growth, mineral composition and ion balance of the potato cultivar Irene, infested with *Globodera pallida*, Nematologica, (in press).

Ferris, H., Turner, W.D. and Duncan, L.W., 1981, An algorithm for fitting Seinhorst curves to the relationship between plant growth and preplant nematode densities, J. Nemat., 13:300.

Greco, N., Brandonisio, A. and Marinis, G. de, (1982a), Tolerance limit of the sugarbeet to *Heterodera schachtii*, J. Nematol., 14:199.

Greco, N., di Vito, M., Brandonisio, A., Giordano, I. and Marinis, G. de, (1982b), The effect of *G. pallida* and *G. rostochiensis* on potato yield, Nematologica, 28:379.

Hellinga, J.J.A., 1942, De invloed van het Bietenaaltje op de opbrengst en de samenstelling van Suikerbieten, Meded. Inst. Suikerbiet. Bergen O.Z., 12:165.

Jones, F.G.W. and Kempton, R.A., 1978, Population Dynamics, Population Models and Integrated Control, in "Plant Nematology", J. Southey, ed, MAFF AS Publication GD/1 HMSO, London.

Kaai, C., 1979, Bietecystenaaltjes (*Heterodera schachtii* en *Heterodera* sp.) in akkerbouw en tuinbouwgewassen, Jaarverslag 1978, Instituut voor Plantenziektenkundig Onderzoek, p.132.

Kaai, C., 1980, Verband tussen bevolkingsdichtheid van bietecystenaaltjes (*Heterodera schachtii*) en de opbrengst van kroten, Jaarverslag 1979, Instituut voor Plantenziektenkundig Onderzoek, p.127.

Kaai, C., 1981), in Jaarverslag 1980, Instituut voor Plantenziektenkundig Onderzoek, pp.135-136.

Kaai, C., 1982, Verband tussen bevolkingsdichtheid van *H. schachtii* en de opbrengst van spruitkool, Jaarverslag 1981, Instituut voor Plantenziektenkundig Onderzoek, p.141.

Meagher, J.W. and Brown, R.H., 1974, Microplot experiments on the effect of plant hosts on populations of the cereal cyst nematode (*Heterodera avenae*) and on the subsequent yield of wheat, Nematologica, 20:337.

Seinhorst, J.W., 1960, Over het bepalen van door aaltjes veroorzaakte opbrengstvermindering bij cultuurgewassen, Meded. Landb. HogeSch. Opzoekstns Gent, 25:1025.

Seinhorst, J.W., 1965, The relation between nematode density and damage to plants, Nematologica, 11:137.

Seinhorst, J.W., 1978, De,invloed van een tweede aaltjesgeneratie op de opbrengst van een gewas, Jaarverslag 1977, Instituut voor Pflantenziektenkundig Onderzoek, pp.121-122.

Seinhorst, J.W., 1979, Nematode and growth of plants: Formalization of the nematode-plant system, in "Root Knot Nematodes (*Meloidogyne* species) Systematics, Biology & Control", F. Lamberti and C.E. Taylor, eds., Academic Press.

Seinhorst, J.W., 1979b, Effect of *Heterodera avenae* on the shoot:root ratio of oats, Nematologica, 25:495.

Seinhorst, J.W., 1979c, De invloed van uitstel van aantasting op het verband tussen aaltjesdichtheid en groeiremming, Jaarverslag 1978, Instituut voor Plantenziektenkundig Onderzoek, p.130.

Seinhorst, J.W., 1981a, Water consumption of plants attacked by nematodes and mechanisms of growth reduction, Nematologica, 27:34.

Seinhorst, J.W., 1981b, Growth and yield of oats at a range of *Heterodera avenae* densities and under different watering regimes, Nematologica, 27:52.

Seinhorst, J.W., 1981c, Achtergronden van aaltjesbestrijding I, Bedrijfsontwikkeling, 12:491.

Seinhorst, J.W., 1982a, The relationship in field experiments between population density of *Globodera rostochiensis* before planting of potatoes and yield of potato tubers, Nematologica, 28:277.

Seinhorst, J.W., 1982b, Achtergronden van aaltjesbestrijding. 3. Economische aspecten van vruchtwisseling en chemische bestrijding, Bedrijfsontwikkeling, 13:494.

Seinhorst, J.W., Been, T. and Schomaker, C., 1986, The relation between initial egg density of *Globodera rostochiensis* and *G. pallida* and the growth, final haulm and tuber weights and dry matter contents of various potato cultivars, (in preparation).

Seinhorst, J.W. and Kozlowska, J., 1977, Damage to carrots by *Rotylenchus uniformis*, with a discussion on the cause of increase of tolerance during the development of the plant, Nematologica, 23:1.

Seinhorst, J.W. and Ouden, H. den, 1971, The relation between density of *Heterodera rostochiensis* and growth and yield of two potato varieties, Nematologica, 17:347.

Seinhorst, J.W. and Ouden, H. den, 1981, The effect of different dosages of aldicarb on the multiplication at small population densities of *Globodera rostochiensis* on potato on rotavated and non rotavated plots, Neth. J. Pl. Path., 86:147.

Trudgill, D.L., Evans, K. and Parrott, D.M., 1975, Effects of potato cyst nematodes on potato plants, Nematologica, 21:169.

Whitehead, A.G., Tite, D.J., Fraser, Janet E. and French, Eileen M., 1980, Control of potato cyst nematode, *Globodera rostochiensis* in a three course rotation, J. Agric. Sc. Camb., 95:293.

DISCUSSION

Trudgill referred to experiments which showed that increasing densities of potato cyst nematodes were associated with an effect on N, P, K nutrients in the potato haulm, and asked if nematode damage is primarily a physiological phenomenon. Seinhorst considered the effects of potato cyst nematodes to be primarily physiological and this applied to other types of nematodes e.g. *Pratylenchus penetrans* on apple trees, where physiological damage was observed at low nematode densities. He surmised that there was a change in the habitus of the plant indicating the introduction of a toxic substance. Stone said there was good evidence of nematodes causing perturbation of the physiology of the host plant and considered this to be similar to effects induced by stresses such as water lack. Seinhorst observed that older plants tend to be more susceptible to water imbalance than younger plants.

Roberts asked if there was evidence of plant growth being stimulated by low densities of plant parasitic nematodes. Seinhorst replied that he had only seen this effect in his experiments with *P. penetrans* on apple, but such an effect was hard to measure. Wallace had reported a positive growth response to *Meloidogyne*.

AGRONOMIC ASPECTS OF POTATO CYST NEMATODE INFESTATION

J. Willem Seinhorst

formerly: Instituut voor Plantenziektenkundig Onderzoek
Wageningen
Netherlands

INTRODUCTION

Potato cyst nematodes, *Globodera rostochiensis* (Woll.) Behr. and *Heterodera pallida* (Stone) Behr., are present in most potato growing areas throughout the world (Evans and Stone, 1977). They seem particularly adapted to Western and Central European conditions. Despite a moderate population increase per season on susceptible potatoes, but because of a very small decrease of populations in the absence of host plants (only 35 per cent per year) infestations are long lived there. In many cases economically desirable rotations are not wide enough to prevent population increase and spread of the nematodes, a particular example being the short rotations customary in potato crops grown in the Netherlands for starch production. Potato cyst nematodes seem to do less well in the Mediterranean area, possibly because of a much greater population decrease in the absence of host plants (Mugniery, 1978). Also the USA has stayed largely free of potato cyst nematodes, which can almost only be due to adverse natural conditions; otherwise, how could it be explained that the beet cyst nematode spread to all beet growing areas in the USA.

Undoubtedly potato cyst nematodes have been, and are, spread mainly with seed potatoes, first from South America to Europe, then throughout Europe and, in recent times, back to South America again (Evans and Stone, 1977). The spectacular spread of the nematodes and their devastating effect on potato yield has induced governments to impose statutory measures on potato growing, ranging from quarantine measures to prevent the import of infested material and the spread from infested to uninfested land, to those regulating the growing of potatoes within a country. As an example of the latter, Dutch regulations prescribe chemical treatment of the soil depending on whether it has been found infested or not and according to the frequency potatoes are grown; the aim is not only to protect the individual potato growers but equally the international trade of seed potatoes, flower bulbs and nursery stock, by keeping overall degree of infestation in the country at a low level. These interests were considered important enough to impose a little more control on potato growers than is, at least, theoretically, strictly economical. However, whether control is left to the initiative of the individual farmer or is prescribed in some detail, it must be based on the same knowledge of the behaviour of the nematode populations and of the effects of the various control methods - crop rotation, chemical control and growing resistant cultivars. The present state of knowledge cannot be considered sufficient for either of the two approaches. Growing resistant cultivars, once population density of the nematode has decreased to levels which are not expected to reduce the yields of these cultivars, is by far the cheapest method as long as suitable high-yielding cultivars are available and long rotations with other crops are not required for reasons other than nematode control. Such resistant cultivars exist for starch production, except for those fields where new pathotypes occur and where sufficient resistance against them is not yet available. A few such fields may be found soon after the introduction of potato cultivars resistant to all the known potato cyst nematode pathotypes. The population densities of the new pathotypes then gradually increase with time, which is an annoying but unavoidable phenomenon associated with the use of resistant cultivars in situations where resistance against all possible pathotypes has not yet been found. The extreme patchiness of such infestations, often only one or a few patches per field, indicates the introduction of one or a few cysts per patch some five or more crops of susceptible potatoes previously, depending on the rate of reproduction of the nematodes on the cultivars grown. One cyst introduced into a hectare means that one cyst was present per 400 to 3,000 kg seed potatoes, harvested from 200 to 1,400m^2. It is not known at what level of infestation of the soil at harvest time such an infestation of well cleaned tubers is to be expected in probably a small proportion of the separate lots of seed potatoes.

Infestations in a new patch, detectable by one or a few square metres of poor growth of the potatoes, have a well defined pattern (Seinhorst, 1982a (and elsewhere in this publication), which makes it possible to define an area with nematode populations exceeding a certain density. Chemical treatment of the area could prevent, at relatively low cost, the rapid spread of the nematode.

Cultivars resistant to all pathotypes and acceptable for human consumption and to the processing industry are not yet in sufficient supply. Therefore, susceptible cultivars are often grown on infested land. Bintje still is the most widely grown cultivar in the Netherlands, and because of its nematode susceptibility crop rotation and soil fumigation are indispensible in its cultivation.

Where cultivars grown for seed are susceptible to the pathotypes present, infestations in the fields must be kept at a low level by crop rotation, fumigation or rotation with resistant cultivars. This is of particular importance because seed potatoes are generally required by quarantine regulations to be "free" of potato cyst nematodes. The suitability of fields for growing seed potatoes is generally checked by the investigation of soil samples of a prescribed size from areas also of a prescribed size. Before export the consignment is inspected for the presence of cysts, either indirectly by checking whether it is sufficiently free from soil or, directly, by investigating a sample of this soil. Such sampling and inspection cannot, of course, given an absolute guarantee of freedom from cysts. Therefore, to the grower the system poses the question - what degree of infestation can be allowed in his field (= what sampling effort must be made) to ensure that the number of cysts found by quarantine inspection will be acceptably small, taking into account the success of efforts to remove the soil adhering to the tubers. Apparently a matching of methods and efforts has been achieved. However, the proportion of the consignments containing cysts that is actually shown to be infested is and will remain unknown. An indication of the spread of potato cyst nematodes by seed potatoes considered to be free of them on the basis of current standards of sampling and inspection could be obtained from the numbers of infestations that became established in previously uninfested areas. As many potato crops have to be grown successively before an introduction becomes apparent, it will hardly be possible to trace its year and source. The best that can be achieved is a guarantee that a technically feasible and economically acceptable amount of effort is made to keep the spread of cysts with seed potatoes at a minimum (which is unknown). This does mean that potato cyst nematode densities in seed potato fields must be kept very small: if a grower wants to keep the chance of detection of cysts in a 250 g soil sample from such a field below one per cent, the density must be kept lower than one cyst per 25 kg soil. As the nematodes increase on the plants to give an infestation level of 10-75 cysts per kg soil, he must also ensure that the quarantine inspector does not collect more than 3.5 (!) to 25 g soil from the harvested consignment if the chance of finding cysts is to be kept below one per cent.

DETERMINING THE NECESSARY POPULATION REDUCTION BY CONTROL MEASURES, WHEN GROWING SUSCEPTIBLE POTATO CULTIVARS

The reduction of potato cyst nematode populations following a crop of susceptible potatoes and required to prevent yield losses in the next potato crop can be based on either an estimate of the average rate of increase of the nematodes on susceptible potatoes, or on nematode counts in a soil sample taken some time between the last and the next potato crop. The statutory measures regulating potato growing in the Netherlands are based on an estimate of the average rate of increase of the nematodes. Using the second method the grower must decide the level to which he requires the nematode density to be reduced. For ware potato crops the density at which maximum returns may be expected seems to be the most sensible level (Seinhorst, 1982b), and for seed crops that at which the chance of detection of cysts in the product is negligible.

AVERAGE RATE OF REPRODUCTION OF POTATO CYST NEMATODES

The variance of the estimate of the nematode density after n susceptible crops, starting from a fixed density, is the sum of the n variances of the individual multiplication rates and the variances of all other changes of nematode density. Therefore, to be a reliable base for determining the necessary amount of reduction of population density after the nth crop of susceptible potatoes, not only the average rate of reproduction of potato cyst nematodes on such potatoes must be know, but rates for individual fields must also vary little from field to field and from year to year. This last requirement is not fulfilled in the Netherlands and probably not elsewhere.

In an investigation, multiplication rates were determined in fairly uniformly infested parts of the thirteen fields with initial nematode densities not larger than about 5 eggs per g soil in seven years (between 1957 and 1971). Egg densities before planting and after the lifting of susceptible potatoes were determined in soil samples of 1 to 1.5 kg from at least 20 adjacent square metre plots in each field. Sampling errors of average egg densities were not larger than 7 per cent and 5 per cent for initial and final densities, respectively. The ratios between the average final and the average initial density in each field are given in Table 1, together with data from the literature. The ratios vary from 2.5 to 150, with considerable differences between different fields in the same year and for one field in different years. There was no correlation with soil type.

Table 1 Maximum rates of reproduction and maximum population densities of potato cyst nematodes (*G. rostochiensis*) in different fields and years.

Year	Soil type	Maximum rate of reproduction	Maximum density eggs/g soil
1976	clay	3	250
1966[1]	humic sand	6	
1969	sandy loam	8	
1967	clay	11	70
1957	light sandy soil	20	75
1967[1]	humic sand	22	328
1971	sandy loam	22	240
1965	peaty sand	25	200
1967	sandy soil	25	300-400
1968	sandy soil	26, 56	234
1968	clay	26	
1970	sandy loam	29	234
1968	clay	41	
1967[1]	humic sand	51	500
1968	sandy loam	72	
1968	clay	79, 158	264
1961[2]	peaty sand		764
1967/68[1]	peaty sand		290
	sandy soil		245
	sandy loam		302
		geom. mean : 24	

[1]) Nollen and Mulder, 1969
[2]) Oostenbrink, 1961

Variation in maximum densities in the same or different fields, and between years, is small (Table 1). With the exception of two examples at unusually low levels (70 and 75) all are between 230 and 500 and 8 of 12 even between 230 and 300. The resulting variation in rates of reproduction at *Pi* = 0.5 to 300 eggs per g soil is given in Fig. 1. It demonstrates why the rates published in the literature, which were mostly measured at fairly large initial population densities, vary less than those in Table 1. However, according to Seinhorst (1982b) the economy of European potato growing requires initial egg densities not larger than about 8 eggs per g soil at which level variation in multiplication rates is of the same order as in Table 1.

To calculate the consequences of this variation it was assumed that the distribution of log reproduction rates is normal with mean 1.398 and standard deviation 0.41 (Fig. 2). The resulting variation in egg densities to be expected after successive crops of susceptible potatoes, followed by control that reduces the population to 24^{-1} times with a standard deviation of log reduction (-1.398) of 0.088 is given in Fig. 3. The variability of nematode densities to be expected after 1, 2, 3 etc. years increases rapidly, also indicating that the difference between necessary and actually executed amount of control, if based on an average multiplication rate (whether too much or too little control) on a large number of randomly chosen fields increases rapidly. The only way to reduce the chance of small to considerable losses in half the fields is to reduce the populations more strongly, at the cost of more unnecessary control. Therefore, it is more efficient and cheaper to measure egg density after each crop of susceptible potatoes.

To demonstrate the possible effect of variation as shown in Table 1, successive nematode densities in a number of years resulting from choosing two series of multiplication rates at random from Fig. 2 (by choosing a series of number at random from a range 1 to 84 and deriving the associated multiplication rate from Fig. 2) and dividing each by 24, are shown in Fig. 4.

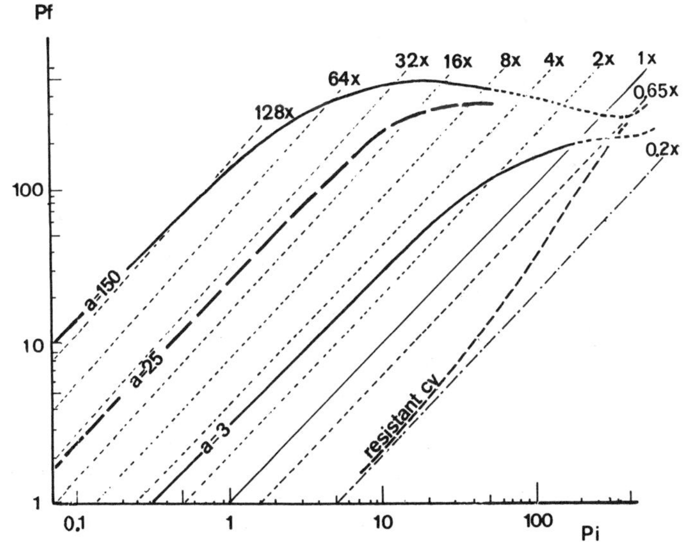

Fig. 1 Variability of multiplication rates and maximum densities of *G. rostochiensis* on susceptible potato cultivars.

Most or all of the data in Table 1 are of *G. rostochiensis*, pathotype Ro 1 populations. Average reproduction rates of *G. pallida*, pathotypes Pa 2 and 3 may be slightly different for cultivars susceptible to all known pathotypes of potato cyst nematodes and distinctly smaller for cultivars resistant to one or more of these (Seinhorst and Oostrom, 1984). There is no reason, however, to suppose that the variation will be smaller than for Ro 1. Obviously, the increase of a small potato cyst nematode density on a susceptible potato cultivar and, certainly, nematode density after two or more crops of susceptible potatoes, cannot be predicted except within wide to very wide limits. Therefore, the determination of the economically optimal amount of control required before the next crop of potatoes is grown in a particular field can only be based on the actual nematode density in the field after the last crop of susceptible potatoes. This density can only be determined by counting the nematodes in a soil sample from the field.

MEANS OF REDUCING POPULATION DENSITIES

Populations of potato cyst nematodes can be reduced by crop rotation, chemical soil treatments and growing resistant cultivars.

Crop rotation

In the absence of host plants a certain percentage of the eggs of potato cyst nematodes hatches spontaneously during a short period in spring (den Ouden, 1960). It most probably is the only cause of practical importance of population decrease in the absence of hosts. Although it seems easy enough to collect information on the magnitude of this population decrease, the available observations, although accurate enough in some cases, do not allow conclusions on variability from field to field and from year to year. Huijsman (1961) calculated an average decrease of 35 per cent per year from observations during six years on the same small plots without potatoes. Cole and Howard (1962) found an average reduction per year over three years of about 20 per cent. Average egg content of cysts decreased by 26 per cent and average number of eggs per g soil by 21 per cent in fourteen fields in 1969 (den Ouden, 1970), but average

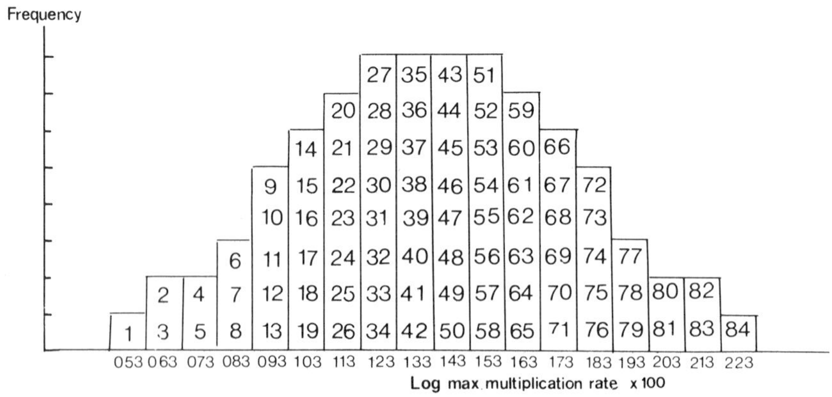

Fig. 2 Theoretical log normal distribution of maximum multiplication rates of *G. rostochiensis* on susceptible potato cultivars.

number of eggs per g soil in six other fields by 49 per cent in 1973 (den Ouden, 1974b). Sampling errors of these averages are, most probably, small enough to consider the differences to be highly significant, but the total information still is too limited to allow generalisation.

Volunteer plants

Absence of hosts also means the absence of volunteer plants. Den Ouden (1967) demonstrated that the effect of the latter depends on their plant density (Fig. 5) according to the equation:

$$Pf = a(1 - q^{Pi})(-{}^{e}\log q)^{-1} + bPi,$$

in which a = the maximum rate of multiplication of the nematodes on the volunteer plants, b = the proportion of the population that survived in the absence of host roots, q = a constant <1 and $a(-{}^{e}\log q)^{-1}$ (= M = the maximum number of eggs produced on the roots of the volunteer plants proportional to volunteer plant density (4 and 16 plants per m^2). Nematode population densities decreased when larger than M but increased when smaller than M. In den Ouden's (1967) experiment the volunteer plants did not induce a noticeable number of eggs to hatch. All multiplication was of juveniles that hatched spontaneously (about 35 per cent). At still larger volunteer plant densities than 16 per m^2 these would certainly have induced some additional hatching.

Chemical control

See A. G. Whitehead elsewhere in this volume.

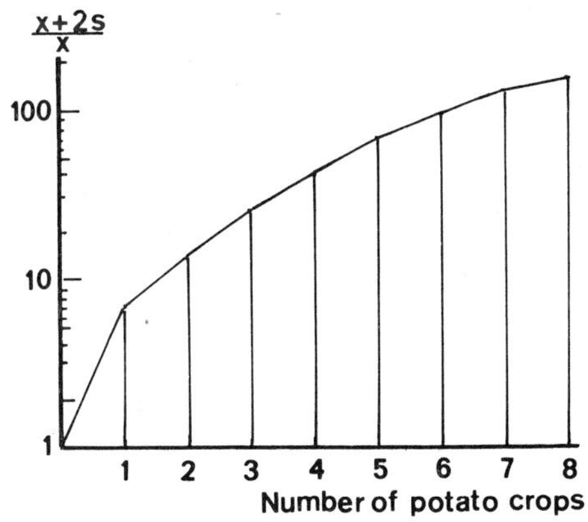

Fig. 3 Variation of density x of *G. rostochiensis* after n crops of susceptible potatoes as estimated at planting of the first crop, assuming a standard deviation of log multiplication rate of 0.41 and one of log population decrease between potato crops of 0.088.

Growing resistant potatoes

Resistant potato cultivars induce eggs of potato cyst nematodes to hatch. Therefore, if a large proportion of the hatched juveniles penetrating into roots die there before becoming adult and producing eggs, there will be far fewer eggs in the soil after harvesting the resistant potatoes than before they were planted. Resistant potato cultivars can, therefore, be used not only to prevent increase of the nematode population but also to reduce it. This is only possible when a stimulus from the plant is necessary to induce hatching. When hatching is induced irrespective of presence or absence of a host or resistant plant, as in *H. avenae* and in *H. trifolii*, growing a resistant plant has the same effect as growing a non-host or fallow. Even susceptible potatoes do not induce more than 70 to 80 per cent of the eggs of potato cyst nematodes in the field to hatch, because the density of roots in the furrows is too small to affect all the eggs there (Seinhorst and den Ouden, 1980). As some 35 per cent of the non-affected eggs hatch spontaneously at least 15 to 20 per cent of the initial population is still present after growing a very resistant cultivar, if its ultimate size is not materially affected by the nematodes. Resistance does not automatically mean a greater tolerance than in susceptible cultivars and reduction of the ultimate size of the root system will also reduce its effect on the nematode population. As in susceptible cultivars, this reduction which is also associated with a reduction in tuber yield, is to be avoided for economic reasons. Therefore, if resistant cultivars are used to reduce nematode densities well above the tolerance limit, two contrary phenomena are combined: the greater need of a strong reduction and the smaller effect of the means used at larger than at smaller nematode

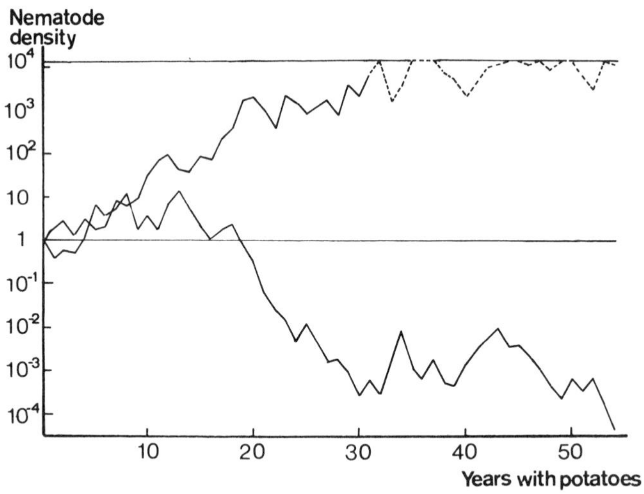

Fig. 4 Population densities of *G. rostochiensis* (eggs per 100 g soil) after successive potato crops, according to two series of reproduction rates chosen at random from Fig. 2, which rates were then divided by the average, 24. Dotted lines: reproduction reduced by intraspecific competition, which is assumed to result in a maximum density of 2 x 10^4eggs per 100 g soil. Similar deviations from the initial population density can be expected in about 5 per cent of the fields if maximum multiplication rates have a frequency distribution as in Fig. 2.

densities. Whenever resistant cultivars are to be grown on infested land not only their degree of resistance against, but also their degree of tolerance (sensitivity to yield reduction) of, the pathotype present in the soil must be known. Unfortunately not only is little known about the tolerance of susceptible cultivars, but also almost nothing about the resistant ones; and what is known applies only at large nematode densities. Tuber yield in a field experiment by Jones et al. (1967) indicated considerable differences between cultivars at nematode densities >50 eggs per g soil. However, as explained by Seinhorst (elsewhere in this volume) yield reductions at large nematode densities are not necessarily a measure of the tolerance at the economically much more important range of 2 - 20 eggs per g soil. Only accurate measurements of growth and yield at a range of small to medium densities of potato cyst nematode can supply the necessary information on tolerance of potato cultivars in general and of resistant cultivars in particular. The available information suggests that little difference in tolerance at small nematode densities is to be expected between susceptible cultivars and between pathotypes, but at least some differences between resistant cultivars. Unless a special test is found (e.g. growing young sprouts for some days in soil with a range of nematode densities after which root lengths are measured, as done by Seinhorst and den Ouden, 1971), the testing will be tedious and time consuming, as at least some forty plants of each cultivar must be grown in pots for the proper time to obtain the necessary data. Field experiments generally are far too inaccurate or too laborious to be of any use.

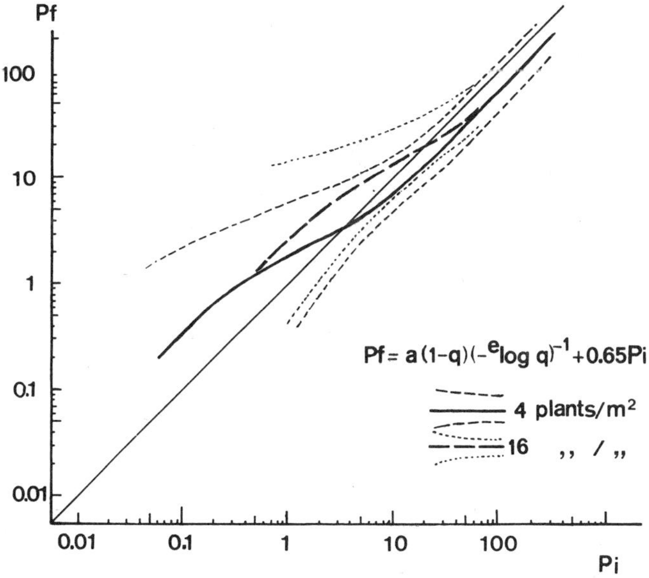

Fig. 5 Effect of two densities of "volunteer plants" (small tubers planted at 20 cm depth) in oats on population density of *G. rostochiensis* in 1 sq.m pots. *Pi* and *Pf*: eggs per g soil before planting the tubers and in the autumn, respectively (Adapted from den Ouden, 1967).

RESISTANCE AND PATHOTYPES

Once potato cyst nematode densities have been reduced to a level which makes growing resistant cultivars economically acceptable, they could be kept at that level or even be reduced further, by growing sufficiently resistant cultivars. There are two stumbling blocks - the lesser is that no suitable resistant cultivars are available for every purpose; the more important is that when growing a resistant potato cultivar repeatedly in the same field the nematode population originally present tends to be replaced by one which can multipy on the resistant cultivar. Only after such a new "pathotype" is found can the search for resistance against it and the breeding of potato cultivars incorporating this new resistance be started. Until there are resistant cultivars available only crop rotation and chemical control can keep potato growing possible.

According to Kort et al. (1977) there are five *G. rostochiensis* and three *G. pallida* pathotypes. The numbers are limited by the means to distinguish them. Nematode populations against which resistance has not yet been found are not necessarily, and until now were not, of one and the same pathotype.

The pathotypes (formerly called biological races and also biotypes) were originally indicated as A, B, C etc., but Dutch and English B appeared to be even different species. Therefore, Kort et al. (1977) proposed a new nomenclature in which all *G. rostochiensis* pathotypes are indicated by the code Ro and all *G. pallida* pathotypes by the code Pa, both followed by a number for each pathotype. Dutch B became Ro 2 and English B became Pa 2. They also gave a list of clones from various cultivated and wild *Solanum* species to be used as a test assortment and suggested that on these either fair to large numbers of cysts are formed or very few to none at all, indicated by a + or a - respectively (Table 2). By planting the test clones in soil containing the nematode population to be identified and, after several weeks, inspecting the roots or the soil for the presence or virtual absence of new females or cysts a list of plusses (+) and minuses (-) can be made and compared with Kort et al. (1977) Table II (here Table 2). The alternative classification of Canto and De Scurrah (1977) does not differ materially from that of Kort et al. (1977). The use of only two classes suggests that susceptible and resistant plants can always be distinguished easily in tests. Kort et al. (1977) Table III contradicts this supposition (it fails to demonstrate a difference between pathotypes Ro 2 and Ro 3 in the way required by their Table II). It suggests instead, that the qualifications "susceptible" (+) and "resistant" (-) could only be assigned to the various test plants with the help of an arbitrary definition of these classes in terms of numbers or ratios of numbers of cysts on these plants.

Resistance against *G. rostochiensis*, pathotype Ro 1, derived from *Solanum tuberosum* subsp. *andigena* is based on a dominant gene, which operates against females but not against males (Toxopeus and Huijsman, 1952, 1953). Resistant plants react to the penetration of Ro 1 juveniles by necrosis of the cells in the immediate vicinity of the nematode instead of the development of a transfer cell (hypersensitivity reaction).

A gene for gene relation has been shown to exist between resistance ex *andigena* against pathotype Ro 1 and the capacity of other pathotypes to reproduce on the resistant plants; and also for resistance against pathotype Pa 1 and the capacity to break it. Only females that are homozygously recessive for the virulence gene are capable of reproduction on resistant plants, as shown by Jones et al. (1981) in experiments on mating of

Table 2 Identification of pathotypes of *G. rostochiensis* (Ro) and *G. pallida* (Pa) (Adapted from Kort et al., 1977, Table II)

	Ro					Pa		
Clone	1	2	3	4	5	1	2	3
S. tuberosum ssp. *tuberosum*	+	+	+	+	+	+	+	+
S. tuberosum ssp. *andigena* CPC 1673 hybr.	-	+	+	-	+	+	+	+
S. kurtzianum hybr. 60-21-19	-	-	+	+	+	+	+	+
S. vernei hybr. 58-1652/4	-	-	-	+	+	+	+	+
S. vernei hybr. 62-33-3	-	-	-	-	±	+	+	+
S. vernei hybr. 65-346/19	-	-	-	-	-	+	+	+
S. multidissectum hybr. P 55/7	+	+	+	+	+	-	+	+
S. vernei hybr. 69-1377/94	-	-	-	-	-	-	-	-

+ : susceptible, - : resistant

pathotypes and virulence of the offspring, and by Seinhorst and Oostrom (1985) who selected a pathotype Ro 3 population from an apparently lightly contaminated Ro 1 population and propagated it for three generations on the Ro 1 resistant cv. Alcmaria (resistance ex *andigena* CPC 1673-20). The relationship between the genetics of resistance of wild *Solanum* species on potato cultivars and the virulence of nematode populations probably is more complicated. Also, resistance against pathotypes other than Ro 1 and Pa 1 and from sources other than *S. tuberosum* subsp. *andigena* against Ro 1 could be a kind other than hypersensitivity to the penetration of juveniles. The hypothesis that all females on a cultivar which is resistant to the bulk of the individuals in the population differ from the latter in having resistance breaking genes, and that, therefore, degrees of resistance between totally resistant and fully susceptible are due to the nematode populations being mixtures of genetically different individuals, is contradicted by the observation that maintaining a pathotype Ro 3 population on cvs Darwina (resistance ex *S. vernei*, $(VT^n)^2$ 62-33-2), Proton and Promesse (relative susceptibilities 4.4 per cent, 1.3 per cent and 0.2 per cent of that of the susceptible cv. Irene, respectively) for three or four years did not result in any increase of reproduction rates nor, therefore, in the selection of individuals with a genetical constitution more suited to development on these cultivars than the bulk of the original Ro 3 population. On the other hand maintaining the same Ro 3 population on cv. Veenster (resistance ex *S. vernei*) resulted in an increase of relative susceptibility from 0.5 per cent of that of cv. Irene in the first generation to 120 per cent in the third, suggesting the presence of a single recessive resistance breaking gene.

If susceptible and resistant plants are not such vastly different groups as suggested by Kort et al. (1977) Table I (here Table 2) and

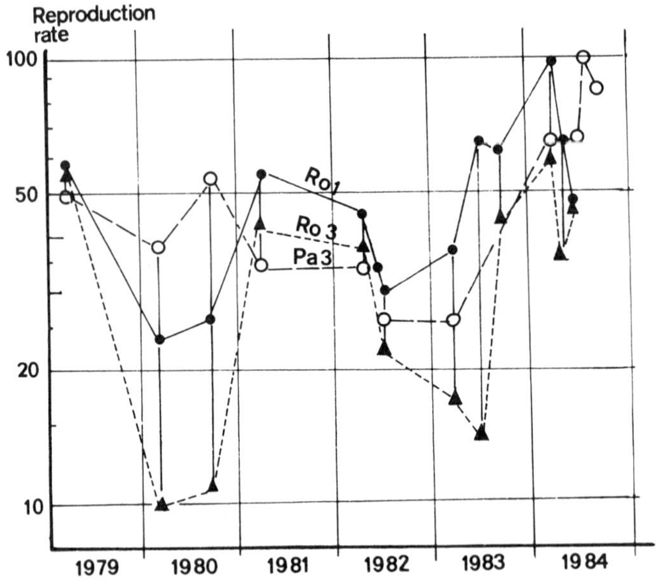

Fig. 6 Reproduction rates of a population of Ro 1, one of Ro 3 and one of Pa 3 on cv. Bintje in 1979 and on cv. Irene in 1980 to 1984. Every year the populations collected in the experiments of the previous year were used. Vertical lines connect rates determined in simultaneous randomised experiments. Variation coefficients about 8 per cent.

finding a nematode population reproducing on a cultivar albeit less than on a truly susceptible one, cannot be regarded as the initial nematode population being a mixture of pathotypes, degrees of susceptibility (resistance) must be measured as genuine characters of cultivars of which practical use can be made. Seinhorst (1983) argues that only a ratio between the rate of reproduction at a small initial nematode density on the clone to be tested and that on a standard susceptible cultivar (i.e. relative susceptibility = r.s.) is sufficiently constant for this purpose under the varying conditions and actual multiplication rates in different experiments. Two or more standards would be better than one. The reproduction rate of Ro 3 on cv. Irene was unreliable as a standard. Cyst contents were, for unknown reasons, much smaller than on other susceptible cultivars in some experiments but not in others. The initial density to be used is open to question, as the above mentioned ratio is density dependent. In very to fairly resistant cultivars it is at Pi = 0.05 eggs per g soil six times as high as at 5 eggs per g soil, but in moderately resistant cultivars (where it is much more important than in the very resistant ones) the difference is considerably smaller if present at all, e.g. for Ro 3 on cv. Darwina (Seinhorst, 1983).

Contrary to the relative constancy of the r.s. ratios, the ratios of reproduction rates of different populations of the same or of different pathotypes may vary considerably (Fig. 6).

Table 3 Reproduction rates of various pathotypes of *G. rostochiensis* and *G. pallida* on four potato cultivars. Densities of all pathotypes at planting: 1 egg/g soil. Reproduction rates on cvs Astarte, Ehud and Multa in percentages of that of the same pathotype as cv. Irene (susceptible to all pathotypes).

	Ro						Pa			
	1	2	3	3 IPO	4	5	1	2	3 (1)	3 (2)
Cv. Irene 100=	102x	54x	74x	36x	124x	109x	112x	59x	64x	120x
Cv. Astarte (resistance ex *Solanum vernei* VT[5] 62-69-5)	0.35	1.5	1.4	6.8	5.0	5.2	40	31	37	44
Cv. Ehud (resistance ex *S. tuberosum* subsp. *andigena* CPC 1673-1)	0.025	0.32	23	48	0.09	24	78	29	78	69
Cv. Multa (resistance ex *S. tuberosum* subsp. *andigena* CPC 1673-1)	(~100)		91					12	35	90

RELATION BETWEEN DEGREES OF RESISTANCE OF ONE CULTIVAR AGAINST DIFFERENT PATHOTYPES

Resistance against one or more pathotypes apparently is often associated with reduced susceptibility to other pathotypes (den Ouden, 1974a). All cultivars with resistance against Ro 1 and against Ro 1, Ro 2 and Ro 3 tested by Seinhorst and Oostrom (1984) were also less susceptible to Pa 3 than cultivars without such resistance, except cv. Multa which although highly susceptible to Ro 1 and Ro 3 has a reduced susceptibility to Pa 2 (r.s. = 12 per cent) and to Pa 3 (r.s. = 35 per cent) (Table 3). Resistance against Ro 1 in cv. Astarte, derived from *S. vernei*, divides the *G. rostochiensis* pathotypes tested into three groups: Ro 1, Ro 2 + Ro 3, and Ro 3 (IPO) + Ro 4 + Ro 5, with ratios of relative susceptibilities of cv. Astarte of about 1 : 3.7 : 3.7^2. Resistance ex *andigena* CPC 1673-1 against Ro 1 in cv. Ehud divides these populations into four groups: Ro 1, Ro 4, Ro 2, and Ro 3 + Ro 5 (+ Ro 3 IPO?) with ratios of relative susceptibilities of cv. Ehud of about 1 : 3.7 : 3.7^2 : about 100. The meaning of the peculiar ratio 3.7 is not clear, but it does suggest fixed relations between the virulences of the populations. This seems to contradict the opinion of Trudgill (1985) that populations of Rp 2 to Ro 5 are a continuum of individuals carrying different proportions of virulence genes. The two populations of Ro 3 in Table 3 may not be as different as the fitures suggest. Probably the reproduction rate on cv. Irene was underestimated for Ro 3 (IPO) because of unusually small number of eggs per cyst on this cultivar in this test.

Both the ex *S. vernei* resistance in cv. Astarte and in the ex *andigena* resistance in cvs Ehud and Multa reduce the relative susceptibility of the three cultivars to all *G. pallida* populations tested. Together they create three groups of relative susceptibilities: Pa.1 + Pa 3(1) + Pa 3(2) on cv. Ehud; Pa 1 + Pa 2 + Pa 3(1) + Pa 3(2) on cv. Astarte + Pa 3(1) + Pa 3(2) on cv. Multa + Pa 2 on cv. Ehud; and Pa 2 on cv. Multa, with ratios 6 : 3 : 1, which again suggests fixed virulence relations.

Relative susceptibilities to the two Pȧ 3 populations, which are the same in Table 3, had a ratio Pa 3(1) to Pa 3(2) of 2 to 1 for cultivars with partial resistance ex *S. vernei* $(VT^n)^2$62-33-3 (Seinhorst and Oostrom, 1984, 1985).

MAXIMUM POPULATION DENSITIES OF DIFFERENT PATHOTYPES ON CULTIVARS WITH DIFFERENT DEGREES OF RESISTANCE

Ratios of maximum densities on a susceptible standard and on resistant cultivars ($M_sM_r^{-1}$) generally are larger than those of maximum rates of multiplication ($a_sa_r^{-1}$) (Seinhorst, 1983). This may also apply to cultivars that differ little in susceptibility, e.g. cv. Ehud versus cvs Multa and Astarte with Pa 3 (Seinhorst, 1983). However, $M_sM_r^{-1}$ was equal to $a_sa_r^{-1}$ for Pa 3 and Ro 3 on cvs Irene and Darwina (Seinhorst, 1983; Seinhorst and Oostrom, 1984; Seinhorst elsewhere in this volume). Maximum densities of different pathotypes on partially resistant cultivars have not been compared in further experiments.

USE OF PARTIALLY RESISTANT CULTIVARS IN ROTATIONS

If a new pathotype builds up in a field against which no high resistance incorporated in a suitable cultivar is available, then a susceptible cultivar could be grown but with the smallest possible relative susceptibility to the pathotype and resulting in the smallest possible maximum nematode density. For example, in the case of an infestation with Pa 3, growing cv. Darwina would result in a smaller population density than growing cv. Irene followed by a treatment with 250 l DD per ha. Therefore as soon as a new pathotype is found not only a search for new high degrees of resistance should be started but also the relative susceptibilities of existing cultivars to the new pathotype should be determined, to find the best ones to bridge the gap until suitable cultivars with high degrees of resistance become available.

THE ECONOMY OF CONTROLLING POTATO CYST NEMATODES IN PRACTICAL POTATO GROWING

To obtain maximum returns from growing susceptible potatoes, potato cyst nematodes are controlled by a combination of the measures discussed above, often called integrated control. Seinhorst (1982) described the information that is required to maximise returns and gave simple examples of how to determine the proper combinations of control measures when growing ware potatoes. Necessary data are nematode density after the last susceptible potato crop as determined in a soil sample (see Seinhorst elsewhere in this volume), population decrease to be expected when growing non-host crops or resistant potato cultivars, parameters of the relation at small nematode densities between the densities at planting and tuber yield to be expected, the relation between dosage of contact nematicides and nematode mortality to be expected under the conditions in the field at the time of the treatment (Seinhorst, 1973, 1981), the cost of the chemicals to be used and of their application, the expected market price of the product and the net expected returns of the potato crop and of other crops. Ideally this information should be available soon after the harvest of the previous crop of susceptible potatoes, but some adaptation is of course possible during the period between two potato crops. "Systemic" nematicides were not included as, despite many experiments, the relation between dosage of these chemicals and tuber yield at different nematode densities is not known. The examples given are based on fixed values of system parameters and variables, except for nematode densities, of which the full range between the tolerance limit T and 250 T is considered. In reality these parameters are variable and only partly predictable within sometimes wide limits (e.g. minimum yields). Calculations and decisions must, therefore, take the frequency distributions of these values over fields and years into account. These frequency distributions still are largely unknown and sometimes difficult to determine (e.g. of the minimum yields). Arbitrary distributions could be introduced into the calculations to determine the importance of variations in the values and the extent of the information needed.

REFERENCES

Canto M. and De Scurrah, M.M., 1977, Races of the potato cyst nematode in the Andean region and a new system of classification, Nematologica, 23:340.

Cole, C.S. and Howard, H.W., 1962, The effect of growing resistant potatoes on a potato-root eelworm population - a microplot experiment, Ann. app. Biol., 50:121.

Evans, K. and Stone, A.R., 1977, A review of the distribution and biology of the potato cyst nematodes *Globodera rostochiensis* and *G. pallida* PANS, 23:178.

Huijsman, C.A., 1961, The influence of resistant potato varieties on the soil population of *Heterodera rostochiensis* Woll, Nematologica, 6:177.

Jones, F.G.W., Parrot, D.M. and Perry, J.N., 1981, The gene for gene relationship and its significance for potato cyst nematodes and their solaneceous hosts, in "Plant Parasitic Nematodes Vol. III", Eds. B.M. Zuckerman and R.A. Rohde, Academic Press, New York.

Jones, F.G.W., Parrot, D.M. and Williams, T.D., 1967, The yield of potatoes resistant to *Heterodera rostochiensis* on infested land, Nematologica, 13:301.

Kort, J., Ross, H., Rumpenhorst, H.J. and Stone, A.R., 1977, An international scheme for identifying and classifying pathotypes of *Globodera rostochiensis* and *G. pallida*, Nematologica, 23:333.

Mugniéry, D., 1982, Diversitiés régionales d'application de lutte intégrée en culture de pomme de terre en fonction de la variabilité des populations de nématodes à kystes, Agronomie, 2:629.

Nollen, H.M. and Mulder, A., 1969, Praktijkonderzoek naar de bestrijdingsmogelijkheden van het aardappelcystenaaltje (*Heterodera rostochiensis* Woll.), Onderzoekcommissie van de Provinciale Raad voor de Bedrijfsontwikkeling in Drente. Assen 1969, 39 p.

Oostenbrink, M., 1961, Nematodes in relation to plant growth. II. The influence of the crop on the nematode population, Neth. J. agric Sc., 9:55.

Ouden, H. den., 1960, Periodicity in spontaneous hatching of *Heterodera rostochiensis* in the soil, Nematologica, Suppl. II, 101.

Ouden, H. den., 1967, The influence of volunteer plants in oats on the population density of *Heterodera rostochiensis*, Nematologica, 13:325.

Ouden H. den., 1970, De afname van de bevolkingsdichtheid van het aardappelcystenaaltje bij afwezigheid van aardappelen. Jaarverslag 1969, Instituut voor Plantenziektenkundig Onderzoek, p.133.

Ouden H. den., 1974a, The multiplication of three pathotypes of the potato root eelworm on different potato varieties, Neth. J. Pl. Path., 80:1.

Ouden, H. den., 1974b, Afname van de bevolkingsdichtheid van het aardappelcystenaaltje bij teelt van resistente aardappelen en bij braak. Jaarverslag 1973, Instituut voor Plantenziektenkundig Onderzoek, p.101.

Seinhorst, J.W., 1973, Dosage of nematicidal fumigants and mortality of nematodes, Neth. J. Pl. Path., 79:180.

Seinhorst, J.W., 1981, Achtergronden van aaltjesbestrijding (2). Voorkomen van opbrengstvermindering door aaltjesaantasting, Bedrijfsontwikkeling, 12:733.

Seinhorst, J.W., 1982a, The relationship in field experiments between population density of *Globodera rostochiensis* before planting potatoes and the yield of potato tubers, Nematologica, 28:277.

Seinhorst, J.W., 1982b, Achtergronden van aaltjesbestrijding (3). Economische aspecten van vruchtwisseling en chemische bestrijding, Bedrijfsontwikkeling, 13:494.

Seinhorst, J.W., 1984, Relation between population density of potato cyst nematodes and measured degrees of susceptibility (resistance) of resistant potato cultivars and between this density and cyst content in the new generation, Nematologica, 30:66.

Seinhorst, J.W. and Oostrom, A., 1984, Comparison of multiplication rates of three pathotypes of potato cyst nematodes on various susceptible and resistant cultivars, Meded. Fac. Landbouww. Rijksuniv. Gent., 49:605.

Seinhorst, J.W. and Oostrom, A., 1985, Selectie in populatie van *Globodera rostochiensis* door teelt van resistente aardappelcultivars. Jaarverslag 1984, Instituut voor Plantenziektenkundig Onderzoek, p.137.

Seinhorst, J.W. and Ouden, H. den., 1971, The relation between density of *Heterodera rostochiensis* and growth and yield of two potato varieties, Nematologica, 17:347.

Seinhorst, J.W. and Ouden, H. den., 1980, The effect of different dosages of aldicarb on the multiplication at small population densities of *Globodera rostochiensis* on potato in rotavated and non rotavated plots, Neth. J. Pl, Path., 86:147.

Toxopeus, H.J. and Huijsman, C.A., 1952, Genetical background of resistance to *Heterodera rostochiensis* in *Solanum tuberosum* var. *andigenum*, Nature, London, 170:1016.

Toxopeus, H.J. and Huijsman, C.A., 1953, Breeding for resistance to potato root eelworm. 1. Preliminary data concerning the inheritance of the nature of resistance, Euphytica, 2:180.

Trudgill, D.L., 1985, Methods and problems of testing for resistance to and pathotypes of potato cyst nematodes, Nematol. News no., 46, 20.

DISCUSSION

Stone commented that Stone and Turner (Proc. CIP Tenth Annual Conf.) have demonstrated selection for virulence in *G. pallida* against polygenically based resistance in potatoes. In the first three generations of the selection experiments there was little evidence of virulence selection and this was made difficult to measure by the high standard errors attached to the counts of the low numbers of cysts produced before virulence began to build up. However, by the fourth and fifth generations the process was becoming more obvious, by the sixth it was quite clear (see EPPO Bull. 18) and by the ninth generation virulence populations were reproducing as well on the resistant clones as on the susceptible controls. Seinhorst replied that his selection for virulence experiments were taken only over the first three generations and were concerned with *G. rostochiensis* rather than *G. pallida*.

Perry asked if there were differences in hatching behaviour between eggs from nematodes raised on susceptible and resistant clones. Seinhorst replied that he had not tested this but reiterated that there was as wide a range of numbers of eggs per cyst produced on a susceptible clone as on a resistant clone. In response to a question about the numbers of generations of PCN observed, Seinhorst responded that a partial second generation may occur in pots; some eggs may hatch immediately while others go into diapause. Whitehead remarked that late maturing cultivars such as Cara, which may be planted in April and not harvested until November could, in theory, give a long enough growth period for a partial second generation.

Andersson asked why $z\text{-}T = 1.05$ is uninversally used by Seinhorst when in an early publication he had defined $z\text{-}T$ as ranging from 1.05 to 1.15. Seinhorst replied that a wealth of subsequent experience had indicated that $z\text{-}T = 1.05$ fits most observations.

LIMITS TO METHODS FOR MANAGEMENT OF POTATO CYST NEMATODES IN WESTERN FRANCE

Catherine M. Balandras

I.N.R.A., Laboratoire de Recherches de la Chaire de Zoologie
Domaine de la Motte-au-Vicomte, B.P. 29, 35650 Le Rheu
France

INTRODUCTION

In France, seed potatoes are produced in three main agricultural areas - the West, North and Centre. The western area, Brittany, has an oceanic climate which limits the incidence of aphids and thus favours the seed potato crop; production from a total of 7,000 ha accounts for 56 per cent of the national production. Most of the seed potato crop is produced on small farms (ca 30 ha) which grow a diversity of crops. Seed production is rarely a main crop and never exceeds 3.5 ha per farm.

The discovery of the potato cyst nematode, *Globodera pallida* (Stone) Behr., in seed producing areas prompted the need for efficient and well adapted methods for its management. According to current statutory legislation the production of seed is prohibited on land that has been found contaminated until, after treatment, *G. pallida* can no longer be detected (Anon, 1978). The legislation is aimed at prevention of further multiplication and dissemination of the nematode. Potatoes may not be grown on infested land but non-host crops such as wheat or corn (maize) can be grown, although soil cultivation, sowing, chemical treatment, harvesting the crop and other activities that disturb the soil, can be a means of disseminating the nematode from one field to another and even from one farm to another one. Although the complete absence of the potato crop for a long period may be efficient as a means of eliminating the potato cyst nematode, it cannot be considered as practical because of the adverse economic impact on the local farming systems in Brittany.

During the past four years management methods have been evaluated for their efficiency and for the way in which they best fit local agricultural requirements. Control strategies including cultural, chemical and genetic methods have to fulfill the objectives of preventing any multiplication of the nematode and drastically reducing the density of populations that are present.

POTENTIAL CONTROL STRATEGIES

Non-Host Crops

Assuming the absence of volunteer potatoes, under a non-host crop there is no significant effect on hatching of *G. pallida* in infested soil. The density of the population present in the field decreases but at a very slow rate. Depending on the type of non-host crop and the climatic conditions (temperature and humidity) in western France the decline in nematode population density never exceeds 28 per cent. Thus although the nematode does not multiply, there is a slow rate of reduction of contamination and land may remain infested for a long period of time, so increasing the risk of further dissemination of the nematode.

Chemical Treatment

In western France agricultural practices in spring and autumn preclude the use of fumigants. Indeed in most cases the soil is under crops at these times of the year, and even if there were no early spring crop fumigation at this time would present too great a risk of phytotoxicity for the following crop.

Carbamates and organophosphates could possibly be used but due to their mode of action they are only lethal on early hatched juveniles. Hatching requires the presence of growing potato plants and this is contradictory to the objective of non-multiplication of the nematode.

Resistant Cultivars

The use of resistant cultivars would seem to be the ideal control measure if they prevent multiplication of the nematodes. Unfortunately there is no cultivar resistant to *G. pallida* currently available to farmers and this is the only species of potato cyst nematode so far encountered in western France. In areas infested with *Globodera rostochiensis* (Woll.) Behr. the improper use of totally resistant cultivars may select more virulent populations (Huijsman, 1961). In the case of mixed species infestations the percentage of *G. pallida* may be increased by growing cultivars with resistance only to *G. rostochiensis*.

In practice some of the so-called resistance of cultivars to *G. rostochiensis* only expresses partial resistance (Mugniéry, 1982). This efficiency is a function of the initial nematode population density: at low populations densities the resistance of cultivars is reduced.

Trap Crops

The use of a trap crop, alas described by Carlson and Videgard (1971), can be of value when included in a control strategy. The presence of growing potato plants induces hatching of the nematodes which then penetrate the roots and start their development. The aim of using a trap crop is to kill the plants before the nematodes have developed to the mature adult stage. Regular lifting of control plants is necessary to determine the development of the nematode populations and the appropriate time for the destruction of the trap crop. There is, of course, no marketable production of tubers.

The trap crop procedure requires caution: the optimal efficiency is obtained by allowing the potatoes to grow until the critical date, i.e. the near-to-maturity stage of the females. If this date is passed it has the effect of increasing nematode populations. Likewise the destuction of potatoes has to be thorough in order to avoid regrowth of tuber pieces in the same year and/or progeny tubers in subsequent years.

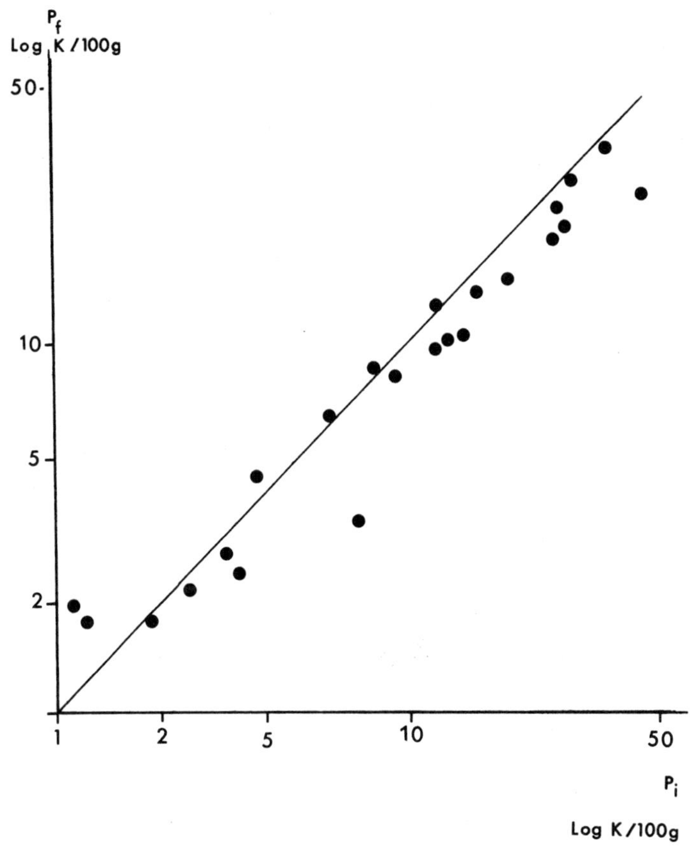

Fig. 1 Effect of chemical treatment followed by a trap crop - Number of cysts Log_{10} (K).

An efficient compromise is the planting of well sprouted small tubers of a late cultivar. Plants tend to grow very rapidly inducing an abundant hatch of juveniles, and using up all the stored products of the mother tuber which makes mechanical destruction easy. Under optimal conditions this procedure provides a 75 per cent decrease of the nematode population. Its efficiency can be increased by incorporating a chemical treatment, the reduction of population then reaching 90 per cent (Mugniéry and Balandras, 1984).

ADOPTED CONTROL STRATEGIES

Overall efficiency of the FOLLOWING control strategy was assessed in two experiments undertaken in western France over a four year period.

The first experiment was at Plouneventer on a 5 ha site; it comprised treatment of the soil with ethoprophos (10 kg a.i./ha) in 1983, followed by

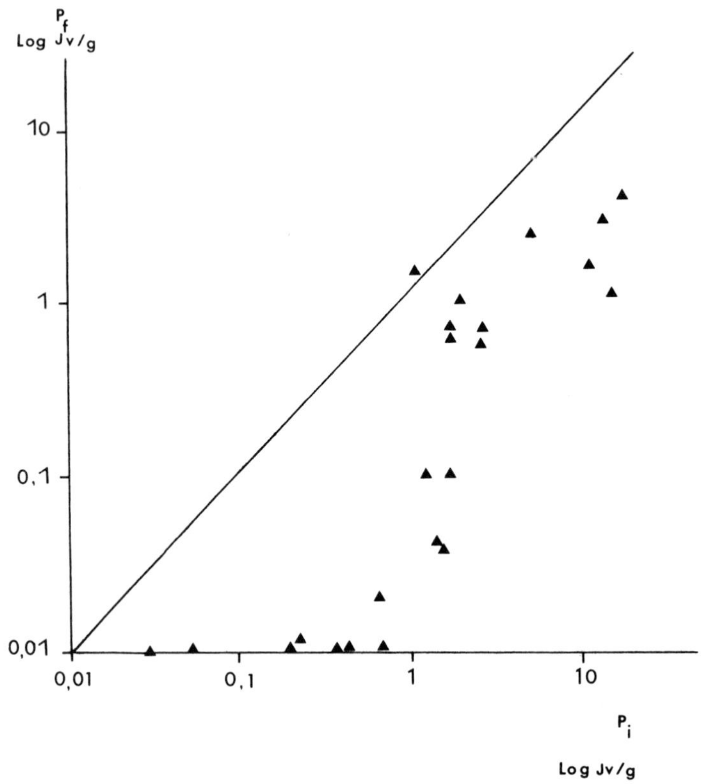

Fig. 2 Effect of a chemical treatment followed by a trap crop - Number of juveniles Log_{10} (Jv).

a trap crop and then by a pea crop sown in June. The soil was sampled before and after trap cropping, forty-sub samples being taken on each occasion with a standard soil corer to provide about 1 kg soil, from which cysts were extracted by the Fenwick can flotation method.

In the absence of a potato crop there was no multiplication of nematodes. The mean number of cysts per 100 g soil, before and after the crop, were not significantly different as compared and tested by the t-test (t=1.01 for 22df, P<.05) (Fig. 1). However, the treatment resulted in a significant decrease in the numbers of juveniles with an average mortality of 90 per cent (Fig. 2).

The second experiment at Plouzévedé comprised chemical treatment with ethoprophos in 1981 followed by a trap crop and then a crop of forage corn (maize). A further trap crop followed by corn was grown in 1982, followed by cauliflower in 1983 and corn in 1984. Soil samples were collected as in the first experiment.

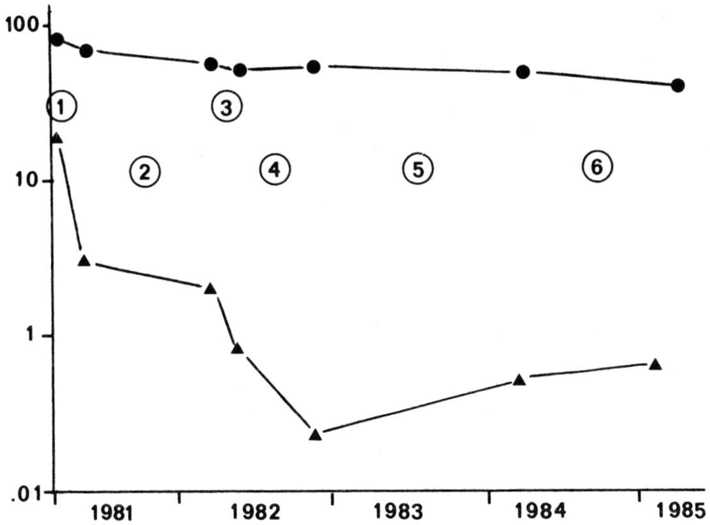

Fig. 3 Population densities of *G. pallida* in relation to chemical treatment and trap crops.
1. Ethoprophos + trap crop; 2. Corn; 3. Trap crop; 4. Corn; 5. Cauliflower; 6. Corn.
●—● K/g (Log.). ▲—▲ Jv/100g (Log.).

The four years of experimentation allow an assessment to be made of the efficiency of the strategy. There was a considerable decrease in the nematode population over the first two years, followed by a continuing progressive decrease (Fig. 3). An analysis of the effect of a non-host crop, i.e. forage corn, shows a dependence between initial population densities and the rate of multiplication of nematodes. Cyst counts significantly dropped ($t=2.38$ for 8df at 5 $P<0.05$); the slight increase in the numbers of juveniles indicates that there had been sporadic multiplication.

CONCLUSIONS

The results of the experiments demonstrate the efficiency of a trap crop preceded by a nematicide treatment and confirm the results obtained by Mugniéry and Balandras (1984), i.e. 90 per cent reduction of the nematode population after one year. This meets one of the requirements of current legislation, which is no multiplication of the nematode with a rapid decrease of population and reduction in the risk of dissemination.

However, continuing observations indicated the difficulty of completely freeing the soil from nematodes and consequently avoiding dissemination from one field to another. They also support Den Ouden's (1967) findings (with *G. rostochiensis*) that nematode densities remain higher under non-host crops with volunteer potato plants than those without, when initial nematode populations are low; high nematode populations are not influenced by volunteer plants. However, it is questioned whether a few volunteer plants that have escaped the farmer's attention do maintain a permanent source of contamination.

Current technical legislation, national as well as international, requires a nil tolerance level for the potato cyst nematode with reference to seed potato production. But however good and accurate sampling methods may be, practice has proved that they do not detect very low population densities whereas a growing potato plant does (and also increases the initial level). Under these circumstances it must be questioned whether the growing of potatoes on land that has previously been found infested but has then been treated does not take the undue risk of increasing nematode populations.

There is clearly a need for improved control methods for potato cyst nematodes to avoid the prospect of an irreversible situation with regard to infestations and seed production. Further, there should be a re-evaluation of the legislation with a view to prohibiting the growing of potato crops on land previously found infested until more efficient and economically feasible methods have been found to solve the problem of volunteer plants.

REFERENCES

Anon, 1978, Règlement technique de la production, du contrôle et de la certification. Tome 2. Certification des plants: plants de pommes de terre, plants de fraisiers, ail et èchalote. Ed. GNIS, Paris, 1978.

Carlson, B. and Videgard, G., 1971, Control of potato cyst nematodes by means of a trap-crop, Potatis, 30.

Den Ouden, H., 1967, The influence of volunteer potato plants in oats on the population density of *Heterodera rostochiensis*, Nematologica, 13:325.

Huijsman, C.A., 1961, The influence of resistant potato varieties on the soil population of *Heterodera rostochiensis* Woll., Nematologica, 6:177.

Mugniéry, D., 1982, Influence de l'hôte sur l'expression du sexe chez un nématode phytoparasite, *Globodera pallida* Stone, à déterminisme sexuel épigénique. C.R. Acad.Sci., Paris, T. 294, Serie III, 294.

Mugniery, D. and Balandras, C., 1984, Examen des possibilités d'éradication du nématode à kystes, *Globodera pallida* Stone, Agronomie, 4(8):773.

INTEGRATED CONTROL OF THE SUGAR BEET CYST NEMATODE

Joachim Müller

Federal Biological Research Centre for Agriculture
and Forestry, Nematology Institut, D-4400 Münster
West Germany

INTRODUCTION

The sugar beet cyst nematode, *Heterodera schachtii* Schmidt, has been known for more than 100 years as a severe pest of sugar beet in Germany. In 1881 Kühn demonstrated that the nematode was the cause of the so called "Rübenmüdigkeit" in the Magdeburg region of Germany. It is now known from almost every country that has had a long tradition of sugar beet production.

Despite many investigations aimed at controlling the pathogen, it remains a limiting factor for beet production in many areas. Chemical treatment of the soil does not completely eradicate infestation and moreover it is expensive and can create problems of environmental pollution. Therefore new strategies are based on integrated control which attempts to combine various factors limiting the pest. Integrated control methods should be adapted to the biological and ecological characteristics of the nematode species in question and for *H. schachtii* seven components are considered to be relevant.

Population thresholds: This is the basis for integrated control methodology. However, routine methods for population assessment used by extension services are incapable of giving reliable results. Predictions of yield losses based upon economic thresholds are questionable. More reliable population estimates for improved yield prediction are possible, but are not economically justifiable at this time.

Natural enemies (antagonists): These are widespread and important, but have not been demonstrated to be useful for active control of the nematode.

Targeted applications: Broadcast nematicide treatments are normally used instead of targeted applications to patches of infection because nematode distribution is unknown or the appropriate technology is not economical.

Selective pesticides: These or products non-toxic to beneficial organisms are not commercially available. Available nematicides are relatively nonspecific in activity. If selective pesticides were to be considered, knowledge relating to the side effects of plant protection products on beneficial parasites and predators would have to be extended.

Cultural practices: Early sowing can be recommended, but trap cropping is too risky. Intercropping of sugar beet with resistant radish has given negative results.

Rotations: Growing sugar beet one year in five would be an effective control measure but this is not economically practical.

Resistance: So far, breeding has not produced commercially acceptable cultivars that are resistant to the sugar beet nematode.

There are disadvantages with each of these seven components for integrated control of *H. schachtii* and therefore a new approach to control is required. In Germany the growing of resistant green manure crops in sugar beet-cereal rotations has produced promising results and these are presented and discussed here.

HISTORICAL REVIEW

Phacelia, clover, grasses, rape, oilseed radish, mustard may be used as green manure crops. In sugar beet-cereal rotations they are sown after winter barley, which is harvested at the end of July in Germany. In this system the cruciferous crops have the advantage of growing quickly and therefore can be sown even late in the season, but they are also hosts of *H. schachtii* and for this reason farmers are advised against sowing rape in sugar beet rotations. There are also doubts about the usefulness of oil seed radish, several authors having found that there was no nematode population increase on the crop (von Horn, 1961; Lücke, 1971; Talatschian, 1974; Hirling, 1976), whereas others indicated the risk of nematode multiplication (Thielemann, 1973; Thomas, 1974; Decker et al., 1977). Probably various factors like sowing date, climatic conditions and soil antagonists contributed to these differing results but the high genetic variability within the species *Raphanus sativus* may also be responsible.

Baukloh (1976) was the first to investigate various cruciferous plant species for their susceptibility to *H. schachtii*. In oilseed radish he found single plants without cysts and from these started breeding for resistance but was unable to select lines that remained completely free from the nematode. For practical reasons he accepted that a plant having up to four cysts is resistant. This limit may be valid for the testing method Baukloh used, but with other environmental conditions, especially in the field with varying population densities, this may have to be changed.

Attack by *H. schachtii* is of no economic importance in cruciferous green manure crops, and breeding for resistance does not aim at preventing damage in these crops. As resistant sugar beet cultivars are not available, the new approach is to reduce sugar beet cyst nematode populations in a long-term rotation system using other sources of resistance.

BIOLOGICAL BASIS OF RESISTANCE

Cyst nematodes withdraw their food from a highly specialised cell complex, and it is known that induction and maintenance of these cell structures is necessary for female development and egg production. The dynamics of giant cell formation, histology, ultrastructure and biochemical reactions are described elsewhere in this book. However, the development of males and females in susceptible and resistant plant tissues is compared here.

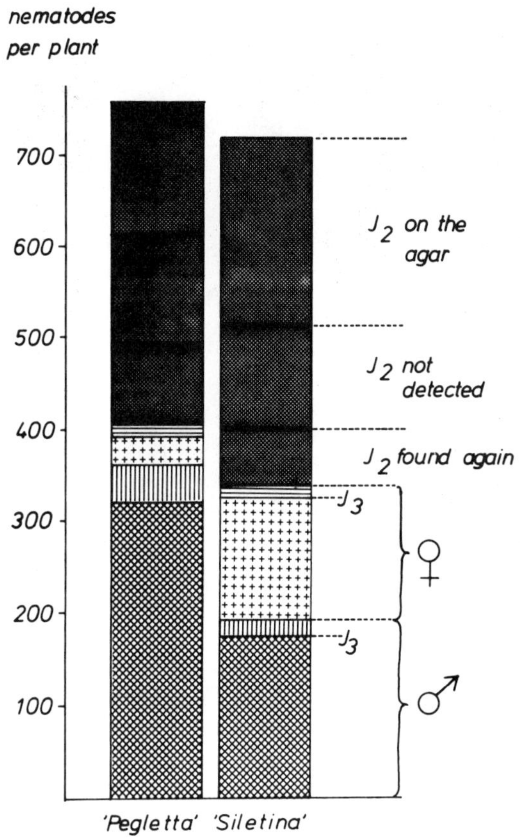

Fig. 1. Frequency of different stages of development on a susceptible and a resistant cultivar (see text).

Molz (1920) found that more males than females were produced under unfavourable conditions and he supposed that this might be due to sex reversal during juvenile development. Other authors favoured the theory of a selective breakdown of female juveniles but this is still disputed. Müller (1985) investigated the sex ratio of *H. schachtii* on susceptible and resistant plants under sterile conditions, as follows.

Plants and nematodes were reared in an axenic system in Petri dishes, which allows observation of the juveniles in the agar and in the roots. Juveniles (1-60) were inoculated to a young seedling (sugar beet, oilseed radish or mustard) growing on nutrient agar, which was transferred after two days into a new Petri dish. After two days about two-thirds of the juveniles had penetrated the roots (Fig. 1). The roots were stained after 15 days and all nematode stages were counted; the results are based on 18,740 inoculated juveniles.

Although it was not possible to account for all the inoculated nematodes it can be assumed that some of the second stage juveniles which penetrated into the root did not continue to develop. However, the same

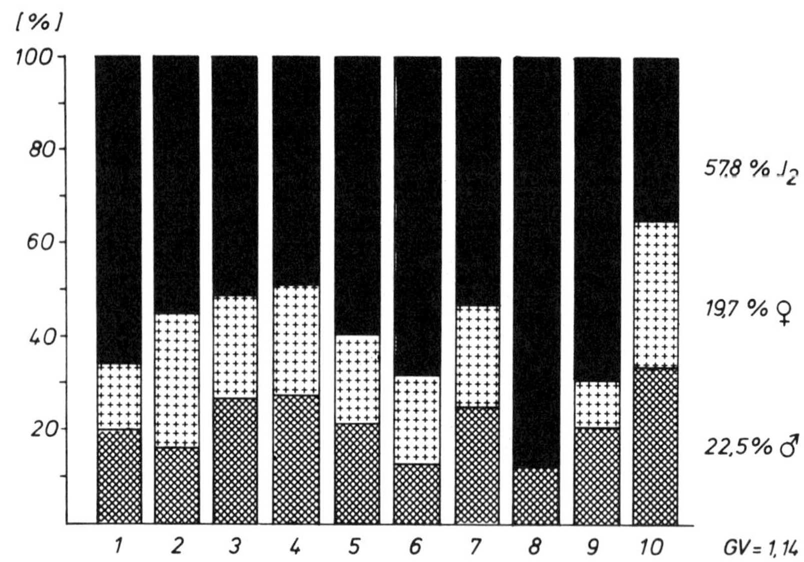

Fig. 2. Relative frequency of males, females and juveniles in susceptible oil seed radish and rape (GV = sex ratio).

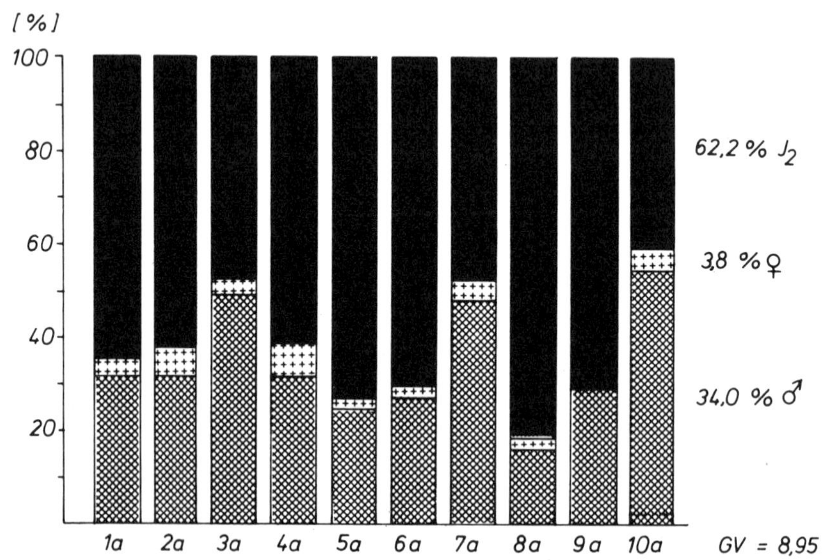

Fig. 3. Relative frequency of males, females, and juveniles in resistant oil seed radish cultivars (GV = sex ratio).

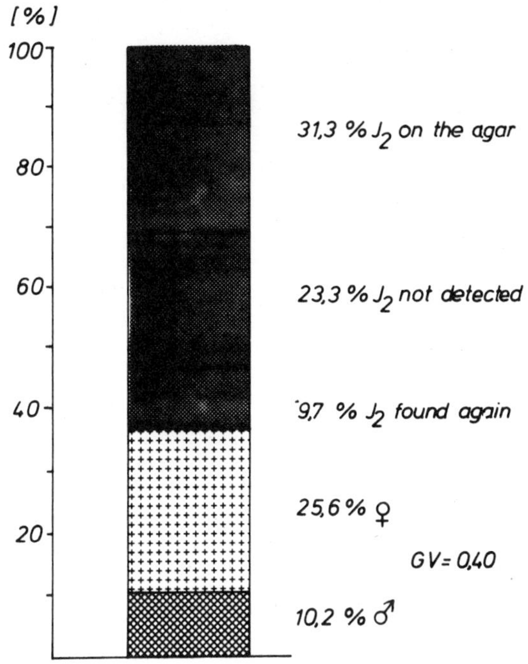

Fig. 4. Relative frequency of males, females, and juveniles on sugar beet inoculated with one juvenile per plant (GV = sex ratio).

percentage of juveniles penetrated the susceptible 'Siletina' as the resistant 'Pegletta' (Fig. 1).

The results of ten experiments with susceptible rape or oilseed radish are summarised in Fig. 2. Of the total inoculated individuals 19.7 per cent had differentiated to immature or adult females, and 22.5 per cent to males, giving a sex-ratio of 1.14. The remaining 57.8 per cent were second stage juveniles.

Corresponding ten experiments with resistant oilseed radish (cvs Slobolt, Pegletta and Nemex) resulted in 3.8 per cent of immature or adult females but 34.0 per cent of males, a sex-ratio of 8.95 (Fig. 3). Of the inoculated juveniles 62.2 per cent did not develop beyond the second stage, this being 4.4 per cent more than in the susceptible plants. The higher sex ratio in resistant plant tissue does not, however, point to a selective breakdown of female juveniles and in a potential female second stage juvenile there must be a sex reversal to male development. This was confirmed in an experiment with juveniles inoculated to sugar beet seedlings (Fig. 4). After two days, 31.3 per cent of the juveniles had not penetrated, and 33.0 per cent did not develop; 25.6 per cent of the inoculated juveniles developed to females, but only 10.2 per cent to males. The sex ratio of 0.40 must be explained by sex reversal in potentially male juveniles to female development under these favourable conditions.

The observations show that not only in parthogenetic species, like *Meloidogyne arenaria* and *M. incognita* but also in the amphimictic *H. schachtii*, sexual differentiation is influenced by environmental factors. The prospective sex of the nematodes is probably decided early in the development of the second stage juvenile soon after it starts to feed within the roots.

TESTING FOR RESISTANCE

A plant variety or cultivar is regarded as resistant, if it does not multiply or increase a particular nematode population. In the case of *H. schachtii* the population in the plant and in the soil has to be considered. Nematode multiplication occurs if the final population density (P_f) after harvesting the crop exceeds the initial density (P_i) before sowing. The official testing for resistance is based on this multiplication rate or P_f/P_i value.

The multiplication rate of a nematode population depends on temperature, soil moisture, plant growth, sowing date, initial population density, nematode antagonists and probably even more factors, and in experiments is impossible to standardise them all. Therefore it was decided to refer all test results to a susceptible standard cultivar, Siletina, the maximum multiplication rate of which was in the range of four under field conditions. This was used as the basis when calculating multiplication rates of the test lines.

Official resistance tests are carried out as pot trials as this ensures a multiplication rate of at least four. In addition, the soil can be mixed thoroughly in order to minimise variation in nematode distribution. The initial population density is adjusted to give 500 - 1,500 eggs plus juveniles per 100 g of soil, as this is the range that occurs most frequently in the field.

FIELD TRIALS

Influence of sowing date

Temperature has a marked influence on the population dynamics of *H. schachtii* and the effect of resistant cruciferous plants was supposed to be correlated with the accumulated day degrees during cultivation. Records of soil temperatures showed October to be the critical period when temperatures dropped below 10°C. The accumulated temperature can reach, however, the level which is necessary to complete a nematode generation. This is particularly true for early sowing in July whereas with late sowing it is doubtful if females will mature and produce eggs. Thus, late sowing may be advantageous. On the other hand I have found that a pronounced population decline can be expected only with relatively warm temperatures. A field trial with different sowing dates provided more information on this topic.

The oilseed radish cultivars Siletina (susceptible) and Pegletta (resistant), a resistant mustard cultivar and Phacelia (a non-host) were sown on four different dates (July 15, August 2, 16 and 30 in 1982). The accumulated day degrees ranged from 218 for the late sowing to 583 for the early sowing. Only the susceptible cv. Siletina showed a response to these differences with a multiplication rate (P_f/P_i) of two in the earliest sowing but no multiplication in the late sowing. When sown on August 30, this cultivar acted as a trap crop and attained almost the effect of a resistant plant. The relatively low multiplication rate of the early sowing indicates that the conditions for nematode development were not optimal as in other fields multiplication may occur even with late sowing. The influence of all other treatments was independent of the accumulative temperature. So even late sown resistant green manure crops may have a beneficial effect and, with the cultivars tested, early sowing does not bring the risk of a population increase. It should be noted that the experiment was in 1982 which was relatively warm in the autumn; other years may give different results.

The effect of resistant cultivars at different soil depths

Oilseed radish and mustard are quickly growing green manure plants, developing a strong tap root which penetrates deeply into the soil. The population dynamics of the sugar beet cyst nematode were investigated at three depths (0-15, 15-30 and 30-45 cm) using the oilseed radish cultivars Siletina (susceptible) and Pegletta (resistant), mustard cv. Maxi (resistant) and the non-host Phacelia (Steudel et al., 1985). The trial was carried out in 1983, in which the summer was hot and also temperatures in autumn were relatively high. Thus the accumulated temperature was higher than in 1981 and 1982. In 1983, 654 day degrees were recorded at 10 cm and 543 at 30 cm depth, with crops sown on July 15; the trial continued until the end of October when the temperature dropped below 10°C.

The nematode multiplied only on cv. Siletina. The highest multiplication rate (P_f/P_i; 3.4) was recorded at 0-15 cm soil depth, slightly less at 15 cm (2.7) and population increase (1.2) was recorded even at 30-45 cm. Nematode populations declined on all the other plants, and this was particularly pronounced in the deepest soil layer. The accumulated temperature of 543 day degrees is misleading as it was not known exactly when plant roots appeared at this depth and probably while intensive hatching occurred, few females matured. At 30-45 cm the contents of cysts (eggs + juveniles) decreased in cv. Siletina during the season, indicating that intensive hatching had occurred (Fig. 5). It is, however, not clear why the same effect was observed in the non-host Phacelia; probably changes of soil humidity play a role in this case.

The trial has shown that resistant cruciferous plants may have a favourable effect especially in deep soil layers. The resistant mustard, with a multiplication rate of 0.25, proved to be particularly efficient. Reducing nematode populations at depth is the main advantage of resistant plants compared with chemical soil disinfestation which is effective only in the upper 20-25 cm.

Host specific population density

The benefit from a resistant cultivar is influenced by the level of the nematode infestation. It is known that at very high initial infestation densities, populations do not increase even when a good host plant is grown. On the other hand it is not possible to eradicate an infestation by growing resistant cruciferous crops, as at least a few new cysts occur on all cultivars. At low infestation levels this will be sufficient to compensate for the population decline caused by hatching.

Schlang (1985) illustrated this situation by plotting the multiplication rates of *H. schachtii* on radish cultivars grown in the field. For cv. Nemex, which is approved as resistant, the equation $y = 23.0 \cdot x^{-0.56}$

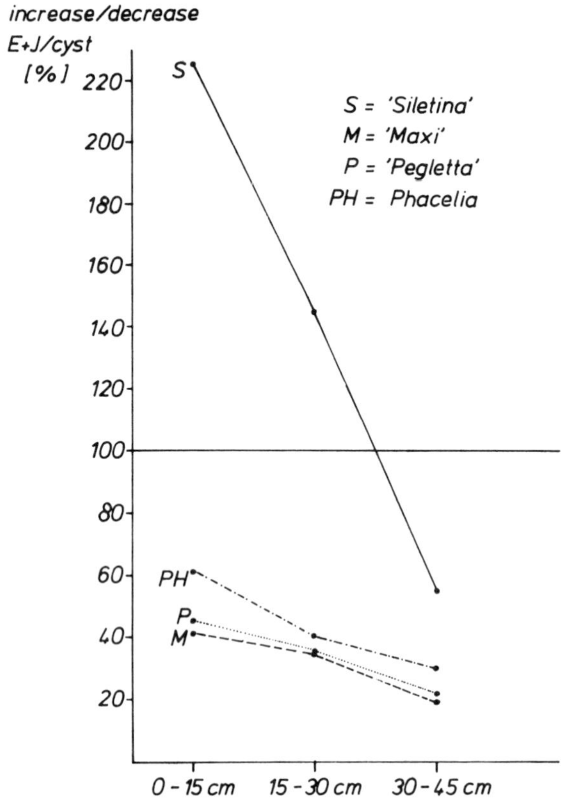

Fig. 5. Percentage increase or decrease of numbers of eggs and juveniles per cyst in a trial with resistant and susceptible cultivars (see text).

was obtained, indicating that at 1,000 eggs and juveniles per 100 g of soil as initial infestation, a multiplication rate of 0.5 can be expected (Fig. 6). For cv. Slobolt he found the equation $y = 94.7 \cdot x^{-0.65}$, and here the same initial density corresponds to a multiplication rate of 1.0 (Fig. 7). Thus, with cv. Slobolt a population reduction can only be achieved in heavily infested fields.

Fig. 8 illustrates the situation for three cultivars in a log-scale graph. The equilibrium density of a *H. schachtii* population will level off at about 3,500 eggs and juveniles per 100 g of soil with cv. Siletina, 1,000 with cv. Slobolt and at 280 with cv. Nemex. It was not possible to reduce populations beyond this level in the fields investigated. Cropping with the resistant Nemex may even increase very low population densities until they reach the equilibrium density. This means that the farmer should not grow resistant cultivars in non-infested fields, for example as a preventative (prophylactic) strategy against new infestations. Such fields should be cultivated with non-host plants. On the other hand, there is no risk on infested land, as the approved resistant cultivars maintain an equilibrium density beyond the damage threshold.

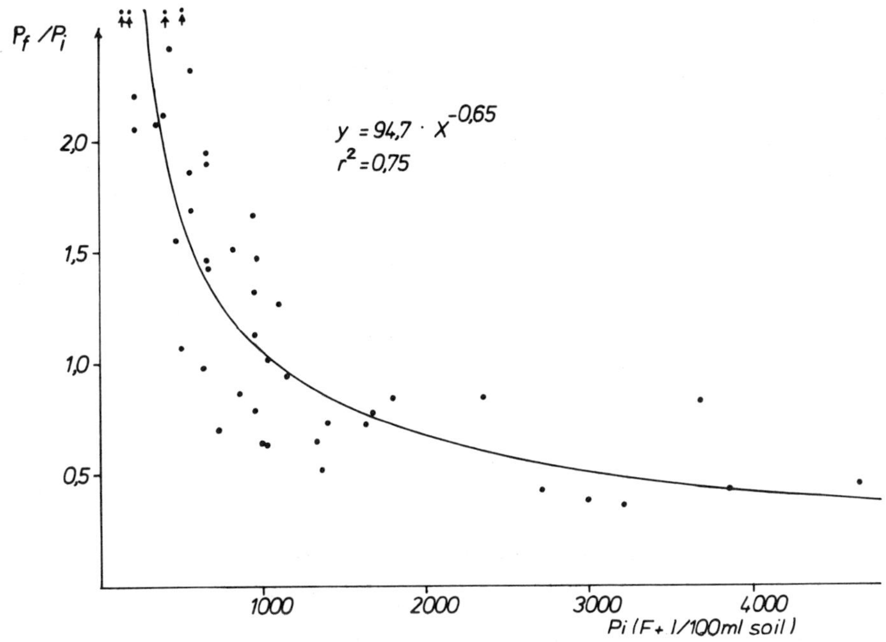

Fig. 6. Correlation between initial population density and multiplication rate of *H. schachtii* for the oilseed radish cv. Nemex (Schlang, 1985).

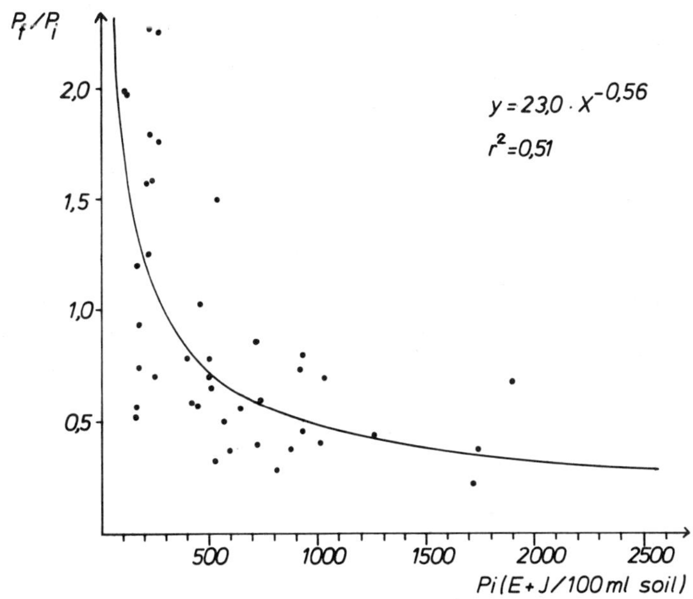

Fig. 7. Correlation between initial population density and multiplication rate of *H. schachtii* for the oilseed radish cv. Slobolt (Schlang, 1985).

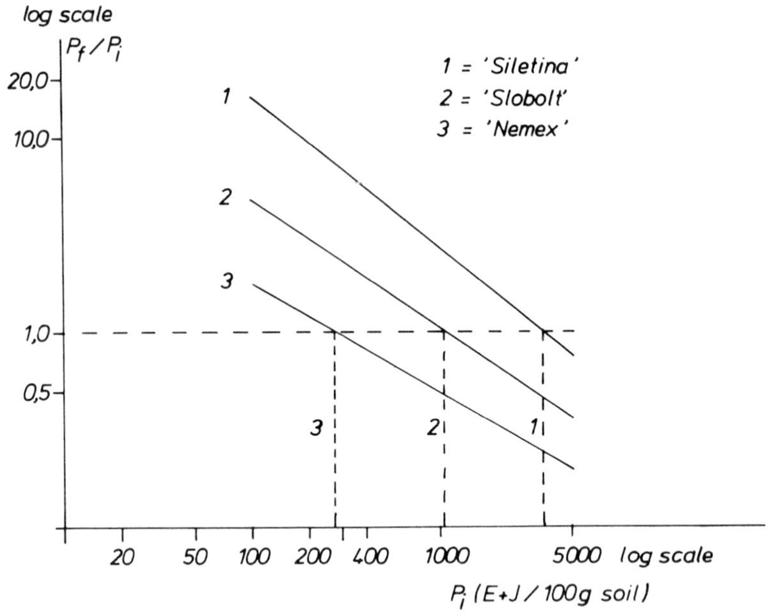

Fig. 8. Equilibrium densities of *H. schachtii* populations for three oilseed radish cultivars (Schlang, 1985).

MODEL CALCULATIONS FOR POPULATION DYNAMICS IN ROTATIONS

Steudel et al. (1981) published the results of 15 years of investigations in the Köln area of Germany on the natural population decline of *H. schachtii* with non-host plants and the multiplication rates on sugar beet. The data provide the basis for the following model calculations (Steudel and Müller, 1983).

Fig. 9 shows that the multiplication rate (P_f/P_i) is dependent on the initial population density (P_i). With this correlation and knowing the population density before growing sugar beet, the prospective final population density can be calculated. The average natural population decline in the crops following sugar beet was determined as 53 per cent in the first year, when winter wheat is usually grown, and 30 per cent in the second year with winter barley. Using these figures, the population dynamics in a three year rotation, which is usual in Germany, is represented in Fig. 10. Starting with a P_f of 1,500 after sugar beet, the population declines during the two years of cereals, but then continually increases every third year to reach about 3,000 after seven years of the three year rotations. The P_i will constantly exceed the damage threshold, which was determined as ca 500 eggs and juveniles per 100 g of soil. This is in agreement with investigations of Heijbroek (1977). A three year rotation on fields infested with *H. schachtii* therefore always has the risk of yield loss.

This traditional rotation system can be modified by including a resistant green manure crop after winter barley (Fig. 11). The population

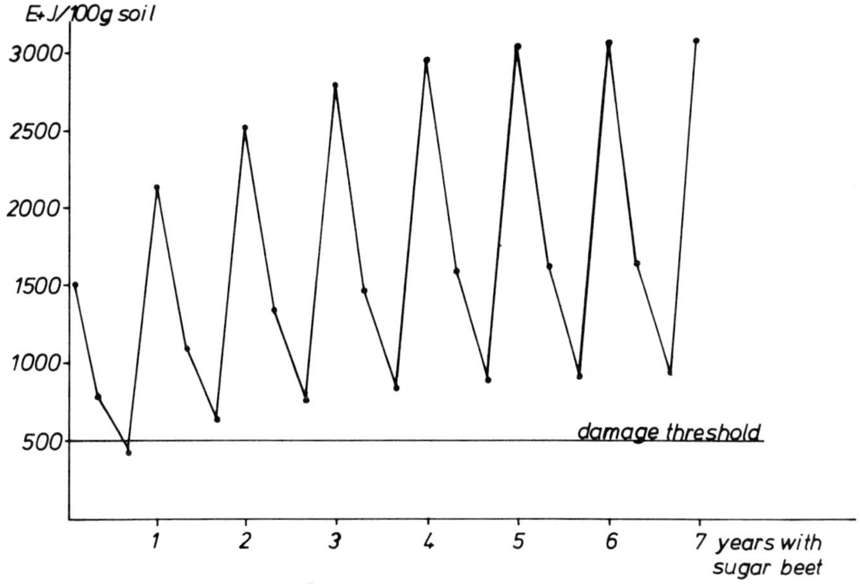

Fig. 9. Correlation between multiplication rate (P_f/P_i) and initial population density (Steudel and Müller, 1983).

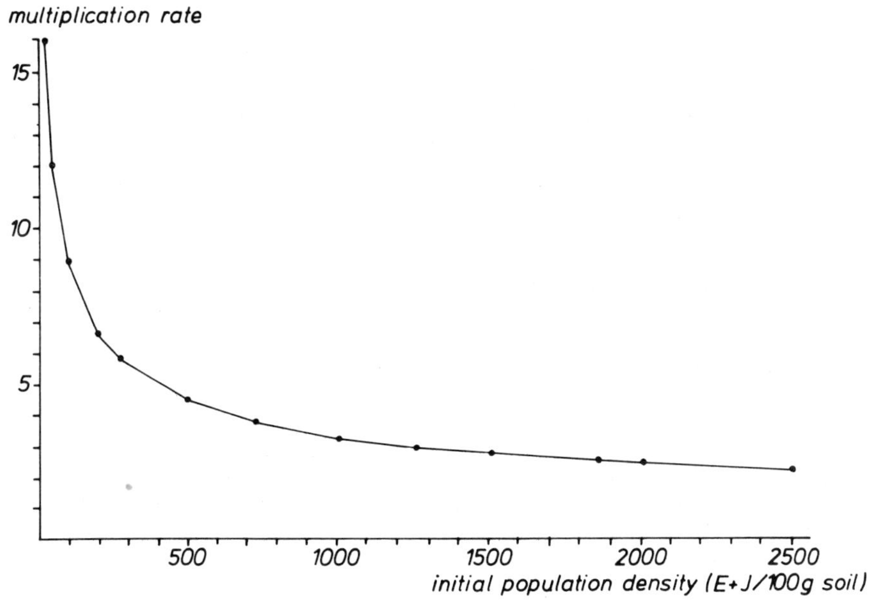

Fig. 10. Population dynamics of *H. schachtii* in a traditional 3 year rotation of beet-cereal cereal (Steudel and Müller, 1983).

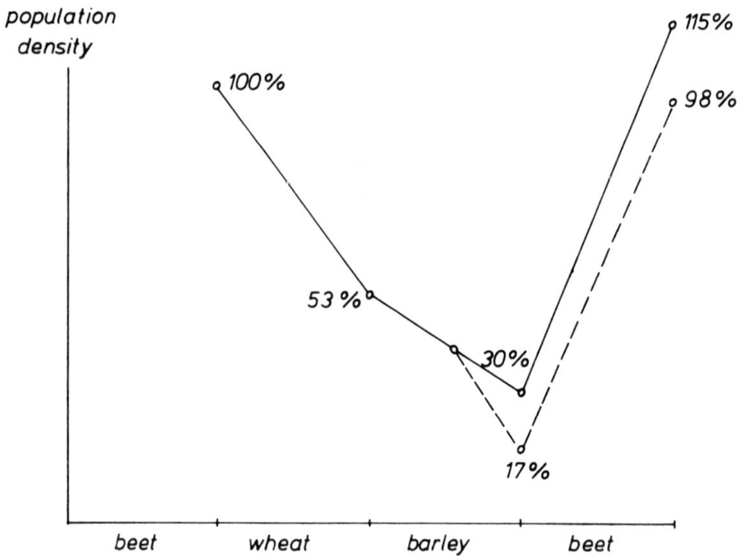

Fig. 11. Population dynamics of *H. schachtii* in a 3 year rotation with and without a resistant green manure crop (Steudel and Müller, 1983).

declines to 17 per cent of the P_f value after sugar beet, instead of 30 per cent without a resistant crop. When sugar beet is grown again, the new P_i would reach 98 per cent of the situation three years before, instead of 115 per cent in the old system. This calculation is based on a multiplication rate of 0.4 using the resistant green manure crop.

Fig. 12 illustrates the situation for four periods of a three year rotation. To make this and the remaining graphs clear, only the population densities before sugar beet (P_i) are indicated. Starting with a high P_i of 5,000, the population declines in both curves (A), but not beyond about 1,000 in the rotation without resistant green manure crops. If a resistant crop is included (dotted line), the P_i drops after three rotations below the damage threshold. In Fig. 12 (B) the situation starts with a low P_i of 500. As explained previously, the population can be kept at a low level with a resistant plant incorporated into the system, but it increases in the conventional rotation.

In some circumstances, e.g. a three year rotation of sugar beet-winter wheat-winter wheat, it should be possible to grow a resistant green manure crop twice in one rotation. Two calculations made for this system are given in Fig. 13: the solid line assumes a resistant cultivar with a multiplication rate of 0.6, which clearly is inadequate. Approved resistant cultivars must attain a multiplication rate of 0.5 or less. The dotted line was calculated with a multiplication rate of 0.4, which can be achieved, and with this the population declines after two rotations below the damage threshold. As explained earlier, however, it will be impossible to eradicate *H. schachtii*; the population will level off at the

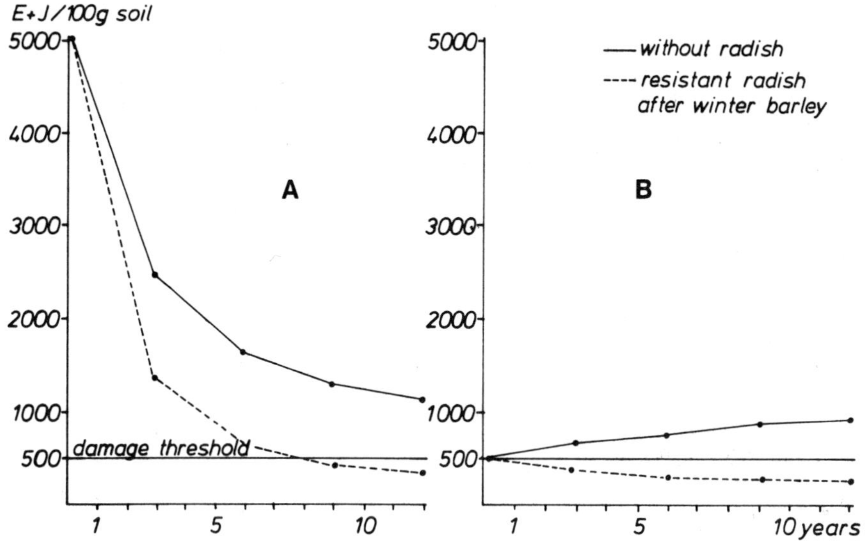

Fig. 12. Population dynamics of *H. schachtii* in 3 year rotations with and without resistant radish (P_f/P_i=0.4), grown once after winter barley, starting with high (A) and low (B) population densities (Steudel and Müller, 1983).

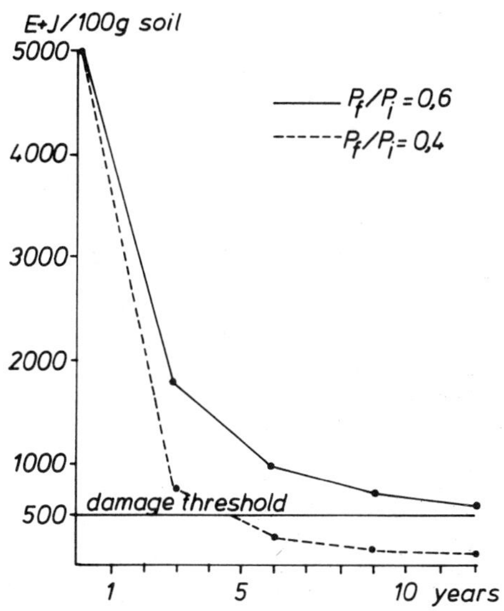

Fig. 13. Population dynamics of *H. schachtii* in a 3 year rotation (beet-winter wheat-winter wheat) with resistant radish (P_f/P_i= 0.6 or 0.4) grown every year after winter wheat (Steudel and Müller, 1983).

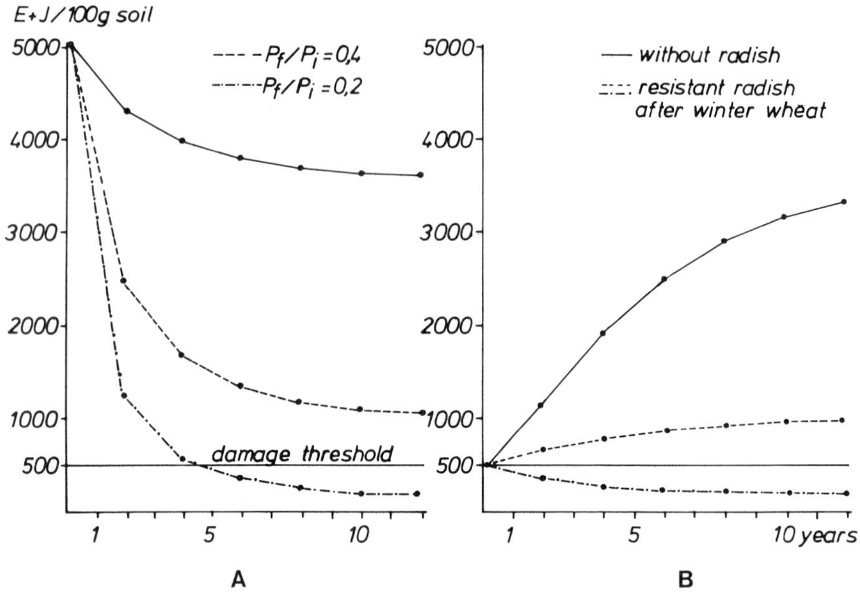

Fig. 14. Population dynamics of *H. schachtii* in a 2 year rotation with resistant radish (P_f/P_i = 0.4 or 0.2), starting with high (A) and low (B) population densities (Steudel and Müller, 1983).

equilibrium density, which depends on the degree of resistance in the cultivar used.

Often the farmer would like to grow sugar beet every second year instead of every third. On fields infested with the sugar beet cyst nematode this type of rotation would quickly lead to serious crop damage. Therefore it was important to investigate whether the situation could be improved by including a resistant cruciferous crop. This is illustrated in Fig. 14, again with a high and a low initial infestation density. The solid line shows the conventional system, which in both graphs ends between 3,000 and 4,000 eggs and juveniles per 100 g of soil, confirming the well authenticated experience of drastic nematode problems in a two year rotation. The dotted lines are calculated with a rotation to include a resistant green manure crop. However, a multiplication rate of 0.4 is too high to bring the population density below the damage threshold. Only a P_f/P_i value of 0.2, which is not achieved with available resistant cultivars, would be sufficient to provide a two year rotation without nematode problems.

The calculations as presented are a model. Under field conditions many factors as soil type, humidity, temperature, extent of root formation, and nematode antagonists will influence the population dynamics of *H. schachtii* and therefore cause considerable deviation from the predicted development. However, the use of resistant green manure crops will be capable of reducing sugar beet cyst nematode populations and are therefore a suitable input to an integrated system of control for this pest.

REFERENCES

Baukloh, H., 1976, Untersuchungen zur Wirtspflanzeneignung der Kruziferen gegenüber dem Rübennematoden *Heterodera schachtii* (Schmidt), unter besonderer Berücksichtigung der Resistenzzüchtung. Diss. Gießen, 1976.

Decker, H., Dowe, A. and Heide, A., 1977, Zur Wirtseignung von Kruziferen-Zwischenfrüchten für das Rübenzystenalchen (*Heterodera schachtii* Schmidt) unter besonderer Berücksichtigung des Ölrettichs, Nachrichtenblatt für den Pflschutz in der DDR, 31:136.

Heijbroek, W., 1977, Mogelijkheden om de schade door het bietezystenaaltje te voorkomen, Bedrijfsontwikkeling, 8:451.

Hirling, W., 1976, Die Wirtseignung des Ölrettichs (*Raphanus oleiferus* L.) für das Rübenzystenälchen (*Heterodera schachtii* Schmidt), Zeitschr. Pflkrankh. Pflschutz, 83:631.

Horn, A. van., 1961, Ölfruchte als Vermehrer des Rübennematoden, Gesunde Pflanzen., 13:248.

Kuhn, J., 1881, Ber. physiol. Lab. Univ. Halle 3, 153 pp.

Lücke, E., 1971, Ölzwischenfruchte in die Rübenfruchtfolge, Hannov. Land- u. Forstw. Ztg., 124:22.

Molz, E., 1920, Versuche zur Ermittlung des Einflusses äußerer Faktoren auf das Geschlechtsverhältnis des Rübennematoden (*Heterodera schachtii* A. Schmidt), Land. Jahrbücher, 54:769.

Muller, J., 1985, Der Einfluß der Wirtspflanze auf die Geschlechtsbestimmung bei *Heterodera schachtii*, Mitt. Biol., Bundesanstalt Land-Forstwirtschaft, Heft 226:46.

Müller, J. and Steudel, W., 1983, Der Einfluß der Kulturdauer verschiedener Zwischenfrüchte auf die Abundanzdynamik von *Heterodera schachtii* Schmidt, Nachrichtenbl. dt. Pflanzenschutzd., (Braunschweig), 35:103.

Schlang, J., 1985, Resistenzverhalten verschiedener Ölrettichsorten gegenüber *Heterodera schachtii*, Gesunde Pflanzen, 37:233.

Steudel, W. and Müller, J., 1983, Untersuchungen und Modellrechnungen zum Einfluß pflanzenverträglicher Nematizide und nematodenresistenter Zwischenfrüchte auf die Abundanzdynamik des Zuckerrübennematoden (*Heterodera schachtii*) in Zuckerrübenfruchtfolgen, Zuckerind, 108:365.

Steudel, W., Thielemann, R. and Haufe, W., 1981, Untersuchungen zur Populationsdynamik des rübenzystenälchens (*Heterodera schachtii* Schmidt) in der Köln-Aachener-Bucht, Mitt. Biol. Bundesanst, Heft., 199.

Steudel, W., Schlang, J. and Müller, J., 1985, Untersuchungen zum Einfluß einiger Zwischenfrüchte auf die Abundanzdynamik des Rübennematoden (*Heterodera schachtii* Schmidt) in verschiedenen Bodentiefen, Mitt. Biol. Bundesanst., Heft, 26:129.

Talatschian, P., 1974, Populationsentwicklung phytoparasitärer Nematoden an Stoppelfrüchten unter besonderer Berücksichtigung von Ölrettich, Zeitschr. Pflanzenkrankh. Pflanzenschutz, 81:538.

Thielemann, R., 1973, Kreuzblütige Zwischenfrüchte in Fruchtfolgen mit Zuckerrüben, Zuckerrübe, 22:15.

Thomas, E., 1974, Einfluß der Bondenbearbeitung auf die Vermehrung des Rübennematoden an verschiedenen Kruziferen, Landw. Zeitschr. Rheinland, 25:1246.

DISCUSSION

A number of questions were asked about the change from males to females in the juvenile stage and whether this could really be called "sex reversal". Müller explained that this may be something of a misnomer as the sex ratio is not fixed and the ratio of males to females depends on the environmental conditions prevailing. With reference to the data on the sugar beet cyst nematode, it should be noted that oil seed radish is not a good host and fewer females are produced than on sugar beet.

Referring to the model of population dynamics Müller explained that because the sugar beet cyst nematode is relatively polyphagous there is less likelihood of selecting pathotypes than with *G. pallida* and *G. rostochiensis*. Sikora observed, however, that root knot nematodes are very polyphagous and yet they have many races.

Sikora asked if the 53 per cent reduction in populations was justified and Muller replied that whilst there was much less variation from year to year he believed the work to be sound. Seinhorst suggested that some of the variation may be due to sampling error and added that from Steudel's data (see references) he had calculated a 53 per cent average reduction in population density.

Whitehead pointed out that according to Steudel (see references) the sugar beet cyst nematode has two generations per year in Germany and yet only a four fold increase in population was illustrated, which seems very low. He asked if any biological factors such as fungal parasites may be involved in suppressing multiplication. Müller said that the role of fungi is difficult to assess, but he thought the multiplication rate to be just average.

Trudgill asked about the type of resistance in oil seed radish and mustard, and Müller replied that it was not known whether it was mono- or polygenic. Andersson asked whether it would be more practical and advantageous when using resistant radish as a trap crop to sow it in the winter barley crop. Müller said that competition between the crops would have to be considered. Sikora added that work with trap crops was being done at Bonn, and it was found that the time of killing the trap was critical, and delay could lead to an increase in nematode numbers. Müller added that intercropping sugar beet with mustard or oil seed radish had no effect on the populations of sugar beet cyst nematode.

CELL RESPONSES TO *HETERODERA SCHACHTII* FEEDING ON SUSCEPTIBLE AND RESISTANT CULTIVARS OF OIL RADISH

Cornelia Stender

Botanisches Institut der Tierärztlichen Hochschule
Bünteweg 17d, D-3000 Hannover 71
Federal Republic of Germany

Ultrastructural changes at feeding sites of the cyst nematode *Heterodera schachtii* Schmidt were examined in roots of a susceptible (Siletina) and three resistant (Pegletta, Nemex and Slobolt) cultivars of *Raphanus sativus* var. *oleiformis* grown in aseptic nutrient agar culture. Apparently identical changes occurred in both cultivars between 1 and 2 days after invasion by the infective second stage juveniles. Induced syncytia (nurse cell systems) derived from parenchymatous cells of the vascular cylinder showed ultrastructural features of high metabolic activity such as a pronounced increase in cytoplasmatic density, reduction of the central vacuole into numerous small vacuoles, abundance of ribosomes, polyribosomes, mitochondria, dictyosomes and enlargement of nuclei.

Distinct ultrastructural differences between modified cells of susceptible and resistant roots first became evident at the second stage of observation, 3 days after nematode invasion. The differences were mainly changes in the nature of the endoplasmatic reticulum (ER) and in the degree of vacuolization. In the susceptible cultivar there was considerable proliferation of smooth-surfaced ER, whereas in the resistant cultivar the ER was rough-surfaced with flattened cisternae. The vacuoles in modified cells of resistant roots were larger than in those of susceptible roots.

Four to 5 days after nematode invasion of the roots of resistant cultivars, the cytoplasm and cell organelles within the syncytial cells were in a state of disintegration and vacuoles had further increased in size. At the same time the syncytia in the susceptible cultivar continued to develop.

Seven days after nematode invasion the developing syncytium in the susceptible cultivar was filled with dense cytoplasm and contained smooth ER. At the same time the syncytial cells of the resistant cultivars disintegrated with nearly complete protoplast degeneration.

The expression of resistance in the three resistant oil radish cultivars seemed to be always the same. In the beginning the syncytial cells were initiated, but degenerated after a few days.

The most striking differences between syncytia within susceptible and resistant cultivars concerned the endoplasmatic reticulum (ER). Three

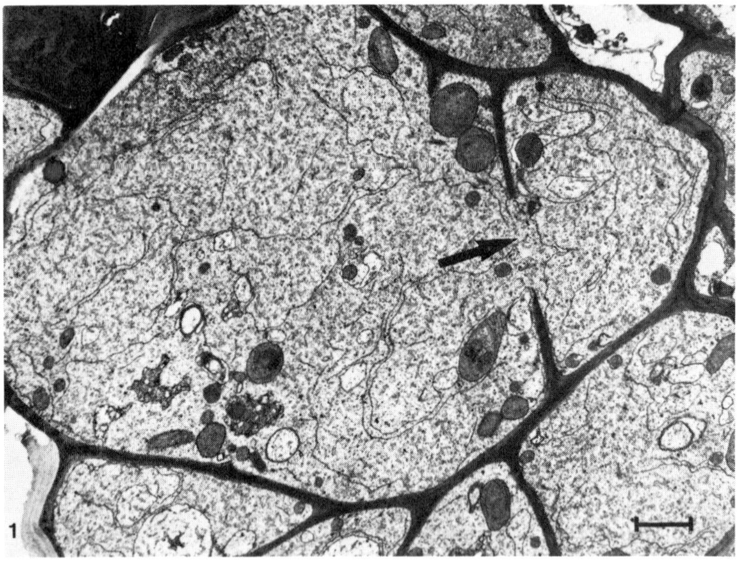

Fig. 1. Cross section through syncytium from root of a resistant cultivar of *R. sativus* var. *oleiformis* induced by a devloping male of *H. schachtii*. The syncytial cells are filled with dense cytoplasm and cell wall dissolutions (arrow) are evident. Bar = 1μm.

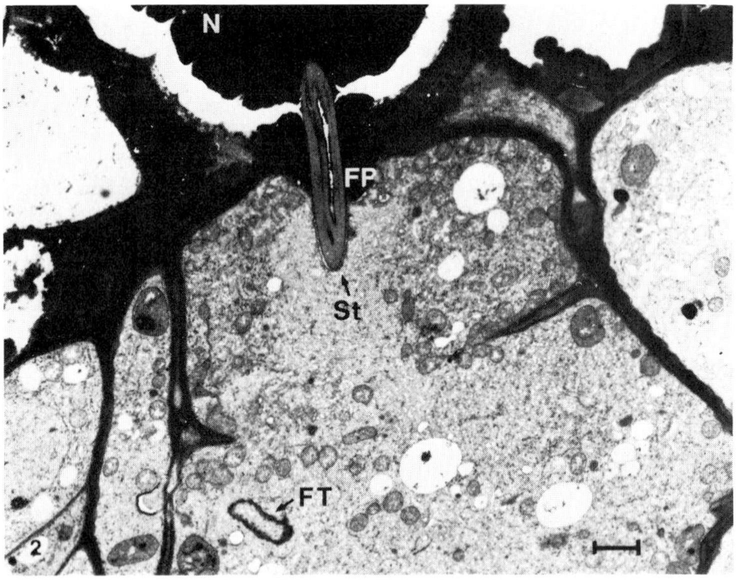

Fig. 2. Cross section through feeding site of *H. schachtii* in the root of a susceptible *R. sativus* var. *oleiformis* cultivar. Nematode stylet tip inserts into a cytoplasmatic zone free of cell organelles and with fewer ribosomes and polyribosomes than in surrounding cytoplasm. NB. electrondense feeding plug and feeding tube. FP, feeding plug; FT, feeding tube; ST, stylet; N, nematode. Bar = 1μm.

days after nematode invasion the smooth-surfaced ER predominated in the susceptible and rough-surfaced ER in the resistant cultivar.

Although just speculation, it is possible that the genetically controlled programming of syncytia degeneration may lead to the synthesis of digestive enzymes at the ribosomes which are attached to the ER. These enzymes are then released into the cisternal cavities of the ER. The lytic enzymes may then be transported into the vacuoles which have been observed to increase in size again in the syncytia of the resistant cultivars. The breakdown of the tonoplast might then lead to the observed rapid lysis of the cytoplasm of the resistant cultivars (Wyss et al., 1984).

These early degenerated syncytia in resistant roots do not provide enough nutrients for juvenile development, probably not even for male development. But as males develop in the resistant cultivars the question arises about the features of syncytial cells induced by males.

Fig. 1 shows a syncytium in the roots of the resistant cultivar Pegletta induced by a developing male, which has just moulted to the J4 stage. At this stage of development the juvenile has just stopped feeding (Müller and Wyss, 1981), and at this time the male induced syncytia in the resistant as well as in the susceptible cultivars are filled with dense cytoplasm, cell wall dissolutions are evident and the cytoplasm does not show any sign of degeneration.

When the juvenile has moulted to a male and has stopped feeding for some time, the syncytium starts to degenerate. Large vacuoles develop again and the cytoplasm starts to degenerate.

Syncytia which developing males are able to initiate in susceptible as well as in resistant varieties, are usually smaller than the syncytia initiated by a developing female.

In some sections the feeding of the nematode on the syncytial cells could occasionally be observed (Fig. 2). Cells directly fed upon by the nematode can easily be recognised by an electron dense feeding plug, which originates from the amphids (Endo, 1978). Also, feeding tubes dispersed within the cytoplasm indicate the cells on which the nematode has fed. These tubes are formed by rapidly hardening nematode saliva secreted through the stylet into the plant cell. The saliva is produced in the very active dorsal oesophageal gland cell (Wyss and Zunke, 1986).

Usually these feeding tubes are surrounded by a cytoplasmatic zone, which is free of cell organelles such as mitochondria and plastids. This zone has less ribosomes, polyribosomes and sometimes less membranous systems than the surrounding cytoplasm (Fig. 2).

The cytoplasm around the feeding tubes is definitely modified, probably by a very subtle action of tube enzymes to promote solute flow from the cytoplasm into the tubes (Wyss et al., 1984).

REFERENCES

Endo, B.Y., 1978, Feeding plug formation in soybean roots infected with the soybean cyst nematode, Phytopathology, 68:1022.

Müller, J. and Wyss, U., 1981, Entwicklung des Zystennematoden *Heterodera schachtii*, Publ.Wiss.Film., Sekt.Biol., Ser.14, Nr. 10/C 1387, 20 pp. Inst.Wiss.Film, Göttingen.

Wyss, U. and Zunke, U., 1986, Observations on the behaviour of infective *Heterodera schachtii* J2-juveniles inside host roots, Rev.Nematol., 9,

Wyss, U., Stender, C. and Lehmann, H., 1984, Ultrastructure of feeding sites of the cyst nematode *Heterodera schachtii* Schmidt in roots of susceptible and resistant *Raphanus sativus* L. var. *oleiformis* Pers. cultivars, Physiol.Pl.Path., 25:21.

ULTRASTRUCTURAL RESPONSES IN SUGARBEET TO *HETERODERA SCHACHTII*

Teresa Bleve Zacheo and Giuseppe Zacheo

Istituto di Nematologia Agraria, C.N.R.
70126 Bari
Italy

Second stage juveniles of *Heterodera schachtii* Schmidt entered the roots of sugarbeet (*Beta vulgaris* L.) by penetrating meristematic cells of the root tip. The juvenile body was often found lying parallel to the root axis in tunnels formed within the cortex during nematode migration. Three days after entering the root, the lip region of the nematode was located beyond the endodermal cells, which were not completely differentiated. Nematode penetration caused mechanical damage to host tissues, evident as a necrotic ring, involving endodermal cells and three layers of the vascular tissue. The empty space around the juvenile permitted the movement of its lip region, allowing it to vary the site of stylet insertion in the feeding area of the vascular parenchyma, which as a result of the feeding was transformed into a syncytium. The occurrence of bundles of tubes at the site of stylet insertion, indicates that they may be produced repeatedly in the syncytial cells (Rumpenhorst, 1984).

Three days after nematode inoculation, numerous cell wall stubs and gaps were present in the syncytium as a result of cell wall dissolution. The absence of the plasma membrane covering these perforations allowed the fusion of protoplasts and movement of nuclei between adjacent cells. The cells in the syncytium showed an increase in the cytoplasmic density in terms of numerous organelles, the amount of membranous material and the many small vacuoles. Clusters of an unusual type of vesicle (Newcomb, 1967) were present in the syncytial cells, very close to the feeding region and associated with profiles of smooth endoplasmic reticulum; they could have a role in transferring proteins from one component to another (Gunning and Steer, 1975). The meristematic tissue outside the syncytium showed an unusual arrangement of endoplasmic reticulum, in the form of concentric cisternae containing secretory material and proplastids with proteinaceous bodies in their stroma. It seems that the meristematic cells, not directly involved in the nematode feeding, respond to the invasion message by producing proteins, which are either transferred into the intracisternal space of rough endoplasmic reticulum and then transported to the exterior, or are accumulated in the stroma of plastids.

The syncytium, twelve days after *H. schachtii* inoculation, was transformed into a large cell, containing residues of the previous cell walls, which had become thickened because of polysaccharide deposition. The cell wall enclosing the syncytium also was thickened and the portions adjacent to vessels, especially xylem vessels, showed numerous wall protrusions.

The ground cytoplasm, without vacuoles and very few organelles, was densely packed with vacuolated plastids near the wall protrusions and an extensive system of membranes, resembling smooth endoplasmic reticulum. Smooth endoplasmic reticulum may be involved in the transport of solutes to the feeding area of the nematode (Wyss et al., 1984). At a later stage of syncytial development the smooth endoplasmic reticulum, produced by a loss of ribosomes on its surfaces, became conspicuous. The transformed reticulum is specialised in transport and secretion (Robards, 1974) suggesting that the syncytium, at this stage, may function as a secretory cell.

REFERENCES

Gunning, B.E.S. and Steer, M.W., 1975, Ultrastructure and the Biology of Plant Cells, Edward Arnold, London.

Newcomb, E.H., 1967, A spiny vesicle in slime-producing cells of the bean root, J. Cell Biol., 35:C17.

Robards, A.W., 1974, Dynamic Aspects of Plant Ultrastructure, McGraw-Hill, UK.

Rumpenhorst, H.J., 1984, Intracellular feeding tubes associated with sedentary plant parasitic nematodes, Nematologica, 30:77.

Wyss, U., Stender, C. and Lehmann, H., 1984, Ultrastructure of feeding sites of the cyst nematode *Heterodera schachtii* Schmidt in roots of susceptible and resistant *Raphanus sativus* L. var. *oleiformis* Pers. cultivars, Physiol. Plant Pathol., 25:21.

THE SOYBEAN CYST NEMATODE

Gregory R. Noel

US Department of Agriculture, Agricultural Research Service
Department of Plant Pathology, University of Illinois
Urbana, IL 61801,

HISTORY AND DISTRIBUTION

The soybean cyst nematode (SCN), *Heterodera glycines* Ichinohe, now infests many soybean producing areas in the USA and will, no doubt, eventually infest every major production area. Ichinohe (1959) reviewed the early history of SCN. It was first reported in Japan in 1915 by Hori (1915), but the nematode had been known for many years prior to that and was believed to be *H. schachtii* Schmidt (Ishikawa, 1916). Ishikawa referred to the soybean disease as "Tsukiyobo" which translates into "moon night disease" in reference to the pale yellow oval patches of diseased soybeans. Ito (1921) later coined the name "Daizu-iwo-byo" or "soybean yellow dwarf disease", the name more commonly used today.

H. glycines occurs in Columbia (Norton, 1983), Egypt (Elmiligy, 1968), Indonesia (Nishizawa, 1984) Korea, (Yokoo, 1936), People's Republic of China (Nakata and Asuyana, 1938), the United States of America (USA) (Winstead et al., 1955) and the Amur region of the Soviet Union (R. Gotlova, personal communication). There are no published reports of the occurrence of SCN in Argentina or Brazil, but agriculturists from South America have told the author that the nematode does infest parts of the extensive soybean production areas in those two countries.

In the USA, SCN is the most important nematode occurring on soybean. In spite of recent advances in chemical control and release of resistant cultivars, the nematode continues to be an ever increasing and serious problem. For instance, in Illinois, the state with the largest production of soybeans in the USA, SCN was reported for the first time in 16 counties during the 1984 and 1985 growing seasons.

Since its discovery in the USA in 1954, SCN has been reported in 24 states, Nebraska being the only state with large-scale soybean production from which it has not been reported. The question "Is *H. glycines* indigenous to the USA or was it introduced?" often arises. Recent two-dimensional electrophoresis studies (Ferris et al., 1985) reported that certain USA populations were more closely related to each other than they were to a Japanese population, indicating that USA populations may be indigenous and derived from an ancient ancestor which was widely distributed in Asia and North America. Some believe that *H. glycines* may have evolved in the United States from *H. schachtii*. However, there are

Table 1 Characterisation of *Heterodera glycines* races.

Race	Reproduction				
	Pickett	Peking	PI88788	PI90763	Lee
1	-	-	+	-	S
2	+	+	+	-	S
3	-	-	-	-	S
4	+	+	+	+	S
5	+	-	+	-	S

Reproduction on a cultivar or line less than 10% of that on Lee is given a negative sign (-) and an amount of 10% or more is given a positive sign (+); S = susceptible host. Source: Golden et al. (1970); Inagaki (1979).

several reports that soil was imported from China and Japan for *Rhizobium* studies during the latter half of the 19th century (Anon., 1968). This soil was then sent to various agriculture experiment stations and scientists in the USA. SCN could have been imported several times and distributed in soil to many locations where it went unnoticed for many years while infecting native weed hosts, lespedeza used for hay, and soybeans. It was not until the intensive soybean production which has occurred in the USA since World War II that SCN became a problem.

Most SCN hosts are in the Leguminoseae, but reproduction occurs on hosts in other botanical families (Riggs and Hamblen, 1962; Riggs and Hamblen, 1966). Hosts of economic importance are soybean (*Glycine max* Merr.), snap bean (*Phaseolus vulgaris* L.), adzuki bean (*Phaseolus anqularis* Wright), certain *Lespedeza* spp., and certain *Melilotus* spp. (crimson or scarlet clovers). Most red clovers (*Trifolium*) spp. and peas (*Pisum sativum* L.) are either highly resistant or immune. No monocotyledonous plants have been reported as hosts.

RACES

Five races of SCN have been reported although many populations exist which do not fit the current race scheme illustrated in Table 1 (Golden et al., 1970; Inagaki, 1979). Riggs et al. (1981) studied the race identification of 33 SCN populations collected from the USA and Japan. They found that when five soybean differentials were used, six "races" were identified, but when 13 differentials were used there were 25 "races". Miller (1966) reported that different SCN races could be identified from the same field. Triantaphyllou (1975) reported that under greenhouse conditions, changes in the degree of parasitism of certain SCN populations occurred rapidly within seven generations. However, one population did not reproduce on the resistant cultivar Pickett. Triantaphyllou (1975) stated that differences in the ability of individual populations to parasitize resistant hosts were attributed to quantitative differences in gene frequency among the populations. In addition to race shifts in field populations, the ability of races to interbreed and produce viable progeny further enhances the possibility for a population to parasitize a resistant cultivar (Triantaphyllou, 1975; Price et al., 1978). Several genes have been reported to confer resistance to *H. glycines* (Ross and Brim, 1957; Caldwell et al., 1960; Matson and Williams, 1965; Hartwig and Epps, 1970; Thomas et al., 1975).

McCann et al. (1982) used a field population of SCN to select several populations which multiplied on different sources of resistance. When the selected populations were challenged with the seven resistant lines on which selection occurred, differences in numbers of white females were observed. The authors concluded that low numbers of females in certain population-resistance combinations were due to low frequency of alleles for parasitism.

To better understand the nature of SCN races and to develop an improved method of race identification which would not rely on differential cultivars, studies have been made on the use of immunology, electrophoresis and scanning electron microscopy (SEM). Variations in serological reaction occurred but SCN races could not be identified (Riggs et al., 1982). In another study using crossed immunoelectrophoresis race 4 had an antigen not present in race 3 (Griffith et al., 1982). Pozdol and Noel (1984) found no differences in native proteins of *H. glycines* races 1, 2, 3 and 4 when separated by polyacrylamide gel electrophoresis, but when proteins were denatured with sodium dodecyl sulphate, races 3 and 4 were consistently separable from races 1 and 2. Radice et al. (1985) reported enzyme polymorphism for aspartate aminotransferase and -esterase in single females from SCN populations. When SEM was used, no differences were found in juvenile cephalic morphology among the five SCN races (Noel and Stanger, 1986). However, similarities in cyst morphology among certain races enabled placement of the races into three groups consisting of race 1, race 3, and races 2, 4 and 5 (Stanger and Noel, 1986).

Most nematologists who study *H. glycines* agree that the present race scheme is inadequate. Nevertheless, it has provided a system of identification useful to growers who want to know what race is predominant in a field and what cultivar to plant to avoid nematode damage.

POPULATION DYNAMICS

Several workers have investigated hatching and emergence of *H. glycines*. Factors studied have included pH, moisture, relative humidity, temperature, inorganic ions, herbicides, root diffusates, seasonal variation, and a chemically defined hatching factor from kidney bean. Greatest emergence from cysts occurred at pH 3.5, using either phosphate or phthalate buffers, which gave nearly identical results (Lehman et al., 1971a). Juveniles emerged from egg masses earlier (2 to 6 days) than from cysts and was unaffected by pH 2.5, 3.5, or 4.5. Emergence from cysts was greatest at pH 3.5 and was maximum at 8 to 11 days. However, hatch from eggs free of cysts was greatest at pH 6.0 (Tefft et al., 1982). Eggs maintained at relative humidity (RH) 8.5% hatched up to 4 days, but hatch at 100% RH continued at a low rate up to 16 days at which time the experiment was terminated (Endo, 1962). Greater hatch occurred when soil moisture was 25% than at 33 or 15% (Tefft et al., 1982). Dessication of cysts reduced juvenile emergence and numbers of viable eggs (Skotland, 1957; Slack and Hamblen, 1961). Slack and Hamblen (1961) also reported that juveniles emerged from cysts at temperatures between 16 and 36°C, but failed to emerge at higher or lower temperatures. The optimum temperature for emergence was 24°C.

Zinc salts and ferrous sulphate highly stimulated *H. glycines* hatch in vitro (Clarke and Shepherd, 1966; Tefft and Bone, 1984, 1985a) and calcium moderately stimulated emergence from cysts (Lehman et al., 1971a). However, compounds containing ammonium or nitrate were inhibitory to juvenile emergence from cysts.

Organic compounds in the form of soil-applied herbicides commonly used in soybean production differ in their ability to stimulate hatch and emergence of juveniles in vitro (Kraus et al., 1986). Vernolate, metribuzin, alachlor and trifluralin stimulated hatch and emergence in vitro but metalachlor did not. The stimulation occurred during 14 days after initiation of the experiment, but there was no effect at 21 days. Vernolate and metribuzin were most active at 7 days while trifluralin was most active at 14 days. Alachlor activity was similar at 7 and 14 days. In general, concentrations of 1.25, 2.5 and 5.0 ppm of the herbicides were most effective in stimulating hatch and emergence whereas 10.0 ppm was ineffective. Bostian et al. (1984a) found that combinations of phenamiphos (0.05 µg/ml) plus alachlor (0.063, 0.125 or 1.0 µg/ml) increased hatch at 18 days when compared with the control and with alachlor plus 1.0 µg/ml phenamiphos. Greater numbers of juveniles were associated with alachlor concentrations of 0.125 µg/ml. *H. glycines* populations in greenhouse and field studies exhibited trends similar to those observed in vitro (Riggs and Oliver, 1982; Kraus et al., 1982; Schmitt et al., 1983). In greenhouse studies five times as many juveniles were recovered from trifluralin treated soil than from the untreated controls (Riggs and Oliver, 1982). The densities of juvenile field populations were twice those of the control at two weeks and six times greater than the control at six weeks. Treatment with metolachlor also resulted in large population increases at six weeks. Schmitt et al. (1983) reported that both egg and juvenile populations were higher in phenamiphos plus alachlor treatments than with phenamiphos in combination with either metribuzin or linuron. The herbicides and phenamiphos had single and interactive effects. These effects might have resulted from hatch stimulation or perhaps altered host response resulting in increased host suitability of the soybeans. Bostian et al. (1984b) found that during early soybean growth, plants treated with alachlor had less nematode infection than did control plants, but plants treated with phenamiphos and alachlor had more nematode penetration than plants treated with phenamiphos alone.

Early results regarding hatch stimulation by root diffusates were often contradictory with most workers reporting a lack of stimulation (Ichinohe, 1955; Skotland, 1957; Slack and Hamblen, 1961; Clarke and Shepherd, 1966). However, Tsutsumi and Sakurai (1966) reported that hatching was greater for diffusate from kidney bean than either soybean or adzuki bean root leachings. Success with root leachings was apparently dependent upon use of diluted leachings (Okada, 1971). Okada (1972b) found that kidney bean diffusate and an unknown factor from ultrasonically ruptured eggs were equal in their ability to stimulate hatch. Okada (1972a) also reported that an unknown factor within cysts inhibited hatch of eggs; this factor was not present in the cyst wall. By unknown mechanisms the factor(s) within eggs and cysts work in concert to control long-term egg survival and hatch.

Ross (1963) surmised that diapause occurred in *H. glycines* when he reported a sharp decline in numbers of juveniles emerging from cysts in field soil beginning in late September. During November no juveniles emerged. In subsequent months, a gradual increase in emergence occurred through mid-April. In the northern USA and other northern production areas such as northern Japan, Korea, and the People's Republic of China, survival is probably greater than what would be expected in the southern USA or warmer temperate climates. In the northern USA the temperature in the autumn usually drops quite drastically in September and October. Nematode activity and that of biocontrol organisms diminishes rapidly, whereas in the southern USA temperatures are more moderate and both nematodes and biocontrol organisms may remain active during the winter. Thus in northern production areas, percent survival of *H. glycines* eggs and juveniles is greater than in southern areas. In cooler soybean production regions only three to four nematode generations are produced. In more temperate regions six or seven generations may occur. However, most of the population increase occurs during the first generation (Lawn and Noel, 1986).

Tefft and Bone (1985b) found that hatch stimulation by root diffusates from intact plants increased during vegetative and reproductive stages of soybean development but declined as plants senesced. Addition of the chelating agent ethylenediamine tetraacetic acid to root diffusates from plants of various ages enhanced the rate of SCN egg hatch. Leucine amino peptidase (LAP) was associated with hatch (Tefft and Bone, 1985a). They reported that zinc chloride at 10^{-4} - 10^{-10} M increased LAP activity of egg supernatant by 36% compared with the controls. Additionally, egg supernatant with zinc hydrolyzed more substrate than nontreated egg supernatant. Egg supernatant with zinc chloride at 10^{-3} M decreased LAP activity by 20%. Root diffusates had no effect on LAP activity in the supernatant from homogenised eggs of *H. glycines* (Tefft and Bone, 1985b).

Masamune et al. (1982) reported the isolation and purification of an *H. glycines* hatching factor, glycinoeclepin A, from kidney beans. Detailed structural analysis was provided by Fukuzawa et al. (1985). Glycinoeclepin A is a dibasic acid with a molecular weight of 446 and a molecular formula of $C_{25}H_{34}O_7$. Glycinoeclepin A is highly active, stimulating hatch <u>in vitro</u> at concentrations of 10^{-11} - 10^{-12} g/ml.

HOST-PARASITE INTERACTIONS

Melton et al. (1986) developed the equation $DS = 5(10^{-6})x^2y^2 - 3(10^{-4})x^2y - 2.8(10^{-3})x^2 - 1.94(10^{-2})y^2 + 0.4288x + 1.0220y - 12.7185$ which defined the relation between developmental stage (DS), time in days after inoculation (x), and constant temperature (y) from 16 to 32°C. The r^2 for this equation is 0.84. The authors also found that the number of males increased at lower temperatures. The necrotic response of resistant snapbeans and soybeans was not affected by temperature. The necrotic response was more extensive in the resistant soybean than in the resistant snapbean. Melton et al. (1986) also reported that resistance in snapbean and soybean did not affect the male:female ratio, but snapbean increased production of males. Juvenile penetration was similar in susceptible and resistant soybeans and in snapbeans. In resistant soybeans mature females were smaller and produced fewer eggs and egg masses than in susceptible soybeans (Noel et al., 1983).

Incompatible reactions in soybean usually resulted in smaller syncytia that degenerated and were replaced by parenchyma cells (Acedo et al., 1984). Syncytia of the few successful females in incompatible reactions were large, located within the stele, and had normal cytoplasm. However, the syncytia were ovoid rather than elongated and apparently did not incorporate adjacent cells as in the compatible reaction.

INTERACTION WITH RHIZOBIUM

H. glycines race 1 can severely reduce *Rhizobium japonicum* (Kirchner) Buchanan nodule weight and nitrogen fixing capacity of the bacterium (Lehman et al., 1971b). Nodule weight was reduced more when soybeans were inoculated with race 1 when compared with soybeans inoculated with races 2 and 4. Also, nitrogen fixation capacity was much less when plants were infected with race 1 when compared to race 4. Nodules from nematode infected plants were poorly organised with less distinct zones of nodular tissue and were characterised by early appearance of vascular elements and sclerenchyma (Ko et al., 1985). Nodules from infected plants also contained less leghemoglobin content per gram of nodule than control plants and had massive accumulation of starch granules and crystalline arrays of phytoferratin in cell plastids (Huang and Barker, 1983). These observations suggest that metabolism of iron-containing compounds and carbohydrates are affected by SCN infection. Huang et al. (1984) reported that binding of *R. japonicum* to soybean root hairs was suppressed by SCN infection and concluded that this was due to SCN interference with soybean lectin metabolism.

PESTICIDE INTERACTIONS

Field studies of interactions between nematicides and soil-applied herbicides used for weed control in soybeans have proven quite complex. Schmitt et al. (1983) reported that *H. glycines* egg and juvenile populations were greater in phenamiphos plus alachlor treatments than with phenamiphos in combination with either metribuzin or linuron. Combinations of aldicarb plus either alachlor, linuron, or metribuzin were much more effective in increasing SCN populations than were combinations of those herbicides and phenamiphos (D. P. Schmitt, personal communication). Soybeans treated with aldicarb and alachlor yielded more than soybeans treated with aldicarb plus linuron or metribuzin (Schmitt et al., 1983). Kraus et al. (1982) reported that trifluralin, metribuzin, and vernolate also increased SCN populations in field experiments but alachlor had no effect. They also reported that aldicarb was ineffective for increasing soybean yields unless used in combination with either vernolate, trifluralin, metribuzin, or trifluralin plus metribuzin. Use of alachlor plus aldicarb did not result in a yield increase.

CONTROL

If soybeans were grown once every four or five years SCN would not be a problem. Even with the present intensive soybean production in the USA, SCN can be managed with integrated control.

A number of cultivars with resistance to SCN have been developed in the USA, Japan and the People's Republic of China (Wrather et al., 1984; Inagaki, 1979: Wu et al., 1982). Most of the cultivars released in the USA have been developed by the USDA Agricultural Research Service and various universities, but private seed companies have become increasingly active in cultivar development during the past few years. Unfortunately, most SCN resistant cultivars do not have the yield potential of their high-yielding susceptible parents when both are planted in noninfested fields. Yields of resistant cultivars are commonly 85 to 90 per cent of those obtained with the susceptible parent. In addition, during the time required to develop resistant cultivars, newer high-yielding susceptible cultivars with greater yield potentials are also ready for release. The end result is that, in many instances, the need to plant a resistant cultivar costs the grower a substantial percentage of his yield potential.

Another approach for controlling SCN is the development of tolerant cultivars which can support large numbers of SCN, but with minimal effects on yield. Boerma and Hussey (1984) found soybean lines that supported high levels of SCN reproduction (were susceptible) but had little suppression of yield (were tolerant). The resistant-intolerant combination was not found. However, D. P. Schmitt (personal communication) indicated that in North Carolina, cultivar Forrest (resistant to races 1 and 3) has become susceptible to SCN in certain areas and is more intolerant to certain SCN populations than standard susceptible lines.

DBCP was the most effective nematicide for SCN control. It was economical and easy to apply. However, environmental considerations resulted in its ban during 1980. Much research effort has been spent to find alternative nematicides, but none is as inexpensive or has provided results equal to those obtained with DBCP. Efficacy of carbamate and organophosphate nematicides differs in different areas of the USA. In Illinois, aldicarb has provided the most consistent control (Edwards and Noel, 1981). Reis et al. (1983) showed that within certain economic considerations, aldicarb used on high-yielding SCN-susceptible soybeans rotated with maize was economically sound and provided more income for the grower than continuous planting of a resistant cultivar or resistant cultivar rotated with maize. The resistant cultivar rotated with maize was nearly as competitive economically as the susceptible cultivar treated with aldicarb and rotated with maize.

Success of any control approach is dependent on knowledge of the damage potential of SCN populations in the field. In Illinois, on silt loam soils with 2% organic matter, economic injury levels are 600 eggs plus juveniles or 12 cysts containing viable eggs/250 cm^3 at planting (Noel, 1984). A linear regression of yield on cyst numbers and a curvilinear (quadratic) regression of yield on numbers of eggs plus juveniles describe the relationship between SCN and yield. In North Carolina linear relationships were found with sandy soils and quadratic relationships in other soil types (D. P. Schmitt, personal communication).

Biological control of SCN is receiving more emphasis. Ownley Gintis et al. (1983) recovered a large number of fungal associates of *H. glycines*. A progressive increase in fungal colonisation occurred with nematode development. Some of the fungi, namely *Chaetomium cochloides* (Palliser), *Paecilomyces* spp., *Phoma terrestris* (Mont.), *Phytophthora cinnamoni* Rands, *Phythium* sp., *Verticillium chlamydosporium* Goddard, and *V. lecanii* (Zimm.) Viegas demonstrated chitinase activity when grown on chitin agar plates. Rodriguez-Kabana et al. (1984) demonstrated control of *H. glycines* using crustacean chitin at rates of 0.5 - 4.0% (w/w).

IN SUMMARY

H. glycines presents many challenges to nematologists and soybean growers. SCN cannot be controlled solely by use of resistance, nematicides, or rotation. Current resistance is not durable; nematicides are not always efficaceous and may pose environmental problems; rotations are frequently unprofitable. Durable resistance or tolerance and development of increased efficacy and safety of nematicides with new technologies are needed. Markets for rotation crops must be developed. When a better understanding of economic injury levels, population dynamics and biological control is obtained, SCN control utilising integrated management can be developed. In the future growers may be able to utilise genetically engineered hatching factors to reduce nematode populations. Whatever SCN management system is developed, its success will be dependent on grower acceptance.

REFERENCES

Acedo, J.R., Dropkin, V.H. and Luedders, V.D., 1984, Nematode population attrition and histopathology of *Heterodera glycines*-soybean associations, J. Nematol., 16:48.

Anon, 1968, Inoculate Soybeans? Soybean News, 20:4.

Boerma, H.R. and Hussey, R.S., 1984, Tolerance to *Heterodera glycines* in soybean, J. Nematol., 16:289.

Bostian, A.L., Schmitt, D.P. and Barker, K.R., 1984a, In vitro hatch and survival of *Heterodera glycines* as affected by alachlor and phenamiphos, J. Nematol., 16:22.

Bostain, A.L., Schmitt, D.P. and Barker, K.R., 1984b, Early growth of soybean as altered by *Heterodera glycines*, phenamiphos and/or alachlor, J. Nematol., 16:41.

Caldwell, B.E., Brim, C.A. and Ross, J.P., 1960, Inheritance of resistance of soybeans to the cyst nematode, *Heterodera glycines*, Agron. J., 52:635.

Clarke, A.J. and Shepherd, A.M., 1966, Inorganic ions and the hatching of *Heterodera* spp., Ann. appl. Biol., 58:497.

Edwards, D.I. and Noel, G.R., 1981, Update: Soybean cyst nematode, Ill. Res., 23:4.

Elmiligy, I.A., 1968, Occurrence of *Heterodera glycines* from the Golden Island, U.A.R., Nematologica, 14:148.

Endo, B.Y., 1962, Survival of *Heterodera glycines* at controlled relative humidities, Phytopathology, 52:80.

Ferris, V.R., Ferris, J.M. and Murdock, L.L., 1985, Two-dimensional protein patterns in *Heterodera glycines*, J. Nematol., 17:422.

Fukuzawa, A., Furusaki, A., Ikura, M. and Masamune, T., 1985, Glycinoeclepin A, a natural hatching stimulus for the soybean cyst nematode, J. Chem. Soc., Chem. Commun., 222.

Golden, A.M., Epps, J.M., Riggs, R.D., Duclos, L.A., Fox, J.A. and Bernard, R.L., 1970, Terminology and identity of infraspecific forms of the soybean cyst nematode (*Heterodera glycines*), Plant Dis. Reptr., 54:544.

Griffith, T.W., Koehler, N.J., Coker, S.G. and Riggs, R.D., 1982, Race comparisons of *Heterodera glycines* using crossed immunoelectrophoresis, J. Nematol., 14:598.

Hartwig, E.E. and Epps, J.M., 1970, An additional gene for resistance to the soybean cyst nematode, *Heterodera glycines*, Phytopathology, 60:854.

Hori, S., 1915, Phytopathological notes. Sick soil of soybean caused by a nematode, (In Japanese). J. Pl. Protect., 2:927.

Huang, J.-S. and Barker, K.R., 1983, Influence of *Heterodera glycines* on leghemoglobins of soybean nodules, Phytopathology, 73:1002.

Huang, J.-S., Barker, K.R. and Van Dyke, C.G., 1984, Suppression of binding between rhizobia and soybean roots by *Heterodera glycines*, Phytopathology, 74:1381.

Ichinohe, M., 1955, Studies on the morphology and ecology of the soybean nematode, *Heterodera glycines*, in Japan, Hokkaido Agr. Exp. Sta. Rep. No. 48:1.

Ichinohe, M., 1959, Studies on the soybean cyst nematode, *Heterodera glycines*, and its injury to soybean plants in Japan, Plant Dis. Reptr. Suppl., 260:238.

Inagaki, H., 1979, Race status of five Japanese populations of *Heterodera glycines*, Jap. J. Nematol., 9:1.

Ishikawa, T., 1916, Principal insect pests and diseases of plants occurring during the fourth year of Taisho. II. Occurrence of "Moon night" and wilt disease of soybean. (In Japanese), J. Pl. Protect., 3:197.

Ito, S., 1921, Studies on "Yellow dwarf" disease of soybean. (In Japanese), Hokkaido Agr. Exp. Sta. Rep., 11:47.

Ko, M.P., Huang, P.-Y., Huang, J.-S. and Barker, K.R., 1985, Accumulation of phytoferritin and starch granules in developing nodules of soybean roots infected with *Heterodera glycines*, Phytopathology, 75:159.

Kraus, R., Noel, G.R. and Edwards, D.I., 1982, Effect of pre-emergence herbicides and aldicarb on *Heterodera glycines* population dynamics and yield of soybean, J. Nematol., 14:452.

Kraus, R., Edwards, D.I. and Noel, G.R., 1986, Herbicides as hatching factors for *Heterodera glycines*, Nematologica, 32: (in press).

Lawn, D.A. and Noel, G.R., 1986, Field interrelationships among *Heterodera glycines*, *Pratylenchus scribneri* and three other nematode species associated with soybean, J. Nematol., 18:98.

Lehman, P.S., Barker, K.R. and Huisingh, D., 1971a, Effects of pH and inorganic ions on emergence of *Heterodera glycines*, Nematologica, 17:467.

Lehman, P.S., Huisingh, D. and Barker, K.R., 1971b, The influence of races of *Heterodera glycines* on nodulation and nitrogen-fixing capacity of soybean, Phytopathology, 61:1239.

Masamune, T., Anetai, M., Takasugi, M. and Katsui, N., 1982, Isolation of a natural hatching stimulus, glycinoeclepin A, for the soybean cyst nematode, Nature, 297:496.

Matson, A.L. and Williams, L.F., 1965, Evidence of a fourth gene for resistance to the soybean cyst nematode, Crop Sci., 5:477.

McCann, J., Luedders, V.D. and Dropkin, V.H., 1982, Selection and reproduction of soybean cyst nematodes on resistant soybeans, Crop Sci., 22:78.

Melton, T.A., Jacobsen, B.J. and Noel, G.R., 1986, Effects of temperature on development of *Heterodera glycines* on *Glycine max* and *Phaseolus vulgaris* and on host resistant reaction, J. Nematol., 18: (in press).

Miller, L.I., 1966, Variations in development of two morphologically different isolates of *Heterodera glycines* obtained from the same field, Phytopathology, 55:1068.

Nakata, K. and Asuyama, H., 1938, Survey of the principal disease of crops in Manchuria. (In Japanese), Bur. Indus. Rep., 32:166.

Nishizawa, T., 1984, The cyst nematode occurring in soybean fields in the tropics, Java Island, Proc. Sixth Biennial Cyst Nematode Workshop, College Park, Maryland, USA.

Noel, G.R., 1984, Relating numbers of soybean cyst nematode to crop damage, Proc. Fifth Cyst Nematode Workshop, p.17.

Noel, G.R. and Stanger, B.A., 1986, Scanning electron microscopy of second-stage juvenile cephalic morphology in *Heterodera glycines*, J. Nematol., 18 (in press).

Noel, G.R., Stanger, B.A. and Bloor, P.V., 1983, Population development, reproductive behavior, and morphology of race 4 of the soybean cyst nematode, *Heterodera glycines*, on resistant soybeans, Plant Disease, 67:179.

Norton, D.C., 1983, *Heterodera glycines* on soybean in Columbia, Plant Disease, 67:1389.

Okada, T., 1971, The hatching response of the soybean cyst nematode, *Heterodera glycines* Ichinohe (Tylenchida: Heteroderidae), Appl. Ent. Zool., 6:91.

Okada, T., 1972a, Hatching inhibitory factor in the cyst contents of the soybean cyst nematode, *Heterodera glycines* Ichinohe (Tylenchida: Heteroderidae), Appl. Ent. Zool., 7:99.

Okada, T., 1972b, Hatching stimulant in the egg of the soybean cyst nematode, *Heterodera glycines* Ichinohe (Tylenchida: Heteroderidae), Appl. Ent. Zool., 7:234.

Ownley Gintis, B., Morgan-Jones, G. and Rodriguez-Kabana, R., 1983, Fungi associated with several developmental stages of *Heterodera glycines* from an Albama soybean field soil, Nematropica, 13:181.

Pozdol, R.F. and Noel, G.R., 1984, Comparative electrophoretic analyses of soluble proteins from *Heterodera glycines* races 1-4 and three other *Heterodera* species, J. Nematol., 16:332.

Price, M., Caveness, C.E. and Riggs, R.D., 1978, Hybridization of races of *Heterodera glycines*, J. Nematol., 10:114.

Radice, A.D., Riggs, R.D. and Huang, F.E., 1985, Protein and enzyme polymorphism of *Heterodera glycines*, J. Nematol., 17:510.

Reis, R.P., Noel, G.R. and Swanson, E.R., 1983, An economic analysis of alternative control methods for soybean cyst nematode in Southern Illinois, Plant Disease, 67:480.

Riggs, R.D. and Hamblen, M.L., 1962, Soybean cyst nematode studies in the family Leguminosae, Ark. Exp. Sta. Rep., Series 110, 18 pp.

Riggs, R.D. and Hamblen, M.L., 1966, Further studies on the host range of the soybean cyst nematode, Ark. Exp. Sta. Bull., 719, 19 pp.

Riggs, R.D. and Oliver, L.R., 1982, Effect of trifluralin (Treflan) on soybean cyst nematode, J. Nematol., 14:466.

Riggs, R.D., Hamblen, M.L. and Rakes, L., 1981, Infra-species variation in reactions to host in *Heterodera glycines* populations, J. Nematol., 13:171.

Riggs, R.D., Rakes, L. and Hamblen, M.L., 1982, Morphometric and serologic comparisons of a number of populations of cyst nematodes, J. Nematol., 14:188.

Rodriguez-Kabana, R., Morgan-Jones, G. and Ownley Gintis, B., 1984, Effects of chitin amendments to soil on *Heterodera glycines*, microbial populations, and colonization of cysts by fungi, Nematropica, 14:10.

Ross, J.P., 1963, Seasonal variation of larval emergence from cysts of the soybean cyst nematode, Phytopathology, 53:609.

Ross, J.P. and Brim, C.A., 1957, Resistance of soybeans to the soybean cyst nematode as determined by a double-row method, Plant Dis. Reptr., 41:923.

Schmitt, D.P., Corbin, F.T. and Nelson, L.A., 1983, Population dynamics of *Heterodera glycines* and soybean response in soils treated with selected nematicides and herbicides, J. Nematol., 15:432.

Skotland, C.B., 1957, Biological studies of the soybean cyst nematode, Phytopathology, 47:623.

Slack, D.A. and Hamblen, M.L., 1961, The effect of various factors on larval emergence from cysts of *Heterodera glycines*, Phytopathology, 51:350.

Stanger, B.A. and Noel, G.R., 1986, Scanning electron microscopy of vulval cones of *Heterodera glycines* and three related cyst nematode species, J. Nematol., 18 (in press).

Tefft, P.M. and Bone, L.W., 1984, Zinc-mediated hatching of eggs of soybean cyst nematode, *Heterodera glycines*, J. Chem. Ecol., 10:361.

Tefft, P.M. and Bone, L.W., 1985a, Leucine aminopeptidase in eggs of the soybean cyst nematode *Heterodera glycines*, J. Nematol., 17:270.

Tefft, P.M. and Bone, L.W., 1985b, Plant-induced hatching of eggs of the soybean cyst nematode *Heterodera glycines*, J. Nematol., 17:275.

Tefft, P.M., Rende, J.F. and Bone, L.W., 1982, Factors influencing egg hatching of the soybean cyst nematode *Heterodera glycines* race 3, Proc. Helminthol. Soc. Wash., 49:258.

Thomas, J.D., Caveness, C.E., Riggs, R.D. and Hartwig, E.E., 1975, Inheritance of reaction to race 4 of soybean cyst nematode, Crop Sci., 15:208.

Triantaphyllou, A.C., 1975, Genetic structure of races of *Heterodera glycines* and inheritance of ability to reproduce on resistant soybeans, J. Nematol., 7:356.

Tsutsumi, M. and Sakurai, K., 1966, Influence of root diffusate of several hosts and non-host plants on the hatching of the soybean cyst nematode, *Heterodera glycines* Ichinohe, 1952, Jap. J. Appl. Ent. Zool., 10:129.

Winstead, N.N., Skotland, C.B. and Sasser, J.N., 1955, Soybean cyst nematode in North Carolina, Plant Dis. Reptr., 39:9.

Wrather, J.A., Anand, S.C. and Dropkin, V.H., 1984, Soybean cyst nematode control, Plant Disease, 68:829.

Yokoo, T., 1936, Host plants of *Heterodera schachtii* Schmidt and some instructions, Korea Agric. Exp. Sta. Bull., 8:47.

Wu, H., Yao, Z., Li, X., Liu, H., Shang, S. and Mao, D., 1982, Studies on the screening of resistant sources to cyst nematode of soybean, China Ag. Sci., 6:19.

DISCUSSION

Noel stated that aldicarb was applied at 8 Kg/ha (1.2 Kg a.i./ha) and of the 4.0 million hectares of soybean in Illinois about 30,000 ha would be treated with the chemical. Jacobsen explained that Illinois was a "two crop" state, maize and soybean, and that small fields were 16-32 ha, large fields 200 ha plus. Trudgill referred to the parallel curves on the figure showing yields from treated (aldicarb) and untreated plots which to him suggested that other pests were being controlled. Noel commented that possibly they did not detect low SCN infestations and possibly in the presence of aldicarb there may have been side effects on other parasites. Jacobsen said that 22 fungi had been identified with the soybean nematode.

B'Chir asked why three years of a non-host in a crop rotation system was considered necessary if populations of SCN dropped by 50-90% in one year. Jacobsen replied that the decision depended very much on the starting level of the infestation; detectable damage to the crop was caused at a level of 5 billion cysts/ha. Sikora asked about the future outlook for soybean in Illinois if there are several generations of SCN per year and soybeans are grown in two year rotations. Noel said it was like fighting a forest fire with a garden hose, but there were only 25 years work done on the problem so far and the outlook for the future was one of confidence and optimism, but he added that finances for research had been cut to the extent that resistant soybean lines could not be examined in the 1985 programme.

Stone said the situation with regard to races and physiological races in SCN was similar to that with the potato cyst nematode and asked if the pathotype system for SCN was satisfactory or is any change in it anticipated. Noel replied that the system for SCN is being examined critically.

SOURCES OF RESISTANCE TO THE SOYBEAN CYST NEMATODE

Satish C. Anand

Department of Agronomy, University of Missouri-Columbia
Portageville, MO 63873

INTRODUCTION

Soybean cyst nematode (SCN), *Heterodera glycines* Ichinohe, was identified on soybean in the USA in 1954 (Winstead et al., 1955). Subsequently, several pathogenic races of the nematode were discovered (Ross, 1962). Golden et al. (1970) described four physiological races of SCN based on their reaction on five soybean lines (Table 1). Another race (race 5) was first reported from Japan (Inagaki, 1979) and it has now been isolated from a number of places in the USA (Anand et al., 1985; Young, 1982). Continuous cultivation of resistant soybean cultivars was found to change the parasitic capabilities of nematode populations (Triantaphyllou, 1975), consequently the field populations of SCN may be a mixture of several races. Currently races 3 and 4 are very prominent in the southeast and southern regions of the USA whereas race 5 has been reported in areas of mid-south, where race 4 resistant cultivars have been planted for a number of years.

The first set of SCN resistant cultivars released were Pickett (Brim and Ross, 1966), Dyer (Hartwig and Epps, 1968), and Custer (Luedders et al., 1968). These cultivars were slightly less productive than the adapted cultivars in SCN non-infested fields. The second set of cultivars included Pickett 71 (Hartwig et al., 1971), Forrest (Hartwig and Epps, 1973), Mack (Caviness et al., 1972) and Centennial (Hartwig and Epps, 1977). All these cultivars carried resistance from 'Peking', a black seeded plant introduction from China. Breeding efforts were reorganised with the recognition of race 4 as a highly virulent race. This led to the development of Bedford in maturity group V (Hartwig and Epps, 1978), Fayette in maturity group III, and Bradley in maturity group VI (Anand and Shannon, 1984). These cultivars possess resistance to race 4 derived from PI 88788, another introduction from China. Currently, all SCN resistant cultivars released in the USA have derived their resistance from either Peking or PI 88788, or both. Thus there is a very narrow genetic base for SCN resistance in our soybean cultivars. Commercial cultivars resistant to SCN races 2 and 5 are not available at this time.

SCREENING METHODS

Ross and Brim (1957) tested the soybean lines in double row method in cyst infested fields. In recent years, screening is done by planting the

Table 1. Reaction of soybean differentials to the races of *Heterodera glycines*[a]

Soybean Differential	Races 1	2	3	4	5
Lee 68	+	+	+	+	+
Pickett	-	+	-	+	+
Peking	-	+	-	+	-
PI 88788	+	+	-	-[b]	+
PI 90763	-	-	-	+	-

[a] Races 1-4 as described by Golden et al. (1970) and Race 5 by Inagaki (1979).

[b] PI 88788 was originally classified as susceptible, but later work showed it was resistant to race 4.

+ number of white females 10% or more of those on Lee 68; -,number of white females less than 10% of those on Lee 68.

cultivars in SCN infested soil in pots in the greenhouse. Resistance is evaluated by counting white females on the roots 25 to 30 days after planting (Anand, 1982). A more accurate method is inoculating each plant with a known number of eggs or juveniles. The reaction of a plant is based on the number of white females per plant compared with those on the susceptible check. A cultivar with less than 10% of the number of white females of those on Lee is considered resistant. Since there was a great variation between cultivars for the level of resistance, the plants with 5% to 10% of the cysts were categorised as moderately resistant. Plants with greater than 10% of the cysts but significantly less than those on the susceptible check were classified as moderately susceptible.

IDENTIFICATION OF RESISTANT GERMPLASM

Plant introductions PI 90763, PI 84751, Ilsoy, and Peking were reported to be resistant to the field populations of soybean cyst nematode in North Carolina (Ross and Brim, 1957). Since the SCN races had not been defined at this time, it was not known to which races these PI lines were resistant. Epps and Hartwig (1972) screened approximately 3000 soybean lines against race 4, and found PI 88788, PI 89772, PI 87631-1, Cloud, Columbia, Peking, 84751, and PI 90763 to have a high level of resistance. Another 2000 PI lines were screened by Anand (1982) who reported PI 416762 to be resistant to both races 3 and 4. The entire USA Soybean World Collection of over 9000 PI lines was screened by Anand and Gallo (1984) against race 3 and they found 34 additional resistant lines. Among these, 14 had a moderate to high degree of resistance to race 4 and 11 were moderately to highly resistant to race 5. PI 437654 was the only line resistant to races 3, 4 and 5 of SCN. Subsequently it was found that PI 437654 carried resistance to races 1 and 2 as well. The reaction of resistant PI lines against the races 3, 4 and 5 is given in Table 2.

Table 2. Reaction of resistant soybean germplasm to races 3, 4 and 5 of *Heterodera glycines*

	Races		
Soybean Line	3	4	5
Ilsoy	R	MR	MS
PI 16790	R	MR	S
PI 17852B	R	S	MR
PI 22897	R	S	S
PI 54591	MR	S	S
PI 70218-2-19-3	MR	S	S
PI 79609	R	MS	S
PI 79693	MR	S	S
PI 84751	R	S	MR
PI 87631-1	R	R	S
PI 88788	R	R	S
PI 89008	MR	S	S
PI 89014	MR	S	S
PI 89772	R	S	R
PI 90763	R	S	R
PI 91138	MR	S	S
PI 92720	R	S	S
PI 200495	MR	S	MS
PI 209332	R	R	S
PI 303652	R	S	S
PI 339868B	R	MS	MR
PI 398680	R	R	S
PI 398682	MR	S	S
PI 404166	R	S	R
PI 404198A	R	MR	R
PI 404198B	R	MR	MS
PI 407729	R	MR	MS
PI 407944	MR	S	S
PI 408192-2	MR	S	S
PI 416762	R	MR	MS
PI 417091	MR	S	MS
PI 417094	MR	S	S
PI 437488	MR	S	S
PI 437654	R	R	R
PI 437655	R	MR	MS
PI 437679	R	S	MR
PI 437690	R	S	MR
PI 437725	R	S	MS
PI 437770	MR	MR	S
PI 438183	MR	S	S
PI 438489B	R	S	R
PI 438496B	R	S	S
PI 438497	R	S	MS
PI 438498	R	S	MS
PI 438503A	R	R	MS

R: Resistant
MR: Moderately resistant
MS: Moderately susceptible
S: Susceptible

The majority of these resistant lines were in maturity groups II, III and IV (Table 3). Although maturity groups 000-I constituted 25% of the total collection, none of the resistant lines were in these early groups. Similarly no resistant line was in maturity groups VI-X. Approximately two thirds of the entire collection also had yellow seed coat whereas 71% of the resistant lines were black seeded. These two factors would indicate that the resistant lines probably originated in the same geographical area and may have a common genetic background.

IDENTIFICATION OF SOURCES OF RESISTANCE

Different soybean lines may carry the same or different genes for resistance to a biotype of SCN. It is important to identify the genetic relationships for resistance among soybean lines. If two soybean lines have the same genes/alleles for resistance, they may be considered as a single source of resistance. Preserving genetic diversity would require utilisation of SCN resistance from several sources.

The relationship between two soybean lines can be recognised by either testing them against the same biotype of the pathogen or by studying the segregation for resistance following hybridisation. Both PI 88788 and PI 90763 were reported to be resistant to race 4 of SCN when tested against the field population (Epps and Hartwig, 1972). Studies involving selected SCN populations indicated that those two PI lines were in fact different (Anand and Brar, 1983). Whereas PI 88788 and PI 90763 had moderate levels of resistance to SCN population developed on Forrest, they differed significantly when tested against populations developed on PI 88788 and PI 90763 (Table 4). The same research showed that PI 87631-1, PI 209332 and Cloud were similar to PI 88788, whereas PI 89772 was closely related to PI 90763 (Anand and Brar, 1983).

After the identification of an additional 34 resistant PI lines (Anand and Gallo, 1984), it was felt necessary to compare them with race 3 resistance in Forrest derived from Peking. Single F_2 plants involving crosses of

Table 3. Resistant lines of *Heterodera glycines* in different maturity groups

	Number of lines screened		Number of Resistant lines	
Maturity Group	Total	% of Total	Total	% of Total
000-I	2257	24.8	0	0
II	1111	12.1	9	20.0
III	1030	11.2	16	35.5
IV	2126	23.2	17	37.8
V	1369	15.0	3	6.7
XI-V	1250	13.7	0	0
Total	9153		45	

Table 4. Number of white females per plant incoulated with populations of *Heterodera glycines*[1]

Inoculated	SCN Population Selected on			
Soybean Line	Essex	Forrest	PI 88788	PI 90763
Essex	104 a[2]	222 a	315 a	221 a
Forrest	95 a	217 a	178 b	257 a
PI 88788	5 b	7 b	97 c	9 c
PI 90763	16 b	51 b	4 d	78 b

[1] From Anand and Brar, 1983.

[2] Values with same letters are not significantly different at 5% level of probability.

Forrest with the new resistant PI lines were screened against race 3. The F_3 progenies were also tested to confirm the results of F_2 plants. Some of the crosses were also screened for race 5 resistance. Although it will take several years to study all the combinations, the results of these experiments which are so far available are presented in Table 5.

Of the eight crosses involving Forrest, four segregated for SCN race 3 resistance. It would indicate that the resistance for race 3 in PI 437665, PI 88788, PI 209332 and PI 90763 is carried on locus or loci different than those in Forrest. Since the SCN resistance in soybean is controlled by several genes (Caldwell et al., 1960), more than one gene differenece may exist between Forrest and the other four soybean lines. No segregation for resistance was observed in crosses involving PI 437654, PI 437679, and PI 438498 which indicated that the resistance for race 3 in these PI lines was located on the same loci as Forrest. Gene difference for resistance were also observed between Peking and PI 438489B for race 3, and between PI 90763 and PI 404198A for race 5.

PI lines which were susceptible to race 3 were not found to be resistant to any other race. Race 3 may lack the ability to parasitise resistant germplasm. This supports the hypothesis of Triantaphyllou (1975) that race 3 is either devoid of genes for parasitism or it may have low frequency of genes for parasitism. Previous work (Anand and Shumway, 1985; Luedders and Dropkin, 1983; Young, 1984) indicated that SCN genes for parasitism to reproduce on PI 90763 and PI 89772 were not independent for ability to reproduce on PI 88788. This was attributed to close linkage or multiple allelism in genes for parasitism.

Plant introduction 437654 was found to be resistant to all the known races of SCN. The discovery of a soybean line with resistance to all known SCN races indicates that PI 437654 either possesses combination of resistance genes from different sources or it has additional genes not present in other soybean lines. More work is required to elucidate genetics of resistance to cyst nematode. Nevertheless, PI 437654 would be a good source of resistance in cultivar development and it could also be used in SCN race identification.

Table 5. Reaction of F_2 plants and F_3 progenies against races 3 and 5 of *Heterodera glycines*

Cross		Segregation for Race 3	Segregation for Race 5
Forrest	x PI 437654	No	-
	x PI 437655	Yes	-
	x PI 437679	No	-
	x PI 438496B	No	-
	x PI 438498	No	-
	x PI 88788	Yes	-
	x PI 209332	Yes	-
	x PI 90763	Yes	-
Peking	x PI 88788	No	-
	x PI 404166	No	No
	x PI 437654	No	No
	x PI 438489B	Yes	No
PI 88788	x PI 438496B	No	-
PI 90763	x PI 404166	-	No
	x PI 404198A	-	Yes
	x PI 438489B	-	No

-Not tested.

REFERENCES

Anand, S.C., 1982, New soybean strain resistant to soybean cyst nematode: PI 416.762, Pl.Dis., 66:933.

Anand, S.C., and Brar, G.S., 1983, Response of soybean lines to differentially selected cultures of soybean cyst nematodes (*Heterodera glycines* Ichinohe), J.Nematol., 15:120.

Anand, S.C. and Gallo, K.M., 1984, Identification of additional soybean germplasm with resistance to race 3 of the soybean cyst nematode, Pl.Dis., 68:593.

Anand, S.C. and Shannon, J.G., 1984, Registration of Bradley soybeans, Crop Sci., 24:998.

Anand, S.C. and Shumway, C.R., 1985, Response of secondary selection on soybean cyst nematode reporoduction on resistant soybean lines, Crop Protection, 4:231.

Anand, S.C., Wrather, J.A and Shumway, C.R., 1985, Soybean genotypes with resistance to races of soybean cyst nematode, Crop Sci., 25:1073.

Brim, C.A. and Ross, J.P., 1966, Registration of Pickett soybeans, Crop Sci., 6:305.

Caldwell, B.E., Brim, C.A. and Ross, J.P., 1960, Inheritance of resistance of soybeans to the cyst nematode, *Heterodera glycines*, Agron.J., 52:635.

Caviness, C.E., Riggs, R.D. and Walters, J.H., 1972, Registration of Mack soybeans, Crop Sci., 12:396.

Epps, J.M. and Hartwig, E.E., 1972, Reaction of soybean varieties and strains to race 4 of the soybean cyst nematode, J.Nematol., 4:222 (Abstr.).

Golden, A.M., Epps, J.M., Riggs, R.D., Duclos, R.D., Fox, L.A. and Bernard, R.L., 1970, Terminology and identity of infraspecific forms of the soybean cyst nematode (*Heterodera glycines*), Plant Dis.Reptr., 54:544.

Hartwig, E.E. and Epps, J.M., 1968, Dyer soybeans, Crop Sci., 8:402.

Hartwig, E.E. and Epps, J.M., 1983, Registration of Forrest soybeans, Crop Sci., 13:287.

Hartwig, E.E. and Epps, J.M., 1977, Registration of Centennial soybeans, Crop Sci., 17:979.

Hartwig, E.E. and Epps, J.M., 1978, Registration of Bedford soybeans, Crop Sci., 18:915.

Hartwig, E.E., Epps, J.M. and Edwards, C.J., 1971, Registration of Pickett 71 soybeans, Crop Sci., 11:603.

Inagaki, H., 1979, Race status of five Japanese populations of *Heterodera glycines*, Jap.J.Nematol., 9:1.

Luedders, V.D. and Dropkin, V.H., 1983, Effect of secondary selection on cyst nematode reproduction on soybeans, Crop Sci., 23:263.

Luedders, V.D., Williams, L.F. and Matson, A.L., 1968, Registration of Custer soybeans, Crop Sci., 8:402.

Ross, J.P., 1962, Physiological strains of *Heterodera glycines*, Plant Dis.Reptr., 46:766.

Ross, J.P. and Brim, C.A., 1957, Registance of soybeans to the soybean cyst nematode as determined by a double-row method, Plant Dis.Reptr., 41:923.

Triantaphyllou, A.C., 1975, Genetic structure of races of *Heterodera glycines* and inheritance of ability to reproduce on resistant soybeans, J.Nematol., 7:356.

Young, L.D., 1982, Reproduction of Tennessee soybean cyst nematode population on cultivars resistant to race 4, Pl.Dis., 60:251.

Young, L.D., 1984, Changes in the reproduction of *Heterodera glycines* on different lines of *Glycine max*., J.Nematol., 16:304.

Winstead, N.N., Skotland, C.B. and Sasser, J.N., 1955, Soybean cyst nematode in North Carolina, Plant Dis.Reptr., 39:9.

HETERODERA AVENAE: VIRULENCE AND RESISTANCE

Kirsten Andersen and Sigurd Andersen

Department of Crop Husbandry and Plant Breeding
Royal Veterinary and Agricultural University, Copenhagen
Denmark

INTRODUCTION

Cereals and grasses are damaged by several species of cyst forming nematodes. One species, *Heterodera avenae* Woll. (the oat nematode = cereal root eelworm), is common in countries in Northwest Europe, and is found in Canada, North Africa, India and Australia. Control is possible by use of resistant cultivars, but different pathotypes are found, and the resistance of a cultivar must correspond to the pathotype of the nematode.

A conference about *H. avenae* and related species was held at Rennes, France in 1981 (see EPPO Bull. 12:307-526). Three years later, another EPPO conference was held at Münster, Germany, where the main interest was the potato root nematode, but a small follow-up meeting discussed *H. avenae*. Conclusions from these meetings recommended the double figure principles for pathotypes, and the use of an international test collection of different cultivars (Anon, 1982, unpublished information).

Although there is much information about *H. avenae*, it is not known why damage to crops is only slightly correlated to level of soil infestation; or why the multiplication rate may be high or low. Also information about the genetics of the nematode and of resistant cultivars is insufficient.

SPECIES AND PATHOTYPES

At the beginning of this century *H. rostochiensis* Woll. *H. schachtii* Schmidt, and *H. avenae* were considered to be pathotypes of one species but later were classified as separate species. The first evidence of pathotypes of *H. avenae* was that of Andersen (1959). During the following 30 years several reports were published about virulence and resistance. The existence of pathotypes is clear, but the situation is complicated. Some of the populations interbreed e.g. pathotypes 11 and 12 (Andersen, 1965), and pathotypes 11 and 31 (Rivoal and Person-Dedryver, 1982). These pathotypes must belong to the same species. Other populations were preliminarily listed as pathotypes, but later classified as true species. Cook (1982) listed seven species of cyst forming nematodes as virulent to cereals and grasses, and Sturhan (1984) identified 15 species. Research on the hybridisation of individual nematodes from different populations is important as it may show whether pathotypes interbreed, or whether they are true species.

The virulence of many populations differ when tested against a sufficient number of cereal cultivars. An example is the reaction of pathotypes 11, 41 and 61 (Andersen and Andersen, 1982). These three pathotypes could not be identified when only 4-5 test cultivars were used, but they showed different virulence when populations were tested against a sufficient number of cultivars. The different results reviewed by Andersen and Andersen (1982) indicate that the nematode have several genes for virulence. If different pathotypes interbreed we may expect to find many different combinations of genes.

TERMINOLOGY OF PATHOTYPES AND GENES FOR RESISTANCE

A simple classification of pathotypes and genes for resistance was suggested at the 1981 EPPO meeting (Andersen and Andersen, 1982) as a useful tool for the barley breeder. Genes from three cultivars: Drost, KVL 191, and Morocco, are used by many breeders. Consequently the symbols for resistance genes in Drost, KVl 191, and Morocco were termed Hal, Ha2, and Ha3 respectively.

The pathotypes of *H. avenae* were classified into three groups. Group 1 includes pathotypes to which resistance was found in cv. Drost (Hal). The pathotypes from this group were given a two-figured number, the last figure being 1, e.g. the numbers 11, 21, 31, etc. The first figure is the number of a pathotype of the group.

KVL 191 is resistant to pathotypes in group 2. The numbers 12, 22, 32 etc. are reserved for pathotypes virulent to cultivars with Hal, but avirulent to cultivars with the gene Ha2. Populations in group 3 are virulent to cultivars with Hal and Ha2, but do not form cysts on cultivar Morocco.

The number of populations in each group is unlimited, according to the system, but we need an international panel of experts to decide if a population may be classified as a new pathotype and consequently given a number.

LINKAGE OF GENES

The gene Hal was found in cv. Drost, and in the years 1977-84 F_2 populations were investigated from crosses between Drost and 19 marker genes, but no linkage was found (unpublished data).

The gene Ha2 from KVL 191 is closely linked to the genes pau and cer g^{10} on chromosome 2. The recombination values to these genes are approximately 3. The recombination value to Vv (2-rowed versus 6-rowed) is about 15 (Cotton and Hayes, 1969; Andersen and Andersen, 1973). The gene Ha3 from cv. Morocco is allelic or closely linked to Ha2 (Andersen and Andersen, 1970a).

Cotten and Hayes (1969) crossed KVL 191 with KVL 14. Both cultivars were considered to have the gene Ha2. They found two susceptible plants, thus indicating the possibility that the gene for resistance in KVL 14 is not Ha2. Nielsen (1982) crossed Bajo Aragon and Martin 403-2 with other cultivars and demonstrated that Bajo Aragon has two genes for resistance and Martin 403-2 one gene.

Cv. Morocco contains the Ha3 gene for resistance, but probably one gene more (O'Brien et al., 1979). A selection from five backcross generations to the susceptible cv. Pallas showed different resistance to Danish and Australian populations (Table 1). This indicates that the selection has only one of the Morocco genes for resistance.

In oats monogenic, dominant resistance was found in cv. CI 3444, whereas resistance in *Avena sterilis* was found to depend on three genes (Clamot and Rivoal, 1984). In wheat, monogenic dominant resistance was found in cv. Loros (for references see Andersen and Andersen, 1982). The resistance in oats and wheat is not complete as some cysts develop on resistant plants, but usually less than 10 per cent of the number of cysts on the susceptible check cultivar.

BREEDING FOR RESISTANCE

The genes Ha1, Ha2, and the gene from the oat cv. CI 3444 are easily transferred to high yielding, commercial cultivars. There are no reports about linkage to undesireable genes. High yielding, resistant cultivars are now available to farmers at least in Skandinavia, United Kingdom, France and Germany. If such cultivars are combined with regular investigations of the nematode infestation of the soil, the nematode should not cause serious yield losses in these countries.

DECREASE AND INCREASE IN POPULATIONS

Heterodera avenae does not, as far as is known, respond to root exudates, but approximately 60 per cent of the eggs inside the cysts hatch each year (Andersen, 1961). The hatching rate may be as low as 40 per cent, or as high as 90 per cent (Andersen and Andersen, 1970b).

The multiplication of the hatched juveniles varies very much from place to place. In some fields a decrease in the population will occur from spring to autumn, even if a susceptible crop is grown. In other fields the multiplication rate will be high.

The average multiplication rate may be high, when the initial population is low, but in fields with a high population a decrease in numbers may be expected (Fig. 1). The ceiling (optimum) level is the optimum of the population density in a specific field, in a specific year (weather conditions), and when a specific crop is grown. It is not known exactly why the multiplication rate will be high or low, but some fungi are harmful to the nematode (Kerry, 1982; Juhl, 1982). Another reason for variation in the multiplication rate may be the ability of the host to supply large numbers of nematodes with food.

The breeder requires a high multiplication to test his breeding lines for resistance. This is best achieved by using an unfested sandy loam soil which is inoculated with cysts.

Some small cysts are without eggs but large cysts often contain 300 or more (average 200 eggs per cyst). If the yearly hatch is 60%, and with 50% females, only 1% of the females have to develop into cysts in order to maintain a stable population; with 2% of the females developing to cysts the population should be doubled. In some of our multiplication trials, 30% of the females developed into cysts but the remarkable fact is that 70% did not. It must be concluded that a high proportion of the females die inside the roots of the plants (Seinhorst, in this volume). The death rate is higher in resistant plants than in susceptible ones, but even in the latter the death rate must be extremely high.

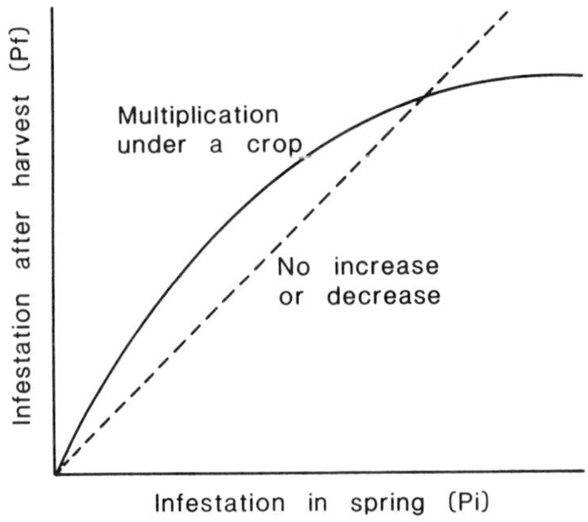

Fig. 1 Change in a nematode population when a susceptible crop of cereals is grown.

THE ACTION OF THE PLANTS

Female nematodes develop for some time inside the roots and the plant may react in terms of susceptibility, tolerance or resistance. Tolerance is the ability of a plant to grow unaffected by the invasion by the nematode (Trudgill, in this volume). Resistance is the reaction of plants which do not allow the nematodes to develop into cysts. Resistant plants have some tolerance, as the vascular system is damaged but not to the extent as in susceptible plants. As an example, results from 20 year old field trials (Table 1) show that the resistant selection outyielded the susceptible cultivar at high levels of infestation, but not in uninfested or lightly infested soil.

A high level of tolerance to *H. avenae* was found in commercial cultivars and breeders lines of wheat by Fischer et al. (1981). Large numbers of cysts developed on the roots, and cultivars and lines in infested soil yielded from 13 to 138 per cent of the yield in uninfested soil. Williams and Beane (1982) reported a yield benefit of 9 to 26 per cent from the use of resistant barley varieties.

Much has been published about resistance (Andersen and Andersen, 1982; Andersson, 1982; Cook and York, 1982; Rivoal and Person-Dedryver, 1982). The cultivars mostly used for breeding and virulence testing are distributed as a collection of test cultivars for pathotypes of *H. avenae* and related species (Anon, 1985). Seed lots and instructions on how to carry out the tests are available on request.

Resistance in most cultivars depends on one dominant gene. The barley cvs. Drost and Zita are completely resistant to pathotype 11 and 12, respectively. The number of cysts found on these cultivars is often less than 0.1 per cent of that on susceptible cultivars.

As the death rate of females in susceptible cultivars is about 95 per cent, it poses the questions:

1. Are environmental conditions important for formation of cysts on resisters? Some results, reported by Cook (1982), indicate an

Table 1 Yield of a susceptible barley cultivar and a resistant barley selection in field trials (Andersen, 1967)

Eggs & Juveniles per kg soil	Number of trials	Kg grain/ha Susceptible cultivar	Kg grain/ha Resistant selection
Less than 1000	23	4630	4370
1800	8	4800	4430
6300	24	4640	4560
30100	23	3690	3810

influence of temperature.

2. Does resistance depend completely on one gene?

3. Is the whole genetical constitution of a cultivar able to improve or impair the resistance?

Australian workers (Brown and Meager, 1970; Ellis and Brown, 1976; O'Brien et al., 1979) tested KVL 191 and other cultivars with the Ha2 gene. Resistance to the Australian nematodes were not complete as was the case in experiments with Danish pathotypes and Danish cultivars with the gene Ha2. Such differences between Denmark and Australia may be due to differences in environment, or the Australian nematodes may be another pathotype.

In 1985, *H. avenae* from Australia (kindly supplied by R.W. Brown, Department of Agriculture, Victoria) were tested in Denmark on several barley cultivars. The test was done in a glasshouse, using sections of plastic pipes which were filled with cyst-free soil and each pipe then infected with 15 cysts. The plants were grown to harvest when grain and straw were cut and burned. The soils were heated for 2 days at 100°C, and cysts were washed from the soil and counted. Preliminary results are given in Table 2.

The Danish pathotype 11 produced very few cysts on cvs. Drost or Ortolan and no cysts on cvs. Siri and Zita. Pathotype 12 was avirulent to Zita and Siri. The Australian *H. avenae* produced cysts on both Zita and Siri, but less than on susceptible cultivars. Thus, it is concluded that the Australian nematodes are not pathotype 12, but as Zita and Siri are less susceptible than Emir and Herta, they may be closely "related" to pathotype 12.

The cysts produced by the Australian population on Zita and Siri contained fewer eggs and juveniles than cysts produced on the susceptible Herta. This indicates that conditions for development of the nematodes were less favourable on Siri and Zita. Although some nematodes were able to overcome the resistance barrier, they lost some vigour.

In this experiment, we tried as far as possible to exclude environmental effects, but the cysts themselves are produced in two very different

Table 2 Reaction of barley cultivars to two Danish pathotypes and an Australian population of *H. avenae* (8 replicates)

	Cultivar	Pathotype 11	Pathotype 12	Australian cysts
Cysts per plant	Emir	326	414	260
	Herta	394	378	463
	Drost	2	486	421
	Ortolan	1	342	272
	Siri	-	1	62
	Zita	-	0 a)	59
	Morocco	-	1	1
	Mor. x P B)	-	2	34
Eggs & Juveniles per plant	Emir	58400	59100	44600
	Herta	63300	37000	84400
	Drost	x	49800	75800
	Ortolan	x	42600	43900
	Siri	0	0	7600
	Zita	x	0	5500
	Morocco	x	x	x
	Mor. x P b)	0	x	4300
Eggs & Juveniles per cyst	Emir	180	143	172
	Herta	161	98	182
	Drost	x	102	180
	Ortolan	x	125	161
	Siri	0	0	122
	Zita	x	0	93
	Morocco	x	x	x
	Mor. x P b)	0	x	128

\- = 1-3 cysts on 8 plants

x = At least one cyst with more than 100 eggs

a) = Average of 7 plants. Plant no. 8 developed 49 cysts

b) = Selection from Morocco x Pallas.

climates. It must also be recognised that each plant was infected with 15 cysts and it is possible that these may have been one pathotype only or two or more pathotypes. The results, considered with many other data from different parts of the world, indicate that the action of a major gene in a plant may depend on other genes, e.g. the whole metabolic processes of a cultivar. Similarly, the efficiency of major genes for virulence may depend on other genes of the nematode. Another explanation of the results is that plants and nematodes have an unlimited number of genes giving resistance or virulence of varying degree.

FUTURE RESEARCH

More information is needed about the genetics of inheritance when resistant cultivars or different pathotypes are crossed. Experiments must be done in isolation within a glasshouse to avoid the risk of infestation with a pathotype which is not a native of the country.

REFERENCES

Andersen, S., 1959, Resistance to various populations of the cereal root Eelworm (*Heterodera avenae*), Nematologica, 4:91.

Andersen, S., 1961, Resistens mod havreål, Dansk Videnskabs Forlag, Diss., 178 pp.

Andersen, S., 1965, Heredity of race 1 and race 2 in *Heterodera avenae*, Nematologica, 11:121.

Andersen, S., 1967, Erfaringer vedrørende havreål, Tidsskr. Landøkonomi, 154:94.

Andersen, S. and Andersen, K., 1970a, Sources of genes which promote resistance to races of *Heterodera avenae* Woll., EPPO Public. ser. A no., 54:29.

Andersen, S. and Andersen, K., 1970b, Nedgang i smitte af havrenematoder efter dyrkning af resistente bygsorter eller græs., Tidskr. Planteavl., 74:559.

Andersen, S. and Andersen, K., 1973, Linkage between marker genes on barley chromosome 2 and a gene for resistance to *Heterodera avenae*, Heteditas, 73:271.

Andersen, S. and Andersen, K., 1982, Suggestions for determination and terminology of pathotypes and genes for resistance in cyst-forming nematodes, especially *Heterodera avenae*, EPPO Bull., 12:379.

Andersson, S., 1982, Population dynamics and control of *Heterodera avenae* - A review with some original results, EPPO Bull., 12:463.

Anon., 1982, Conclusions of the EPPO colloquium on cereal cyst nematodes, EPPO Bull., 12:525.

Anon., 1983, International test collection for *Heterodera avenae* and related species (Distributed by Sortssamlingen, Landbrugets Plantekultur, Agrovej 10, 2630 Taastrup, Denmark).

Brown, R.H. and Meagher, J.W., 1970, Resistance in cereals to the cyst nematode (*Heterodera avenae* Woll.) in Victoria, Austr. J.expl.Agric. Anim.Husb., 10:360.

Clamot, G. and Rivoal, R., 1984, Genetic resistance to cereal cyst nematode *Heterodera avenae* Woll. in wild oat *Avena sterilis* I. 376, Euphytica, 33:27.

Cook, R., 1982, Cereal and grass hosts of some gramineous cyst nematodes, EPPO BULL., 12:399.

Cook, R. and York, P.A., 1982, Resistance of cereals to *Heterodera avenae*: Methods of investigations, sources and inheritance of resistance, EPPO Bull., 12:423.

Cotten, J. and Hayes, J.D., 1969, Genetic resistance to the cereal cyst nematode (*Heterodera avenae*), Heredity, 24:593.

Ellis, S.E. and Brown, J.A.M., 1976, The reaction of resistant cultivars of barley to some Australian populations of *Heterodera avenae*, Nematologica, 22:87.

Fischer, J.M., Rathjen, A.J. and Dube, A.J., 1981, Tolerance of commercial cultivars and breeders lines of wheat to *Heterodera avenae* Woll., Aust.Jour.Agric.Res., 32:

Juhl, M., 1982, Cereal cyst nematode. Soil heating and fungal parasitism, EPPO Bull., 12:505.

Kerry, B.R., 1982, The decline of *Heterodera avenae* populations, EPPO Bull., 12:491.

Nielsen, C.H., 1982, Heredity of *Heterodera avenae* resistance originating from two barley cultivars and one spring wheat cultivar, EPPO Bull., 12:457.

O'Brien, P.C., Sparrow, D.H.B. and Fischer, J.M., 1979, Reaction of cereals to populations of *Heterodera avenae* in South Australia, Nematologica, 25:348.

Rivoal. R. and Person-Dedryver, F., 1982, Caracterisation des pathotypes d'*Heterodera avenae* en France: influence de periode de la culture sur le pouvoir discriminant de cultivars d'*Avena sativa* et differences dans la capacite a former des femelles, EPPO Bull., 12:387.

Sturhan, D., 1984, Report EPPO Panel on Ceral Cyst Nematodes, 9-10.

Williams, T.D. and Beane, J., 1982, Variation in cereal yield losses associated with *Heterodera avenae* in England and Wales, EPPO Bull., 12:485.

DISCUSSION

Referring to the occurrence of partial resistance in the Australian *H. avenae*, Stone suggested that instead of the concept of modifying genes it might be postulated that another gene was present in about 30% of the Australian population. Andersen responded that he considered that dominant genes would govern the whole physiological process in nematodes but this would be modified by several other "modifying" genes. Perry commented that he found that the Australian nematode did not hatch easily and considered that the storage of cysts seemed to be an important factor. He asked if experiments could be standaridsed to take into account the physiological differences between populations. Andersen had no problems in Denmark with the Australian *H. avenae* which took about two weeks to hatch from the cysts. However, it should be kept in mind that pathotypes 11 and 12 were produced under different environmental conditions from those in Australia.

Asked how he would identify a resistant plant, Andersen replied that this depended on the eggs and juveniles present and that differences in population curves differentiation between cultivars. In simple terms, a resistant plant is one on which the nematodes does not develop to the cyst stage. Ireholm considered that such a definition was at variance with others and Seinhorst commented that resistance is a matter of definition by the research worker. Asked about the genetics of virulence, Andersen said this was complicated and would involve the interbreeding of pathotypes and the testing of subsequent generations. Andersson pointed to the tolerance shown by resistant barley cultivars. Resistant oat cultivars were usually very susceptible. Andersen agreed that there was low tolerance in oats.

Concerning *H. avenae* in Spain, Romero said that there was only one generation per year and hatching occurred at low (2-10°C) temperatures.

In a general discussion on *H. avenae*, it was observed that in many countries *H. schachtii* and *H. avenae* occur together and this presents difficulties of distinguishing between the cysts of the two species. Stone and Seinhorst considered that with experience it was possible to distinguish between cysts of the species using a microscope. Heinicke said that in tests undertaken in Germany, it was found that approximately 80% of *H. avenae* occurring in samples extracted for *H. schachtii* were determined correctly, simply on the appearance of the cysts. Antoniou remarked that the problem does not present itself in Greece because of the absence of *H. schachtii*. Greco said that the distribution of *H. schachtii* is geographically limited in Italy.

On the occurrence of *Punctodera*, Stone commented that it can usually be found in wet, permanent pastures but not in dry, tilled fields. There was no information about the damage to grasses, but Golden referred to investigations on the occurrence of *P. punctata* on golf courses in New Jersey, USA. Miller added that under some conditions in Texas, wheat was damaged by *P. punctata*.

Widening the discussion, Stone referred to the similarity of appearance between *G. pallida* and *G. achillae*, the latter species being harmless to potatoes but sometimes found in statutory samples and thus a problem if not identified correctly. Mugniery said that separation of the two species can easily be done by a biotest in Petri dishes; the juveniles of *G. pallida* immediately penetrate potato roots. Also, cysts of *G. achillae* contain few eggs compared with *G. pallida*. Andersson commented that *G. achillae* was found relatively frequently in Sweden; Miller said that Sturhan had found it only once in Germany.

THE BIOLOGY OF *HETERODERA AVENAE* IN SPAIN

Alicia Valdeolivas and Maria Dolores Romero

Instituto de Edafología y Biología Vegetal
Serrano 115 dup. 28006
Madrid

INTRODUCTION

During 1983-84 and 1984-85 a detailed study was made of four populations of *Heterodera avenae* Woll. from three cereal growing areas of Spain. The populations were selected from three different climatic zones (Elias Castillo and Ruiz Beltran, 1977) (Table 1).

The experiments were done at "La Higueruela" Experimental Farm in 1983-84 and in the garden of the Instituto Español de Entomologia, Madrid in 1984-85, starting at the end of October and terminating in mid-June in each year. Infested soil from each location was placed in plastic tubes (20 cm long, 6 cm diameter) which were then placed on top of steam-sterilised 50 cm deep, plastic-lined holes. A wheat seed of a cultivar susceptible to *H. avenae* was sown in each.

Table 1

Locality Nematode Population	Province	Climate	Mean monthly temp. Dec-Jan °C	Jul-Aug
La Higueruela Experimental Farm	Toledo	cont. medit.[ɸ]	4.6	27.8
Alcala de Guadaira	Sevilla	subtr. medit.[ɸ]	8.7	27.0
Bello (X)	Teruel	cont. medit.	0.6	21.6
Concud	Teruel	temp. medit.[ɸ]	3.5	21.1

(X) Discarded after first year

[ɸ] cont. medit. = continental mediterranean; subtr. = subtropical; temp. = temperate

Every 15 days nematodes were extracted from two cylinders of soil from each locality and counted. Cysts were extracted using a Fenwick apparatus; males and J2s were extracted by centrigufation (De Grisse, 1969); roots were stained with lactophenol-cotton blue and examined with a stereomicroscope. Mean temperatures were calculated from records taken every 6 hours at a depth of 15 cm.

OBSERVATIONS

Specimens from the Teruel population are morphologically different from those from the other two populations; they are in the range of the British pathotype III or Götlandstrain.

Second stage juveniles first appeared in early November, coinciding with lowering temperatures and the start of the winter rains. Third stage juveniles were present one month later. Fourth stage juveniles, males, females and cysts appeared first in the Sevilla population (and the males from this population also disappeared first) but otherwise development of populations from the different localities were similar (Fig. 1).

The relationship between emergence of juveniles at the three locations and temperature and rainfall is shown in Fig. 2. Juvenile emergence increased from November to January-February and then decreased until the end of March/beginning of April, fluctuations in numbers of juveniles being related to variations in temperature and rainfall. Juvenile emergence also increased near the middle of March in populations from Teruel and in 1983-84 from Toledo.

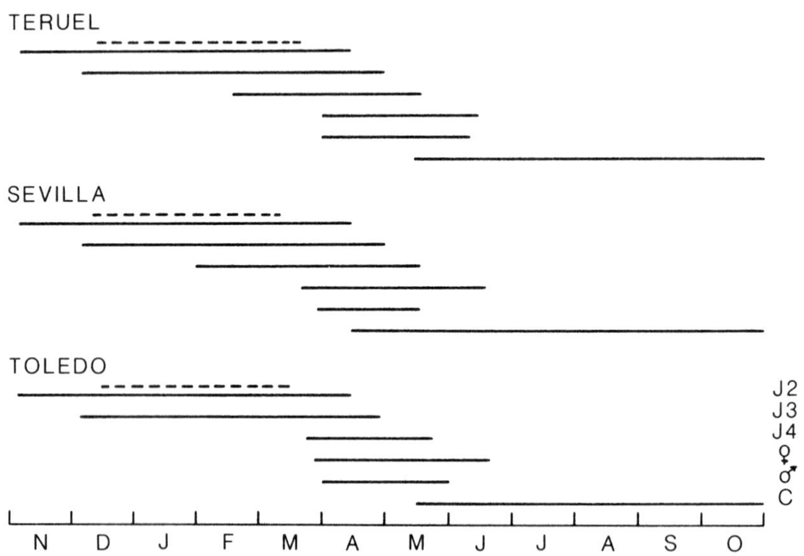

Fig. 1. Life cycle of *H. avenae* populations from three localities.

----- = maximum emergence; J = juveniles; C = cysts.

In the first year of the study juvenile emergence began at 13 and reached a maximum as the temperature decreased to 5-7℃. In the second year emergence began at 14℃ for nematode populations from all three locations but there were two peaks, at 13℃ and at 7℃, and with the Teruel population there was a third peak at 11℃ at the end of March.

All stages of the Spanish *H. avenae* populations appeared earlier in the year than those reported from Greece and Italy (Kyrou, 1976; Mezzeti, 1952). The J2 and J3 stages were earlier than the FR_1 population from

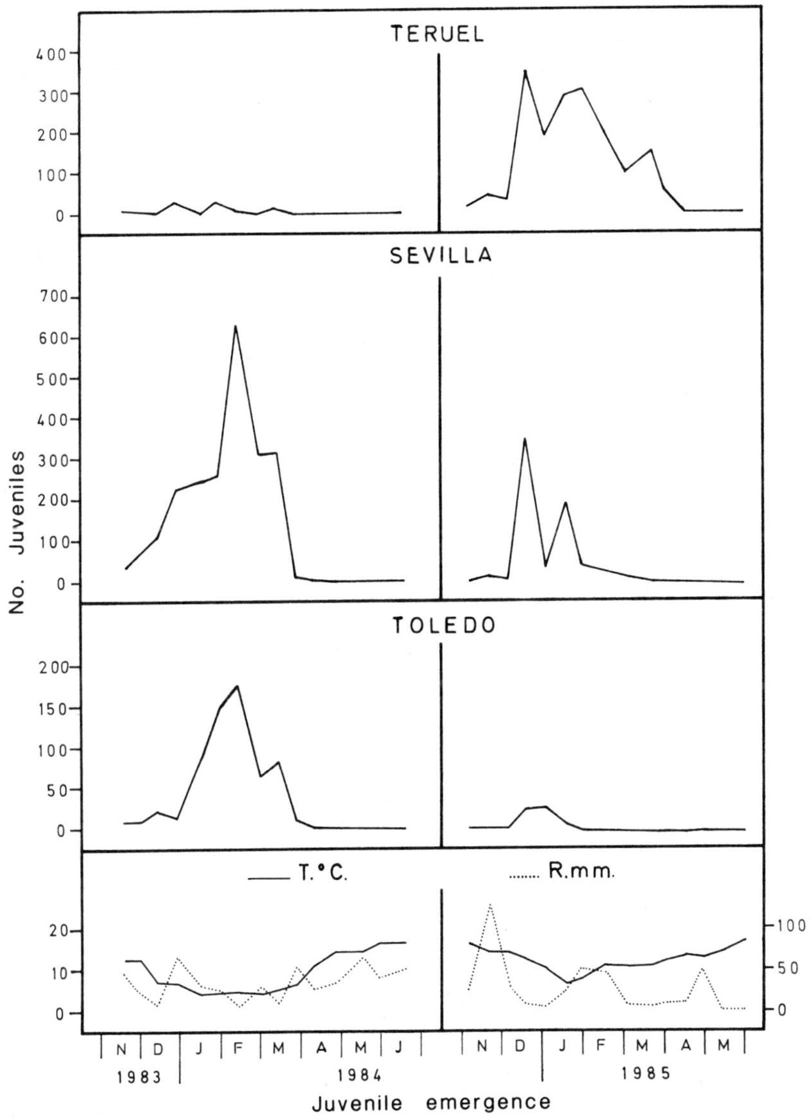

Fig. 2. Relationship between juvenile emergence, temperature and rainfall.

Castelnaudary, South of France and other stages appeared similarly to populations from Toledo and Teruel (Rivoal, 1978). The earlier emergence of J2 in the Spanish populations is due to ability to emerge at higher temperatures than the French population reported by Rivoal (1979).

Although based on only two years work, all the populations of *H. avenae* studied had a life cycle of the mediterranean type with juveniles active during the winter. However, in populations from Toledo and Teruel there was a second emergence of juveniles in March, which is typical of populations from Northern Europe and Canada. If these observations are confirmed, it is probably explained by cold winter temperatures in these regions inducing a second diapause for part of the juvenile populations which emerge in spring as the temperatures rise. The observed cycle would then be a combination of the two cycles previously described for *H. avenae*.

ACKNOWLEDGEMENTS

The authors thank Carlos Lacasta and Antonio Duce for their technical assistance.

REFERENCES

De Grisse, A., 1969, Redescription on modification de quelques techniques utilisées dans l'étude des nematodes phytoparasitaires, Meded. Rijksfaculteit Land b.Wetensch. Gent, 34:351.

Elias Castillo, F. and Ruiz Beltran, L., 1977, Agroclimatologia de España, Cuadernos INIA, no.7.

Kyrou, N.C., 1976, Biological notes on *Heterodera avenae* Woll., 1924 atudied on wheat in Central Macedonia, Ann. Inst. Phytopath. Benaki, 2:187.

Mezzetti, A., 1952, Observazioni sull' anguillulosi radicale dei cereali in Italia, Ann. Speriment. Agr. NS, 7:743.

Rivoal, R., 1978, Biologie d'*Heterodera avenae* Wollenweber en France. I. Differences dans les cycles d'eclosion et de developpement de deux races Fr_1 et Fr_4, Revue Nématol., 1:171.

Rivoal, R., 1979, Biologie d'*Heterodera avenae* Wollenweber en France. II. Etude des differences dans les conditions thermiques d'eclosion des races Fr_1 et Fr_1, Revue Nématol., 2:233.

INCIDENCE OF *HETERODERA AVENAE* IN SPRING BARLEY FIELDS IN DENMARK

Lars M. Hansen

Danish Research Centre for Plant Protection,
Institute of Plant Pathology, Lottenborgvej 2, DK-2800,
Lyngby, Denmark

In the period 1959-61 spring barley fields in Denmark were examined for the occurrence of the cereal cyst nematode, *Heterodera avenae*. More than 1,000 eggs per kg were found in about 40 per cent of the fields, and in 3 per cent more than 10,000 eggs per kg soil. In a corresponding survey in 1972-73 the numbers were 54 and 10 per cent, respectively, and in 1984, 41 and 9 per cent.

Thus there has been little change in the distribution of the cereal cyst nematode in Denmark in the last 25 years, with a similar percentage of heavily infested fields in 1984 as in 1972-73 (Hansen, 1986). This may be associated with the limited availability of resistant cultivars (Jakonsen, 1978).

Only pathotypes 1 and 2 of *H. avenae* are of frequent occurrence in Denmark. Resistant cultivars were grown in only 5 per cent of the fields examined in 1984. An increase in the use of resistant cultivars would result in a considerable reduction of the number of infested fields.

In the last five years the area of spring barley has decreased but there has been an increase in the area of winter barley. Spring barley now occupies an area of about 1 million ha. It is estimated that a nematode density of 1,000 eggs per kg soil would cause a substantial yield loss, and at densities of more than 10,000 egg per kg soil considerable yield losses would result in most years. Thus in Denmark the cereal cyst nematode causes yield losses of 30-75,000 tons of barley a year. In financial terms this represents an annual loss of income for the Danish farmers of 42-100 million Danish kroner or 5-12 million ECU.

REFERENCES

Hansen, L.M., 1986, Incidence of *Heterodera avenae* in spring barley fields in Denmark, Danish J. of Pl. Soil Sci., (in preparation)

Jakonsen, J., 1978, Experience in Denmark with cereal cultivars resistant to cereal cyst nematode (*Heterodera avenae* Woll.), EPPO Bull., 8:29.

AN ANALYSIS OF THE *HETERODERA TRIFOLII* COMPLEX AND OTHER SPECIES IN THE *SCHACHTII* GROUP ATTACKING LEGUMES

Richard A. Sikora and Paul W. Th. Maas

Institut für Pflanzenkrankheiten der Universität, Bonn
Federal Republic of Germany and Plantenziektenkundige Dienst
Wageningen, The Netherlands

INTRODUCTION

Information is limited on the many species and host races in the *Heterodera trifolii*-complex and the *schachtii*-group attacking legumes. This review is based primarily on literature published between 1974 and 1985. Mulvey and Anderson (1974) should be referred to for information on *H. trifolii* published before this date.

THE *HETERODERA TRIFOLII* COMPLEX

When the first cyst nematode, *Heterodera schachtii* Schmidt, was described in 1871 host races of *H. schachtii* existed, because of differences in host ranges between different populations (Schmidt, 1871). These host races later appeared to be different species with some subsequently being placed in different genera. For example, the clover cyst nematode, *H. trifolii* Goffart, 1932 was first described as a race of *H. schachtii* (Goffart, 1932).

In 1972, Mulvey divided the cyst nematodes into five groups on the basis of cyst morphology and Stone (1975) divided them into six groups on the basis of second stage juvenile (J2) lip characters. Utilising both sets of characters and additional criteria, Stone (1979) established eleven groups with *H. trifolii* in the *schachtii* group.

The *schachtii* group is characterized by lemon-shaped cysts with a long vulval slit (>35 mμ); under bridge and bullae generally strongly developed; ambifenestrate; and J2 with a greatly elongated oral disc, with a partial median division and incomplete boundaries dorsally and ventrally; sub-median lips absent and lateral field with four incisures. Eleven nominal species, some undescribed species and four species inquirendae can be placed in the *schachtii* group (Table 1).

Detailed studies of *H. trifolii* in comparison with *H. schachtii* (Franklin, 1951; Oostenbrink and Den Ouden, 1954; Wouts and Weischer, 1977) and *H. glycines* Ichinohe (Hirschmann, 1956) showed *H. trifolii* to be a distinct biological entity. Mulvey (1958) demonstrated that *H. trifolii* was a parthenogenetic species without males. Triantaphyllou and Hirschmann (1978) made a cytological study of 24 populations of cyst nematodes belonging to the *H. trifolii* parthenogenetic species complex including 18 popula-

Table 1. List of species in the *Heterodera schachtii* group

Species	Males
H. daverti Wouts & Sturhan, 1978	+
H. glycines Ichinohe, 1952	+
H. lespedezae Golden & Cobb, 1963	-
H. medicaginis Kir'janova, 1971 (Gerber & Maas, 1982)	+
H. rosii Duggan & Brennan, 1966	+
H. rumicis Pogosyan, 1961	-
H. salixophila Kir'janova, 1969	+
H. schachtii Schmidt, 1871	+
H. sonchophila Kir'janova, Krall & Krall, 1976	+
H. trifolii Goffart, 1932	-
H. ciceri Vovlas, Greco & Di Vito, 1985	+
Undescribed species	
H. trifolii-18 : *Dianthus caryophyllus* (Cuany & Dalmasso, 1975)	+
H. sp. : *Rumex crispus* (Triantaphyllou & Hirschmann, 1978)	-
H. daverti : *Trifolium subterraneum* (Sikora, 1977)	+
H. sp. : *Cicer arietinum* (Sikora & Maas, unpublished)	-
Synonyms	
H. paratrifolii Kir'janova, 1963 = *H. trifolii* (Krall, 1977)	
H. galeopsidis Goffart, 1936 = *H. trifolii* (Maas et al., 1982)	
Species inquirendae	
H. graduni Kir'janova, 1971 → *H. cruciferae* group (Vovlas et al., 1981)	
H. limonii Cooper, 1955	
H. oxiana Kir'janova, 1962	
H. polygoni Cooper, 1955 → *H. cruciferae* group (Mulvey, 1972)	
H. scleranthii Kaktina, 1957	
H. tadshikistancia Kir'janova & Ivanova, 1966	

tions of *H. trifolii* from white clover, two of *H. lespedezae* Golden & Cobb from *Lespedeza striata*, two of *H. galeopsidis* Goffart from white clover and two of *Heterodera* sp. from *Rumex crispus*. They found mitotic parthenogenesis (apomixis) in all populations with chromosome numbers of 3n=24-28 and 4n=32-35 and concluded that these polyploid forms probably have evolved along the same parthenogenetic line. They concluded that the taxonomic treatment of this group should be based on comparative morphology, cytology, behaviour and any other source of information that provides characteristics for convenient differentiation.

From a detailed morphological and morphometric comparison and limited host tests of six populations with chromosome numbers of 24, 26, 27, 28, 32, and 34, Hirschmann and Triantaphyllou (1979) concluded that the chromosomal forms of the *H. trifolii* complex can be distinguished morphologically. However, the morphological differences are not striking and can hardly be considered as adequate for normal species characterization. They proposed recognition of chromosomal forms as subspecies indicating their close relationship in this parthenogenetic complex of apparently monophyletic origin. Subspecies, however, implies a potential for interchange of genetic information and the possible recombination into one species not expected through apomixis. Since the usual subspecies concept is based on the biological species concept, subspecies cannot be applied to the parthenogenetic members of the *H. trifolii*-complex.

HOST RACES

Maas et al. (1982) studied the *H. trifolii* complex by comparing five different host races including *H. trifolii* and *H. galeopsidis* collected in their type localities. They found statistically significant morphometric differences between some of the populations, tending to be more pronounced when each host race was grown on its specific host plant. The morphometric differences between the populations, however, showed a continuous gradient "Cline" and were not always characteristic for the host races.

As long as there is no evidence of how chromosomal forms, morphotypes and host races are related, this agamic *H. trifolii* complex cannot be split into morphologically separate taxonomic units or subspecies. For practical plant nematology these populations are best classified as host races by suitable host range tests (Table 2). A host race is a clearly defined term used for example with *Ditylenchus dipsaci* (Kühn) Filipjev and species of *Meloidogyne*.

Maas et al. (1982) proposed the term forma specialis, commonly used in mycology as a more acceptable term for international notation of host races. This is a latin term which is accepted in botanical nomenclature but not yet recognised in zoology.

The only host races of *H. trifolii* currently of economic importance are the yellow beet cyst nematode, damaging beets, cabbages and peas in the Netherlands (Maas and Heijbroek, 1982) and of unknown importance in Sweden (Andersson, 1984) and Switzerland (Vallotton, 1985) and the clover host race that causes serious decline in pasture legumes. The host range differences between the other races of *H. trifolii* have almost no practical meaning. However, from a phylogenetic and co-evolutionary point of view they are intriguing.

Table 2 Reproduction of five *Heterodera trifolii* populations on ten test plants.

Test plants	Nematode population from (locality) Clover (Wag.)	Hebe (Wag.)	Hemp-nettle (Wag.)	Spurrey (Wouw)	Beet (Overloon)
Spinacia oleracea 'Viroflex'	+ +	+ +	+ +	+ +	+ +
Rumex crispus	+ +	+ +	+ +	+ +	+ +
Trifolium repens 'Retor'	+ +	+ +	+ +	+ +	-
T. pratense 'Robina'	+ +	+ +	+ +	+ +	-
Dianthus deltroides	+	+	+ +	-	+ +
Hebe andersonii	-	(+ +)	+	-	+
Galeopsis tetrahit	-	-	(+ +)	-	+
Spergula arvensis	-	-	-	(+ +)	-
Beta vulgaris var. *rubra*	-	-	-	-	(+ +)
Brassica oleracea var. *sabauda* 'Savoy'	-	-	-	-	+ +

HETERODERA SCHACHTII GROUP

Similar problems exist when attempting to distinguish between species in the *H. schachtii* group associated with legumes. A major difficulty is that in most host tests an attempt is made to characterize variable nematode populations, under artificial conditions, with test plant combinations often containing different cultivars that are genetically heterogeneous within themselves. Nevertheless, differences in host range between species may be striking. In Table 3, *H. glycines* is the only species reproducing on soybean; on the other hand, lucerne is the only known host of *H. medicaginis* Kirjanova (Gerber & Maas).

An unsatisfactory situation appears when *H. daverti* (Wouts & Sturhan) from its type localities (Germany and the Netherlands) is compared with *H. daverti* (Nordmeyer, 1979) from Tunisia. Although they are morphologically indistinguishable, the Dutch population reproduces on white clover and maiden pink carnation but not on subterranean clover, whereas the Tunisian population reproduces on subterannean clover but not on the other two test plants. In the European as well as in the Tunisian populations males were present, although this does not necessarily mean that nematode reproduction is heterosexual (amphimixis).

Likewise, Cuany and Dalmasso (1975) comparing a diploid (n=9) amphimictic population, *H. trifolii*-18, with other nematode populations of the *H. schachtii* group, concluded that *H. trifolii*-18 is distinct from the amphimictic *H. schachtii* and *H. glycines* by morphometrics and host range but is morphologically close to *H. rosii* (Duggan & Brennan) from which it differs slightly in host range.

We found a monosexual *H. trifolii*-like population reproducing on chickpea in Syria to be morphologically close to *H. rosii* but distinct in host range and by the absence of males. *Heterodera ciceri* (Volvas, Grecor & Di Vito) a new species from chickpea in Syria, which is also similar to *H. rosii* is bisexual (Vovlas et al., 1985). There is insufficient information on *H. rumicis* (Pogosyan), *H. salixophila* (Kirjanova) and *H.sonchophila* (Kirjanova, Krall & Krall) to be included in Table 3.

The presence of small morphological differences and distinct host ranges between some of the species in the *schachtii*-group indicates that some may not be true species, but host races. This is especially true for populations of *H. daverti* and *H. rosii*.

It is concluded that the *H. trifolii*-complex consists of more or less morphologically distinct amphimictic and apomictic populations, more or less adapted to special host plants. For practical purposes, identification of nematode populations in the *H. schachtii* group should be based on both morphology and host preferences, because of the occurrence of host races in *H. trifolii* and pathotypes in *H. glycines*. It should be stressed that pathotypes can exist within the host race concept (Maas et al., 1982). For example, a pathotype of *H. trifolii* f. sp. *beta*, reproduces well on an otherwise resistant cultivar of oil radish, *Raphanus sativus* (Maas unpublished data).

DISTRIBUTION AND ECONOMIC IMPORTANCE

Information on the distribution of the various species in the *schachtii*-group attacking legumes and host races in the *H. trifolii*-complex, with the exception of the clover race, is almost nonexistent. *H. trifolii*, a cosmopolitan parasite has been detected in North America, western Europe,

Table 3 Reproduction of species of the *Heterodera schachtii* group on test-plants, presence of males and length of J2

Nematode species \ Test plant	*Trifolium repens*	*T. subterraneum*	*Glycine max*	*Medicago sativa*	*Beta vulgaris*	*Brassica oleracea*	*Dianthus*	Males	Length of juveniles II (µm)
H. glycines	-	?	++	+	-	-	?	+	375 - 490 (440)
H. medicaginis	-	?	-	++	-	-	-	+	417 - 512 (462)
H. ciceri	-	?	-	+	-	-	?	+	440 - 585 (525)
H. sp. (Sikora and Maas)	-	-	-	++	-	-	+	-	465 - 700 (640)
H. rosii	++	-	-	-	-	-	-	+	430 - 662 (550)
H. trifolii - 18 (Cuany and Dalmasso, 1975)	+	?	-	?	-	+	?	+	515 - 638 (584)
H. daverti	++	-	-	-	-	-	++	+	494 - 535 (517)
H. daverti (Nordmeyer, 1979)	-	++	-	-	-	-	-	+	444 - 496 (460)
H. trifolii									
*f,*sp. *trifolium*	++	-	-	-	-	-	+	-	462 - 544 (503)
*f.*sp. *hebe*	++	-	-	-	-	-	+	-	469 - 539 (504)
*f.*sp. *galeopsis*	++	-	-	-	-	-	++	-	458 - 554 (506)
*f.*sp. *spergula*	++	?	-	-	-	-	-	-	493 - 599 (546)
*f.*sp. *rumex*	++	?	?	?	?	?	?	-	459 - 551 (508)
*f.*sp. *lespedeza*	++	?	-	-	-	-	?	-	400 - 510 (457)
*f.*sp. *beta*	-	-	-	-	++	++	++	-	504 - 624 (564)
H. schachtii	-	-	-	-	++	++	±	+	404 - 492 (448)

USSR, Australia, New Zealand, Israel and India (Mulvey and Anderson 1974). In addition to typical movement of cysts by wind, water and farm implements, *H. trifolii* has been shown to be transported over long distances adhering to potato tubers and by bird migration (Kimpinski and Platt, 1983; Lobb and Wood, 1971).

The impact on agricultural production of the various species and host races has in many cases not been determined by exact field or greenhouse experimentation. In many instances their role in reducing plant growth has been estimated by survey work, greenhouse tests with sterilized soil or studied by measuring unspecific growth response after soil fumigation. The results of recent studies on crop loss are summarised in Table 4.

Controversy has existed concerning the importance of *H. trifolii* in decline of white clover in established pastures in the Netherlands. The development of this debate is very interesting to follow (Ennik et al., 1965; Hidding et al., 1963; Seinhorst and Sen, 1966). Results obtained during the past 15 years, however, have demonstrated that the nematode can be a primary cause of clover decline in pastures (Norton, 1967; Healy et al., 1972; Yeates et al., 1976; Korteweg and s'Jacob, 1980; Cook et al., 1983). More recently a host race of *H. trifolii* has been shown to cause economic damage to sugar beet, cabbage and peas in the Netherlands (Maas and Heijbroek, 1982). *H. daverti* is also considered a serious problem on subterranean clover in some areas of northern Tunisia (Sikora, 1977; Nordmeyer et al., 1978), damage being more severe on sandy soils. Norton (1967) also reported that *H. trifolii* reproduced best in loamy clay sand.

We are of the opinion that crop loss is a complex interaction often involving environmental factors, farming practices, and interactions with other organisms. Yeates (1973, 1974) for example, indicated that the combination of *H. trifolii* invasion and low soil moisture was responsible for reduced root growth and lack of vigour of white clover. Differences in economic importance may also relate to host susceptibility as described by Kuiper (1960) for *H. trifolii*, and Nordmeyer (1978) for *H. daverti*. The often reported occurrence of soil borne fungi in the roots of plants affected by this cyst nematode group indicates that complex interactions are important; these observations are discussed later.

BIOLOGY AND LIFE CYCLE

Mankau and Linford (1960) and Mulvey and Anderson (1974) have thoroughly covered this topic for *H. trifolii*. Data on the development of the other species are presented in Table 5. Comparisons are difficult, because of differences in host plant and techniques used to study development and the duration of the life cycle.

The only comparisons of development times for each life stage of the first generation are those for *H. trifolii* on *Trifolium repens* (Mulvey, 1959a, b) and *H. lespedezae* on *Lespedeza striata* (Bhatti et al., 1972). The development of these two species do not vary significantly (Table 6).

Subbotin (1984) made observations on *H. trifolii* development in the field. Using weather data he estimated that 3 generations could be produced in one year. Yeates and Risk (1976) reported two generations per year. Dzhunusov (1982) calculated the life cycle of *H. medicaginis* on lucerne by taking plants from a naturally infested field. He stated that J2s began penetrating the root at temperatures of 6°C in the third week of March. J3s were found 20 days later. The time for one generation was 86

Table 4 Crop loss associated with the presence of nematodes in the *Heterodera trifolii* - complex and other species in the *H. schachtii* - group attacking legumes

Nematode	Plant	Effect	Technique Ø	Reference
H. daverti	*Trifolium subterraneum*	severe losses	S	Sikora 1977
H. daverti	*T. subterraneum*	-0-56% dry weight	GH	Nordmeyer 1978
H. galeopsis	*T. pratense*	-15 and 50% dry weight	GH	Solov'eva et al., 1979
	T. repens	and reduced flowering	GH	Solov'eva et al., 1979
H. galeopsis	*Ballota bifida*	severe damage	S	Solov'eva et al., 1979
H. lespedezae	*Lespedeza striata*	stunting	GH	Bhatti and Sasser, 1973
H. lespedezae	*L. striata*	reduced shoot height, shoot and root weight	GH	Fagbenle et al., 1986
H. medicaginis	*Medicago sativa*	decline	S	Alpat'ev et al., 1980
			S	Dzhunusov 1982
			S	Terent'ev 1982
H. trifolii	*T. pratense*	-15 and 50% dry weight	GH	Stelter and Meinl, 1972
	T. repens	-15 and 50% dry weight	GH	Stelter and Meinl, 1972
H. trifolii	*T. pratense*	-50% dry weight	GH+N	Healy et al., 1973
	T. repens	-50% dry weight	GH+N	Healy et al., 1973
H. trifolii	*T. repens*	sign. losses	GH	Mowat 1974
H. trifolii	*Dianthus caryophyllus*	growth reduction	GH	Cuany and Delmasso 1975
H. trifolii	*T. repens*	decreased clover pop.	F+N	Yeates and Risk, 1976
H. trifolii	*T. repens*	-50% dry weight	GH	Kortewege and s'Jacob, 1980
		disappearance from pastures		
H. trifolii	*Beta vulgaris*	increased branched beets	F+GH	Maas and Heijbroek, 1982
f.sp beta	*Pisum sativum*	decreased root weight and	F+GH	Maas and Heijbroek, 1982
	Brassica spp	yield 40%	F+GH	Maas and Heijbroek, 1982
	T. pratense	reduced growth	F+GH	Maas and Heijbroek, 1982
H. trifolii	*T. repens*	severe decline	S	Subbotin 1984
H. trifolii	*T. repens*	decreased yield	F+N	Cook et al., 1983

Ø F = Field experiment; S = Survey or general field observation; N = Nematicide; GH = pot experiment in greenhouse or field

Table 5 Duration of the life cycle and number of generations per year of species of *Heterodera* attacking legumes

Nematode	Host	Temp. ℃	Life cycle duration	Day degrees	Gen/yr	Reference
H. lespedezae	*Lespedeza striata*	22-26	26 days	624	-	Bhatti et al., 1972
H. medicaginis	*Medicago sativa*	Field	30 days	-	2	Dzhunusov, 1982
H. trifolii	*Trifolium repens*	10-20	56 days	840	8	Mulvey, 1959a, b
H. trifolii	*T. repens*	21-27	17 days	408	-	Gerdemann and Linford, 1960
H. trifolii	*T. repens*	Field	42 days	-	3	Subbotin, 1984

Table 6 Development of *H. trifolii* on *Trifolium repens* and *H. lespedezae* on *Lespedeza striata*

	H. trifolii Ø 10-20°C	*H. lespedezae* Ø 14-18°C
Juvenile III	7	9-18
Juvenile IV	17	18-21
Premature adults	22	24-30
Egg formation	30	30
Matrix deposition		33
Yellow females	60	42
Eggs + Juvenile II	56	48
Hatching		51
Brown cyst stage	64	72

Ø Number of days after penetration of the roots by J2s, at the temperature ranges specified.

days in the field. A second generation was produced 74 days after inoculation of noninfested plots with first generation cysts. Norton (1967) estimated the optimum temperature for *H. trifolii* development to be 25°C. A population of *H. lespedezae* from Illinois had an optimum temperature for cyst production of 22-26°C (Fagbenle, 1973) whereas Bhatti et al. (1972) working with an isolate from North Carolina found the optimum to be 26°C.

Conflicting reports on the number of generations per year are probably due to the overlap of different stages of the life cycle (Gerdemann and Linford, 1953). The use of "day-degrees" (Ladygina, 1961) should be included in future work on population development to help in making comparisons, especially in field studies. Day-degrees are calculated by adding the average daily temperatures above the threshold temperature of the nematode i.e. that at which it begins to become active.

It would also be advantageous if life cycle studies started with J2s as inoculum and were measured until the next generation of J2s was observed in the eggs. Inoculation of nematode-free plants of the same age with cysts of the first generation should be used to determine the length of the second generation. With these two parameters an estimate of the number of generations in any single year can be determined from weather data and comparisons made between different studies.

HOST REACTION

Damage to young roots caused by *H. trifolii* penetration is often extensive. Multiple invasion of a root tip kills it, even in nonhosts. The syncytium forms quickly after penetration and if initiated in the stele, it remains in this tissue with greatest direction of growth away from the root tip. The nematode failed to survive if a syncytium was initiated in the cortex and did not extend into the stele (Mankau and Lindford, 1960). Bhatti and Sasser (1974) working with *H. lespedezae* noted that the juveniles began feeding on the stele after 24 hr, and that synctia formation began 72 hr after penetration in the pericycle and extended into the phloem and xylem elements. One month after infection the degeneration of the syncytia began and at 40 days there were signs of cytoplasmic breakdown. Mankau and Lindford (1960) were the first to observe the presence of feeding tubes in the giant cells of a species of *Heterodera*. Rumpenhorst (1984) made detailed observations of the feeding tube.

Seinhorst (1981) reported that *H. trifolii* infection did not affect water consumption per gram plant per day or dry matter content of plants directly at low densities. At higher population levels, however, the nematode reduced water uptake and increased dry matter levels.

MALES

Males are common in some species, and in others a few have been reported to occur occasionally. Wouts (1978) found that the only true *H. trifolii* males appear to be those collected by Norton (1967) from a population reared from a single *H. trifolii* cyst; they are male intersexes. The other males reported as being *H. trifolii* were considered to belong to unidentified, contaminating species. He stated that the male-producing populations reported to exist on carnations and clover (Cuany, 1970; Oostenbrink, 1961) respectively, are undescribed species. He later described the Oostenbrink population as *H. daverti* (Wouts and Sturhan, 1978).

Fagbenle et al. (1986b) detected an extremely low proportion of juveniles that developed into males under good growing conditions with an Illinois isolate of *H. lespedezae*. Males were found more often on red clover than on the type host *Lespedeza striata*, and only at 22 and 26°C ambient temperature, where large numbers of cysts were produced. No males were found in an isolate from North Carolina.

The production of males under stress conditions in the heteroderids is a recognized fact. Green and Plumb (1970) found low numbers of males of *H. trifolii* on *Trifolium incarnatum* but no females, and considered this to be an extreme manifestation of the host resistance/sex determining mechanism which operates with *G. rostochiensis* (Woll.) Behr. However, Wouts (1978) reported that the males in this population belonged to a contaminating species. The presence of males in many instances indicates that in naturally occurring populations more than one species is often present or that stress factors are important controlling factors. Tests on the influence of stress factors using populations from single cysts would clarify this situation.

HATCHING

Most of the eggs of *H. trifolii* are retained within the cyst, but some occur in the gelatinous matrix. Emergence has been reported to be cyclic and influenced by a diapause and temperature, with emergence possible over a temperature range of 4.2 - 31.4°C with 17.2 being considered optimum. It was concluded that hatching coincides with the early start and long growing period of clover roots (Oostenbrink, 1967). This effect was not observed by Yeates and Risk (1976). Winslow (1956) found that in many *Heterodera* species, in vitro hatch was lower in winter and that this condition was induced by earlier environmental conditions. This effect could be by-passed by extracting cysts earlier in the season (Fenwick and Reid, 1953). This effect, "removal from the field prevents reduced hatch", has been named "facultative diapause" by Shepherd and Cox (1967).

Winslow (1955) found hatching to be stimulated by pea root extracts, but not with extracts from clover. *H. trifolii* and *H. galeopsidis*, which have large host ranges, are less dependent on stimulation by root exudates and hatch more readily in water than species with narrow host ranges (Williams, 1978).

Maas and Heijbroek (1982) working with *H. trifolii* f. sp. *beta* found that maximum hatch occurred within 4 weeks at 25-30°C, 70% of the juveniles hatching in the absence of a host plant at temperatures above 15°C. When cysts were treated with root leachate or picric acid hatch increased to 90%. *H. daverti* emergence increased in zinc chloride (Nordmeyer and Sikora, 1983a).

THRESHOLD AND POPULATION DENSITIES

Wardojo et al. (1963) and Hidding et al. (1963) found that *H. trifolii* caused no damage to white clover during the first 4-8 weeks after sowing in pot experiments with populations of up to 50 juveniles or eggs per g of soil. Growth reduction occurred only after the clover had been cut one or more times. According to Seinhorst and Sen (1966) *H. trifolii* will only damage well developed white clover plants in pastures at densities of more than 200 eggs per g soil, this level being seldom reached in the Netherlands where the equilibrium density seems to be well below 50 per g soil. Ennik et al. (1964) for example, found fewer than 10 eggs per g soil in pastures at Wageningen, the Netherlands.

Yeates (1973) reported that during a typical summer drought in New Zealand *H. trifolii* densities in a clover pasture were 600-800 cysts/350ml sample compared with 200-300 cysts in the autumn and spring. Pasture growth in the spring corresponded well with the invasion of roots by the nematode. In pot trials Yeates (1978) estimated that 600 cysts/kg soil corresponded with the equilibrium population density for field soil. He found that low *H. trifolii* infestation levels stimulated plant growth.

Korteweg and s'Jacob (1980) found that *H. trifolii* and nitrogen reinforced each other in suppressing white clover in a grass-clover mixture. *H. trifolii* diminished the competative power of white clover in a mixture and was considered to be a serious factor in the disappearance of white clover from grasslands, especially when high amounts of nitrogen fertilizer are applied.

Freckman and Chapman (1972) found that only 25-30% of the initial *H. trifolii* inoculum entered the root system of red clover seedlings. Nordmeyer and Sikora (1983b) noted that *H. daverti* juvenile attraction to the root and penetration increased when the plant was previously attacked by root-rotting fungi. Root penetration seems to be influenced by initial population density, environmental factors and the presence of rhizosphere microorganisms.

Sugar beet and red clover growth were negatively correlated with the density of the beet host race of *H. trifolii*. The threshold level was set at 5 eggs/g soil (Maas and Heijbroek, 1982).

Populations of *H. medicaginis* causing thinness of lucerne stands ranged from 15 cysts and 1429 eggs and juveniles to 271 cysts and 33216 eggs and juveniles/100ml of soil on different lucerne varieties (Aplat'ev et al., 1980). Decreased yields were reported at densities of 500 eggs and juveniles per 100ml of soil (Dzhunusov, 1982).

Population densities in the field are often strongly influenced by environmental factors and the presence of parasites and antagonists. A large measure of variation is also due to extraction techniques. For these reasons the population densities cited are not of particular significance or usefulness in estimating possible crop losses.

INTERACTIONS

Interactions between most species and host races in the complex and other soil microorganisms have not been studied in great detail. Those that have been studied often produce unique effects.

Work by Freckman and Chapman (1972) showed that when *H. trifolii* and *M. hapla* were present simultaneously they developed independently, often directly next to one another. The effect of dual inoculations of the same two species on white clover yield was studied by Healy et al. (1973) and Widdowson et al. (1973). They reported the presence of heavy mixed infestations in unsterilised soil, but did not mention any direct interaction between the two species. The high rates of superphosphate associated with farming on yellow-brown loams in New Zealand, usually attributed to the capacity of these soils to "fix" phosphate, may in part be a consequence of nematode activity. When the nematodes were present, more than five times as much phosphate was needed to give yields comparable to those obtained in soil free of nematodes.

Taha and Raski (1969) showed that the introduction of the nematode one week before, simultaneously, or one week following inoculation with *Rhizobium* bacteria did not hinder nodule formation. Nodule size on white clover was not altered although *H. trifolii* showed a preference for nodulated tissue (6 *H. trifolii*/nodule). The nitrogen-fixation efficiency of nematode-infected nodules was not impaired. They concluded that the reduction in the total N in *H. trifolii* infected plants was the result of stunting of the entire plant. Yeates et al. (1977), conversely, detected a reduction in rhizobia number.

The existence of interactions between nematodes in the *H. trifolii* - complex with soil borne fungi has been suggested as important in clover decline for: *H. trifolii* (Ennik et al., 1964; Yeates, 1973; Anon, 1977), *H. medicaginis* (Alpt'ev et al., 1980), and *H. daverti* (Sikora, 1977).

Ennik et al. (1964) isolated *Rhizoctonia* and species of *Fusarium* from tissue adjacent to dead lateral roots of *Trifolium repens* in pots with untreated field soil. They also noted that *H. trifolii* populations were higher in these pots; a phenomenon similar to that observed by Nordmeyer and Sikora (1983b). Poor growth could not be explained by the presence of the nematode alone.

Sikora (1977) reported the presence of the root rotting fungi *Fusarium oxysporum* and *F. avenaceum* associated with *H. daverti* in pastures of *Trifolium subterraneum* in northern Tunisia. The diseased plants were severely stunted and had a deformed and rotted primary and secondary root system. Several varieties exhibited leaf reddening and to a lesser extent leaf chlorosis. Nordmeyer et al., (1978) reported that *F. oxysporum* and *F. avenaceum* were present in 54 and 41% of diseased plants. *Pythium irregulare* was also present in the tissue.

Simultaneous inoculation of *H. daverti* with either species of *Fusarium* did not reduce the yield of two *Trifolium subterraneum* clones over the additive effect of the organisms inoculated separately (Nordmeyer and Sikora, 1983b). Of great interest was the fact that inoculation of the nematode 1 or 2 weeks after the fungi caused synergistic reductions in plant dry weights. Furthermore, nematode penetration was increased when the fungi were previously established in the root system. Leaf reddening, a symptom observed under field conditions, developed only on plants inoculated with both the nematode and *F. avenaceum*.

CONTROL

A number of the techniques used for the control of other species of *Heterodera* are applicable to nematodes in the *H. trifolii* - complex. Varieties of white clover resistant to *H. trifolii* have been reported (Kuiper, 1960) with inheritance of resistance determined by more than one gene (Dijkstra, 1971). Some cultivars of *Medicago polychroa* were found to be resistant to *H. medicaginis* (Alpat'ev 1981).

Differences in host susceptibility to *H. trifolii* (Widdowson et al., 1972; Yeates et al., 1973) and to *H. lespedezae* (Edwards and Malek, 1971) have been reported and could be used as alternative crops. Thirtyfour out of forty plant cultivars or breeding lines of different plant families were highly susceptible to the beet host race of *H. trifolii* (Steele et al., 1983).

Cauny (1970) recommended the use of systemic nematicides to control *H. trifolii* on carnation. Hadar et al. (1983) found that preplant soil treatment with methyl bromide, sodium methyldithiocarbamate or soil solarization by mulching with polyethylene sheets controlled *H. trifolii* on carnation.

FUTURE CHALLENGES

1. More information is needed on the correlation between morphological, cytological and host range variability to further elucidate the phylogenetic relationships as a basis for taxonomic classification.

2. Cross-breeding experiments may contribute to our understanding of genetic relationships.

3. The effect of environmental factors and host plant on the presence of males should be looked at more closely, and where possible with populations developed from single cysts.

4. Nematode and host plant protein-electrophoresis and DNA probe technology may bring us even closer to nematode phylogeny and the basic principles of host specificity.

5. Work on this group has to a certain extent been curtailed by a reduction in the importance of pasture legumes caused by the increased use of grasses for animal feeding. The detection of host races on sugar beet and a new species attacking chickpea may give the group a new level of economic importance.

6. Hatching and seasonal effects on population dynamics should be looked

at more closely. The presence of a high level of water hatch in some species and host races could be utilized in a pest management programme by crop rotation.

7. The many observations involving interactions with other microorganisms deserves more study. The interaction between the nematode, rhizobia and nitrogen fertilizer in clover-grass mixtures deserves more attention.

REFERENCES

Alpat'ev, N.M., 1981, Effectiveness of evaluating lucerne varieties under conditions of infection. (Rus). Selektsiya i Sememovodstvo, USSR, 3:16.

Alpat'ev, N.M., Terent'eva, T.G. and Shestiterova, I.Z. 1980, The causes of the thinness of lucerne stands in the northern Causasus. (Rus). Zaschita kornovykh kul'tur ot vreditelei, bolez ei i sornyakov. "Kolos", 61.

Andersson, S., 1984, First record of a yellow beet cyst nematode *(Heterodera trifolii)* in Sweden, Växtskyddsnotiser, 48:11.

Anon, 1977, Ann.Rpt on Research and Technology of Dpt.Agric. Northern Ireland, 47:102.

Bhatti, D.S. and Sassar, J.N., 1973, Pathogenicity of *Heterodera lespedezae* on 'Kobe' lespedeza, Indian J.Nematol., 2:125.

Bhatti, D.S. and Sassar, J.N., 1974, Histopathology of 'Kobe' lespedeza roots infected with *Heterodera lespedezae*, Indian J.Nematol., 4:20.

Bhatti, D.S., Hirschmann, H. and Sassar, J.N., 1972, Post-infection development of *Heterodera lespedezae*, J.Nematol., 4:104.

Cook, R., Plowright, R.A., Rushton, W.H. and Bowling, P.J., 1983, Ann.Rpt. Welsh Pl.Br.Stn., 1982, 67.

Cuany, A., 1970, A study of the action of aldicarb against *Heterodera* infestations of carnations in the south of France, Phytiatrie-Phytopharmacie, 19:177.

Cuany, A. and Delmasso, A., 1975, Characters and specificity of two biological races of *Heterodera* on *Dianthus caryophyllus*, Nematol. medit., 3:11.

Dijkstra, J., 1971, Breeding for resistance to *Heterodera* in white clover, J.Euphytica, 22:36.

Dzhunusov, K. K., 1982, The biology of *Heterodera medicaginis*, (Rus). Sbornik Nauchnykh Trudov Vsesoyuznogo Nauchno-Issledovatel'skogo Instituta Zashchity Rasteneii Leningrad, USSR, 45.

Edwards, D.I., and Malek, R.B., 1971, Cyst production by *Heterodera lespedezae* on selected legumes, Plant Dis.Reptr., 55:927.

Ennik, G.C., Kort, J. and Luesink, B., 1964, The influence of soil disinfection with DD, certain components of DD and some other compounds with nematocidal activity on the growth of white clover, Neth.J.Pl. Path., 70:117.

Ennik, G.C., Kort, J. and van de Bund, C.F., 1965, The clover cyst nematode (*Heterodera trifolii* Goffart) as the probable cause of death of white clover in a sward, J.Br.Grassld.Soc., 29:258.

Fagbenle, H.H., 1973, The lespedeza cyst nematode, *Heterodera lespedezae* Golden & Cobb: Temperature effects on histopathology on two hosts and infraspecific physiological variation, Ph.D. Thesis, Univ.Illinois, USA, Diss.Abst Intern Nr 34B, 5277.

Fagbenle, H.H., Edwards, D.I. and Malek, R.B., 1986a, Comparative cyst production by two isolates of *Heterodera lespedezae* on selected legumes, Plant Dis., (in press).

Fagbenle, H.H., Edwards, D.I. and Malek, R.B., 1986b, Effects of temperature on pathogenicity and population development of two *Heterodera lespedezae* isolates, J.Nematol., (in press).

Fenwick, D.W., Reid, E., 1953, Population studies on the potato root eelworm (*Heterodera rostochiensis* Woll.), J.Helminth., 27:119.

Franklin, M.T., 1951, The cyst forming species of *Heterodera*, Commn.Bur. Agr.Tech.Commun. No.29, 147pp.

Freckman, D.W. and Chapman, R.A., 1972, Infection of red clover seedlings by *Heterodera trifolii* Goffart and *Pratylenchus penetrans* Cobb, J.Nematol., 4:23.

Gerber, K. and Mass, P.W.Th., 1982, A redescription of *Heterodera medicaginis* Kirjanova, Nematologica, 28:94.

Gerdemann, J.W. and Linford, M.B., 1953, A cyst-forming nematode attacking clovers in Illinois, Phytopathology, 43:603.

Goffart, H., 1932, Studies on the cereal nematode *Heterodera schachtii* Schm. with particular emphasis on the conditions in Schleswig-Holstein. III. Race study of *Heterodera schachtii* Schm., Ab.Biol. Reichsanst., Berlin, 20.

Green, C.D. and Plumb, S.C., 1970, The interrelationships of some *Heterodera* spp. indicated by the specificity of the male attractants emitted by their females, Nematologica, 16:39.

Hadar, E., Sofer, S., Brosh, S., Mordechai, M., Cohn, E. and Katan, Y., 1983, Control of clover cyst nematode on carnation (Hebrew). Hassadeh, 63:1698.

Healy, W.B., Widdowson, J.P. and Yeates, G.W., 1972, The effect of nematodes on the growth and utilization of phosphorus by white clover on a yellow-brown loam, N.Z. Soil News, 20:125.

Healy, W.B., Widdowson, J.P. and Yeates, G.W., 1973, Effect of root nematodes on the growth of seedling and established white clover on a yellow-brown loam, N.Z. J.Agric., 16:70.

Hidding, J., Hijink, J. and Oostenbrink, M., 1963, Opbringst en kwaliteitsverlies von witte klaver en ein gras-klaver-mendsel door let klavercystenaalje, *Heterodera trifolii*, Meded.Landb.hogesch. OpzoekStns. Gent, 28:679.

Hirschmann, H., 1956, Comparative morphological studies on the soybean cyst nematode, *Heterodera glycines* and the clover cyst nematode, *Heterodera trifolii* (Nematoda: Heteroderidae), Proc.Helm.Soc.Wash., 2:140.

Hirschmann, H. and Triantaphyllou, A.C., 1979, Morphological comparison of members of the *Heterodera trifolii* species complex, Nematologica, 25:458.

Kimpinski, J. and Platt, H.W., 1983, Washing of potatoes to remove nematodes and to observe the effect on storage rot diseases, Can.Pl. Dis. Survey, 63:45.

Korteweg, C.R. and S'Jacob, J.J., 1980, Influence of *Heterodera trifolii* and nitrogen on white clover in a clover-grass mixture, Med.Fac. Landbouww. Rijksuniv. Gent, 45:761.

Krall, E., 1977, Compendium of cyst nematodes in the USSR, Nematologica, 23:311.

Kuiper, K., 1960, Resistance of white clover varieties to the clover cyst-eelworm *Heterodera trifolii* Goffart, Nematologica Suppl. II, 95.

Ladygina, N.M., 1961, Reaction of the beet nematode to temperature and humidity (Rus), in "Voprosy fitogel'mintologii", K.I. Skryabin and E.S. Turlygina, eds., Moscow.

Lobb, W.R. and Wood, J., 1971, Insects in the food supply of starlings in mid-Canterbury, N.Z. Entomologist, 5:17.

Maas, P.W.Th. and Heijbroek, W., 1982, Biology and pathogenicity of the yellow beet cyst nematode, a host race of *Heterodera trifolii* on sugar beet in the Netherlands, Nematologica, 28:77.

Maas, P.W.Th., Du Bois, E. and Dede, J., 1982, Morphological and host range variation in the *Heterodera trifolii* complex, Nematologica, 28:263.

Mankau, R., and Linford, M.B., 1960, Host-parasite relationships of the clover cyst nematode *Heterodera trifolii* Goffart. Univ.Illinois Agr. Exp.Sta.Bull., 667.

Mowat, D.J., 1974, The host range and pathogenicity of some nematodes occurring in grassland in northern Ireland, Rec.Agric.Res., 22:51.

Mulvey, R.H., 1958, Parthenogenesis in a cyst-forming nematode, *Heterodera trifolii* (Nematoda: Heteroderidae), Can.J.Zool., 36:91.

Mulvey, R.H., 1959a, Susceptibilities of plants to the clover cyst nematode, *Heterodera trifolii*, and the period required to complete a life cycle, Nematologica, 4:132.

Mulvey, R.H., 1959b, Investigations on the clover cyst nematode *Heterodera trifolii* (Nematoda: Heteroderidae), Nematologica, 4:147.

Mulvey, R.H., 1972, Identification of *Heterodera* cysts by terminal and cone top structures, Can.J.Zool., 50:1277.

Mulvey, R. A. and Anderson, R.V., 1974, *Heterodera trifolii*, C.I.H. Descriptions of Plant Parasitic Nematodes Set 4, No.46, 4pp.

Nordmeyer, D., 1978, Studies on the root rot complex in *Trifolium subterraneum* in north-west Tunisia, with special reference to interactions among the causal agents, (Ger). Ph.D. Thesis, Univ.Bonn, Germany.

Nordmeyer, D., and Sikora, R.A., 1983a, Effect of a culture filtrate from *Fusarium avenaceum* on the penetration of *Heterodera daverti* into roots of *Trifolium subterraneum*, Nematologica, 29:88.

Nordmeyer, D. and Sikora, R.A., 1983b, Studies on the interaction between *Heterodera daverti*, *Fusarium avenaceum* and *F. oxysporum* on *Trifolium subterraneum*, Rev.de Nematol., 6:193.

Nordmeyer, D., Sikora, R.A., and Jaritz, G., 1978, The importance of nematodes and soil-borne fungi as disease agents in *Trifolium subterraneum* pastures in northwest Tunisia, (Ger). Med.Fac.Landbouww. Rijksuniv. Gent, 43:747.

Norton, D.C., 1967, Relationships of *Heterodera trifolii* to some forage legumes, Phytopathology, 57:1305.

Oostenbrink, M., 1961, The genus *Heterodera*, in "Nematology", J.N. Sasser and W.R. Jenkins, eds., University of N. Carolina Press, Chapel Hill.

Oostenbrink, M., 1967, Studies on the emergence of encysted *Heterodera* larvae, Meded.Rijksfac.LandbWet. Gent, 32:503.

Oostenbrink, M. and Den Ouden, H., 1954, De structuur van de kegeltop als taxonomisch kenmerk bij *Heterodera*-soorten met citroenvormige cysten, Tijdschr.Plantenziekten, 60:146.

Rumpenhorst, H.J., 1984, Intracellular feeding tubes associated with sedentary plant parasitic nematodes, Nematologica, 30:77.

Schmidt, A., 1871, Über den Rubennematoden, Z.Ver.Rübenzuckerindustr. Zollver, 21:1.

Seinhorst, J.W., 1981, Water consumption of plants attacked by nematodes and mechanisms of growth reduction, Nematologica, 27:34.

Seinhorst, J.W. and Sen, A.K., 1966, The population density of *Heterodera trifolii* in pastures in the Netherlands and its importance for the growth of white clover, Neth.J.Pl.Path., 72:169.

Shepherd, A.M. and Cox, P.M., 1967, Observations on periodicity of hatching of eggs of the potato cyst nematode *Heterodera rostochiensis* Woll. Ann.appl.Biol., 60:134.

Sikora, R.A., 1977, *Heterodera trifolii* associated with *Fusarium* root rot of *Trifolium subterraneum* in northern Tunisia, Nematol.medit., 5:319.

Sikora, R.A. and Maas, P.W.Th., 1986, A new species of *Heterodera* attacking *Cicer arietium*, Nematologica (in press).

Solov'eva, G.I., Krall, E.L. and Vasil'eva, A.P., 1979, The hemp nettle cyst nematode-*Heterodera galeopsidis* (Goffart, 1936) in Karelia, (Rus)., Parazetolissled Karelsk. ASSP, Petrozavodsk, 151.

Steele, A.E., Toxopeus, H. and Heijbroek, W., 1983, Susceptibility of plant selections to *Heterodera schachtii* and a race of *H. trifolii* parasitic on sugarbeet in the Netherlands, J.Nematol., 15:281.

Stelter, H. and Meinl, G., 1972, The effects of the infestation of red and white clover by *Heterodera trifolii* and *Heterodera galeopsidis*. (Ger)., Archiv. fur Pflanzenschutz, 8:463.

Stone, A.R., 1975, Head morphology of second-stage juveniles of some Heteroderidae (Nematode: Tylenchoidea), Nematologica, 21:81.

Stone, A.R., 1979, Co-evolution of nematodes and plants, Symb.Bot.Upsal., 22:46.

Subbotin, S.A., 1984, *Heterodera trifolii* Goffart, 1932 in the Moscow region, (Rus)., Byulleten Vbsesoyuznogo Instituta Gel'mintologii im. K.I. Skryabina, 36:45.

Taha, A.H.Y. and Raski, D.J., 1969, Interrelationships between root-nodule bacteria, plant-parasitic nematodes and their leguminous host, J.Nematol., 1:201.

Terent'eva, T.G., 1982, Parasitic nematodes of lucerne, (Rus)., in "Sbornik Nauchnykh Trudov Vsesovuznogo Nauchno-Issledovatel'skogo instituta Zashchity Rasteneii", Leningrad, USSR.

Triantaphyllou, A.C. and Hirschmann, H., 1978, Cytology of the *Heterodera trifolii* parthenogenetic species complex, Nematologica, 24:418.

Vallotton, R., 1985, Preliminary observations in Switzerland on formae specialis *beta* of the cyst nematode *Heterodera trifolii* Goffart, (Fr)., Rev.suisse Agric., 17:137.

Vovlas, N., Inserra, R.N. and Stone, A.R., 1981, *Heterodera mediterranea* n. sp. (Nematoda: Heteroderidae) on *Pistacia lentiscus* in southern Italy, Nematologica, 27:129.

Vovlas, N., Greco, N. and Di Vito, M., 1985, *Heterodera ciceri* n. sp. (Nematoda: Heteroderidae) on *Cicer arietinum* L. from northern Syria, Nematol. medit., (in press).

Wardojo, S., Hijink, M.J. and Oostenbrink, M., 1963, Damage to white clover through inoculation with *Heterodera trifolii*, *Meloidogyne hapla* and *Pratylenchus penetrans*, (Du), Meded.Landb.hogesch. OpzorkDynd. Gent, 28:672.

Widdowson, J.P., Healey, W.B. and Yeates, G.W., 1972, The effect of nematodes on the growth and utilization of phosphorus by white clover on a yellow brown loam, Proc.Agr.Soc.N.Z. (Abstr.), 125.

Williams, T.D., 1978, Cyst Nematodes: Biology of *Heterodera* and *Globodera* in "Plant Nematology", J.F. Southey ed., H.M.S.O., England.

Winslow, R.D., 1955, The hatching responses of some root eelworms of the genus *Heterodera*, Annl.appl.Biol., 43:19.

Winslow, R.D., 1956, Seasonal variations in the hatching responses of the potato-root eelworm, *Heterodera rostochiensis* Wollenweber, and related species, J.Helminth., 39:157.

Wouts, W.M., 1978, On the males of *Heterodera trifolii* Goffart, 1932 (Nematoda: Heteroderidae), Nematologica, 24:115.

Wouts, W.M. and Weischer, B., 1977, A classificiation of fifteen species of Heteroderinae commonly occurring in western Europe based on the larval characteristics, (Ger)., Nematologica, 23:289.

Wouts, W.M. and Sturhan, D., 1978, The identity of *Heterodera trifolii* Goffart, 1932 and the description of *H. daverti* n. sp. (Nematoda: Tylenchida), Nematologica, 24:121.

Yeates, G.W., 1973, Annual cycle of root nematodes on white clover in pasture. I. *Heterodera trifolii* in a yellow grey earth, N.Z. J.Agric. Res., 16:569.

Yeates, G.W., 1974, Effects of *Heterodera trifolii* on the growth of clover on a yellow-grey earth under greenhouse conditions, N.Z. J.Agric.Res., 17:379.

Yeates, G.W., 1978, Influence of *Heterodera trifolii* cyst inoculum on white clover growth and final cyst count in pots of a yellow-brown pumice soil, N.Z. J.Agric.Res., 21:149.

Yeates, G.W. and Risk, W.H., 1976, Annual cycle of root nematodes on white clover in pasture. III. *Heterodera trifolii* in a yellow-brown earth, N.Z. J.Agric.Res., 19:393.

Yeates, G.W., Healy, W.B. and Widdowson, J.P., 1973, Screening of legume varieties for resistance to the root nematodes *Heterodera trifolii* and *Meloidogyne hapla*, N.Z. J.Agric.Res., 16:81.

Yeates, G.W., Healy, W.B., Widdowson, J.P., Thomson, N.A. and McDiarmid, B.N., 1976, Effects of a soil fumigant on the establishment and growth of a grazed pasture on a yellow-brown loam, N.Z. J.Agric.Res., 19:397.

Yeates, G.W., Ross, D.J., Bridger, B.A. and Visser, T.A., 1977, Influence of the nematodes *Heterodera trifolii* and *Meloidogyne hapla* on nitrogen fixation by white clover under glasshouse conditions, N.Z. J.Agric. Res., 20:401.

DISCUSSION

Commenting extensively on definitions relating to the *H. trifolii* complex, Stone outlined his objection to the use of the nomenclatorial forma specialis in preference to host race. He considered that the meaning of host race was clearly established in plant nematology and can be defined as "a race of a nematode species distinguished by inherited ability or ability to reproduce on designated species of plants". Stone said that introducing a latin name for host race introduces an illusion of stability, which is spurious and undesirable. It is spurious because f.sp. is not nomenclatorially acceptable for zoological species. It is undesirable because it differs from the Linnean system in not being concerned with groups of organisms believed to be stable and natural. A host race is defined on the ability or inability of a nematode phenotype to reproduce on an arbitrary selection of plant species. Cyst nematodes have, frequently, large host ranges within their host families, and host races may only become apparent when attention is focussed on a few of the potential hosts.

Stone continued that the use of f.sp. or similar designation had been considered and rejected by the International Committee for Zoological Nomenclature and was botanical, not zoological, in the context of nomenclature. He argued that f.sp. serves only to provide a new jargon for host race and that there was a need to keep nomenclatorial descriptions as simple as possible. If necessary "host race" could be left untranslated in other languages, such as English speaking zoologists have accepted "artenkreis" and other similar examples. Stone acknowledged that the biological species concept (ref. interbreeding) cannot be applied to all races and taxa in the *H. trifolii* complex but observed this is not a unique problem. In zoology parthenogenetic species are defined by using morphological differences, as are most if not all of the amphimictic species. The *schachtii* group itself is only an example of a species complex in nematology; it, too, needs special nomenclatorial treatment.

In discussing various aspects of the biology of the *H. trifolii* complex Vallotton said that *H. trifolii* f.sp. *beta* (yellow beet cyst nematode) had recently been found in Switzerland, but that its distribution and economic importance still had to be assessed; host tests readily separated it from related forms. Jacobsen asked if other forms of the *H. trifolii* complex, other than *H. trifolii*, readily hatched in water and Seinhorst responded they do, at room temperature, but just need a few days of cold treatment (in the field or the refrigerator) to obtain the best hatch. Miller reported that he had found that *H. glycines* reproduced on beets and *H. schachtii* on soybean, and considered that *H. glycines* should be a subspecies of *H. schachtii*. Golden commented that the two species had destinct juvenile and cyst characters and that they should continue to be regarded as separate species.

OBSERVATIONS ON A POPULATION OF CYST FORMING NEMATODES FROM CARNATION

Laura Ambrogioni[1], Francesco P. D'Errico[2] and
Anna Marinari-Palmisano[1]

[1]Istituto Sperimentale per la Zoologia Agraria, Cascine del Riccio, Firenze, Italy, [2]Istituto di Entomologia Facolta di Agraria, Portici, Napoli, Italy

Since 1983 a disease of carnation has become widespread in greenhouses in the Naples area. Young plants are stunted, chlorotic and wilt under moisture stress. They do not produce normal flower stalks, the sepals sometimes have reddish streaks and buds usually fail to open. Patches of poorly growing diseased plants appear and then rapidly increase in size. Carnation mottle virus has been found associated with these symptoms (V. Lisa and A. Ragozzino, personal communication). Cyst forming nematodes of the *Heterodera schachtii* Schmidt group have been observed on the roots of diseased plants.

Two populations of the nematode, from Torre del Greco and Castellammare di Stabia (Naples), have been reared in a greenhouse on *Dianthus caryophyllus* L. and *Trifolium repens* L. and large numbers of cysts and males have been obtained on both plants.

Characters and measurements of the nematode were compared with those of the species belonging to the *schachtii* - group for which sufficient data are available (Table 1).

Among the bisexual species of the group, the specimens from Naples are most similar to the *Heterodera daverti* Wouts & Sturhan but there are some differences in the length of the cyst underbridge and in the shape and length of the male tail. The Italian population can be distinguished from *H. glycines* Ichinohe by the length and depth of the underbridge. It is also clearly differentiated from *H. glycines, H. rosii* Duggan & Brennan and *H. trifolii 18* which damages carnation (Cuany and Dalmasso, 1974) by several juvenile and male characters (Table 1).

In comparison with the *H. trifolii* complex, a group of monosexual taxonomic units (nominal species: *H. trifolii* Goffart, *H. lespedezae* Golden & G. S. Cobb, *H. galeopsidis* Goffart), our populations differ chiefly by the presence of true males; the only males certainly belonging to *H. trifolii* appear to be intersexes (Norton, 1967; Wouts, 1978). The male intersexes of *H. trifolii* are very similar to our specimens in tail shape, but are stouter (<u>a</u> smaller) and have a longer stylet. Other differences are related to the cyst underbridge depth which is greater in comparison with populations of *H. trifolii* complex, and to several measurements of the juveniles. Our specimens are distinguished from *H. rumicis* Poghossian, by some characters and measurements of the juveniles (Table 1).

Table 1. Characters and measurements (μm) of cysts, juveniles and males of some *Heterodera* spp and a closely related Italian population.

CYSTS

Species or Population	Main Host Plant	Vulval slit length	Under-bridge length	Under-bridge depth
1) Naples, Italy	*Dianthus caryophyllus*	49.6[a] (47-56)	127 (96-145)	69.7 (54-82)
2) Naples, Italy	*Trifolium repens*	50.2[a] (48-54)	115 (103-127)	62.0 (54-73)
3) *H. daverti* (1)	*T. repens*	47 (34-55)	82 (70-102)	
4) *H. glycines* (2,3,4)	*Glycine max*	(45-60)	(70-100)	(35-37)
	Trifolium spp	40		
5) *H. rosii* (3,5)	*Rumex crispus* *Trifolium* spp	(45-53)	(110-130)	(43-62)
6) *H. trifolii* 18 (6)	*D. caryophyllus* *Brassica* spp *Trifolium* spp *R. acetosella*		present	
7) *H. trifolii* (1,3,4,7,8, 12)	*Trifolium* spp	(39-50)	(80-100)	(33-38)
	R. crispus			
	Lespedeza sp	(39-62)		
	D. caryophyllus			
8) *H. lespedezae* (3,4,7)	*Trifolium* spp	(50-55)	(90-115)	(30-35)
	R. crispus	47		
	Lespedeza sp	(42-53)		
9) *H. galeosidis* (3,7)	*Galeopsidis tetrahit*	(40-50) 49.1 (42-58)	(80-100)	(37-40)
10) *H. rumicis* (9, 10)	*Rumex* spp	55.7 (47-64)	100	

[a]n = 20; [b]n = 10; measurements from various authors: (1) Wouts and Sturhan, 1978; (2) Hirschmann, 1956; (3) Mulvey, 1972; (4) Mulvey and Golden, 1983;

SECOND-STAGE JUVENILES

	Body length	*a*	Stylet length	Head tip to median bulb valve	Tail length	Hyaline terminal length	Knobs shape	Head annules lip region
1)	461.4[a] (394-593)	20.0 (16-23)	25.5 (24-27)	69.9 (57-77)	54.7 (48-62)	31.8 (25-36)	anterior face deeply concave	1
2)	427.8[a] (399-466)	21.5 (21-23)	25.0 (24-25)	64.2 (54-73)	48.1 (43-54)	25.5 (19-30)	ant. face deeply concave	1
3)	457 (400-480)	24.0 (22-25)	25 (24-26)	70 (62-76)	55 (49-60)	33.3 (28-39)	ant. face slightly concave	1
4)	439 (375-490) 471 (437-504)		*23* *(22–24)*		50.4 (42-59) 45.0 (42-47)	26.6 (20-33)	anterior face concave	*3–4*
5)	*549.7* *(430–661)*	*28.3* *(27–31)*	*31.4* *(27–34)*		*66* *(58–77)*	*40.4* *(37–45)*	*rounded*	
6)	*584* *(515–638)*		*28* *(25–30)*		*61* *(55–68)*	*39* *(30–44)*	anterior face concave	1
7)	510 (490-530) *(459–678)*		*27* *(25–29)* *(24–30)*	*80.7* *(76–86)*	*59* *(55–65)* *(51–78)*	*36* *(30–38)* (27-45)	anterior face concave	*1–4*
8)	457 (400-510) 481 (401-531)		25.2 (24-26) 24.3 (23-26)		55.9 (47-66) 53.5 (46-61)	29.6 (23-36) 26.3 (20-38)	slightly concave	
9)	*553* *(495–620)*		*27.1* *(26–28)*		*68.5* *(60–75)*	*40.3* *(33–45)*	anteriorly concave amalgamated	
10)	*[illegible]21* *(430–555)*		*29* *(28–30)*			*equal to stylet length*	anchor shaped	*4–5*

(continued)

(5) Duggan and Brennan, 1966; (6) Cuany and Dalmasso, 1974; (7) Hirschmann and Triantaphyllou, 1979; (8) Norton, 1967; (9) Krall, 1977; (10) Poghossian, 1961; (12) Wouts, 1978.

Table 1. (continued)

MALES

Species or Population	Body length	*a*	Stylet length	Head lip to excretory pore	Tail length	Head annules knobs shape	Tail shape
1) Naples, Italy	1120[b] (907-1391)	45.8 (39-57)	28.0 (27-29)	136 (109-157)	9.5 (7-11)	4 anteriorly concave	constricted posterior extension of body
2) Naples, Italy	1166[b] (956-1403)	49.4 (42-58)	28.4 (27-30)	145 (127-181)	7.8 (6-9)	4 anteriorly concave	constricted posterior extension of body
3) *H. daverti*	1043 (715-1240)	42 (28-50)	27.6 (26-29)	144 (122-177)	5	4/anteriorly slight concave	*slight extension*
4) *H. glycines*	*1265 (1035–1400) 1330 (1200–1400)*		26.9 (25-28) (26-27)	144	5.3 *(4–8)* 3.5 *(2–6)*	*6–7 4–5 massive rounded*	posterior extension
5) *H. rosii*	*1210 (840–1630)*	*38.5 (28–40)*	*29.8 (28–31)*	*(61–63)*		*3–4 rounded*	*rounded 4 lateral indentations*
6) *H. trifolii* 18	*1500 (990–1800)*		*31.5 (29–34)*			*3–4 rounded*	
7) *H. trifolii*	1200 (950-1290)	*36 (32–39)*	*(32–33)*			4 anteriorly concave	constricted posterior extension of body

Differential characters and measurements of the species related to the population from Naples are printed in italics.

The taxonomic units of the *H. trifolii* complex and the bisexual species closely related to *H. trifolii* are difficult to separate by characters of the cyst or the juvenile. Moreover they have many host plants in common and growing conditions can have significant effects on their morphometric characters (Maas et al., 1982). Some differences between individuals grown on carnation and on white clover also have been found in our population (Table 1).

Hirschmann and Triantaphyllou (1979) proposed the adoption of subspecies as a taxonomic unit in their review of the parthenogenetic *Heterodera trifolii* complex. Sturhan (1983) suggested a category of superspecies for *H. trifolii*, which would include the bisexual species *H. daverti*.

The morphological differences between the Naples population and *H. daverti* can hardly be considered as adequate for characterisation of a separate species. Further information is needed for both forms on the range of variability of some characters, reproductive compatibility and host range to determine the position of the population from Naples. At present the latter may conveniently be regarded as a subspecies of *H. daverti*.

REFERENCES

Cuany, A. and Dalmasso, A., 1974, Caractères et specificité de deux espèces biologiques d'*Heterodera* se developpant sur *Dianthus caryophyllus*, Nematol. medit., 3:11.

Duggan, J.J. and Brennan, P.A., 1966, *Heterodera rosii* (Heteroderidae), a new species of cyst-forming nematode from curled dock (*Rumex crispus* L.), Ir. J. Agric. res., 5:113.

Hirschmann, E., 1956, Comparative morphological studies on the soybean cyst nematode, *Heterodera glycines* and the clover cyst nematode, *H. trifolii* (Nematode: Heteroderidae), Proc. helminthol. Soc., Wash., 2:140.

Hirschmann, E. and Triantaphhyllou, A.C., 1979, Morphological comparison of members of the *Heterodera trifolii* species complex, Nematologica, 25:458.

Krall, E., 1977, Compendium of cyst nematodes in the USSR, Nematologica, 23:311.

Maas, P.W.Th., Du Bois, E. and Dede, J., 1982, Morphological and host range variation in the *Heterodera trifolii* complex, Nematologica, 28:263.

Mulvey, R.H., 1972, Identification of *Heterodera* cysts by terminal and cone top structures, Can. J. Zool., 50:1277.

Mulvey, R.H. and Golden, A.M., 1983, An illustrated key to the cyst-forming genera and species of Heteroderidae in the Western Hemisphere with species morphometrics and distribution, J. Nematol., 15:1.

Norton, D.C., 1967, Relationships of *Heterodera trifolii* to some forage legumes, Phytopathology, 57: 1305.

Poghossian, E.E., 1961, A new nematode species *Heterodera rumicis* sp. n. (Nematoda: Heteroderidae) from Armenian SSR, (in Russian), Dokl. Akad. Nauk. Arm. SSR., 32: 171.

Sturhan, D., 1983, The use of the subspecies and the superspecies categories in nematode taxonomy, in "Concepts in nematode systematics", A.R. Stone, H.M. Platt and L.F. Khalil, eds., Academic Press, London and New York, 41.

Wouts, W.M., 1978, On the males of *Heterodera trifolii* Goffart, 1932 (Nematoda: Heteroderidae), Nematologica, 24: 115.

Wouts W.M. and Sturhan, D., 1978, The identity of *Heterodera trifolii* Goffart, 1932 and the description of *H. daverti* n. sp. (Nematoda: Tylenchida), Nematologica, 24: 121.

THE PEA CYST NEMATODE

Mauro Di Vito and Nicola Greco

Istituto di Nematologia Agraria del C.N.R.
70126 Bari
Italy

INTRODUCTION

Liebscher (1890) reported a disease of pea (*Pisum sativum* L.) caused by nematodes in an experimental field at the Agriculture Institute at Göttingen. At first he attributed the damage to a strain of *Heterodera schachtii* Schmidt, but further studies on morphology and biology led to the conclusion that the nematode was a new species which was named *Heterodera goettingiana* (Liebscher, 1892). Damage to pea crops by *H. goettingiana* was reported a few years later in Great Britain (Theobald, 1912) and in the district of Gironde in France (Capus, 1917; 1918). However, investigations on the biology and pathogenicity of the nematode were not undertaken until much later (Goffart, 1941; Franklin, 1951).

GEOGRAPHIC DISTRIBUTION AND HOST RANGE

The pea cyst nematode occurs world wide. It has been reported from Germany (Goffart, 1941), the Netherlands (Oostenbrink, 1951), Great Britain (Jones, 1965), Belgium (D'Herde, 1966), USSR (Kir'janova and Krall, 1971), Spain (Tobar Jimenez, 1962), Portugal (Macara, 1963), France (Cayrol, 1961), and from other countries of the Mediterranean Basin, such as Israel (Cohn et al., 1975), Algeria and Malta (Lamberti and Dandria, 1979). The nematode also occurs in some limited areas in the USA (Thorne, 1961). In Italy *H. goettingiana* has been reported in many regions, such as Piedmont, Tuscany, Umbria, Marche, Abruzzo, Campania, Apulia, Basilicata, Calabria, Sardinia (Garofalo, 1964a; Di Vito et al., 1973; D'Errico et al., 1975; Ambrogioni and Marinari-Palmisano, 1979), and probably is also present in Sicily (Marinari-Palmisano and Cavalli, 1982).

H. goettingiana can infest some leguminous species, such as garden and field pea, (*Pisum arvense* L.), broad and field bean (*Vicia faba* L.), several species of *Vicia*, soybean (*Glycine max* (L.) Merr), lentil (*Lens culinaris* Medic.), and species of *Lathyrus*, including wild plants (Jones, 1950; Winslow, 1954; Goodey et al., 1965). Beside leguminosae a few females and cysts of the nematode have been observed on the roots of *Asperula arvensis* L. (Di Vito, 1976).

Di Vito et al. (1980) tested the reaction of some leguminous species to six Italian populations of *H. goettingiana*, under greenhouse conditions. They found that chickpea (*Cicer arietinum* L.) sainfoin (*Onobrychis viciaefolia* Scop.), soybean (*Glycine hispida* Moench), sweet vetch (*Hedysarum coronarium* L.), lentil, lupin (*Lupinus albus* L.), alfalfa (*Medicago sativa* L.), bean (*Phaseolus vulgaris* L.), Egyptian clover (*Trifolium alexandrinum* L.), alsike clover (*T. hybridum* L.), crimson clover (*T. incarnatum* L.), red clover (*T. pratense* L.), and white clover (*T. repens* L.) were resistant, while field and garden peas, broad bean, grosspea (*Lathyrus cicera* L.) and vetch (*Vicia sativa* L.) were susceptible to the nematode.

FIELD SYMPTOMS

Like other nematodes, infestations of *H. goettingiana* are first limited to small circular areas and then may extend to the entire field. Heavy infestations can result in total loss of yield. Infested plants are stunted, with pale green leaves at an early stage of growth which later turn yellowish. In infested crops symptoms are very evident and characteristic at the flowering stage. Pods are small, contain few seeds and may not be produced on heavily infested plants. Infested roots are poorly developed and *Rhizobium* nodulation is greatly reduced or completely inhibited (Oostenbrink, 1955; Moriarty, 1962). Further, infested roots are often invaded by fungi such as *Fusarium oxysporum* (Schil.) Synder & Hansen (Oostenbrink, 1955; Garofalo, 1964b), which usually leads to the early death of the plants.

LIFE CYCLE

The life cycle of the pea cyst nematode is similar to that of other species of cyst nematodes. After embryogenesis, the first moult occurs within the eggs and therefore second stage infective juveniles emerge from cysts, penetrate the host roots and after three further moults become adults. Females remain attached to the root from which they partially emerge. When roots of the host plants have hard tissues, as in broad bean, females are detached with difficulty. Under suitable environmental conditions - low temperatures (10-13°C), adequate soil moisture and sufficient food supply - females produce eggsacs in which variable numbers of eggs are laid, occasionally more than 100 per sac (Greco et al., 1986). With inadequate soil moisture and at high soil temperatures, the host is unable to supply the neamtode with sufficient nutrients and then the females soon transform to cysts but no eggsac is formed. Males are very common in this species and thus reproduction is amphimictic.

The sex ratio is affected by environmental conditions. When the nematode invades the roots of chickpea, infested tissues react causing many necrotic areas and only males reach the adult stage (Varo Alcala et al., 1970). The initial population density also affects the sex ratio; Jones et al. (1965) reported that on peas grown in soil infested with four *H. goettingiana* eggs/g soil, 50 per cent of the specimens within the roots are females, while at 359 eggs/g less than 1 per cent females were found. Similar results were observed in Spain (Guevara Benitez et al., 1970).

The number of generations per year varies according to the host species and environmental conditions. Jones (1950) found that in England one or two generations per year are completed on pea and broad bean, respectively. In Southern Italy three generations per year may occur (Di Vito et al., 1974; Greco et al., 1986). These authors studied the effect of sowing time on the life cycle of *H. goettingiana* on pea cv. Progress 9. In crops sown at fortnightly intervals from the autumn through to the spring of the following year, females, eggsacs and cysts were observed at 7, 12 and 15 weeks respectively from root invasion in autumn sown crops; lesser development

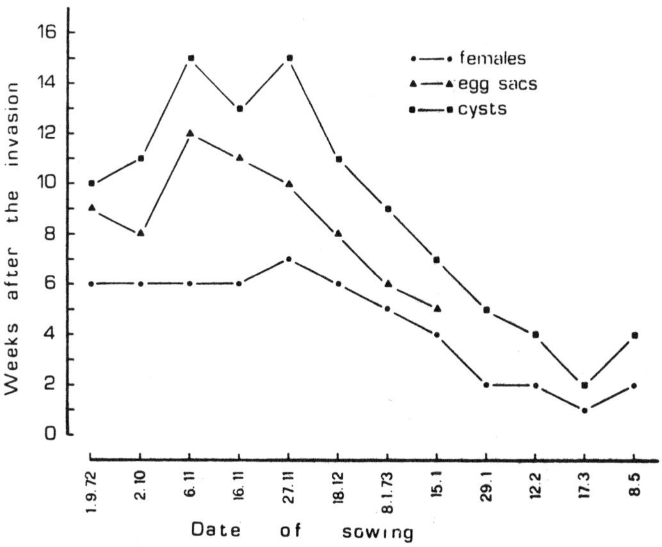

Fig. 1 Time from second stage juveniles invasion required to develop females, females with eggsacs and cysts, on roots of pea cv. Progress 9 at different dates of sowing in soil infested with *H. goettingiana*. (After Greco et al., 1986).

times were recorded for crops sown in January and later (Fig. 1). In Spain females and cysts of *H. goettingiana* occurred on roots of vetch after 58-98 days and 117-129 days from sowing, respectively (Guevara Benitez et al., 1970).

Soil temperature affects the length of the life cycle and egg sac formation. Above 15°C egg sacs are small and eggs are few or absent. From May to August when soil temperatures can reach 25°C juvenile invasion of the roots was suppressed, even in the irrigated crops (Greco et al., 1986). Jones (1975) and Beane and Perry (1984) estimated that the basal temperature for development of the nematode is 4.4°C and that lower temperatures inhibit the fourth moult.

HATCHING

Second stage juveniles emerge from the eggs only in the presence of root exudates from host plants or of some chemicals. Shepherd (1963) tested 179 substances, root leachates from 15 plant species, and some extracts of fungi isolated from the rhizosphere of pea. She obtained an egg hatch of more than 30 per cent in 5 per cent calcium hypochlorite solution, probably because the egg shell was weakened or dissolved. More than 40 per cent hatch was obtained from cysts incubated in root leachate from 4-6 week old plants of pea cv. Kelvedon Wonder. The hatching was lower when root leachates were collected from younger than from older pea plants (Perry et al., 1980). In broad bean root leachate the maximum hatch (87 per cent) was achieved with 16 week old plants but there was 50 per cent hatch in root leachate from 12-28 week old plants (Beane and Perry, 1983). These studies indicate that a high rate of hatching of *H. goettingiana* is usually obtained at 15°C. In southern Italy emergence of juveniles is suppressed in the summer when the temperature exceeds 25°C and resumes when it drops in autumn (N. Greco, unpublished).

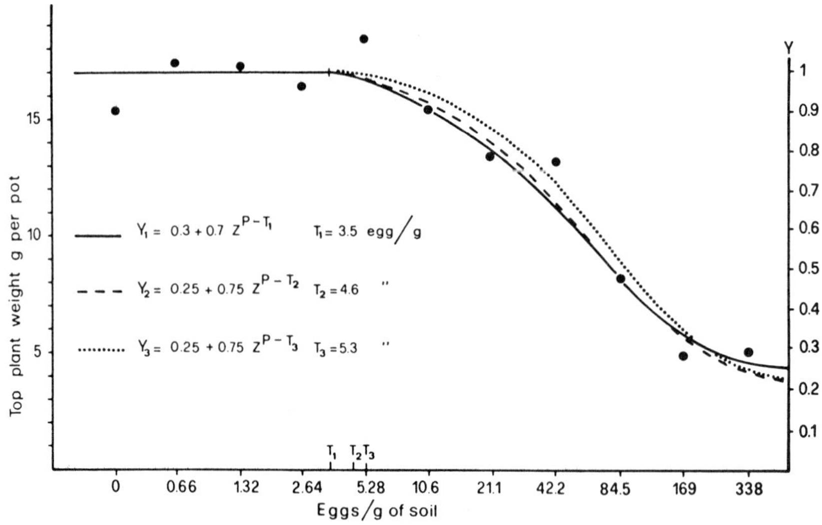

Fig. 2 Relationship between initial population densities of *H. goettingiana* and relative fresh top weight of pea cv. Verdone Fulminante. (After DiVito et al., 1978).

HISTOPATHOLOGY

Once emerged, second stage juveniles penetrate the roots of the host plant. The nematode embeds its cephalic region near the endodermal cells (Varo Alcala et al., 1970) and a syncytium forms (Vovlas and Inserra, 1978). The latter investigators observed plurinucleate and multiple syncytia formed from hypertrophied cells; also plurinucleate syncytia which apparently were independent but had interruptions in their cell walls which allowed continuity of the cytoplasm of contiguous cells. Syncytia have thick cell walls, dense and granular cytoplasm, and hypertrophied nuclei and nucleoli. Usually syncytia are located on the periphery of the central cylinder, and do not affect the structure of the cells of the inner xylem. However, some interruption of vascular elements may occur and thus limit the efficiency of lateral roots because of syncytium formation in the differentiating zone. Cells surrounding syncytia become arranged irregularly; cell walls next to the nematode are necrotic and are more intensively coloured compared to those of healthy cells.

YIELD LOSSES AND POPULATION DYNAMICS

As already mentioned, *H. goettingiana* infestations cause considerable reductions in crop yields. A pot experiment conducted in a screenhouse, using soil infested with *H. goettingiana* at population densities of 0.7-338 eggs/g (Di Vito et al., 1978), showed a tolerance limit of pea to the nematode of 3.5-5.3 eggs/g soil and a relative minimum yield of plant fresh weight above ground of 25 per cent (Fig. 2). Similar results were obtained by Winfield (1965) under field conditions. Broad bean, on the other hand, is more tolerant of nematode attack because reproduction is less than on pea (Winslow, 1955; Moriarty, 1963a), on which the population density (eggs/g soil) of *H. goettingiana* may increase 50 times (Di Vito et al., 1978).

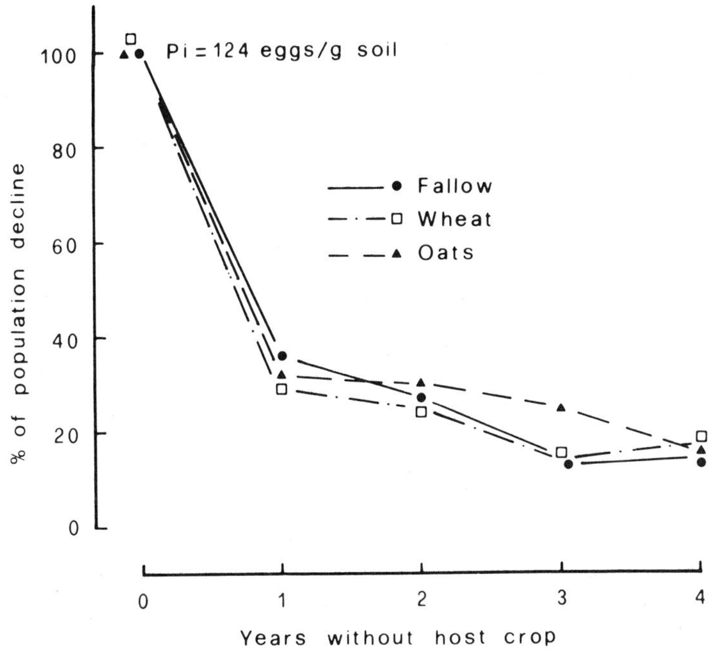

Fig. 3 Effect of fallow, wheat and oats crop on the population decline of *H. goettingiana*.

The pea cyst nematode can persist for several years in the soil in the absence of the host plants and damage to pea crops has occurred after an interval of 10-12 years (Brown, 1958). A population decline of 50 per cent has been observed during the first three years and every year thereafter (Moriarty, 1963b). A similar decline has been reported from the Netherlands (Stemerding, 1960), thus confirming the slow decline of nematode population densities in the absence of a host. A more marked decline of the pea cyst nematode was, however, observed in field experiments conducted in southern Italy where in fallow plots or cultivated with wheat or oats, populations declined by about 68 per cent in the first year and by about 85 per cent after four years (Fig. 3). Different environmental conditions may account for discrepancy of results obtained by independent investigators. In southern Italy soil temperature is rather high and the soil moisture is low from May to October, which can greatly affect the viability of eggs within the cyst.

CONTROL

Depending on climate and on the population density of *H. goettingiana*, a 3-6 year rotation without host plants would reduce nematode levels to non-damaging levels (Fig. 4). Weed control is also suggested in the absence of host crops as several wild leguminous plant species have been reported to be hosts of *H. goettingiana*.

Provided irrigation is possible, cultivating spring pea would allow a good control of the nematode in warm countries, such as southern Italy and other Mediterranean countries, where high soil temperatures, occurring

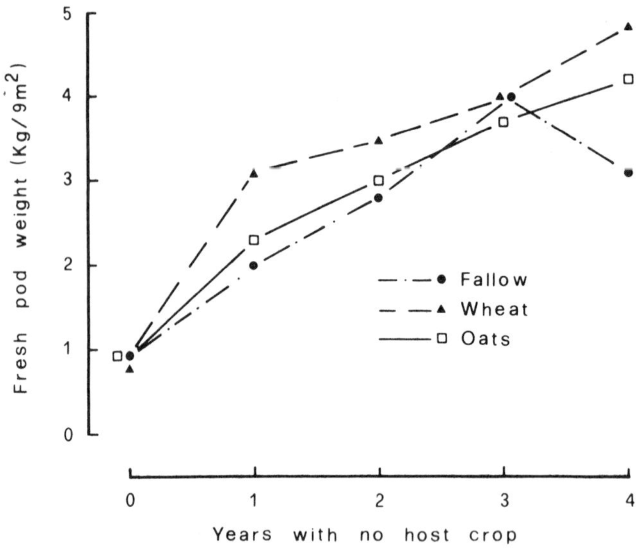

Fig. 4 Effect of fallow and crop rotation with either wheat or oats on yield of fresh pods of pea cv. Verdone Fulminante cultivated in plots infested with *H. goettingiana*.

from late spring onward, suppress egg hatching (Greco et al., 1986) and therefore no substantial damage would be expected.

Chemical control of *H. goettingiana* with consequent crop yield increases has been obtained with both liquid and granular fumigants, such as DD (300 l/ha), Di-Trapex (300 l/ha), Telone (300 l/ha), EDB (150 l/ha) and dazomet (300 kg/ka) (Di Vito et al., 1973; Di Vito and Lamberti, 1976a). To avoid phytotoxicity these chemicals should be applied at least three weeks before sowing, and even longer intervals with dazomet.

Good control of the pea cyst nematode has also been achieved with non-volatile nematicides, such as fenamiphos, aldicarb, carbofuran, thionazin and oxamyl (Di Vito and Lamberti, 1976b); Di Vito and Lamberti, 1977; Whitehead et al., 1979; Green et al., 1981). In alkaline soil oxamyl gave better control when applied as a spray than as a granular formulation which was rapidly deactivated in the soil.

The slow hatching of *H. goettingiana* eggs prevents the satisfactory control of the nematode with very low rates of nematicides, especially fumigants, contrary to what has been shown with *H. avenae* Woll. (Brown et al., 1982).

Growing resistant cultivars is the simplest and probably the cheapest way to prevent nematode damage and to avoid environmental pollution. Unfortunately no cultivar resistant to *H. goettingiana* is available so far. However, investigations by Di Vito and Perrino (1978) have shown that some accessions of *P. abyssinicum* Brown, *P. arvense* L. and *P. elatius* Ster, were moderately resistant to the pea cyst nematode (Table 1). Hybrids and lines obtained by crossing the susceptible *P. sativum* cv. Progress 9 and the resistant line MG101791 of *P. abyssinicum* were moderately resistant. to *H. goettingiana* in the F_2 generation (22 per cent of plants resistant

Table 1 Reaction of *Pisum* spp. lines to infestations of *H. goettingiana* (from Di Vito and Perrino, 1978).

Host species	Number of lines and variety tested	Range of infestation (% respect to control)	Resistant	Susceptible	% of resistant lines
Pisum abyssinicum	14*	6 - 66	5	9	35.7
P. arvense	91	15 - 194	0	91	0
P. elatius	2	12 - 46	1	1	50
P. sativum	5	28 - 91	0	5	0
P. sativum (Control)	1	100**	0	1	0

* Inoculum level 6,000 egg and juveniles/plant

** 100 = 810 females and cysts/plant

compared with 90 per cent for *P. abyssinicum*), indicating that the resistance in *P. abyssinicum* is recessive. An increase in ascorbic free radical (AFR) reductase and ascorbic acid (AA), has been observed in some of the *P. abyssinicum* accessions compared with the susceptible pea (Zacheo et al., 1981). Changes of peroxidase and superoxide dismutase (SOD) were also reported in these lines (Arrigoni et al., 1981).

The wide distribution of *H. goettingiana* and the damage it causes to some host plants, such as pea and broad bean, necessitate the use of control measure to limit yield losses. Crop rotations and chemical nematicides can reduce the nematode population to the level of the tolerance limit of the host crops, and should be used until resistant cultivars become available.

ACKNOWLEDGEMENTS

The authors thank Mr V. Radicci for assistance with the figures.

REFERENCES

Ambrogioni, L. and Marinari-Palmisano, A., 1979, Nematodi del genere *Heterodera* su colture ortensi ed ornamentali in Italia, Proc. Giornate Nematologiche S.I.N., Florence 28-29 Nov. 1979, 89.

Arrigoni, O., Zacheo, G., Bleve-Zacheo, T., Arrigoni-Liso, R. and Lamberti, F., 1981, Changes of superoxidedismutase and peroxidase activities in pea roots infested by *Heterodera goettingiana*, Nematol. medit., 9:189.

Beane, J. and Perry, R.N., 1983, Hatching of the cyst nematode *Heterodera goettingiana* in response to root diffusate from bean (*Vicia faba*), Nematologica, 29:360.

Beane, J. and Perry, R.N., 1984, Observations on the invasion and development of the pea cyst-nematode *Heterodera goettingiana* in winter field beans (*Vicia faba*), Ann. appl. Biol., 105:503.

Brown, E.B., 1958, Pea root eelworm in the eastern countries of England, Nematologica, 3:257.

Brown, R.H., Pye, D.L. and Stratford, G.T., 1982, A comparison of low volume, in row applications of nematicides at seeding, for control of the cereal cyst nematode (*Heterodera avenae*) in wheat, Nematol. medit., 10:9.

Cayrol, J.C., 1961, Les nématodes des coltures maraichéres, "Les Nématodes", Journées d'etude et information, A.C.T.A., FNGPC, Paris.

Capus, J., 1917, Sur un deperissement des cultures du pois en Gironde du à la double action de l'*Heterodera schachtii* et du *Fusarium vasinfectum*, Bull. Soc. Zool. Agric., Bordeaux, 16(7):70.

Capus, J., 1918, *Fusarium* des cultures des pois en Gironde par l'*Heterodera schachtii* Schmidt, Ann. Epiphyt, 5:239.

Cohn, E., Sher, S.A., Bell, A.A. and Minz, G., 1973, Soil nematodes occurring in Israel, Spec. pubbl. N. 22, Divn. Scient. Publ., Bet Dagan, Israel.

D'Errico, F.P., Furia, A. and Mancini, G., 1975, L'*Heterodera goettingiana* Liebscher in Campania e sua patogenicità, Boll. Lab. Entomologia Agraria "Filippo Silvestri", Portici, 32:1.

D'Herde, J., 1966, Het erwtencysterraaltje *Heterodera goettingiana* een rienw parasiet van de erwtenteelt voor België, Landen Tuinb Jaarb, Year 1965-1966, 427.

Di Vito, M., 1976, *Asperula arvensis* L., nuovo ospite di *H. goettingiana* Liebscher, Nematol. medit., 4:93.

Di Vito, M. and Lamberti, F., 1976a, Prove di lotta chimica contro *Heterodera goettingiana* su pisello in Puglia, Nematol. medit., 4:121.

Di Vito, M. and Lamberti, F., 1976b, Efficacia di nematocidi sistemici nella lotta contro *Heterodera goettingiana* su pisello, Nematol. medit., 4:63.

Di Vito, M. and Lamberti, F., 1977, Ulteriori saggi di lotta chimica contro *Heterodera goettingiana* su Pisello, Inf. Fitopatol., 27(10):17.

Di Vito, M. and Perrino, P., 1978, Reazione di *Pisum* spp. agli attacchi di *Heterodera goettingiana*, Nematol. medit., 6:113.

Di Vito, M., Lamberti, F. and Inserra, R.N., 1973, Prove preliminari di lotta chimica contro *Heterodera goettingiana* Liebscher, grave parassita del pisello nell'Italia Meridionale, Nematol. medit., 1:125.

Di Vito, M., Lamberti, F. and Greco, N., 1974, The life cycle of *Heterodera goettingiana* Leibscher under field conditions in Southern Italy (Abstr.), Proc. XII Int. Symp. Nematol. Eur. Soc., Granada (Spain).

Di Vito, M., Lamberti, F. and Inserra, R.N., 1978, Effects of *Heterodera goettingiana* on weight of foliage of pea plants in pots, Nematologica, 24:209.

Di Vito, M., Greco, N. and Lamberti, F., 1980, Comportamento di popolazioni di *Heterodera goettingiana* su specie diverse di leguminose, Inf. Fitopatol., 30:7.

Franklin, M.T., 1951, The cyst-forming species of *Heterodera*, Commonw. Agric. Bur. Farnham Royal, 147 pp.

Garofalo, F., 1964a, *Heterodera goettingiana* Liebscher e *Fusarium oxysporum* (Schl.) Sny. et Hans. nell'avvizzimento di piante di pisello e di lupino, Boll. Lab. Sper. Oss. Fitopatologico, Torino, 27:34.

Garofalo, F., 1964b, Fenomeni di correlazione tra *Heterodera goettingiana* e *Fusarium oxysporum* (Schl.) Sny. et Hans. nell'avvizzimento delle piante di pisello e di lupino, Boll. Lab. Sper. Oss. Fitopatologico, Torino, 27:33.

Goffart, H., 1941, Untersuchungen am Hafernematoden (*Heterodera schachtii*, Schmidt) under besonderer Berucksichtigung der Schleswigholtteinischen Verhaltuisse II, Arb. biol. Reichsanst. Berlin, 23:141.

Goodey, J.B., Franklin, M.T. and Hooper, D.J., 1965, T. Goodey's "The nematode parasites of plants catalogued under their hosts", (3rd edit.), Commonw. Agric. Bur., Farnham Royal, 214 pp.

Greco, N., Di Vito, M. and Lamberti, F., 1986, Studies on the biology of *Heterodera goettingiana* in Southern Italy, Nematol. medit., 14:23.

Green, C.D., Williamson, K., Dennis, E..B. and McBurney, T., 1981, The effect of oxamyl on the growth of peas attacked by pea cyst nematode, Ann. appl. Biol., 97:303.

Guevera-Benitez, D.C., Tobar-Jimenez, A. and Guevara Pozo, D., 1970, Estudio cuantitativo del ciclo vital de *Heterodera goettingiana* Liebscher, 1892 y de la possibilitad de su control mediante cultivos cepo, Revta. Iber. Parasit., 30:229.

Hashim, Z., 1979, A preliminary report of plant-parasitic nematodes in Jordan, Nematol. medit., 7:177.

Jones, F.G.W., 1950, Observations on the beet eelworm and other cyst forming species of *Heterodera*, Ann. appl. Biol., 37:407.

Jones, F.G.W., 1965, Pea, cabbage and carrot root eelworms, in "Plant Nematology", J. F. Southey, ed, Tech. Bull. No. 7, Min. Agric. Fish. Fd., London.

Jones, F.G.W., 1975, Accumulated temperature and rainfall as measures of nematode development and activities, Nematologica, 21:62.

Jones, F.G.W., Meaton, V.H., Parrott, D.M., Shepherd, A.M. and King, J.M., 1965, Population studies on pea cyst-nematode *Heterodera goettingiana* Liebs., Ann. appl. Biol., 55:13.

Kir'yanova, E.S. and Krall', E.L., 1971, "Parasitic nematodes of plants and their control", Vol. II, Leningrad, Izdatel'stvo "Nauka".

Lamberti, F. and Dandria, D., 1979, Plant-parasitic nematodes in the Maltese islands and the problems they cause, Phytopathol. medit., 18:71.

Lamberti, F., Greco, N. and Zaouchi, H., 1975, A nematological survey of date palms and other major crops in Algeria, F.A.O. Pl. Prot. Bull., 23:156.

Liebscher, G., 1890, Eine Nematoden als Ursache der Erbsennündigklit des Bodens, Dtsch. Landw. Pr., 17:477.

Liebscher, G., 1892, Beobachtungen über das eines Nematoden au Erbesen, Jour. Landw., 40:357.

Macara, A.M., 1963, Contribuiçao para o estudo morfologico e biologico dos nematodos *Heterodera göttingiana* Liebscher, 1892 e *H. rostochiensis* Wollenweber, 1923 encontrados en Portugal, Brotéria, 32:25.

Marinari-Palmisano, A. and Cavalli, M., 1982, Contributo alla conoscenza dei generi *Heterodera* e *Globodera* (Nematoda: Heteroderidae) in Italia, Redia, 65:21.

Moriarty, F., 1962, The effect of sowing time and nitrogen on peas, *Pisum sativum* and on pea root eelworm, *Heterodera goettingiana* Liebscher, Nematologica, 8:169.

Moriarty, F., 1963a, A population experiment with *Heterodera goettingiana* Lieb. and a relatively inefficient host, broad bean, Nematologica, 9:152.

Moriarty, F., 1963b, The decline of a pea root eelworm (*Heterodera goettingiana* Liebscher) population in the absence of host plant, J. Agric. Sci., 61:135.

Oostenbrink, M., 1951, Het erwtencystenaaltje, *Heterodera göttingiana* Liebscher, in Nederland, Tijds. Pl.ziekten, 57:52.

Oostenbrink, M., 1955, Ein inoculatienproef met het erwtencystenaaltje *Heterodera goettingiana* Liebscher, Tijds. Pl.ziekten, 61:65.

Perry, R.N., Clarke, A.J. and Beane, J., 1980, Hatching of *Heterodera goettingiana* in vitro, Nematologica, 26:493.

Shepherd, A.D., 1963, The emergence of larvae of *Heterodera goettingiana* Liebs. in vitro and a comparison between field populations of *H. goettingiana* and *H. rostochiensis* Woll., Nematologica, 9:143.

Stemerding, S., 1960, The influence of different rotations on a population of pea cyst eelworm, *Heterodera göttingiana* Liebscher, Nematologica (suppl. II), 97.

Theobald, F.V., 1912, Animals injurious to pulse-eelworm (*Heterodera schachtii*) on peas, J.S.E. Agric. Coll. Wgl., 21:173.

Thorne, G., 1961, Principles of Nematology, New York, McGraw-Hill.

Tobar-Jimenez, A., 1962, La *Heterodera goettingiana* Liebscher, 1892, parasito de las habas (*Vicia faba*L.) gramadimas, Rev. Iber. Parisit, 22:323.

Varo Alcala, J., Tobar-Jimenez, A. and Muñoz Medina, J.M., 1970, Lesiones causadas y reacciones provocadas par algunos nematodes en las raices de ciertas planta, Revta. Iber. Parasit, 30:547.

Vovlas, N. and Inserra, R.N., 1978, Osservazioni comparative delle alterazioni istologiche indotte da *Heterodera goettingiana* e *Meloidogyne javanica* in radici di Pisello, Nematol. medit, 6:187.

Whitehead, A.G., Bromilow, R.H., Tite, D.J., Finch, P.H., Fraser, J.E. and French, E.M., 1979, Incorporation of granular nematicides in soil to control pea cyst-nematode, *Heterodera goettingiana*, Ann. app.. Biol., 92:81.

Winfield, A.L., 1965, Microplot experiments with pea cyst nematode, Pl. Path., 14:158.

Winslow, R.D., 1954, Provisional lists of host plants of some root eelworm (*Heterodera* spp.), Ann. appl. Biol., 41:591.

Winslow, R.D., 1955, The effects of some leguminous crops on the soil population level of pea eelworm, Pl. Path., 4:86.

Zacheo, G., Arrigoni-Liso, R., Bleve-Zacheo, T., Lamberti, F., Perrino, P. and Arrigoni, O., 1981, Changes of ascorbate free radical reductase in pea roots infested by *Heterodera goettingiana*, Nematol. medit., 9:181.

DISCUSSION

Whitehead asked if it was necessary to reinoculate dazomet treated plots with rhizobium. Di Vito replied that it was not necessary as enough survived the treatment. Di Vito explained that only some of the chemicals used in his trials were officially approved for use in Italy, e.g. aldicarb can be used only on sugar beet. Replying to Valloton, he said against the pea cyst nematode only ethylene dibromide (EDB) at 80 litres a.i. per hectare was used, but rarely. Roberts commented that EDB has been universally withdrawn from use on crops in the USA.
Di Vito said it was still available in Italy and was also used on wheat in Australia.

THE CARROT CYST NEMATODE

Nicola Greco

Istituto di Nematologia Agraria del C.N.R.
70126 Bari
Italy

INTRODUCTION

In 1931 Triffit referring to females of a cyst nematode found on roots of carrots (*Daucus carota* L. ssp. *sativus*), stated: "These white stage females were rather small and of the lemon-shaped type, and were furnished with a rather larger egg sac than is usual for *H. schachtii*. This infection is of particular interest in that in appears to be the first known case of *H. schachtii* attack carrots". At that time most of the species of lemon shaped cyst forming nematodes were assumed to be strains of the well known species *Heterodera schachtii* Schmidt, and so was the cyst nematode infesting carrots.

In 1944 Jones (1950a) again found roots of carrots infested with a cyst nematode in several fields at Chatteris, Isle of Ely, England. Further studies convinced Jones that he was dealing with a new species, which he named *H. carotae* and described its morphology and biology (Jones, 1950a, b).

GEOGRAPHIC DISTRIBUTION AND ECONOMIC IMPORTANCE

H. carotae is generally distributed in Europe where carrots are grown (Fig. 1). In the United Kingdom, in addition to its occurrence in England (Jones, 1950 a, b) it has been recorded from Northern Ireland (Anon, 1971) and from Scotland (Osborne, 1971). In the Netherlands, *H. carotae* was associated with carrot sickness but not considered to be a major nematode problem (Oostenbrink, 1955). Severe damage to carrot crops caused by *H. carotae* was reported in Brittany and Normandy, France (Oudinet et al., 1962), where 60 per cent of the fields of carrots were found to be infested (Oudinet, 1968). In Italy, three of the four major carrot areas are heavily infested with *H. carotae*. So far the nematode occurs in the regions of Lazio (Ambrogioni, 1969), Puglia (Lamberti, 1971), Veneto (Tacconi, 1976), and recently has been found in Trentino Alta Adige (Greco, unpublished information). In Switzerland heavy infestations of *H. carotae* were recorded in 1973 (Vallotton, 1980) and since then the nematode has been associated with stunted carrot crops in the area between Martigny and Saillon. Severe infestations of the nematode were also reported from the Federal Republic of Germany (Sturhan, 1960) and the Democratic Republic of Germany (Decker, 1968, cited in Stelter, 1973).

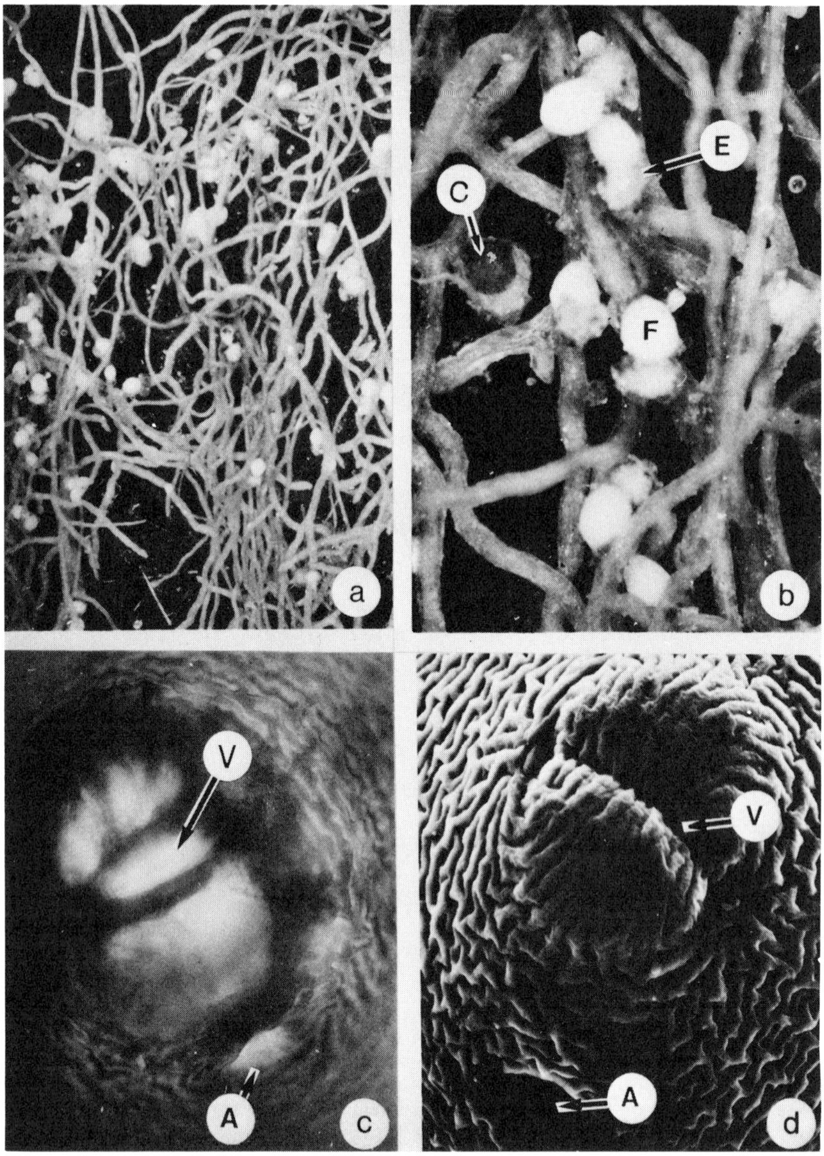

Fig. 1 (a) Distribution of *H. carotae.* (b) White females of *H. carotae*, on feeder roots of carrot. (c) White females (F), brown cyst (C), and eggsacs (E) full of eggs of *H. carotae.* (d) Light microscopy (A) and scanning electron (B) microphotography of a cyst of *H. carotae.*

V = vulval slit; A = anus. Courtesy of Dr N. Vovlas.

H. carotae has been reported from several other countries, but its effect on carrot yield was generally negligible. The nematode was found in southern Sweden in 1964 (Andersson and Nyberg, 1964) in a heavily infested carrot crop, but it was not detected during a later survey. Only two soil samples, out of 600 checked for cyst nematodes in Poland (Brzeski, 1970), contained a few cysts of *H. carotae*, and so far the nematode has not assumed economic importance there. *H. carotae* has also been reported from East Czechoslovakia (Sabova' and Valotska, 1976). Hungary (Javor, 1968, cited in Stelter, 1973) and Russia (Kir'yanova and Krall, 1980), but there is no information about its importance in these countries. Outside Europe *H. carotae* has been reported only from Cyprus (Philis, 1976) and India (Swarup et al., 1964). The Indian population was obtained from a potato field but no morphometric information was published with the record of occurrence.

Movement of contaminated farm machinery and carrots may account for the spread of *H. carotae* to neighbouring areas, but spread of the nematode to different countries is less obvious. However, although stocks of exported carrots are considered to be free of adhering soil, foci of infestation may be established in new areas or countries by cysts adhering to the tap roots or to seed.

HOST RANGE

The host range of *H. carotae* is rather narrow compared with most other cyst forming nematodes. Jones (1950) observed that only cultivated and wild carrot (*D. carota*), out of 106 plants tested, including 20 species of the *Umbelliferae*, were hosts of *H. carotae*. Winslow (1954) also reported *D. pulcherrimus* as a host of the nematode. Vallotton (1980) found that the nematode was able to complete its development on *Torilis arvensis*, a wild *Umbelliferae* occurring in the Valais region of Switzerland, and stated that Mugniery also observed reproduction on *T. leptophylla*. These wild species are common in several European countries and may have played an important role in maintaining populations of the nematode.

BIOLOGY

The biology of *H. carotae* is essentially similar to that of other cyst forming nematodes, but differs in some particulars.

Hatching

Eggs are contained within cysts and eggsacs but whereas the hatching of eggs in cysts is usually delayed, those contained in the eggsacs hatch as soon as soil temperature and soil moisture are suitable. Winslow (1955) and Greco (1981) showed that root exudates of carrots are a prerequisite for a substantial emergence of juveniles from cysts of *H. carotae* and that root leachates from other plant species, including members of *Umbelliferae*, did not significantly stimulate the hatching (Winslow, 1955). Although juveniles may emerge from eggs even at 5°C, the highest rate of hatching is in the range 15-20°C, and negligible at 25°C (Greco et al., 1982). The hatching activity of carrot root leachates is affected by the age of the plants and most juveniles emerge in leachates from 5-7 week old carrots (Winslow, 1955; Greco et al., 1982).

Practically no hatching occurred from newly formed cysts, i.e. less than two months old, but thereafter there was a regular emergence of juveniles provided environmental conditions were suitable (Greco, 1981).

Juveniles emerged from cysts collected at different times of the year from infested fields of southern Italy, but fewer juveniles emerged from cysts collected in the late autumn or in winter (N. Greco, unpublished information). Hatching of embryonated eggs contained in eggsacs in prompt, but it can be suppressed by high summer temperatures and resumed when the temperature drops (Greco, 1981). Carrot root leachate, however, is not necessary to stimulate the emergence of juveniles from eggsacs, but emergence is affected by soil moisture and temperature.

Root invasion and life cycle

Second stage juveniles of *H. carotae* invade carrot roots when the temperature is in the range 5-30°C, but no development occurs below 10°C (N. Greco, unpublished information), which can be considered as basal temperature for the development of the nematode. At 20°C a large number of juveniles had invaded the roots of carrot two days after inoculation, and that third stage juveniles, fourth stage females, adult females and cysts, had developed after 12, 21, 26 and 36 days from the incolation, respectively, when 120, 210, 260 and 360 days degrees had accumulated above the development basal temperature (N. Greco, unpublished information). Fourth stage males and adult males were observed a few days earlier than female stages, i.e. 16 and 21 days from the inoculation, respectively, equivalent to 160 and 210 day degrees. Soon after reaching the adult stage, females extruded eggsacs with eggs, 31 days after inoculation. The eggsacs continued to enlarge and were full of eggs (average content 160 eggs) some two weeks later (460 days degrees). At 20°C, cysts developed in 41 days (410 days degrees). Newly produced eggs soon undergo embryogenic development and second stage juveniles were observed at 51 in eggsacs and 57 days in cysts. These are similar to results obtained at 18-20°C by Ambrogioni (1971).

Females and cysts of *H. carotae* are rather small compared to most other cyst nematodes, but eggsacs are relatively large and may contain as many eggs as the cysts (Fig. 1A, B). The cysts have characteristic perineal patterns (Fig. 1).

Several authors (Jones, 1950a, b); Oostenbrink, 1955; Sturhan, 1960; Ambrogioni, 1969, 1971; Matthews, 1975) have provided measurements of the developmental stages of the nematode. Oostenbrink (1955) has also designated types.

Second stage juveniles soon emerge from the eggsacs and initiate a new generation but the pattern of the life cycle then depends on the availability of the carrot host crop and the soil environment. In most of the European countries, the carrot crop is usually sown in late winter or early spring but along the Mediterranean coast it may also be sown from the end of August to early spring. Jones (1950b) found that in England the development of the nematode was slow in early spring, but increased with rising temperature and reached its maximum in July, and again was slow from November to March. In the field two generations of the nematode were completed during a carrot growing season. In Switzerland 1-4 generations per year can be developed if carrots are sown twice a year on the same land (Vallotton, 1980, 1983). More than one generation was also assumed to be possible by Stelter (1969) under controlled conditions.

In Italy when carrots were sown at the end of August or early September, juveniles invaded the carrot rootlets in September and the first adults were observed towards the end of October (Ambrogioni, 1971; N. Greco, unpublished information). Females soon extruded eggsacs which

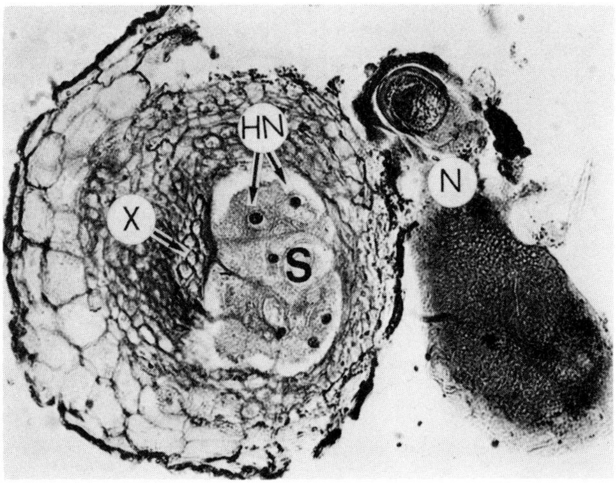

Fig. 2 Syncytium (2) caused by penetration of *H. carotae* in feeder roots of carrot. N = nematode; HN = hypertrophied nuclei; X = compressed vascular elements. (From N. Vovlas, 1978).

became full of eggs from the end of November through to January. Cysts were seen at the end of November. When temperatures are high in late summer and early autumn there is usually a delay in the juvenile invasion of the carrot roots and then mature females with large eggsacs and cysts are not present before January. Heavily infested carrots usually die by early December and most of the nematodes in the roots have by then turned to cysts, but with light infestations the nematode population consists mainly of white females in the feeder roots of the carrots at harvest (January - March); this is also because of the occurrence of a second generation which starts from juveniles emerged from newly formed eggsacs.

When carrots were sown in January - February roots were invaded by juveniles after plant emergence and adults had developed by the first week of April with cysts appearing 10 days later (N. Greco, unpublished information). If the spring is warm and dry, no second stage juveniles invade the roots in May and most of the specimens found on the roots at the end of May are brown cysts. Nematodes fail to invade the roots when carrots are sown during warm summer months, even if the crop is irrigated. Therefore two generations of *H. carotae* may develop when carrots are sown in late summer but only one generation if the crop is sown early in the year. However, in central and northern Italy the nematode populations seem to behave like those in central and northern European countries.

H. carotae juveniles invade only carrot feeder roots and mainly at the tips. Although the nematode can develop in the early growth stage of the main root, no invasion nor development occurs in tap roots.

HOST REACTION AND CROP LOSSES

After the second stage juveniles penetrate the roots, they move parallel to the central cylinder for a few millimetres, and then become sedentary with their cephalic region embedded in the pericycle cells and

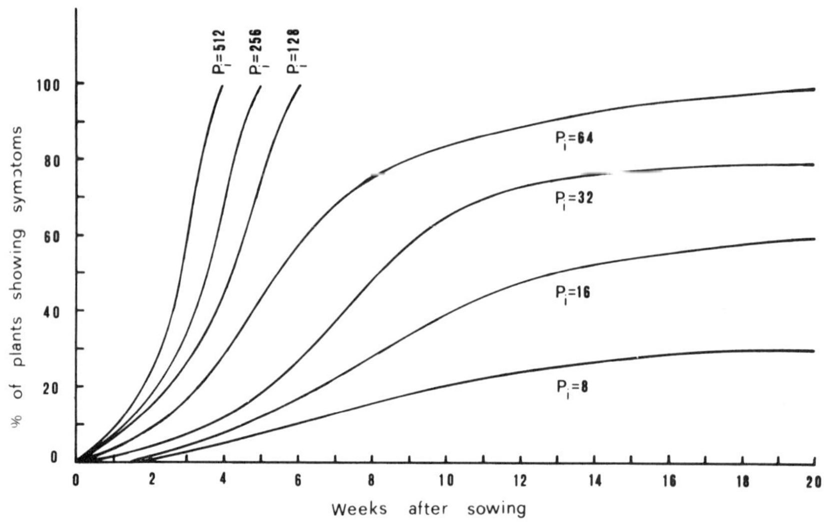

Fig. 3 Relationship between initial densities of *H. carotae* and appearance of symptoms of the nematode attack. Numbers next to the curves indicate initial population densities in egg/ml of soil (From: N. Greco and A. Brandonisio, 1980).

the rest of the body parallel to the central cylinder. On their pathway through the roots the juveniles pierce cells, which become necrotic. Once the nematodes become sedentary, the reaction of the root tissues is similar to that caused by other cyst forming nematodes and a syncytium forms near the cephalic region (Fig. 2), on which the nematode feeds (Ambrogioni, 1972; Vovlas, 1978). At first only a few cells are involved in the syncytium, but after 21 days it consists of about ten cells of the cortical parenchyma and in size is about one fourth of the root diameter and exerts pressure on the vascular elements. When exhausted the syncytium becomes necrotic and with heavy infestations several necrotic areas join and may kill the root. Because carrot rootlets are thin, death may also occur when they are massively invaded by juveniles. The plant reacts by producing new rootlets which in turn may also be invaded by juveniles. Thus the tap roots of infested plants have a heavily bearded appearance.

Field infestations are usually patchy, with stunted plants on which the foliage becomes reddish before they die. Tap roots are greatly reduced in size and unmarketable. Complete failure of the crop may occur in heavily infested soil.

The intensity and timing of the development of the infestation with the appearance of symptoms depends on the nematode population density at sowing. Greco and Brandonisio (1980) found that at 128 eggs/cm^3 of soil all plants showed clear symptoms of nematode attack six weeks from sowing (Fig. 3), while at 8 eggs/ml of soil only 30 per cent of the plants appeared stunted and reddish at harvest. No symptoms were observed at lower population densities.

There is relatively little information on the affect of increasing nematode population densities on crop yield. In Switzerland, Vallotton

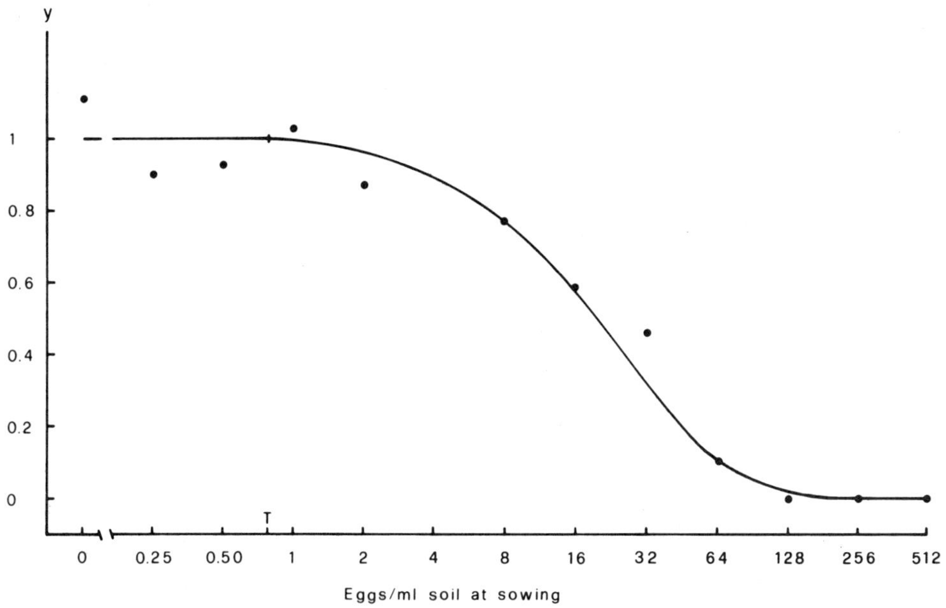

Fig. 4. Initial densities of *H. carotae* on yield of carrots. $y = z^{P-T}$; tolerance limit (T) = 0,8 eggs/ml of soil (From: N. Greco and A. Bremdomisio, 1980).

(1980) estimated a tolerance limit of carrot to *H. carotae* of 40 cysts/250 ml of soil, which he found to be exceeded in about 30 per cent of the sampled fields. In Italy, tolerance limits of 0.19 and 0.8 eggs/ml of soil were estimated from a field trial (Ambrogioni and Marinari-Palmisano, 1976) and a microplot experiment (Greco and Brandonisio, 1980). Yield losses of 40 and 70 per cent occurred at 16 and 32 eggs/cm^3 of soil, respectively (Fig. 4). At the latter nematode density most of the carrots were unmarketable and at higher densities complete failure of the crop can be expected (Greco and Brandonisio, 1980).

POPULATION DYNAMICS

Population dynamics of the nematode are affected by cropping methods, environmental conditions and length of the growing season. In Italy and Switzerland carrot is sometimes cropped twice a year on the same land. Therefore several generations (2-4) of the nematode can be completed leading to high densities. When carrots are sown early in August and harvested at the end of November the population density of the nematode may be drastically reduced because the nematode may not have been able to complete its life cycle. However, a reproduction rate of X30 has been obtained in microplots (Greco and Brandonisio, 1980) (Fig. 5) and X72 in pots (Stelter, 1969). In field crops in Switzerland, Vallotton (1980) calculated a population increase of X10 time. Greco and Brandonisio (1980) estimated an equilibrium density of about 50 eggs/ml of soil.

In a late harvested carrot crop a population was found to consist of 63 per cent cyst eggs and 37 per cent of eggsac eggs (N. Greco, unpublished information). The decline of the population was faster for the eggsac eggs than for the cysts, and 22 and 60 per cent of them, respectively, were still present after eight months.

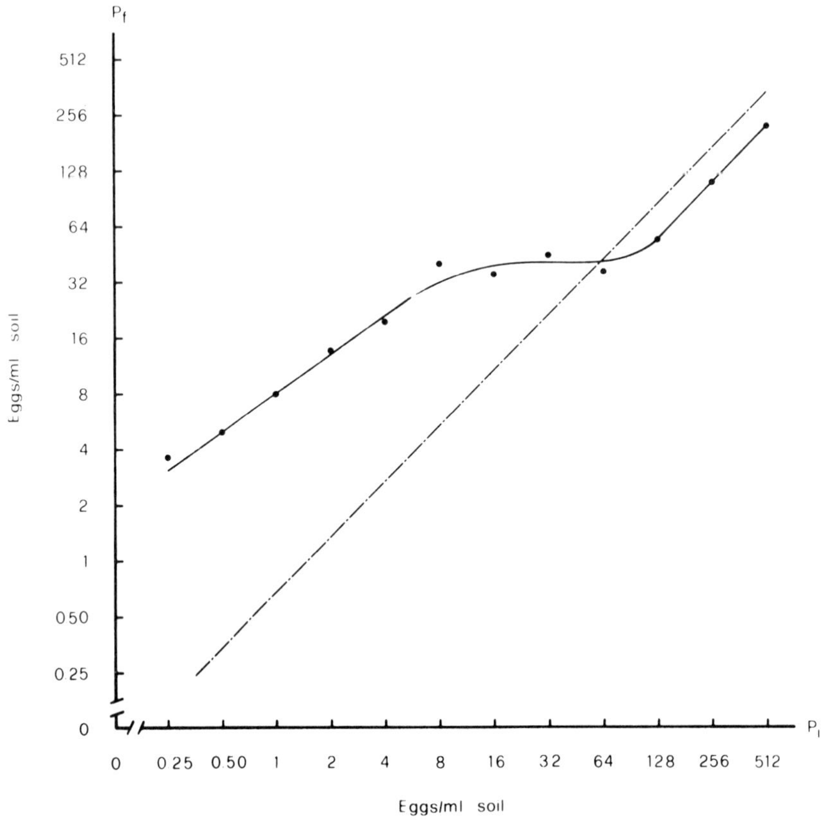

Fig. 5 Relationship between initial (Pi) and final (Pf) populations of *H. carotae*. -.-.-. proportion of unhatched eggs in the absence of host (0.66Pi) (From N. Greco and A. Brandonisio, 1980).

CONTROL

Chemical

Chemical control can be economic when early crops are cultivated. Nematicides used against other cyst nematodes are also effective against the carrot cyst nematode. In France (Oudinet et al., 1962; Oudinet, 1968; Ferron and Mimaud, 1971) soil applications of 200-600 l/ha of DD, 200-240 l/ha of 1,3D or 700 kg/ha of dazomet, gave satisfactory control of the nematode and doubled carrot yields. Good control of the nematode was also achieved with dazomet in Switzerland (Vallotton, 1983b).

Several nematicides have been tested against *H. carotae* in Italy. Good nematode control and increased carrot yields were obtained by soil fumigation with DD, dazomet, and a mixture of DD (80 per cent) and methylisothiocyanate (MIT) (20 per cent) (Greco and Lamberti, 1977a; Greco et al., 1974). Increased rates of DD (22-500 l/ha) or of a mixture of DD + MIT did not improve carrot yields.

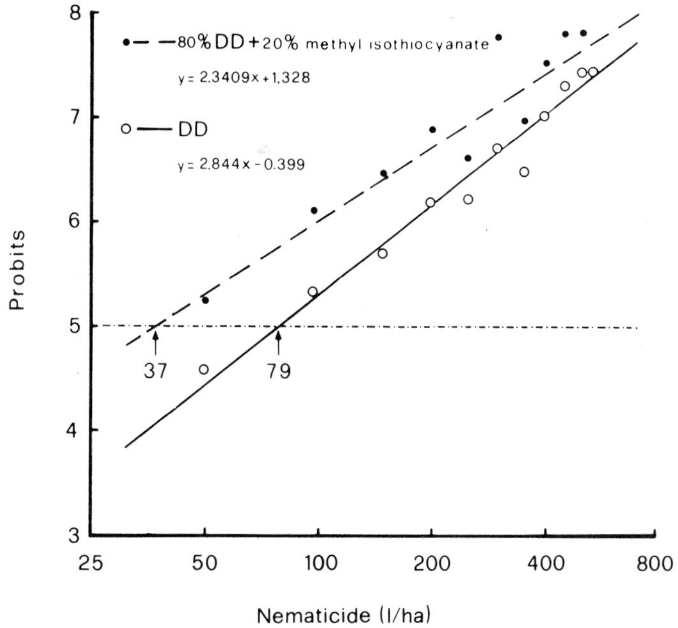

Fig. 6 Relationship between the log dosage of DD + methyl-isothiocyanate and probit mortality of eggs within cysts of *H. carotae* (From: N. Greco et al., 1984).

The rate of a nematicide required to provide nematode control and to obtain satisfactory crop yields depends on the density of nematodes in the soil at sowing and the effectiveness of the chemical under the given conditions. Greco et al. (1984) found that 79 or 37 l/ha of DD or DD + MIT, respectively, were required to kill 50 per cent of the eggs within cysts in the top 30 cm of a sandy soil (Fig. 6). On the basis of these findings Greco and Brandonisio (1980) calculated the rates of the nematicides required to reduce the populations of the nematode to the tolerance limit level (Fig. 7). They also found that rates greater than 250 l/ha DD + MIT or 370 l/ha DD did not give an increased kill of nematodes and hence it was not economic to exceed these rates..

Non-volatile nematicides, such as aldicarb, fenamiphos and oxamyl, have also been tested against *H. carotae* (Greco and Lamberti, 1977a, c), but generally they are less effective than fumigants. Usually the best control with non-volatile nematicides is achieved when they are applied at sowing and again 20-30 days after plant emergence. However, application to established crops would lead to higher residues in the harvested crop.

<u>Cultural methods</u>

In Italy, carrot crops grown every three years on the same land resulted in a doubling of the nematode population (Ambrogioni and Marinari-Palmisano, 1976). Vallotton (1983) reckoned that a two to three year rotation is insufficient to limit yield losses caused by *H. carotae* to an acceptable level. Most probably carrot rotations of four to five years,

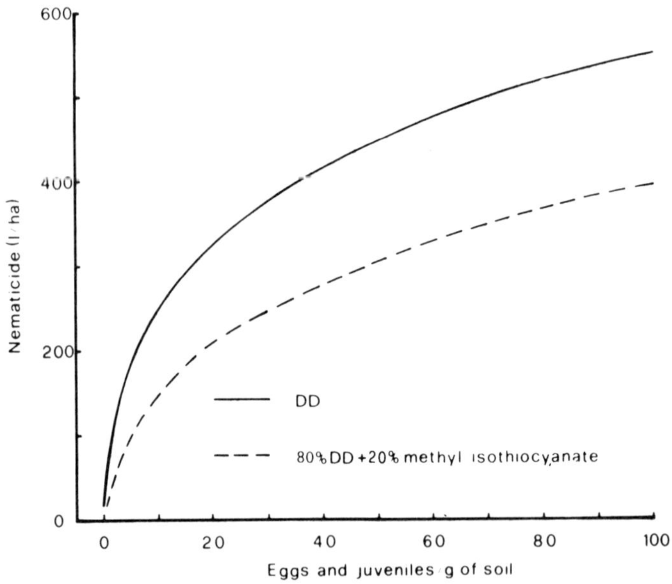

Fig. 7 Relation between numbers of eggs of *H. carotae* and amounts of DD or DD + methylisothiocyanate required to reduce the nematode density at the carrot tolerance limit level (From N. Greco et al., 1984).

as suggested for other cyst forming nematodes, could provide satisfactory control of the carrot cyst nematode.

H. carotae is inactive at high soil temperatures and therefore sowing carrots early in August and harvesting in November or early in December could reduce yield losses in warm countries; at high temperatures juveniles do not invade the roots and thus the nematode may not complete its life cycle by harvest time, thus leading to lower soil populations (N. Greco, unpublished information). In central and northern Europe, when early carrots are cultivated the nematode would complete only one generation per season, thus limiting the increase of soil populations.

Unfortunately no carrot cultivar is resistant or tolerant to the nematode (Greco and Lamberti, 1977b; Greco and Brandonisio, 1982). However, early cultivars appeared to be less susceptible than late ones under field conditions.

Other methods of control

Biological control has shown promise against nematodes (Cayrol and Frankowski, 1979; Stirling, 1984), but only preliminary investigations have been made with regard to *H. carotae* (Vallotton, 1983a). Soil solarisation gives satisfactory control of some fungi and nematodes in the top layers of the soil (Katan, 1981; Porter and Merriman, 1983); in Italy 35-76 per cent of *H. carotae* eggs were killed by mulching the soil with polyethylene sheets for 1-8 weeks from June to August (Greco et al., 1985). However, increase of soil temperatures at greater than 25 cm depth is negligible in solarised soil and it is therefore suggested that a mulching period of 4-8 weeks during the summer could provide satisfactory control of *H. carotae* when the nematode soil population is below 20 eggs/g of soil.

ACKNOWLEDGEMENT

The technical assistance of Mr V. Radicci for preparing drawing and photographs is acknowledged.

REFERENCES

Ambrogioni, L., 1969, Due casi di infestazioni miste da nematodi dei gen. *Heterodera* e *Meloidogyne*, Redia, 51:159.

Ambrogioni, L., 1971, Osservazioni morfologiche e biologiche su *Heterodera carotae* Jones, 1950, Redia, 53:241.

Ambrogioni, L. and Marinari-Palmisano, A., 1976, Effetto di avvicendamenti colturali su *Heterodera carotae* (Nematoda: Heteroderidae) e sulla produzione di carota in terreno infestato, Redia, 59:355.

Andersson, S. and Nyberg, A., 1964, Vaxtsjukdomar och skadedjur i Skane och Halland under vegetationsperioden 1963, Vaxtskyddsnotiser, 28:58.

Anon, 1971, Annual report on research and technical work, Min. Agric. N. Ireland, 35pp.

Brzeski, M.V., 1970, Plant parasitic nematodes associated with carrot in Poland, Roczniki Nauk Rolniczych Seria E, 1:93.

Cayrol, J.C. and Frankowski, H.P., 1979, Une méthode du lutte biologique contre les nematodes à galles des racines appartenant au genre *Meloidogyne*, P.H.M, - Revue Horticole, 193:15.

Ferron, J. and Mimaud, J., 1971, Résultats des essais de produits antiparasitaires effectués en 1968 et 1969 par le service de la Protection des Vegetaux, Phytoma, 225:32.

Greco, N., 1981, Hatching of *Heterodera carotae* and *H. avenae*, Nematologica, 27:366.

Greco, N. and Brandonisio, A., 1980, Relationship between *Heterodera carotae* and carrot yield, Nematologica, 26:497.

Greco, N. and Brandonisio, A., 1982, Comportamento di cultivar di carota in presenza di infestazioni di *Heterodera carotae*, Redia "Appendice", 65:111.

Greco, N. and Lamberti, F., 1977a, Ricerca di dosi ottimali di alcuni nematocidi per la lotta contro *Heterodera carotae*, Nematol. medit., 5:25.

Greco, N. and Lamberti, F., 1977b, Suscettibilita di varieta di carota agli attacchi di *Heterodera carotae*, Nematol. medit., 5:103.

Greco, N. and Lamberti, F., 1977c, Ulteriori saggi di lotta chemica contro *Heterodera carotae* in Puglia, Inf. tore Fitopatol., 27:15.

Greco, N., Lamberti, F. and Inserra, R., 1974, Prove di lotta chemica contro *Heterodera carotae* Jones in Puglia, Nematol. medit., 2:13.

Greco, N., Di Vito, M. and Brandonisio, A., 1982, Effect of temperature and plant age on the hatching of *Heterodera carotae* and H. fici, J. Nematol., 14:443.

Greco, N., Brandonisio, A. and Elia, F., 1984, Effetto di varie dosi di nematocidi fumiganti sulla mortalita delle uova di *Heterodera carotae*, La Difesa delle Piante, 7:331.

Greco, N., Brandonisio, A. and Elia, F., 1985, Control of *Ditylenchus dipsaci*, *Heterodera carotae* and *Meloidogyne javanica* by solarization, Nematol. medit., 13:191.

Jones, F.G.W., 1950a, A new species of root eelworm attacking carrots, Nature, Lond., 165:81.

Jones, F.G.W., 1950b, Observations on the beet eelworm and other cyst-forming species of *Heterodera*, Ann. appl. biol., 37:407.

Katan, J., 1981, Solar heating (solarization) of soil for control of soil borne pests, Ann. Rev. Phytopathol, 19:211.

Kir'yanova, E.S. and Krall', E.L., 1980, Plant-parasitic nematodes and their control, Vol. II, Amerind Publishing Co. Pvt. Ltd., New Delhi.

Lamberti, F., 1971, Nematode-induced abnormalities of carrot in southern Italy, Pl. Dis. Reptr., 55:111.

Oudinet, R., 1968, Le nématode de la carotte, Phytoma, 20:33.

Oudinet, R., Cherblanc, G., Schneider, J. and Deloustal, J., 1962, Quatre années d'essais de traitments contre le nématode le la carotte, Phytoma, 14:11.

Oostenbrink, M., 1955, Nematologische Waarnemingen, IV. *Heterodera carotae* Jones, 1950, op peen *Daucus carota* L. PD Wageningen, Vers. en Meded. Nr. 127, Jaarboek 1954/55:238.

Osborne, P., 1971, First record of *Heterodera carotae* in Scotland, Pl. Path, 20:148.

Philis, J., 1976, Occurrence and control of nematodes affecting carrot crops in Cyprus, Nematol. medit., 4:7.

Porter, I.J. and Merriman, P.R., 1983, Effect of solarization of soil on nematode and fungal pathogen at two sites in Victoria, Soil Biol. Biochem., 15:39.

Sabova, M. and Valotska, B., 1976, Helminth biocenoses of vegetables under different types of cultivation, Third Int. Symp. Helminth. Inst., Kosice, Czechoslovakia, 12-15 October.

Stelter, H., 1969, Zur Biologie der Möhrennematoden, *Heterodera carotae* Jones, 1950, Biol. Zbl., 68:365.

Stelter, H., 1973, Die Arten der Gattung *Heterodera* (Nematoda: Tylenchidae) und ihre Verbreitung, Pedobiologia, 13:40.

Stirling, G.R., 1984, Biological control of *Meloidogyne javanica* with *Bacillus penetrans*, Phytopathology, 74:55.

Sturhan, D., 1960, Der Möhrennematode, *Heterodera carotae*, in Deutschland, Z. Pflanzenkrankh. Pflanzensch., 67:543.

Swarup, g., Sethi, C.L. and Gill, J.S., 1964, Some records of plant parasitic nematodes in India, Curr. Sci., 33:593.

Tacconi, R., 1976, Sulla presenza di *Heterodera carotae* Jones, 1950 e *Heterodera latipons* Franklin, 1969 nel veneto, Redia, 59:305.

Triffit, M.J., 1931, On the occurrence of *Heterodera radicicola* associated with *Heterodera schachtii* as a field parasite in Britain, J. Helminth., 9:205.

Vallotton, R., 1980, Le nématode à kystre *Heterodera carotae*, un nouveau ravageur de la carotte en Suisse romande, Le maraicher, 43:259.

Vallotton, R., 1983a, Le nematode a kyste *Heterodera carotae*, un dangereux ravageur de la carotte, Terre Valaisanne, 32:3.

Vallotton, R., 1983b, La lutte biologique contre les nématodes phyto-parasites, Revue Suisse Agric., 15:263.

Vovlas, N., 1978, Studio istopatologico di radici di carota infestate da *Heterodera carotae* e *Meloidogyne javanica*, Inf. tore Fitopatol., 28:25.

Winslow, R.D., 1954, Provisional lists of host plants of some root eelworms (*Heterodera* spp.), Ann. appl. Biol., 41:591.

Winslow, R.D., 1955, The hatching responses of some root eelworms of the genus *Heterodera*, Ann. appl. Biol., 43:19.

DISCUSSION

In response to a question by Miller about the chemicals that are available to farmers, Greco replied that fumigants such as DD and DiTrapex are used but he did not recommend it. Aldicarb is not used for control of the carrot cyst nematode. Meredith asked if the use of polythene mulches is economic. Greco said they were expensive but convenient, the cost being similar to the use of chemicals at US $400-500/ha. Mulching gave control of the nematode to a depth of about 25 cm; the best thickness of the plastic is considered to be 50 microns. B'Chir asked if the controlling effect of mulching was related to an increase in antagonists or just the effect of solarisation. Greco thought that nematode control was due to effects other than just temperature.

Sikora referred to the absence of root tip deformation and asked if this was typical for the carrot cyst nematode, as this is unlike the situation with other cyst nematodes. Greco replied that in the literature there are indications that deformation can occur, but in these instances other nematodes were probably present. In response to a question Greco said that the threshold temperature for development of the carrot cyst nematode is 10°C.

HATCHING OF THE CARROT CYST NEMATODE*

Vincent Aubert

Federal Agricultural Research Station, CH-1260 Nyon
and University of Neuchâtel
Switzerland

A study of the host status of wild and cultivated Umbelliferae growing in the Valais region of Switzerland revealed that the carrot cyst nematode, *Heterodera carotae* Jones, infested only some of the species (Winslow, 1954; Aubert, 1985). Winslow (1955) reported that among five species of Umbelliferae tested, only the leachate from the cultivated carrot markedly stimulated the hatching of the nematode.

Root diffusates collected from two month old plants of the cultivated carrot and wild Umbelliferae varied in their effect on the emergence of juveniles from cysts of *H. carotae* (Fig. 1). Some of the root diffusates were neutral in their effect, but others stimulated or inhibited hatching (Clark and Perry, 1977). As would be expected, hosts for *H. carotae* (wild and cultivated *Daucus carota, Torilis arvensis* and *T. japonica* produced a normal emergence of juveniles from the cysts. However, diffusate from *Chaerophyllum temulum*, which is not a host for the nematode nor does it act as a trap crop, produced the highest recorded hatching of juveniles.

Rates of hatching from cysts and egg masses were compared in water and carrot root diffusate. Juveniles virtually failed to hatch from cysts immersed in tap water. In carrot root diffusate they did not emerge from cysts in the first 55 days but then there was a rapid hatch over a 3 day period, followed by a slow rate of hatch to reach a mean cummulative hatch of 15 juveniles per cyst at 82 days. Juveniles emerged promptly from egg masses in water and a maximum cumulative hatch of 6 per egg mass was reached at 20 days. In carrot root diffusate the hatch from egg masses reached a cumulative hatch of 8.4 juveniles at 20 days; no further hatching occurred for the next 20 days but then a further rapid hatching followed to reach a cumulative maximum hatch of 40 juveniles per egg mass at 60 days. The mean content of egg masses (390 eggs) is almost twice that of the parental cyst (210 eggs) but per cent, the emergence of juveniles from egg masses (1.6% in tap water; 10.6% in carrot root diffusate) is higher than from parental cysts (0.1% in tap water and 7.1% in carrot root diffusate).

In southern Italy, eggs in cysts of *H. carotae* began to hatch only after 2 months immersion in carrot root diffusate (Greco, 1981). This is not due to uncompleted embryony as I observed that 70% of the eggs in 4 month old *H. carotae* cysts contained infectious juveniles. However, partial dormancy

*Work supported by the F.N.S.R.S. credit No. 3.365-0.82.

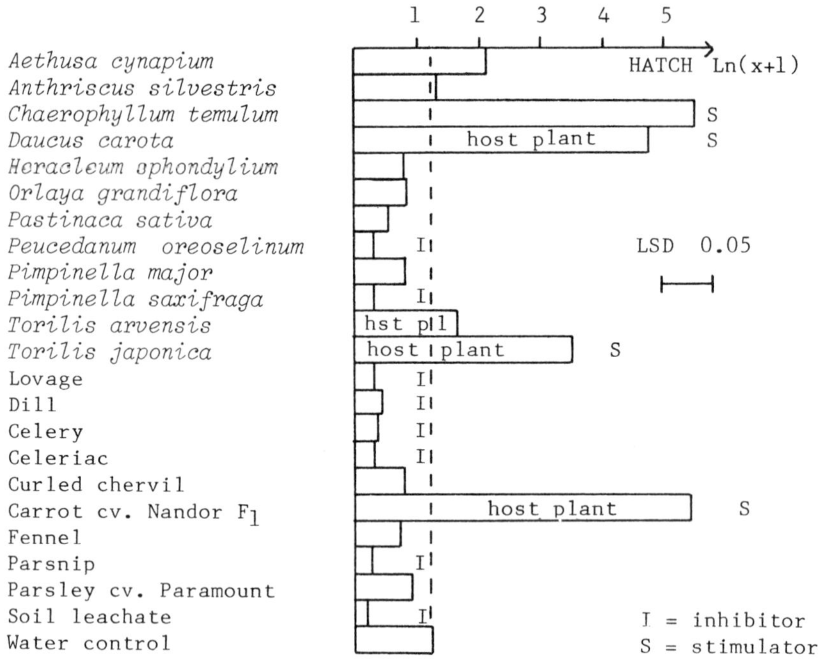

Fig. 1. Emergence of juveniles from cysts immersed in root diffusates (mean of 5 replications with 6 cysts per replication = X).

may occur in egg masses as some eggs may hatch promptly while others, like the cysts, require a relatively long immersion in carrot root diffusate before achieving a high rate of hatching. Such variation in hatching behaviour may ensure the maintenance of an infectious stock of cysts in the soil over a long period. Egg masses would provide a rapid primary infection of crops as well as a high level of infectious juveniles for later infections. Cysts would provide a reservoir of infection and provide a later and moderate level of hatching.

REFERENCES

Aubert, V, 1985, Biologie du nématode *Heterodera carotae* de la carotte, 1ère partie: Etude de la qualité d'hôte de diverses ombellifères, sauvages et cultivées, Revue suisse Vitic. Arboric. Hortic., 17:293.

Clark, A.J. and Perry, R.N., 1977, Hatching of cyst-nematodes, Nematologica, 23:350.

Greco, N., 1981, Hatching of *Heterodera carotae* and *H. avenae*, Nematologica, 27:366.

Winslow, R.D., 1954, Provisional lists of host plants of some root eelworms *Heterodera* spp.), Ann.appl.Biol., 41:591.

Winslow, R.D., 1955, The hatching responses of some root eelworm of the genus *Heterodera*, Ann.appl.Biol., 43:19.

OBSERVATIONS ON THE POPULATION DYNAMICS AND CONTROL OF *HETERODERA CAROTAE* IN WESTERN FRANCE

Michel Bossis

I.N.R.A., Laboratoire de Recherches de la chaire de Zoologie
Domaine de la Motte-au-Vicomte, B.P. 29, 35650 Le Rheu
France

INTRODUCTION

The carrot is an important vegetable crop in western France, accounting for about half of the French production. The main areas of production are near the coast, on sandy or sandy silt soil, where the mild, moist climate is particularly favourable for this crop.

There are three types of production: early-lifted carrots marketed in May-June, carrots for canning harvested in the summer and the autumn, and carrots for long-term storage marketed in the autumn and throughout the winter. Early-lifted carrots are produced in the Nantes area and in the vicinity of the large cities. Carrots for canning are grown in the department of Morbihan. Carrots for long-term storage are produced in the department of Manche (20 per cent of the French production) and in the northern part of the department of Finistère.

Heterodera carotae Jones is a major problem in carrot production, particularly in crops grown for long-term storage (Oudinet et al., 1962;

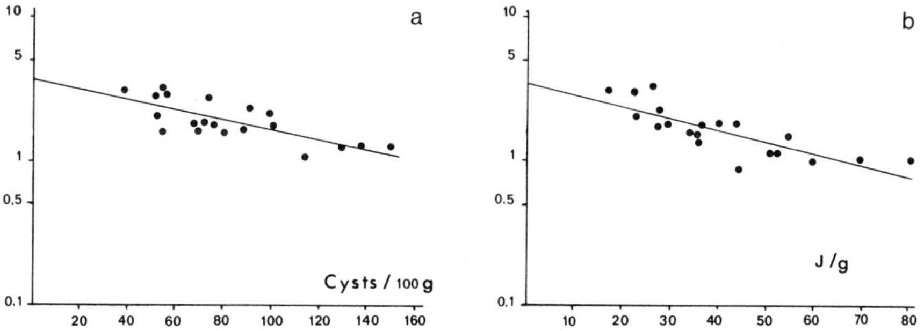

Fig. 1 Population dynamics of *H. carotae* on sandy silt soil, with one year crop of carrots. Relationship between population density of cysts (A) and juveniles (B) and the multiplication rate.

Harranger, 1971). Oudinet (1978) considers that in some intensive crop areas of the department of Manche, 60 per cent of the cultivated land is infested. About 25 per cent of the fields are treated with fumigants, in most cases without first obtaining information about nematode populations. Experiments were carried out to assess the damage caused by *H. carotae* and the efficiency of chemical treatments in an attempt to establish a more suitable control system.

EXPERIMENTS

Field experiments were carried out in the department of Manche on sandy silt soil at "Les Pieux" and on sandy soil at "Creances". Cv. Nandor was sown at both localities at the beginning of June and the crops were harvested in December.

Population densities of *H. carotae* were estimated in 4 m^2 plots by collecting soil samples at 40 loci in each plot (20 cm core depth, 20 - 30 g at each locus); after sieving (2 mm diam.), cysts were extracted from a 300 g sub-sample by flotation in a Kort's elutriator; the cysts were sorted, counted and crushed to estimate the numbers of live juveniles. Initial population densities were estimated just before sowing, and final population densities just after ploughing that followed the harvest.

To test the effectiveness of chemical treatment, 400 l/ha of dichloropropane-dichloropropene (D-D) were applied to nine plots on sandy silt soil one month before sowing (immediately after soil sampling to assess initial population densities). The effect of a non-host crop on nematode populations was assessed on nine plots on sandy silt soil.

The rate of multiplication of *H. carotae* on carrot was investigated in plots providing a progressive gradient from low to high initial densities. The regression curve of multiplication rate was calculated from the equation:

$$\log y = a + bx$$

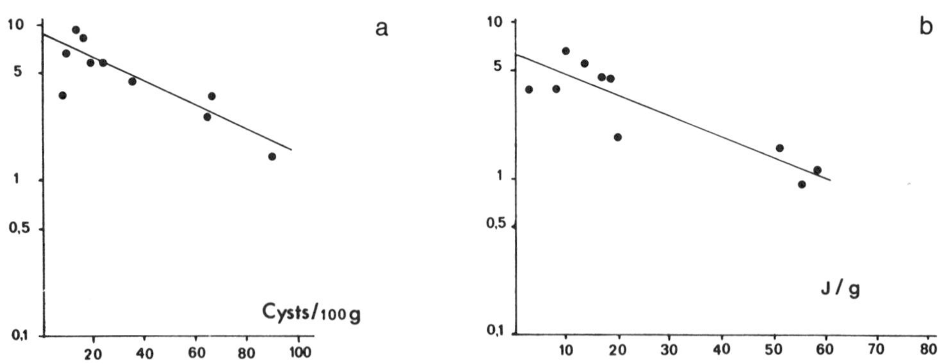

Fig. 2 Population dynamics of *H. carotae* on sandy soil, with one year crop of carrots. Relationship between population density of cysts (A) and juveniles (B) and the multiplication rate.

(Mugniery, 1976), in which x is the initial population (cysts and juveniles) and y the multiplication rate.

RESULTS

The multiplication rate of *H. carotae* in a crop of carrots cv. Nandor grown on sandy silt soil is indicated in Fig. 1A, B.

Correlation coefficients of -0.764 for cysts and -0.808 for juveniles are significant at P = 0.01 with 18 d.f.

Corresponding results for *H. carotae* on sandy soil (Fig. 2A, B) give correlation coefficient of -0.839 for cysts and - 0.879 for juveniles, which are significant at P = 0.01 with 8 d.f.

Multiplication rates of *H. carotae* in plots treated with D-D before sowing are shown as regression curves for cysts and juveniles in Fig. 3A, B. Correlation coefficients of -0.822 for cysts and -0.859 for juveniles are significant at P = 0.01 with 7 d.f.

In plots in which a non-host crop was grown instead of a carrot crop the cyst and juvenile mean populations were respectively 111.3 cysts/100 g of soil and 56.6 juveniles/g. Initial and final density means differed significantly at P = 0.01; t = 3.53 for cysts and t = 3.62 for juveniles with 8 d.f. The multiplication rates were 0.62 for cysts and 0.67 for juveniles.

DISCUSSION

Growing a host crop on sandy or sandy silt soil heavily infested by *H. carotae* generally leads to an increase in the nematode population density. For cysts and for juveniles, the lower the initial population, the higher was the multiplication rate.

At high population densities there is stabilisation at the equilibrium level, the explanation being that at high nematode densities, the root system is heavily damaged by the first generation of the nematode;

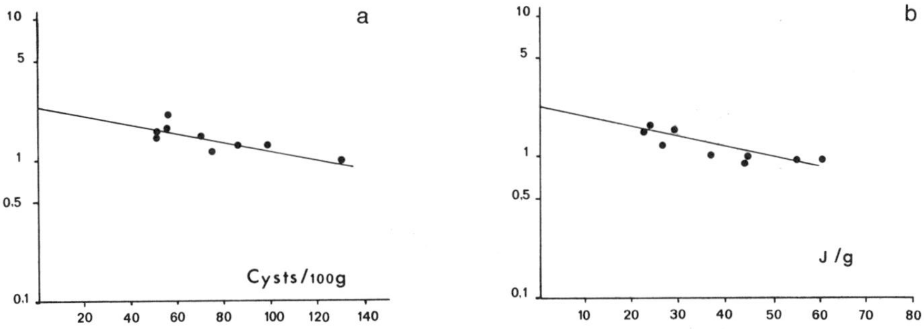

Fig. 3 Population dynamics of *H. carotae* on sandy silt soil, with one year crop of carrots after a D-D treatment. Relationship between population density of cysts (A) and juveniles (B) and the multiplication rate.

consequently carrots remain small and do not sustain more than a moderate increase in nematode numbers. Moreover, intense competition between juveniles probably induces a greater production of male nematodes since sex determination is partially epigenic (Mugniéry and Bossis, 1986). Equilibrium is established at 50 - 70 juveniles/g. At low population densities the multiplication rate of juveniles is 2 - 3 on sandy silt soil and about 7 on sandy soil. Such multiplication rates are similar to those observed in polyvoltine species such as *H. schachtii* Schmidt (Steudel and Müller, 1983) although these are low in comparison with *Globodera rostochiensis* (Woll.) Behr. which can exceed a multiplication rate of 100 (Seinhorst, 1967).

After treatment with the fumigant D-D, which can be expected to achieve a decrease in population of 90 to 95 per cent, it was observed that in most of the treated plots the initial infestation level was reached or even exceeded after one year of growing a carrot crop. These results agree with those of Greco and Lamberti (1977) who had observed high infestations on roots after a D-D treatment at the same dosage.

With a non-host crop, mean population densities decreased by 33 per cent after one year. In western France, it seems that populations decrease less than those observed by Ambrogioni and Marinari Palmisano (1976) in Italy who recorded 49 per cent decrease after a year with a non-host crop. With low populations, a rotational break of just one year with a non-host crop is inefficient, with the tolerance limit being particularly low at less than one juvenile/g soil (Ambrogioni and Marinari Palmisano, 1976; Greco and Brandonisio, 1980).

CONCLUSIONS

The development of populations of *H. carotae* in carrot crops grown for long-term storage, were similar in the two types of soil investigated. High populations remained at the equilibrium level of about 60 juveniles/g and the lowest populations increased to very high levels in one year on a host crop. There are two generations of *H. carotae* on a crop grown for long-term storage, and sometimes a third when the harvest is delayed until the early spring, 10 months after sowing.

Soil treatment with a fumigant such as D-D decreased nematode populations to a level where young plants did not suffer any damage. However, with the development of a second generation, nematode populations reached the initial density level, or greater. In practical terms, this means that with short rotations, every carrot crop should be protected by chemical treatment.

REFERENCES

Ambrogioni, L. and Marinari Palmisano, A., 1976, Effetto di avvicendamenti colturali su *Heterodera carotae* (Nematoda: Heteroderidae) e sulla produzione di carota in terreno infestato, Redia, 59:355.

Greco, N. and Lamberti, F., 1977, Ricerca delle dosi ottimali di alcuni nematocidi per la lotta contro *Heterodera carotae*, Nematol. medit., 5:25.

Greco, N. and Brandonisio, A., 1980, Relationship between *Heterodera carotae* and carrot yield, Nematologica, 26:497.

Harranger, J., 1971, Les nématodes des cultures maraichères, in "Les Nématodes des Cultures", Journées d'études et d'information ACTA FNGPC Paris.

Mugniéry, D., 1976, Etablissement d'un modèle de dynamique de population d'*Heterodera pallida* Stone : Applications à un cas pratique de lutte intégrée, Ann. Zool. - Ecol. anim., 8:315.

Mugniéry, D. and Bossis, M., 1986, Influence de l'hôte, de la compétition et de l'état physiologique des larves infestantes sur la pénétration, le developpement et le phénotype sexuel d'*Heterodera carotae* Jones, Nematologica, (in press).

Oudinet, R., 1978, La carotte de normandie dans le département de la Manche, Phytoma, 208:3.

Oudinet, R., Cherblanc, G., Schneider, J. and Deloustal, J., 1962, Quatre années d'essais de traitements contre le nématode de la carotte, Phytoma, 135:11.

Seinhorst, J.W., 1967, The relationships between population increase and population density in plant parasitic nematodes, Nematologica, 13:157.

Steudel, W. and Müller, J., 1983, Untersuchungen und modellrechnungen zum einfluss pflanzenverträglicher nematizide und nematodenresistenter swischenfrüchte auf die abundanz-dynamik des zucherrübennematoden (*Heterodera schachtii*) in zucherrübenfruchtfolgen, Zuckerind, 108:365.

CYST NEMATODES IN EQUATORIAL AND HOT TROPICAL REGIONS

Michel Luc

Musieum national d'Histoire naturelle
Laboratoire des Vers, 61 rue de Buffon, 75005 Paris
France

INTRODUCTION

In the present communication, the term "cyst nematodes in equatorial and tropical regions" is used instead of the usual "tropical cyst nematodes" for two reasons:

- in addition to those species of cyst nematodes that are closely associated with hot tropical crops and areas, there are others that are common in temperate areas which may also be occasionally found in the tropics;

- the topographical meaning of "intertropical" encompasses some mountainous areas, namely in Central and South America, where climate, vegetation and crops are quite similar to those of temperate regions: these are the "cold tropics".

This paper deals only with the "hot tropics", and the term "tropics" and "tropical" refer here only to these climatic regions.

Regarding the earliest records of "*Heterodera*" in equatorial and tropical areas, it must be kept in mind that prior to Chitwood's (1949) resurrection of the ancient generic name *Meloidogyne* Goeldi, 1877, the genus *Heterodera* contained both cyst forming species of Heteroderidae (or *Heterodera* sensu lato) and various root knot species under the name *Heterodera marioni* (Cornu). As most of the early records of "*Heterodera*" in the tropics were given without morphological detail or description of symptoms on the host plant, it is not possible to determine whether the species in question were *Meloidogyne* or cyst-forming species.

Also, there are numerous records of "*Heterodera*-like juveniles" in the literature concerning tropical crops but such juveniles may belong to species of non-cyst forming Heteroderinae genera which also occur in the tropics. For example, in the Ivory Coast *Heterodera*-like juveniles were frequently found in some areas of primary or ancient forests but extensive investigation failed to find cysts. Eventually it was discovered that these juveniles belonged to a non-cyst forming species of Heteroderinae present on some trees and that they represented a new genus and species, subsequently described under the name of *Hylonema ivorense* by Luc et al. (1978). Therefore, records only of "*Heterodera*" juveniles are not considered here.

The first report of a cyst nematode attacking a tropical crop in a tropical area concerns sugar cane in the Hawaiian Islands (Muir and Henderson, 1926). It is clear that there was no confusion with *Meloidogyne* because the authors stated "It is only recently that the two species * [of "*Heterodera*"] have both been recognised in Hawaiian Islands". A description is given of the damage symptoms caused to the host plant by the "two species" and the distinction between *Heterodera* and *Meloidogyne* is clearly stated. Concerning *Heterodera* they say "In sugar cane this nemaode does not show a preference for the tip of roots but attacks them at any point; ... it does not produce a gall on the root". The "brown cyst stage" is also cited.

Muir and Swezey (1926) also recorded "*Heterodera schachtii*" on pineapple, in the Hawaiian Islands, but here the text is not as clear and doubts remain. These two records are cited again in a later paper by Henderson (1926).

The report of a cyst nematode on sugar cane was not considered seriously by subsequent authors because of the prevailing thought that cyst nematodes were to be found only in temperate areas. For example, in the compilation by Goodey (1956), this record is cited as "*Het. schachtii = ? Mel.* sp.".

This "*Heterodera*" has not been reported again in Hawaiian sugar cane plantations and thus the species, and perhaps the genus, remain unknown. However, I consider that it probably was not *H. schachtii*, as sugar cane has been shown to be a non-host for this species (Skarbilovich and Kurramov, 1973).

Thirty five years later, Luc (1961) reported two tropical *Heterodera* species, one parasitising swamp rice in the Ivory Coast and the other sugar cane in the Congo. They were described as *H. oryzae* Luc & Berdon - Brizuela, 1961 and *H. sacchari* Luc & Merny, 1963 and this clearly established that cyst nematodes could occur in the tropics. Subsequently several other species have been described from tropical crops and weeds. Nevertheless, it appears that the cyst-forming nematodes are not as common nor as numerous, and probably not as economically important, in the tropics as they are in temperate areas.

GENUS *HETERODERA* Schmidt, 1871

Heterodera oryzae Luc & Berdon-Brizuela, 1961

H. oryzae was reported from swamp rice fields at Bokakouamekro in the central part of the Ivory Coast (Luc and Berdon-Brizuela, 1981). This species, of the *schachtii* group, is characterised mainly by the presence of a voluminous egg-mass, attached to the cyst, which often contains more eggs than the cyst itself. The original description has been supplemented by Luc and Taylor (1977).

*Prior to the beginning of 1928, only two species of "*Heterodera*" were recognised: *H. schachtii* Schmidt (the cyst-forming species) and *H. marioni* (Cornu) (the root knot species). A third species, *H. javanica* Treub, 1885, also existed but as it was described from an exotic host plant (sugar cane) and from a tropical country (Indonesia) little attention was given to it at that time.

The role of the egg-mass in relation to population dynamics has been studied (Merny, 1966; Merny, 1972a, Reversat, 1975d). Juveniles in the egg-mass hatch immediately without need of a stimulation by root exudates and this enables two or three generations to develop on the same rice plant (Merny, 1972a). This ready hatching of egg-masses facilitates the culture of the nematode, even under monoxenic conditions (Reversat, 1975a). Thus *H. oryzae* has been used to examine various aspects of the biology of *Heterodera* such as ovigenesis and reproduction, which does not differ from other amphimictic species of the genus (Netscher, 1969); histopathology of attacked roots (Berdon-Brizuela and Merny, 1964); factors influencing the sex ratio and sex determination (Cadet et al., 1975; Cadet and Merny, 1978a); genetic factors related to the length of juveniles (Netscher and Pernes, 1971); production of artificial mutants (Cadet, 1976); factors influencing penetration by the juveniles (Reversat and Merny, 1973; Cadet and Merny, 1976; Merny and Cadet, 1978); chemical composition of juveniles (Reversat, 1977a); oxygen uptake by juveniles (Reversat, 1975c; 1977b) and their survival (Reversat, 1980), particularly in anaerobic conditions (Reversat, 1975b).

The species is known only from some swamp areas in the Ivory Coast (Merny, 1970), and in Senegal where it parasitises banana trees (Taylor, 1978). Reports of *H. oryzae* from Japan (Hashioka, 1964; Shimizu, 1971, 1972, 1973; Nishizawa et al., 1972) refer to a different species, described by Ohshima (1974) as *H. elachista*. However, the two species appear to be closely related and their relationships require further study.

In addition to rice and banana, few other plants have been shown to be hosts. Maize and *Mariscus umbellatus* are susceptible (Merny and Cadet, 1978) whereas sugar cane, *Sorghum vulgare*, *Pennisetum typhoides* and *Panicum maximum* are not, and large differences between various rice cultivars have been observed (Merny and Cadet, 1978). This relatively narrow host range could explain the limited distribution of the species. It is also limited in flooded rice fields by competition with *H. sacchari*; experiments with juveniles marked with ^{35}S have shown that *H. oryzae* is much less aggressive than *H. sacchari* (Reversat and Bois, 1982).

Heterodera sacchari Luc & Merny, 1963

H. sacchari, of the *schachtii* group, was originally described from sugar cane in the Niari Valley, Congo (Luc and Merny, 1963). It has subsequently been reported from the Ivory Coast where it has frequently been found in flooded rice fields in the northern and central regions (Merny, 1970), in Nigeria on sugar cane (Jerath, 1968) and wild grasses (Odihirin, 1975), in flooded rice fields of Casamance Province, Senegal and in Gambia (Fortuner and Merny, 1973) and on sugar cane in Burkina Faso (Cadet and Merny, 1978b). The species is also reported to have been found on *Saccharum spontaneum* in India (Swarup et al., 1964).

In addition to sugar cane and rice, various wild Graminaceae are considered as hosts by Odihirin (1975): *Paspalum conjugatum*, *Axonopus compressus*, *Mariscus umbellatus*, *Cynodon dactylon*, *Eleusine indica* and *Brachiaria brizantha*. Odihirin (1975) considers that *H. sacchari* is indigenous in Nigeria and whilst I agree I also think that the area of distribution is probably wider than indicated by present reports.

The morphology of *H. sacchari* has been described and the original description was supplemented by Luc (1974). The major diagnostic character of the species is a very dark, strongly developed underbridge of the female cyst which has finger-like projections. The apparently non-functional male has been described by Netscher et al. (1969). Netscher (1969) has shown that *H. sacchari* is a triploid parthenogenetic species.

The hatching of juveniles from cysts is similar in sugar cane root diffusate and tap water (Garabedian and Hague, 1984) but Riversat (1981) has found potassium permangate to be useful as a hatching agent. Aldicarb, oxamyl and carbofuran prevent hatching from cysts but this effect is decreased by the addition of sugar cane root diffusate (Garabedian and Hague, 1984).

Jerath (1968) observed that infested sugar cane plants are stunted and thin, and secondary roots are less abundant than healthy plants. In sugar cane plantations, cysts of *H. sacchari* may be transported by the water in irrigation canals for at least 5 to 8 km (Odihirin, 1977).

On rice, experiments show (Babatola, 1983a) that infested plants are chlorotic and their growth retarded, roots are necrotic and blackened, tiller numbers are reduced and the grain yield is lower. *H. sacchari* attacks both swamp and upland rice but the latter appears more susceptible to damage (Babatola, 1983b). Rice cultivars react very differently to infestation by the nematode (Babatola, 1983b).

In western areas of Africa, *H. sacchari* may constitute a danger for sugar cane, and both upland and swamp rice. Foliar applications of oxamyl have been shown experimentally to prevent invasion of young sugar cane plants (Garabedian and Hague, 1982) but field trials have not been reported.

H. cyperi Golden, Rau & Cobb, 1963

H. cyperi was first found infesting *Cyperus esculentus* at Sanford, Florida, USA, and was described by Golden et al. (1963). No other host was known and its distribution was thought to be limited to two countries in Florida and one in Georgia (Mulvey and Golden, 1983). Recently, however, Kumar (1980) recorded it on *Cynodon dactylon* and *Bambusa* sp. in a coffee estate, in Karnataka, India. Eroshenko and Kazachenko (1972) considered it to be a species inquirenda, but there does not seem sufficient reason for this.

Heterodera mothi Khan & Husain, 1965

H. mothi, found on *Cyperus rotundus*, was the first cyst nematode described from India (Khan and Husain, 1965). Cysts of this species, which is of the *goettingiana* group, are characterised by the thinness of their cuticle, the prominence of the anal area, and the slenderness of the vulval bridge. Mulvey (1972) emphasised the fact that the underbridge is rarely present. An egg-sac has been observed (Khan and Jairajpuri, 1975).

Sharma and Swarup (1984) state that this species reproduces only on the type host whereas other records are from *Chrysanthemum carinatum* and *Hibiscus syriacus* in India (Yaqoob and Khan, 1969); *Cyperus esculentus*, *Glycine max*, *Gossypium hirsutum*, in Georgia, USA (Minton et al., 1973); in soil from sugar-beet fields in Iran (Talatschian and Achyani, 1976); sugar cane (Maqbool, 1981) and onion (Maqbool et al., 1983) in Pakistan; and cysts in soil at Viti Levu, Fiji Islands (Kirby et al., 1978).

Sharma and Swarup (1984) gave information on the morphometrics and life cycle of *H. mothi* but nothing is known about its pathogenicity.

H. indocyperi Husain & Khan in Khan et al., 1964 (cited by Hesling, 1978) is considered by Hesling to be a nomen nudum and a synonym of *H. mothi*.

Heterodera cajani Koshy, 1967 (= *H. vigni* Edward & Misra, 1968)

This species, of the *schachtii* group, was first reported under the name of *H. trifolii* in a pigeon-pea field of the IARI at New Delhi, India (Swarup et al., 1964). It was described later as *H. cajani* by Koshy (1967). *H. vigni* described from roots of *Vigna unguiculata* by Edward and Misra (1968) has been recognised (Kalha and Edward, 1979) as a junior synonym of *H. cajani*.

H. cajani is the most widespread cyst-forming nematode in India and has been reported in the states of Andhra Pradesh, Bihar, Delhi, Haryana, Rajasthan, Tamil Nadu and Uttar Pradesh (Sharma and Swarup, 1984). The species has also been reported on cowpea in Egypt (Aboul-Eid and Ghorab, 1974).

On the basis of field observations and experiments by various authors (Koshy, 1967; Koshy and Swarup, 1972; Bhatti and Gupta, 1973; Janarthanan, 1972; Verma and Yadav, 1975; Gaur and Sing, 1977) the following host plants have been established: *Cajanus cajan, Cicer arietinum, Cyamopsis tetragonoloba, Dolichos biflorus, D. lablab, Glycine max, Phaseolus atropurpureus, P. aureus, P. calcaratus, P. lathyroides, P. lunatus, P. mungo, P. vulgaris, Pisum sativum, Sesamum indicum, S. orientale, Vivia narabenesis* cv. *sativa, Vigna sinensis* and *Zea mays*. With the exception of the last species, all these host plants belong to Leguminosae or Pedaliaceae. Koshy and Swarup (1972) established that 82 crop and weed species were non-hosts including *Cyamopsis tetragonoloba, Glycine max, Phaseolus calcaratus, Pisum sativum* cv. Early Wisconsin and several *Vicia* species. The original description of the species has been supplemented by Koshy et al. (1971) and Sharma and Swarup (1984). The biology of *H. cajani* is similar to other *Heterodera* species. Host reactions have been described by Aboul-Eid and Ghorab (1974) and the life cycle by Koshy and Swarup (1971a). More than nine generations per year have been reported (Koshy and Swarup, 1971b). The hatching of juveniles from cysts and egg-masses is optimal at 29°C; root leachates of *Cajanus cajan* and *Dolichos lablab* have a positive effect; hatching of juveniles from egg-masses at 20 days is 100 per cent in contrast to only 16 per cent from brown cysts (Koshy and Swarup, 1971d). The presence of egg-masses, and the rapid hatching of juveniles, explains the high number of generations observed. It has also been observed that *H. cajani* may attack the aerial parts of young seedlings of pigeon-pea (Koshy and Swarup, 1971e).

Although there is no precise data on the damage to crop plants caused by *H. cajani*, it must be considered a dangerous species. Its presence in Egypt shows that it could become established in tropical, subtropical and even Mediterranean countries.

Heterodera graminis Stynes, 1971

H. graminis was described from roots of *Cynodon dactylon* on a bowling green at New Castle, New South Wales, Australia (Stynes, 1971). The species has recently been recorded on undetermined species of grasses on a golf course at Delhi, Rajasthan, India (Sharma and Swarup, 1984) and has been found in soil at Vanua Levu, Fiji Islands (Kirby et al. 1978) and on *Cynodon dactylon* in a sugar cane plantation in Trinidad (Stone, in litt.).

H. graminis is of the *goettingiana* group, and appears to be closely related to *H. cyperi*.

In pot experiments *H. graminis* reproduced on *Setaria italica, Secale cereale* and maize (Sharma and Swarup, 1984). Nothing is known about its pathogenicity but host plants were described as yellowish and stunted (Kirby et al., 1978; Sharma and Swarup, 1984). Data on the life cycle are given by Sharma and Swarup (1984).

Heterodera zeae Koshy, Swarup & Sethi, 1971

This species of the *schachtii* group was described from specimens found in association with maize at Chapli, Rajasthan, India (Koshy et al., 1971). Surveys showed that *H. zeae* is widely distributed throughout India and in the following states: Andhra Pradesh, Bihar, Delhi, Gujarat, Madhya Pradesh, Punjab, Rajasthan, Uttar Pradesh (Koshy and Swarup, 1971c) and Karnataka (Narayanaswamy, et al., 1982). Subsequently it was recorded in Pakistan (Maqbool, 1981), in Egypt (Oteifa, unpublished report, 1978), and in the USA (Sardanelli et al., 1981). *H. zeae* is considered to be one of the most economically important nematodes attacking cereals in India (Sharma and Swarup, 1984).

In addition to the original description, data on morphology and morphometrics have been reported from Indian populations by Sharma and Swarup (1984) and for American populations by Golden and Mulvey (1983). Hutzell (1984) has recently described the male. Sharma and Swarup (1984) provide some information on the biology of the species but there has been no published information on pathogenicity. Hosts recorded for *H. zeae* include several cultivars of *Zea mays*, (Misra et al., 1977), *Echinochloa colona, Panicum* spp., *Setaria italica* and *Triticum aestivum*. Experimentally (Srivastava and Swarup, 1975; Sharma and Swarup, 1984) the following plants have been found susceptible to *H. zeae: Oryza sativa; Secale cereale, Sorghum vulgare, Zea mexicana, Avena sativa* and *Hordeum vulgare*. *Sorghum vulgare* was, however, considered resistant by Srivastava and Swarup (1975) as were *Paspalum* spp. and *Pennisetum typhoideum*. Maqbool (1981) reported non-graminaceous crops (gram, citrus, pear, garlic) as hosts, but it is not clear if cysts were found on the roots of these plants or in the surrounding soil.

Heterodera gaembiensis Merny & Netscher, 1976

Heterodera gambiensis, belonging to the *schachtii* group, has been reported on *Sorghum vulgare* and *Pennisetum typhoideum* in various locations in Gambia, West Africa (Merny and Netscher, 1976). Surprisingly, in the Senegalese regions surrounding Gambia, extensive surveys failed to locate this species. It has, however, recently been recorded on *Eleusine caracana* in Karnataka, India (Narayaswamy et al., 1982). There is no information on the biology, host range or pathogenicity of the species.

Heterodera oryzicola Rao & Jayaprakash, 1978

This species of the *schachtii* group is known only from upland rice in Kerala State, India (Rao and Jayaprakash, 1978). Although it has been observed to penetrate the roots of several graminaceous plants, according to Sharma and Swarup (1984) cysts are produced only on rice, mainly on the cultivars Jaya, Ratna and T (N)-1.

H. oryzicola causes stunting and chlorosis of leaves (Rao and Jayaprakash, 1977) and field experiments indicate yield losses of 21-42 per cent (Usha and Kurien, 1981).

Morphologically *H. oryzicola* is very close to *H. oryzae*, from which it differs mainly in various measurements of the stylet and spicules, and length of the hyaline terminal part of tail of the second stage juvenile (Rao and Jayaprakash, 1978). Sharma and Swarup (1984) provide additional morphometric data and information on the life cycle.

Heterodera delvii Jairajpuri, Khan, Setty & Govindu, 1979

This species, belonging to the *goettingiana* group, was described from "ragi" (*Eleusine coracana*) in Bangalore, Karnataka, India (Jairajpuri, 1979). Other hosts reported by Krishna Prasad et al. (1980) include various *Echinochloa* species, *Sorghum vulgare*, maize, pearl millet, finger millet, *Setaria italica*, wheat and *Panicum miliaceum*.

H. delvii is close to *H. cyperi*, *H. mothi* and *H. graminis*, but differs mainly in the structure of the underbridge which shows a subcircular mass attached to its centre; this is absent in the other species. No information is available on the biology and the pathogenicity of this species. It is not known outside the type location, and attempts to find it again at this site have been unsuccessful (Sharma and Swarup, 1984).

Heterodera sorghi Jain, Sethi, Swarup & Srivastava, 1982

This species, of the *schachtii* group, has been found on *Sorghum vulgare* at Raispur, Utah Pradesh, India (Jain et al., 1982); it has also been recorded, without indication of host, at Hayatpur, Haryana State (Dhawan et al., 1983). In pot experiments *H. sorghi* reproduced on *Echinocloa colona*, *Eleusine caracana*, *Setaria italica*, *Paspalum scrobiculatum*. *Zea mays*, *Cyperus rotundus* and *Pennisetum typhoides* (Sharma and Swarup, 1984).

H. sorghi is characterised by a very thick, prominent and dark brown underbridge. Data on morphometrics and life cycle have been provided by Sharma and Swarup (1984) but nothing is known about its pathogenicity.

Heterodera raskii Basnet & Jayaprakash, 1984

H. raskii was recently described from Hyderabad, India, on *Cyperus bulbosus* Vab., which is presently the only known host (Basnet and Jayaprakash, 1984). It is in the *goettingiana* group and considered to be close to *H. cyperi* Golden et al., 1962. Cysts are very elongated, somewhat resembling those of *Dolichodera* Mulvey & Ebsary, 1980. Nothing is known about pathogenicity or host-parasite relationships.

Other *Heterodera* species

Heterodera schachtii Schmidt, 1871: This typically temperate species has been recorded in Western Tropical Africa in two locations close to each other. In Senegal, *H. schachtii* was found in a small vegetable garden in the centre of Dakar, where it caused damage to beets and cabbage (Luc and Netscher, 1974); extensive surveys in the surrounding area were negative.

It was recorded later in Banjul, Gambia, also on beets (Bridge and Manser, 1980). In both cases, *H. schachtii* was probably introduced from Europe with rooted plants and is not considered to present a danger to crops in Senegal and Gambia.

Heterodera graminophila Golden & Birchfield, 1972: Although found by Golden and Birchfield (1972) outside of a strictly tropical area (Louisiana, USA), *H. graminophila* is cited here as Birchfield (1973) stated that the species "may not be native to the USA since the host plant, *Echinochloa colona* and other closely related forms, are cultivated in tropical Asia and Africa for food". Other hosts are rice and *Sorghum halepense*. *H. graminophila* was said by Birchfield (1973) to be atypical of the genus because of the absence of giant cells at the feeding point within the host root: the females feed in the pericycle and phloem tissues of the host; no galling of root tissues and no foliage symptoms occurred. Birchfield (1973) considered it an example of a well-balanced host-parasite system, possibly indicative of a very old parasite relationship.

H. graminophila belongs to the *goettingiana* group, and is close to *H. cyperi* and *H. graminis*.

Heterodera avenae Wollenweber, 1924: This is a cosmopolitan species present in some warm parts of the world. It has been associated with severe diseases in India known as "molya". Many Indian publications refer to *H. avenae* but it occurs only on temperate cereals such as barley and wheat. On the contrary, tropical cereals such as *Echinochloa colona, Sorghum vulgare, Eleusine coracana, Oryza sativa, Setaria italica, Zea mays* are non-hosts (Gill and Swarup, 1971; Sharma and Swarup, 1984). Thus strictly *H. avenae* does not qualify as a tropical cyst nematode.

Heterodera chaubattia Gupta & Edward, 1973: This species was found in the soil of an orchard at Chaubattia, Uttah Pradesh, India (Gupta and Edward, 1973). This is the only report and Sharma and Swarup (1984) failed to rediscover the nematode at the type locality. Moreover types are apparently not available for study and thus taking into account the discrepancies and lack of information in the original description, Sharma and Swarup (1984) considered *H. chaubattia* to be a species inquirenda.

GENUS *CACTODERA* Krall & Krall, 1979

Cactodera cacti (Filipj. & Sch. Stekh. , 1941) Krall & Krall, 1979

This species was reported on *Echinopsis* spp. in Mysore, India (Kumar, 1964); the nematode could have been introduced with the host plant, which is an ornamental.

Cactodera amaranthi (Stoyanov, 1972) Krall & Krall, 1969

This species was described by Stoyanov (1972) as *Heterodera amaranthi* from roots of *Amaranthus viridis* growing in experimental fields at Santiago de las Vegas, near Hayana, Cuba. Heavy infestation of the host was reported. It was found earlier in Florida by G. J. Rau (unpublished report, 1957). Details of this and subsequent finding are given by Golden and Raski (1978) who also studied the different stages of the nematode.

Hosts are limited to species of *Amaranthus (A. viridis, A. spinosus, A. retroflexus)*. The distribution includes Cuba, Mexico, and the USA

(Florida, Colorado, Illinois, Kansas, Missouri, Tennessee) according to Mulvey and Golden (1983). However, Reis (1982) reported a cyst forming species "closely resembling" *G. amaranthi* and heavily parasitising *Amaranthus patulus* in a citrus orchard near Maputo, Mozambique. Thus, the distribution of this species may be rather wide.

GENUS *GLOBODERA* Skarbilovich, 1959

Globodera rostochiensis (Wollenweber, 1923) Behrens, 1975; *Globodera pallida* (Stone, 1973) Behrens, 1975

The potato cyst nematodes have been reported in tropical countries but only on rare occasions. The first report is that of Tiago (1956) in Mozambique; but at that time the report was considered questionable. In India, *G. rostochiensis* and/or *G. pallida* are considered to be serious pests of potatoes in the Nilgiris region, which is a hilly part of Madras State and does not have a tropical climate (Seshadri and Sivakumar, 1962; Prasad and Chawla, 1965; Logisvaran and Menon, 1969). Sharma and Swarup (1983) report the presence of both species in Tamil Nadu State, the southernmost state of India and tropical. *G. rostochiensis* had also been recorded in South Africa, where its distribution appears to be restricted to the Transvaal (Kleynhans, personal communication).

Apparently, *G. rostochiensis* and *G. pallida* are not adapted to the tropics; they may survive in some locations (as does *H. schachtii*) but they do not pose as severe a threat to potato crops in hot tropical areas as do various root-knot nematode species.

GENUS *AFENESTRATA* Baldwin & Bell, 1985

Afenestrata africana (Luc, Germani & Netscher, 1973) Baldwin & Bell, 1985)

A. africana was described by Luc et al. (1973) as the second species of *Sarisodera* Wouts & Sher, 1971, as the two species apparently shared several important characters including the presence of cysts devoid of fenestra and a peculiar vulval cone formed by the hypertrophied and very protuberant vulval lips. Restudy of the type species, *Sarisodera hydrophila* Wouts & Sher, 1971, by Baldwin and Bell (1985) led these authors to consider that the females do not transform into cysts; consequently, they erected the genus *Afenestrata* (with cysts), distinct from *Sarisodera* (without cysts).

Morphology, caryology (n = 9, as in many Heteroderinae) and histology of the syncytium produced by *A. africana* are well documented (Luc et al., 1973; Taylor and Luc, 1979; Baldwin and Bell, 1985). The only reported host is Guinea grass *Panicum maximum*, and the only location is the type locality - Experimental Fields, ORSTOM Centre, Adiopodoume near Abidjan, Ivory Coast. However, *A. africana* came from Kenya, East Africa and the location in the Experimental Fields where the nematode was found had been planted with rooted seedlings of Guinea grass obtained by plant breeders during an expedition in Kenya.

Because of this no further investigations of *A. africana* have been undertaken and the original field was treated with a nematicide, then planted with a non-graminaceous crop. A population of *A. africana* was transferred to a greenhouse at the Nematology Laboratory, ORSTOM Centre, to be used only for taxonomic or anatomical studies.

CONCLUSION

This paper somewhat resembles a catalogue and it is difficult to separate into sections. Nevertheless, the following tentative collective conclusions can be made:

- with a few exceptions, among the genera of cyst-forming Heteroderinae found in temperate climates, only species pertaining to the genus *Heterodera* have been reported in the hot tropical climates;

- the twelve *Heterodera* species considered herein as tropical belong to only two groups of the genus: the *schachtii* group (*oryzae*, *sacchari*, *cajani*, *zeae*, *gamgiensis*, *oryzicola* and *sorghi*), and the *goettingiana* group (*cyperi*, *mothi*, *graminis*, *delvii* and *raskii*);

- all these species with the exception of *H. mothi* and *H. cajani* show a clear host preference for graminaceous plants; several of them have been described from, or experimentally raised on, rice but it must be kept in mind that rice is the most cultivated cereal in the tropics, and consequently the most frequently sampled;

- these species, with a few exceptions, have been reported only in West Africa and in India but this could not be considered conclusive in relation to geographical distribution because there are no negative reports from other tropical parts of the world;

- currently, only the genus *Afenestrata* can be considered as strictly tropical;

- with the limited information on the biology of these species, it can be stated that the optimal temperature for hatching is rather high, and there are several generations per year, this being correlated in several species with the presence of a voluminous egg-mass attached to the cyst;

- as far as present research observations permit, it may be assumed that only *H. cajani*, *H. oryzicola*, *H. zeae* and to some extent *H. sacchari* are dangerous species;

- cyst forming Heteroderinae appear less important in tropical climates than in temperate areas but research on the group should not be neglected because of their presence on various important crops, particularly cereals.

REFERENCES

Aboul-Eid, H.Z. and Ghorab, A.I., 1974, Pathological effects of *Heterodera cajani* on cow pea, Pl. Dis. Reptr., 58:1130.

Babatola, J.O., 1983a, Pathogenicity of *Heterodera sacchari* on rice, Nematol. medit., 11:21.

Babatola, J.O., 1983b, Rice cultivars and *Heterodera sacchari*, Nematol. medit., 11:103.

Baldwin, J.G. and Bell, A.H., 1985, *Cactodera eremica* n.sp., *Afenestrata africana* (Luc et al., 1974) n.gen., n.comb., and an emended diagnosis of *Sarisodera* Wouts and Sher , 1971 (Heteroderidae), J. Nematol., 17:187.

Basnet, C.P. and Jayaprakash, A., 1984, *Heterodera raskii* n.sp. (Heteroderidae: Tylenchina), a cyst nematode on grass from Hyderabad, Indian J. Nematol., 16:213.

Berdon-Brizuela, R. and Merny, G., 1984, Biologie d'*Heterodera oryzae* Luc & Berdon-Brizuela, 1961. I. Cycle du parasite et reactions histologiques de l'hôte, Revue Pathol. veg. Entom. agric. Fr., 43:43.

Bhatti, D.S. and Gupta, D.C., 1973, Guar, an additional host of *Heterodera cajani*, Indian J. Nematol., 3:160.

Birchfield, W., 1973, Pathogenesis and host parasite relations of the cyst nematode *Heterodera graminophila* on grasses, Phytopathology, 63:38.

Bridge, J. and Manser, P.D., 1980, The beet cyst nematode in tropical Africa, Pl. Dis., 64:1036.

Cadet, P., 1977, Obtention de mutants morphologiques chez le nématode *Heterodera oryyzae* parasite du riz, Cah. ORSTOM Sér. Biol., 11 (1976): 143.

Cadet P. and Merny, G., 1977, Methode d'étude de la pénétration et du développement des deux sexes chez les *Heterodera* (Nematoda: Tylenchida), Cah. ORSTOM Sér. Biol., 11 (1976): 151.

Cadet P. and Merny, G., 1978a, Influence of some factors on sex-ratio in *Heterodera oryzae* and *H. sacchari* (Nematode: Heteroderidae), Revue Nématol., 1:143.

Cadet, P. and Merny, G., 1978b, Premiers essais de traitements chimiques contre les nématodes parasites de la canne a sucre en Haute Volta, Revue Nématol., 1:53.

Cadet P., Reversat, G. and Merny, G., 1975, Facteurs affectant le déterminisme du sexe chez *Heterodera oryzae* (Nematode: Tylenchoidea), Cah. ORSTOM Sér. Biol., 10:207.

Chitwood, B.G., 1949, Root-knot nematodes. Part 1. A revision of the *Meloidogyne* Goeldi, 1887, Proc. helminth. Soc. Wash., 16:90.

Dhawan, S.C., Kaushal, K.K. and Srivastava, A.N., 1983, Record of occurrence of *Heterodera sorghi* from Haryana State, India, Indian J. Nematol., 13:241.

Edward, J.C. and Misra, S.L., 1968, *Heterodera vigni* n.sp. and second stage larvae of *Heterodera* species in Uttar Pradesh, India, Allahabad Fmr., 42:155.

Eroshenko, A.S. and Kazachenko, I.P., 1972, [Key to the species of *Heterodera* Schmidt, 1871], Biol.-Pochvenn. Inst., Dal'nevost. Nauchnyi Tsentr AN SSSR, 11:205.

Fortuner, R. and Merny, G., 1973, Les nematodes parasites des racines associés au riz en Basse-Casamance (Senegal) et en Gambie, Cah. ORSTOM Sér. Biol., 21:3.

Garabedian, S. and Hague, N.G.M., 1982, comparison of methods of application of oxamyl for control of *Heterodera sacchari* on sugar cane plants, Nematropica, 12:39.

Garabedian S. and Hague, N.G.M., 1984, The affect of weekly exposure to non-volatile nematicides and sugar cane root diffusate on the hatching of *Heterodera sacchari*, Revue Nematol., 7:95.

Gaur, H.S. and Singh, I., 1977, Pigeon-pea cyst nematode, *Heterodera cajani*, associated with the mung crop in the Punjab State, J. Res. Punjab agric. Univ., 14:509.

Gill, J.S. and Swarup, G., 1971, On the host range of the cereal cyst nematode, *Heterodera avenae* Woll. 1924, the causal organism of "Molya" disease of wheat and barley in Rajasthan, Indian J. Nematol., 1:63.

Golden, A.M. and Birchfield, W., 1978, *Heterodera graminophila* n.sp., (Nematoda: Heteroderidae) from grass with a key to closely related species, J. Nematol., 4:147.

Golden, A.M. and Mulvey, R.H., 1983, Redescription of *Heterodera zeae*, the corn cyst nematode, with SEM observations, J. Nematol., 15:60.

Golden, A.M. and Raski, D.J., 1972, *Heterodera thornei*, n.sp., (Nematoda: Heteroderidae) and a review of related species, J. Nematol., 9:93.

Golden, A.M., Rau, G.J. and Cobb, G.S,. 1962, *Heterodera cyperi* Heteroderidae) a new species of cyst-forming nematode, Proc. helminth. Soc. Wash., 29:168.

Goodey, T., 1956, The nematodes parasites of plants catalogued under their hosts. Rev. J.B. Goodey and M.T. Franklin. Farnham Royal, England, Commonw. agric. Bur., 140pp.

Gupta, P. and Edward, J.C., 1973, A new record of a cyst forming nematode (*Heterodera chaubattia* n.sp.) from the hills of Uttar Pradesh, Curr. Sci., 42:618.

Hashioka, Y., 1964, Nematode diseases of rice in the world, Riso, 13:139.

Henderson, G., 1926, Remarks on the genera of spear-bearing nematodes found in Hawaii, with a table for their identification, Hawai. Plant. Rec., 30:454.

Hesling, J.J., 1978, Cyst nematodes: morphology and identification of *Heterodera*, *Globodera* and *Punctodera*, in "Plant Nematology", J.F. Southey ed., HMSO.

Hutzell, P.A., 1984, Description of males of *Heterodera zeae*, J. Nematol., 16:83.

Jain, K., Sethi, C.L., Swarup, G. and Srivastava, A.N., 1982, *Heterodera sorghi* sp.n., a new cyst forming nematode parasitizing sorghum, Revue Nematol., 5:201.

Jairajpuri, M.S., Khan, W.U., Setty, K.G.H. and Govindu, H.C., 1979, *Heterodera delvii* n.sp. (Nematoda: Heteroderidae). a parasite of ragi (*Eleusine coracana*) in Bangalore, India, Revue Nematol., 2:3.

Janarthanan, R., 1972, Occurrence of the pigeon-pea cyst nematode in Tamil Nadu, Indian J. Nematol., 2:215.

Jerath, M.L., 1968, *Heterodera sacchari*, a cyst nematode pest of sugar cane new to Nigeria, Pl. Dis. Reptr., 52:237.

Kalha, C.S. and Edward, J.C., 1979, *Heterodera vigni* Edward & Misra, 1968, a synonym of *H. cajani* Koshy, 1967, Allahabad Fmr., 50:143.

Khan, A.D. and Husain, S.I., 1965, *Heterodera mothi* n.sp. (Tylenchida: Heteroderidae) parasitizing *Cyperus rotundus* L. in Aligarh, U.P., India, Nematologica, 11:167.

Khan, U.V. and Jairajpuri, M.S., 1975, A report on the occurrence of an egg sac in *Heterodera mothi* (Nematoda: Tylenchida), Curr. Sci., 44:590.

Kirby, M.F., Kirby, M.E., Siddiqi, M.R. and Loff, P.A.A., 1978, Plant parasitic nematodes from Fiji: nematode distributions and host associations. UNDP/FAO Survey agric. Pests & Dis. S. Paficic Reg. - Coord. Comittee Meet., 28 Feb.-2 March, 1978 Rep. mimeogr. 12 pp.

Koshy, P.K., 1967, A new species of *Heterodera* from India, Indian Phytopath 20:272.

Koshy, P.K. and Swarup, G., 1971a, Investigations on the life history of the pigeon-pea cyst nematode, *Heterodera cajani*, Indian J. Nematol., 1:44.

Koshy, P.K. and Swarup, G., 1971b, On the number of generations of *Heterodera cajani* the pigeon-pea cyst nematode in a year, Indian J. Nematol., 1:88.

Koshy, P.K. and Swarup, G., 1971c, Distribution of *Heterodera avenae, H. cajani, H. zeae* and *Anguina tritici* in India, Indian J. Nematol., 1:106.

Kishy, P.K. and Swarup, G., 1971d, Factors affecting emergence of larvae from cysts of *Heterodera cajani* Koshy, 1967, Indian J. Nematol., 1:209.

Koshy, P.K. and Swarup, G., 1971e, Susceptibility of aerial parts of pigeon-pea seedlings to *Heterodera cajani* larvae, Indian J. Nematol, 1:245.

Koshy, P.K. and Swarup, G., 1972, Susceptibility of plants to pigeon-pea cyst nematode, Indian J. Nematol., 2:1.

Koshy, P.K., Swarup, G. and Sethi, C.I., 1971, Further notes on the pigeon-pea cyst nematode *Heterodera cajani*, Nematologica, 16:477.

Krall, E.L. and Krall, Kh. A., 1979, [The revision of plant nematodes of the family Heteroderidae (Nematoda, tylenchida) by using a comparative ecological method of studying their phylogeny], in "Printsipy i metodyi zucheniya uzaimootnoshenii mezhdu paraziticheskimi nematodami irasteniyami", Tartu, USSR (1978, publ. 1979).

Krishna Prasad, K.S., Krishnappa, K., Setty, K.G.H., Reddy, B.M.R. and Reddy, H.R.C., 1980, Susceptibility of some cereals to ragi cyst nematode *Heterodera delvii*, Curr. Res., 9:114.

Kumar, A.C., 1964, A note on the occurrence of *Heterodera cacti* from Mysore, Curr. Sci., 33:534.

Kumar, A.C., 1980, Studies on nematodes in coffee soils of south India. 4. Occurrence of *Heterodera cyperi*, J. Coffee Res., 10:77.

Logiswaran, G. and Menon, P.P.V., 1969, Preliminary observations on the spread of the golden nematode (*Heterodera rostochiensis* Woll.) of potatoes in the Niligiris, Madras State, All India Nematol Symp. New Delhi:7.

Luc, M., 1961, Nématodes du genre *Heterodera* parasites de cultures tropicales en Afrique, C.r. Seanc. Acad. Agric. Fr., 47:940.

Luc, M., 1974, *Heterodera sacchari*, C.I.H. Descript. Pl. paras. Nematodes, Set 4, No. 48.

Luc, M. and Berdon-Brizuela, R., 1961, *Heterodera oryzae* n.sp. (Nematoda: Tylenchoidae) parasite du riz en Côte d'Ivoire, Nematologica, 6:272.

Luc, M. and Merny, G., 1963, *Heterodera sacchari* n.sp. (Nematoda: Tylenchoidea) parasite de la canne a sucre au Congo-Brazzaville, Nematologica, 9:31.

Luc, M. and Netscher, C., 1974, Presence of the sugar-beet nematode at Dakar, FAO Pl. Protect. Bull., 22:24.

Luc M. and Taylor, D.P., 1977, *Heterodera oryzae*, C.I.H. Descript. Pl. paras. Nematodes, Set 7, No. 91.

Luc M., Germani, G. and Netscher, C., 1973, Description of *Sarisodera africana* n.sp. and observations on the relationships between the genera *Sarisodera* Wouts Sher, 1971 and *Heterodera* A. Schmidt 1871 (Nematoda: Tylenchida), Cah. ORSTOM. Sér. Biol., 21:35.

Luc, M., Taylor, D.P. and Cadet, P., 1978, Description of a new tropical Heteroderidae, *Hylonema ivorense* n.gen., n.sp., and a new outlook on the family Heteroderidae (Nematoda: Tylenchida), Revue Nématol., 1:73.

Maqbool, M.A., 1981, Occurrence of root-knot and cyst nematodes in Pakistan, Nematol. medit., 9:211.

Maqbool, M.A., Hashmi, S. and Begum, Z., 1983, Onion a new host of *Heterodera mothi* from Pakistan, Nematol. medit., 11:213.

Merny, G., 1966, Biologie d'*Heterodera oryzae* Luc & Bredon, 1961. II. Rôle des masses d'oeufs dans la dynamique des populations et la conservation de l'espèce, Annls Epiphyt., 17:445.

Merny, G., 1970, Les nématodes des rizières inondées de Côte d'Ivoire. I. Les especes observées, Cah. ORSTOM Sér. Biol., 11:3.

Merny, G., 1972a, Les nématodes phytoparasites des rizières inondées de Côte d'Ivoire. III. Etudes sur la dynamique des populations de deux endoparasites: *Hirschmanniella spinicaudata* et *Heterodera oryzae*, Cah. ORSTOM Sér. Biol., 16:31.

Merny, G., 1972b, Les nématodes phytoparasites des rizières inondées de Côte d'Ivoire. IV. Essai d'interprétation mathematique de l'intensité de l'infestation des racines par les nématodes endoparasites en fonction de l'inoculum, Cah. ORSTOM Sér.Biol., 16:129.

Merny, G. and Cadet, P., 1978, Penetration of juveniles and development of adults of *Heterodera oryzae* on different plants, Revue Nématol., 1:251.

Merny, G. and Netscher, C., 1976, *Heterodera gambiensis* n.sp. (Nematoda: Tylenchida) parasite du mil et du sorgho en Gambie, Cah. ORSTOM Sér. Biol., 11:209.

Minton, N.A., Tucker, E.T. and Golden, A.M., 1973, Occurrence of *Heterodera mothi*, a cyst nematode in the United States, Pl. Dis. Reptr., 57:946.

Misra, S.M., Edward, J.C. and Gupta, P., 1977, Studies on the occurrence of white/brown cysts of *Heterodera zeae* in a year in pot culture and on emergence of larvae from these cysts, Allahabad Fmr., 48:115.

Muir, F. and Henderson, G., 1926, Nematodes in connection with sugar cane root rot in the Hawaiian Islands, Hawaii Plant. Rec., 30:233.

Muir, F. and Swezey, O.H., 1926, Entomology: Nematodes, Rep. Hawaii Sug. Exp. Stn., 1926:25.

Mulvey, R.H., 1972, Identification of *Heterodera* cysts by terminal and cone top structures, Can. J. Zool., 50:1277.

Mulvey, R.H. and Golden, A.M., 1983, An illustrated key to the cyst-forming genera and species of Heteroderidae in the Western Hemisphere with species morphometrics and distribution, J. Nematol., 15:1.

Narayanaswamy, B.C., Gowda, D.N. and Setty, K.G.H., 1982, *Heterodera gambiensis* (Nematoda: Heteroderidae), a new cyst forming nematode on fingermillet from Karnataka, India, J. Soil Boil. Ecol., 2:90.

Netscher, C., 1969, L'ovogenèse et la reproduction chez *Heterodera oryzae* et *H. sacchari* (Nematoda: Heteroderidae), Nematologica, 15:10.

Netscher, C. and Pernes, J., 1971, Etude concernant l'influence de la constitution génétique sur la longueur des larves d'*Heterodera oryzae*, Nematologica, 17:336.

Netscher, C., Luc, M. and Merny, G., 1964, Description du mâle d'*Heterodera sacchari* Luc & Merny, 1963, Nematologica, 15:156.

Nishizawa, T., Shimizu, K. and Nagashima, T., 1972, [Chemical and cultural control of the rice cyst nematode, *Heterodera oryzae* Luc & Berdon-Brizuela, and hatching response of the larvae to some root extracts], Jap. J. Nematol., 2:27.

Odihirin, R.A., 1975, Occurrence of *Heterodera* cysts nematode (Nematoda: Heteroderidae) on wild grasses in Southern Nigeria, Occ. Public. Nigerian Soc. Pl. Protect., 1:24.

Odihirin, R.A., 1977, Irrigation water as a means for the dissemination of the sugar cane cyst nematode *Heterodera sacchari* at Bacita Sugar Estate, Occ. Publ. Nigerian Soc. Pl. Protect., 2:58.

Ohshima, Y., 1974, *Heterodera elachista* n.sp., an upland rice cyst nematode from Japan, Jap. J. Nematol., 4:51.

Prasad, S.K. and Chawla, M.L., 1965, The potato root eelworm *Heterodera rostochiensis*, on the Indian Agricultural Research Institute, Indian J. Entomol., 23:127.

Rao, Y.S. and Hayaprakash, A., 1977, Leaf chlorosis due to infestation by a new cyst nematode, Int. Rice Res. Newsl., 2:5.

Rao, Y.S. and Jayaprakash, A., 1978, *Heterodera oryzicola* n.sp. (Nematoda: Heteroderidae a cyst nematode on rice (*Oryza sativa* L.) from Kerala State, India, Nematologica, 23:341.

Reis, L.G.L., 1982, Plant parasitic nematodes associated with Citrus in Mozambique, Agronomia lusit., 41:213.

Reversat, G., 1975a, Elevage monoxenique du nematode *Heterodera oryzae* sur le riz, Ann. Sool. Ecol. anim., 7:81.

Reversat, G., 1975b, Etude préliminaire de la survie en anaérobiose des juvéniles du nématode *Heterodera oryzae* (Tylenchida Heteroderidae), C.R. Acad. Sc. Paris, 280 sér. D:2865.

Reversat, G., 1975c, Méthodes pour la mesure de la consommation d'oxygène des nématodes par la technique du ludion, Cah. ORSTOM Sér. Biol., 10:169.

Reversat, G., 1975d, Influence d'une inhibition préalable par la pression osmotique sur l'éclosion des masses d'oeufs du nématode *Heterodera oryzae*, Cah. ORSTOM Ser. Biol., 10:189.

Reversat, G., 1977a, Etude de la composition biochimique globale des juvéniles des nematodes *Meloidogyne javanica* et *Heterodera oryzae*, Cah. ORSTOM Sér. Biol., 11 (1976) : 225.

Reversat, G., 1977b, Influence of external factors on the rate of oxygen uptake by second-stage juveniles of *Heterodera oryzae*, Nematologica, 23:369.

Reversat, G., 1981, Potassium permanganate as a hatching agent for *Heterodera sacchari*, Revue Nematol., 4:174.

Reversat, G. and Bois, J.F., 1982, Marquage par le 35_S d'*Heterodera oryzae* et d'*Heterodera sacchari* et application a l'étude de la compétition pour l'infestation de la plante-hôte, Bull. OEPP., 12:371.

Reversat, G. and Merny, G., 1973, Influence de quelques facteurs sur la pénétration du nematode *Heterodera oryzae* dans les racines du riz, Cah. ORSTOM Ser. Biol., 21:111.

Sardanelli, S., Krusberg, L.R. and Golden, A.M., 1981, Corn cyst nematode, *Heterodera zeae*, in the United States, Pl. Dis., 65:622.

Seshadri, A.R. and Sivakumar, C.V., 1962, The golden nematode of potatoes (*Heterodera rostochiensis* Woll., 1923), - a threat to potato cultivation in the Madras State, Madras. agric. J., 49:281.

Sharma, S.B. and Swarup, G., 1983, Identification scheme for *Heterodera* species in India, Indian J. Nematol., 13:171.

Sharma, S.B. and Swarup, G., 1984, Cyst forming nematodes of India, New Delhi, India Cosmo Publication.

Shimizu, K., 1971, [Influence of the rice cyst nematode on the field of paddy rice], 15th Ann. Meet. Jap. Soc. appl. Entom. Zool., Fûchu, Tokyo, Apr. 7-9 1971, 32.

Shimizu, K., 1972, [Relationship between growth and yield of paddy rice and the time inoculation with the rice cyst nematode], 16th Ann. Meet. Jap. Soc. appl. Entom. Zool., Shizuoka, Apr. 6-8 1972, 114.

Shimizu, K., 1973, [Influence of different sowing times of rice on seasonal fluctuations of the rice cyst nematode], 17th Ann. Meet. Jap. Soc. appl. Entom. Zool., Nagano, Apr.3-5 1973, 130.

Skarbilovich, T.S. and Kurramov, Sh. Kh., 1973, [Nematode infections in sugar cane and soy bean], Byull. Vsesoyuzn. Inst. gel'mint. K.I. Skryabina, 11:80.

Srivastava, A.N. and Swarup, G., 1975, Preliminary studies on some graminaceous plants for their susceptibility to the maize cyst nematode, *H. zeae* Koshy et al., Indian J. Nematol., 5:257.

Stoyanov, D., 1972, *Heterodera amaranthi* n.sp. (Tylenchida: Heteroderidae); un nemátodo formador de quistes en Cuba, Poeyana, 97:1.

Stynes, B.A., 1971, *Heterodera graminis* n.sp. a cyst nematode from grass in Australia, Nematologica, 17:213.

Swarup, G., Prasad, S.K. and Raski, D.J., 1964, Some *Heterodera* species from India, Pl. Dis. Reptr., 48:236.

Talatschian, P. and Achyani, A., 1976, *Heterodera*-Arten in Zückeruben - Feldern verschiedener Anbaugebiete des Iran, Z. angew. Zool., 63:497.

Taylor, D.P., 1978, Parasitism of banana by *Heterodera oryzae*, Revue Nematol., 1:165.

Taylor, D.P. and Luc, M., 1979, Observations on *Sarisodera africana* (Nematoda: Heteroderidae): Redescription of anterior end of females and occurrence of juveniles with aberrant tails, Revue Nématol., 2:111.

Tiago, J.S., 1956, Un grave inimigo da cultura de batata, Gazetta Agric. Moçamb, 8:334.

Usha, K. and Kurien, K.J., 1981, Cyst nematode, *Heterodera oryzicola*, on rice in Kerala. I. Estimation of loss in rice due to *H. oryzicola* infestation in field conditions, Indian J. Nematol., 11:106 [Abstr.].

Verma, A.C. and Yadav, B.s., 1975, Occurrence of *Heterodera cajani* in Rajasthan and susceptibility to certain sesame varieties, Indian J. Nematol., 5:235.

Wouts, W.M. and Sher, S.A., 1971, The genera of the subfamily Heteroderinae (Nematoda: Tylenchoidea) with a description of two new genera, J. Nematol., 3:129.

Yaqoob, M. and Khan, A.M., 1969, Seasonal fluctuation in the population of *Heterodera mothi*, All India Nematol. Symp. 1969, 20.

DISCUSSION

Noel stated that, according to data of Dr W. Hadisoeganda, *H. glycines* is present in Indonesia as are other species of the genus and he asked in what way *H. glycines* is related to *H. cajani*. Golden responded that although the two species are closely related they are distinct, largely on the basis of structure of the cyst wall. Stone stated that populations of *H. oryzicola* he is studying with Dr A. K. Koshy are very close, if not identical, to *H. oryzae;* they attack both rice and banana-trees, as does *H. oryzae*. Stone also stated that *H. chaubattia* is probably a junior synonym of *Globodera mali*.

B'Chir reported that cyst nematodes are not economically important in Tunisia, as compared with *Meloidogyne* or *Pratylenchus*, reflecting the general situation in the tropics, but *Globodera* species can be a problem when potatoes are grown continuously. This led to a discussion about the effect of temperature on the behaviour of potato cyst nematodes from which it was concluded that, based on observations in various countries e.g. New Zealand, Tunisia, Pakistan, they become inactive when soil temperatures exceed 30°C but they may nevertheless constitute a problem in cooler seasons.

Edongali stated that the *H. trifolii* reported in Libya does not attack subterranean clover and in response to a question by Luc he said that *H. latipons* has not yet been recorded in Libya, whereas it is present in Tunisia.

ECONOMIC IMPORTANCE OF CYST NEMATODES IN NORTH AMERICA

Lawrence I. Miller

Department of Plant Pathology, Physiology and Weed Science
Virginia Polytechnic Institute and State University
Blackburg, VA 24061, USA

INTRODUCTION

Globodera rostochiensis (Woll.) Behr., *G. pallida* (Stone), *G. tabacum tabacum* (Lownsbery & Lownsbery), *G. t. solanacearum* (Miller & Gray),, *Heterodera glycines* Ichinohe, *H. schachtii* Schmidt, *H. cruciferae* Franklin, *H. trifolii* Goffart and *H. avenae* Woll. are the most important cyst nematodes of commercial crops in North America (USA and Canada), or have the potential to be important pests if distribution were not restricted. *H. zeae* Koshy, Swarup & Sethi parasitises corn (*Zea mays*) on a few farms in Maryland but its potential economic importance as a pathogen in North America has not yet been determined. *Cactodera cacti* (Filipjev & Schuurmans Stekhoven) on cactus, *Punctodera punctata* (Thorne) Mulvey & Stone on wheat, *H. lespedeza* Golden and G.S. Cobb on lespedeza, *H. leucilyma* Di Edwardo & Perry on St. Augustine grass, *H. fici* Kirjanova on fig, *H. humuli* Filipjev on hop and *H. carotae*Jones on carrot can be important pests in particular situations but they are not pests of general economic importance in North America.

POTATO CYST NEMATODES

Two potato cyst nematodes, *Globodera rostochiensis* (PCN-G) and *G. pallida* (PCN-P) are pests of potato in North America. PCN-G occurs in New York in the USA, on Vancouver Island in western Canada, and in gardens on the island of Newfoundland in Canada, which are also infested with *G. pallida* (Anon, 1984a). PCN-G was first discovered in the USA as a pathogen of potato in Nassau County, Long Island, New York in 1941 (Mai, 1977). According to D. G. Gunther (personal communication) PCN-G has been recovered from 5,115 ha on Long Island and after its discovery in Steuben County, it has been found on 27 farms (412 ha) in western New York. PCN-G is a pathogen of tobacco in Venezuela (Meredith-C, 1976) but it is not a pest of this crop in North America.

The PCN-G is undoubtedly one of the most important pests threatening the potato industry of the USA and one of the most difficult to control. In addition to suppressing potato yields, the potato cyst nematodes cause indirect losses by interacting with the fungus *Verticillium dahliae* which can result in yield losses even higher than caused by the nematodes alone (Evans and Brodie, 1980).

As described by Mai (1977), the PCN-G control programme in the USA represents one of the most extensive and expensive pest management programmes ever attempted, since the principle objective is to keep nematode populations below the discovery level and not just to maintain the populations below the damaging level. According to W. F. Brown (personal communication), the use of highly resistant potato cultivars is the principle control mechanism supported by surveys and regulatory activities. The use of nematicides for control of PCN-G is currently not permitted on Long Island and they are only prescribed and used on infested farms under special conditions in western New York. Based on information in a report prepared by the New York Department of Agriculture and Markets in 1983 (W. F. Brown, personal communication), about $1 million per year is spent by all agencies of the USDA in New York State involved in survey, control, regulatory practices and research on the PCN-G control programme. In New York, 17,000 ha of potatoes were harvested in 1983 at a production value of $76.3 million (Anon, 1985). Potatoes are grown commercially in all the contiguous United States and Alaska and in 1982 the crop area was 529,000 ha with a production value of $1.6 billion (Anon, 1983).

In 1965, PCN-G was discovered parasitising potato in the Sannich Peninsula, Vancouver Island, British Columbia (Orchard, 1965). Soon afterwards all farms in British Columbia on which potatoes were grown commercially were assayed for PCN-G (R. F. Froese, personal communication). Only fourteen fields in the vicinity of the initial infestation were infested. The area was quarantined but the mandatory production restrictions and nematicide treatments designed to control and/or eradicate the nematode were ineffective and by 1981 there were 40 infested fields on Sannich Peninsula. So, in 1982 a ban was placed on commercial production of potatoes, tomatoes, and eggplants in the quarantined area. Seven growers who annually planted 200 ha of potatoes were affected by the ban. During the period 1965-1975 about twelve man-years were utilised annually in soil surveys and assaying for the presence of PCN-G. Since 1976 this activity has been reduced to about two man-years annually and about one-third of a man-year is devoted to extension work in the quarantined area.

G. Greenslade (personal communication) reports that PCN-G was first detected as a pest of potatoes in Newfoundland in 1962 and the quarantine in force for the potato wart disease in 1909 was extended to include this nematode and PCN-P when it was found in 1977 (Anon, 1984a). The quarantine for the cyst nematodes was designed to prevent their spread from Newfoundland to other parts of Canada and included surveys and assays for the nematodes, inspection, regulatory prohibition, restrictions, control treatments and enforcement at an estimated annual cost of $(Can.) 0.9 million. According to K. G. Proudfoot (personal communication), an average of $(Can.) 0.8 million has been spent annually since 1962 in Newfoundland on research on potato cyst nematodes related to the biology and control by chemicals and development of resistant potato cultivars. None of the 400 ha commercially planted to potatoes in Newfoundland are infested with either species of potato cyst nematodes, infestations being confined to home gardens in the eastern and northern coastal areas of the island portion of the Province (Morris, 1971). Nevertheless, their presence restricts the development of export markets for agricultural products from Newfoundland. Potato cyst nematodes do not infest nearby Maritime provinces where the value of table, processing and seed potatoes was worth $(Can.) 214 million in 1984 (Anon, 1984a). If the potato cyst nematodes were to infest the Maritimes, it would severely deflate the sales of all except processing potatoes. In 1983, 113,000 ha were planted to potatoes in Canada worth $(Can.) 400.2 million (Anon, 1983b).

TOBACCO CYST NEMATODES

The Virginia tobacco cyst nematode (VTCN), *Globodera tabacum solanacearum* was first found in Amelia County, Virginia, in 1961 (Miller and Gray, 1972). In 1983, 148 farms (1,230 ha) in an area extending over 10 counties were infested causing an estimated average reduction in yield of tobacco of 15 per cent or $0.7 million (Komm et al., 1983). VTCN populations can be kept at low levels by long rotations with non-solanaceous crops, chemicals, and/or resistant cultivars. VTCN-resistant tobacco cultivars are available but they are highly intolerant to injury at high VTCN population levels (Komm et al., 1983). Tobacco is the most important cash crop grown in Virginia and in 1982 there were 24,900 ha harvested at a production value of $220.7 million (Anon, 1983a).

The Connecticut tobacco cyst nematode (CTCN), *G. tabacum tabacum*, was first found parasitising tobacco on a farm in Hartford County, Connecticut in 1951 (Lownsbery and Lownsbery, 1954). According to G. S. Taylor (personal communication), most of the farmland in which shade type 61 tobacco is grown, and about half the area planted to broad-leaf type 51, in the Connecticut River Valley in Connecticut and Massachusetts is infested with CTCN. High populations of CTCN cause severe losses in yield and quality of tobacco in this area and most of the infested land is treated with a nematicide. There are no CTCN-resistant shade or broadleaf tobacco cultivars. It has been estimated that even with the use of nematicides, CTCN infestation results in a 0.4 per cent annual loss at a value of $50,000 (G. S. Taylor, personal communication). The cost of chemical control treatments are estimated to be about $60,000 a year. In 1982, tobacco was harvested from 1,081 ha (76 farms) in Connecticut at a value of $14.2 million (Anon, 1982) and from 142 ha (46 farms) in Massachusetts at a value of $2.9 million (Anon, 1983a).

In the USA, VTCN and CTCN have been found only in Virginia and the Connecticut River Valley. Their potential importance is indicated by the fact that in 1982 there were 368,000 ha of tobacco harvested in the USA at a production value of $3.5 billion (Anon, 1983a). In Canada, tobacco was harvested from about 40,500 ha in 1983 at a value of $(Can.) 400.2 million (P. W. Johnson, personal communication).

SOYBEAN CYST NEMATODE

Brewer (1981) and Riggs (1977) reviewed the economic importance of the soybean cyst nematode (SCN), *Heterodera glycines*, to soybean production in the USA from its first discovery in 1954 at Castle Hayne, North Carolina up to 1976. The total economic consequences of crop losses due to SCN cannot be assessed accurately, nor is it accurately known how many dollars have been spent on taxonomy, quarantine, survey and regulatory services in addition to research aimed at devising chemical, host plant resistance, cultural and biological control measures.

Sixty-five counties in eight soybean-producing states were infested with SCN in 1965 (Anon, 1965); 229 counties in 15 states in 1976 (Anon, 1976); 426 counties in 22 states in 1980 (Brewer, 1981; Moore, 1984); and 560 counties in 24 states by January, 1985 (Fig. 1). The production loss for 16 southern states in 1984 was estimated to be 611,000 tonnes (2.48 per cent average loss/state) worth $131 million (Mulrooney, 1984). Loss estimates in 1984 from northern states infested with SCN were 547,000 tonnes worth $117 million (personal communications from several sources). No infestations of SCN have been reported in Canada nor the northern USA

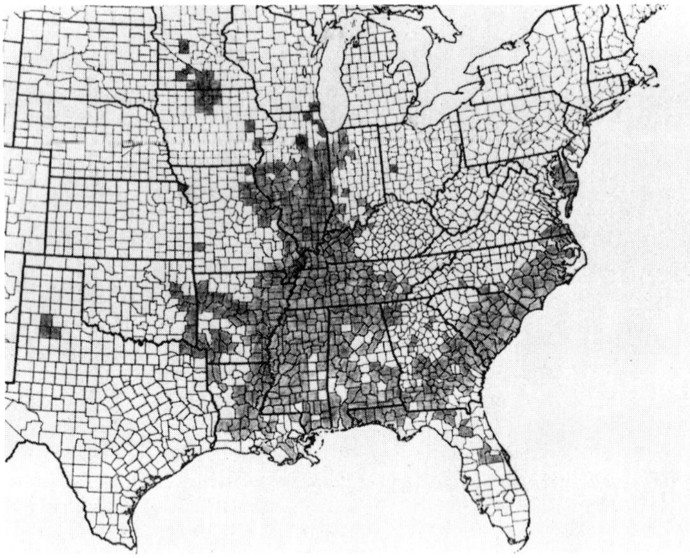

Fig. 1 Counties (shaded) in the USA infested with *H. glycines* in which soybeans are grown, January 1985.

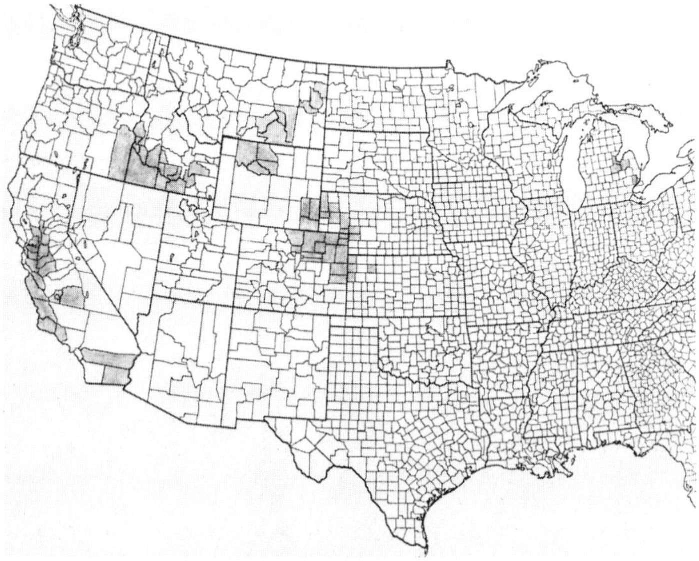

Fig. 2 Counties (shaded) in the USA infested with *H. schachtii* in which sugarbeet is grown, January, 1985.

soybean-producing states of Nebraska, North Dakota, Pennsylvania and South Dakota.

The dollar loss estimates associated with damage by SCN of $352 million in 1979 (Moore, 1984) is considerably higher than the total loss of southern and northern states of $248 million for 1984, even though more land was infested and soybean prices were higher. The degree of accuracy of the loss estimates is not known (Noel, 1982), but the trend towards the lower loss estimation is encouraging. It could indicate that growers are now more careful in management of the SCN by using longer rotations of non-host crops, judicious use of nematicides and selection of the soybean cultivars to be planted, dependent on nematode populations and the SCN race present (Wrather et al., 1984; Edwards and Noel, 1985). The occasional use of a high yielding susceptible cultivar may be advisable when the nematode populations are low in order to reduce the selection pressure on the SCN to form or select new races for which no resistance may be available.

Bradley and Duffy (1982) estimated that yield loss prevented by the use of SCN race 3 resistant hybrid Forrest soybean in the period 1975-1980, in a seven-state region of the USA, accounted for $405 million (1980 dollar value). Even at an estimated cost of $1 million for the development of Forrest, the results justify continuing research to develop new SCN-resistant cultivars. Since no soybean cultivar or breeding line is known to be resistant to all isolates of SCN, resistance probably should be regarded as a depletable reserve.

Soybean is one of the most important crops in the USA and it ranks third in the area planted. There were 26.8 million ha harvested in 1983 in the USA yielding 51 million tonnes with a value of $12.7 billion (Anon, 1983a). Soybean ranked tenth in 1983 in farm value in Canada (Anon, 1983b) with 364,000 ha harvested yielding 848,000 tonnes worth $(Can.) 212.1 million.

Several varieties of red kidney bean and bush-type snapbean *Phaseolus vulgaris*) are parasitised by SCN in Illinois and may prove to be an important factor in production of these vegetables (Noel et al., 1982).

SUGARBEET AND CABBAGE CYST NEMATODES

The sugarbeet cyst nematode (SBCN), *Heterodera schachtii*, is one of the most important pest of sugarbeet in the USA and ranks as one of the most economically important crop pests in that country (Caveness, 1957; Altmanand Thomason, 1971). In 1982, there were about 428,000 ha planted to sugarbeet in 16 states of the USA with a production value of about $803.6 million, not counting the value of sugarbeet seed (Anon, 1983). Sugarbeet is the 18th most important crop in farm value in Canada with 28,700 ha harvested and a production value of $(Can.) 39.5 million (B.E. Hopper, personal communication).

In 1984, SBCN infestations were recorded in nine states in 76 counties of the USA that grow sugarbeet (Fig. 2). The numbers of counties infested in each state were: California - 19, Oregon - 1, Idaho - 14, Montana - 7, Wyoming - 7, Colorado - 12, Kansas - 4, Nebraska - 8, and Michigan - 4, (personal communications from various sources).

SBCN was first discovered in Alberta, near Taber, in 1961 where it caused considerable damage (Lilly et al., 1964). It is assumed that SBCN has infested many fields in the sugarbeet-producing area of southeast Alberta, but nematode populations are very low and do not limit production since a compulsory rotation including sugarbeets only once every four years was initiated about 1979 (P. Bergen, personal communication). A few fields planted to sugarbeets were found to be lightly infested with SBCN about 1975 in Manitoba, and this was followed by a compulsory four-year rotation for all sugarbeet growers in that province (P. Bergen, personal communication).

The SBCN in sugarbeet growing areas is controlled by cultural and chemical methods. Crop rotation with sugarbeet and other susceptible crops planted only once in four or five years minimises yield losses from this nematode (Altman and Thomason, 1971). Reducing SBCN populations by rotating sugarbeet with non-host crops, frequently of lower value, is one important component of measuring economic loss to growers (Roberts et al., 1981). Some reduction of yield loss is usually obtained when sugarbeet is planted at temperatures below that optimal for nematode activity (Cooke and Thomason, 1978). Agronomically acceptable cultivars of SBCN-resistant sugarbeet are not yet available (Altman and Thomason, 1971).

Nematicides are most important in reducing yield loss in sugarbeet production where land area is limited and long rotations are not practical (Altman and Thomason, 1971). The most commonly used nematicides for the control of SBCN in the USA are aldicarb and 1,3-dichloropropene. Cooke and Thomason (1978) estimated that, in 1978, there were 148,347 ha infested with SBCN in sugarbeet growing areas of California and that 2 per cent annual loss in root yield of sugarbeet is due to SBCN. Nematicides are frequently used to control the rootknot nematode pest of sugarbeet in California on the lighter soils, but with the heavier soils infested with very high populations of SBCN, nematicides are not very effective (L.M. Burtch, personal communication). T.K. Schwartz (personal communication) reports that in Colorado, Wyoming, Montana and Nebraska in 1985 about 30 per cent of the sugarbeet producing area was treated with nematicides for SBCN control. Y.K. Yum (personal communication) in 1978 surveyed 81,000 ha of sugarbeet grown in Colorado, Kansas, Nebraska, Wyoming and Montana and found that about 24 per cent of the area was treated with a nematicide for SBCN control. E.D. Kerr (personal communication) estimated that of the 22,680 ha planted to sugarbeet in Nebraska in 1985 about half were treated with a nematicide for SBCN control. O.J. Dickerson (personal communication) reports that SBCN was an important pest of sugarbeet in Finney County, Kansas in the 1960s when sugarbeet was grown in the southwest district, but that long rotation practices prevented severe losses.

The SBCN was first discovered as a pest of cabbage in Florida on a farm in Seminole County in 1969 and by 1975 was known to infest about 166 ha on this farm which also extended into Volusia County (Rhoades, 1971). Treatment of infested areas with nematicides, long rotations of non-host crops, transplanting of only cyst-free seedlings and the expansion of urban development on infested fields near Disneyworld have apparently helped keep the spread of SBCN in check, even although 60 per cent of the 6,000 ha of cabbage grown in Florida in 1982 are in adjacent Flagler, Putnam and St. Johns Counties. Florida growers produced about 6.4 million crates of cabbage during the 1982-83 season (October through June) which had a production value of $22.7 million (H.L. Rhoades, personal communication.

SBCN was first reported as a pest of sugarbeet in New York in 1961, but the crop is no longer grown in this state (Abawi and Mai, 1980). According to D.C. Crosier (personal communication) SBCN is widely distributed throughout the red tablebeet and cabbage producing counties of Ontario and Yates of central New York and causes significant yield losses on a few farms with high populations of SBCN, although most of the populations of infested fields are very low (Abawi and Mai, 1980). Chemical control of SBCN with fenamiphos is approved for cabbage but not for red tablebeet. A cropping sequence involving two or more consecutive years of corn, beans, wheat, oats or other non-hosts lowers the density of SBCN below the economic threshold. SBCN-resistant and/or tolerant cultivars of economically important crops of the Chenopodeaceae or Cruciferae grown in New York have not yet been discovered (Abawi and Mai, 1980). In 1984, there were 1215 ha of red tablebeet harvested in New York at a production value of $1.6 million; and 6,000 ha of cole crops (Brussel sprouts, cauliflower, broccoli, cabbage) with a production value of $28.2 million (Anon, 1985).

SBCN is an economically important pest of cole crops in San Mateo, Santa Cruz, Monterrey, San Luis Obispo and Santa Barbara counties in California (personal communications from various sources). Yields of Brussel sprouts, cabbage, broccoli and cauliflower can be reduced by 50 per cent or more if high populations of SBCN occur. Unless infested soil is rotated out of cole crops to non-host crops for two to three years it is fumigated prior to planting with 1,3-dichloropropene (Lear et al., 1966). In North America the cabbage cyst nematode (CABCN, *H. cruciferae*), is only known to occur in Yolo, San Mateo, Santa Cruz, Monterrey and Santa Barbara counties of California (Siddiqui et al., 1973) and although not as common as SBCN it is recognised as economically important in San Mateo, Santa Cruz, and Monterrey counties (Lear et al., 1965). According to G. Miyao (personal communication) CABCN is not an economically important problem on crucifers in Yolo country. It is not known whether CABCN still occurs in Santa Barbara county or is present in adjoining San Luis Obispo country, as only limited surveys have been made to detect the exact distribution of this cyst nematode, probably since yields of crucifers are reduced about equally at high infestation levels of SBCN and/or CABCN (R.H. Sciaroni, personal communication). In 1983 there were 52,320 ha of cole crops harvested in California at a production value of $325 million. Cole crops were harvested from 43,000 ha in those five counties in which SBCN and/or CABCN was a pest at a production value of $276 million (R. Borton, personal communication).

SBCN was an economically important pest of commercially grown rhubarb in the mid to late 1960s in seven counties of Ontario: Halton, Lincoln, Oxford, Peel, Simcoe, Wentworth and York (Townshend and Olthof, 1967). Little rhubarb is currently grown in Ontario but some damage still occurs to vegetables of the Chenopodiaceae and Cruciferae in Ontario if populations of SBCN reach damaging levels because of absence of rotation or short rotations with non-host crops at seven locations in Halton, Lambton, Oxford, Peel and York counties (J.W. Potter, personal communication). SBCN is not currently a major problem to Canadian agriculture as a whole. The potential importance of the pest, however, is suggested by the area and the 1983 farm value of the following crops which are hosts of SBCN: rapeseed - 1.8 million ha, $(Can.) 604.5 million; cabbage - 4,968 ha, $(Can.) 35.7 million; mustard seed - 63,500 ha, $(Can.) 20.6 million; cauliflower - 2,834 ha, $(Can.) 20.5 million; rutabaga - 3,500 ha, $(Can.) 16.0 million; red tablebeet - 1,043 ha, $(Can.) 3.1 million; spinach - 567 ha, $(Can.) 2.2 million (R.V. Anderson and K.W. Lievers, personal communication).

CLOVER CYST NEMATODE

The clover cyst nematoda (CLCN), *H. trifolii*, is probably the most common cyst in North America and it occurs in southern Canada and at least thirty of the contiguous states of the USA as a pathogen of white (*Trifolium repens*) and red clovers (*T. pratense*) and as a parasite and possible pathogen of certain other forage legumes (Norton and Isley, 1967; Norton, 1984). Isolates of CLCN vary morphologically and in their ability to reproduce and cause injury to different species and cultivars of clover (Chapman, 1964; Singh and North, 1970). The clovers are grown for hay, pasture and soil improvement, usually in mixtures with grasses and are among the most economically important crops grown in North America (Anon, 1983). But research to clearly evaluate the role that CLCN plays in longevity and health of forage legumes and the formulation of control measures is complicated because of difficulties inherent in studies with genetically variable perennial plant clones and the extremely variable populations of *H. trifolii* (Siddiqui et al., 1973; Hirschman and Triantaphyllou, 1979; Norton and Isley, 1984).

CEREAL CYST NEMATODE

Jensen et al. (1975) were the first to report the occurrence of the cereal cyst nematode (CCN), *H. avenae*, in the USA as a parasite of oats on a farm near Cornelius in Washington County, Oregon in 1974. Surveys completed in 1976 indicated that CCN infested seven farms, ca 50 ha, in close proximity to one another in the Cornelius area. The infested area in Oregon drains into the Tualatin river basin, farmland scheduled soon to receive irrigation from the Scoggin reservoir and into two plant nurseries. Should soil in the nurseries become infested with CCN, it would pose a problem in intra- and inter-state shipment of nursery stock (H.W. Jensen, personal communication). Hafez and Golden (1984) discovered CCN in two fields in a wheat and pea rotation in Whitman County, Washington in 1983 and CCN was found in two fields in a barley and potato rotation in Freemont County, Idaho in 1984 (Hafez and Golden, 1984). L.S.O. Graney and G.W. Bird (personal communication) reported that CCN was found infesting four fields planted to cereals in Tuscola County, Michigan in 1983.

The CCN is currently not an economically important pathogen of cereals in the USA, but if it spreads to uninfested sites where cereals are grown, it might become an important pest where the ability to rotate to profitable non-cereal crops for extended periods is limited. Holdeman's (1978) treatise on CCN evaluates the potential economic importance of this cyst nematode in the USA. The area and value of economically important cereals which are hosts and/or damaged by CCN in the USA in 1982 were corn - 29.6 million ha, $19.4 billion; wheat - 31.9 million ha, $9.5 billion; barley - 3.7 million ha, $1.1 billion; oats - 4.3 million ha, $896.2 million; rye - 289.6 thousand ha, $51.7 million (Anon, 1983a).

The CCN in Canada at present is known to occur only in Ontario. It is widespread in the cereal producing areas, with about 25 per cent of the barley and mixed grains (mixture of oats and barley) grain farms infested in Oxford, Waterloo, Durham, and York counties; 10 per cent infested in Dufferin, Huron, Perth, Simcoe, and Wellington; and 1 per cent or less in Kent, Niagara, Bruce, Grey and Victoria counties (J.W. Potter, personal communication). Oats, barley, wheat and rye are attacked in that order of preference, with oats the most susceptible to injury. Wheat, grown in winter, is little affected even although it is a host of CCN. Frequently losses in oats are 50 per cent or higher when grown in soils highly infested with CCN. Many grasses are good hosts of CCN but apparently suffer little injury (Martens et al., 1984). Corn is not a host of CCN in Canada but is very susceptible to injury (delayed maturity) from invasion of the nematode juveniles in roots which do not develop to maturity. Corn is severely injured (stunted) by a combination infestation of CCN and *Pratylenchus penetrans* (Martens et al., 1984). J.W. Potter (personal communication) estimated a loss of $(Can.) 1.2 million annually from damage by the CCN to oats and barley in Ontario, but he did not count losses to wheat, rye, corn or include the possible reduced total income from using the effective Ontario control recommendations (Fushtey, 1978). If the CCN infests and is adapted to other cereal growing provinces of Canada, its potential economic importance in Canada is indicated by the area and value, for 1982, of the following cereals (including corn): wheat - 12.5 million ha, $(Can.) 4.5 billion; barley - 5.1 million ha, $(Can.) 1.4 billion; corn for grain, fodder and sweet corn - 1.5 million ha, $(Can). 1.0 billion; oats - 1.6 million ha, $(Can.) 325.3 million; and mixed grain - 5.3 million ha, $(Can.) 162.8 million (R.V. Anderson and K.W. Lievers, personal communication).

MAIZE CYST NEMATODE

The maize cyst nematode (MCN), *Heterodera zeae*, was first discovered in the USA in Kent County, Maryland in 1981 (Sardanelli et al., 1981). According to L.R. Krusberg (personal communication), surveys for the MCN completed throughout 1984 in 15 counties in Maryland on 2,765 fields (ca 30,590 ha) indicated that MCN infested 27 farms (ca 1,169 ha) in Cecil, Harford, Kent and Queen Anne's counties. A quarantine was enacted by the State of Maryland in 1984 in the infested areas of these counties to control, retard, or eradicate this potentially important pest and regulate movement of infected or infested plant material or non-plant material likely to transfer the MCN from the quarantined area to a non infested area (Anon, 1984b).

Preliminary tests in Maryland by Sardinelli et al. (1981) showed that many corn cultivars and barley are hosts of MCN, but not wheat and oats. Experiments conducted for several years in Kent County, Maryland on a fertile fine silty loam infested with relatively high levels of MCN have not demonstrated that this nematode population is pathogenic and causes yield reduction of corn (L.R. Krusberg, personal communication). At present it is not known whether the high fertility level of the test site, the possible high tolerance level of the cultivars tested or the possible weak virulence of the MCN population are factors to be considered in interpreting the results. This contrasts with the situation in India where MCN is considered to be of economic importance, with corn as a good host but wheat, oats and barley as poor hosts (Koshy et al., 1970; Strivastana and Swarup, 1975).

Corn is the most important cash crop in Maryland and it is the chief provider of feed for the important broiler industry. It is the most important crop in the USA (Anon, 1983a) and the fourth most important agronomic crop in Canada in 1984 (B.E. Hooper, personal communication). The potential importance of the MCN as a pathogen of corn in North America is currently not known but, because of the economic importance of corn and the pernicious nature of other cyst-forming nematodes of other important crops, it is important that intensive studies be continued in Maryland in cooperation with the USDA and Agriculture Canada.

REFERENCES

Anon, 1965, Soybean cyst nematode, USDA, ARS - Plant Pest Control Div., Washington, D.C., 1p.

Anon, 1976, Soybean cyst nematode infested area, N. Carolina Dept. Agric., Pest Control Div., Raleigh, 1p.

Anon, 1982, Geographical Area Series - Connecticut, US Dept. Comm., Bureau of Census ACB2-A-7, Washington, D.C., 176pp.

Anon, 1983a, Agricultural Statistics, US Gonv. Printing Office, Washington, D.C., 558pp.

Anon, 1983b, Selected agricultural statistics Canada and the Provinces, Agric. Canada. Min. Supply and Services, A1-7, Ottawa, 90pp.

Anon, 1984a, Social and economic impact on spread of potato wart disease and golden cyst nematode to the Maritimes, Appendix II, Agric. Canada, Newfoundland Task force on Plant Health, St. John's, 2pp.

Anon, 1984b, In the matter of the corn cyst nematode quarantine in the infested areas of Cecil, Harford, Kent, and Queen Anne's Counties, Maryland, Quarantine Orders 84-1, Annapolis, 2pp.

Anon, 1985, New York Agricultural Statistics, 1984, N. York Crop. Rep. Serv., Albany, 84pp.

Abawi, G.S. and Mai, W.F., 1980, Effects of initial population densities of *Heterodera schachtii* on yield of cabbage and table beets in New York State, Phytopathology, 70:481.

Altman, J. and Thomason, I.J., 1971, Nematodes and their control, in "Advances in Sugar Production: Principles and Practice", R.J. Johnson, T. T. Alexander, G.R. Rush and G.R. Hawkes, eds., Iowa State Univ. Press, Ames.

Bradley, E.B. and Duffy, M., 1982, The value of plant resistance to the soybean cyst nematode: a case study of Forrest soybean, USDA-ERS Staff Rpt. AGES820929, Washington, D.C.

Brewer, F.L., 1981, Special assessment of the soybean cyst nematode *Heterodera glycines* problem, USDA-SEA Joint Planning and Evaluation Staff Paper, Washington, D.C., 40pp.

Brown, E.B., 1969, Assessment of the damage caused to potatoes by potato cyst eelworm, *Heterodera rostochiensis* Woll., Ann. appl. Biol., 63:493.

Caveness, F.E (ed.), 1957, A study of nematodes associated with sugar beet production in selected Northwest and Northcentral States, Beet Sugar Development Foundation, Fort Collins, 157pp.

Chapman, R.A., 1964, Effect of cover cyst nematode on growth of red and white clovers, Phytopathology, 54:417.

Cooke, D.A. and Thomason, I.J., 1978, The distribution of *Heterodera schachtii* in California, Plant Dis. Reptr., 62:989.

Edwards, D.I. and Noel, G.R., 1985, Economic benefits of nematological research, Nematol. Newsletter, 31:7.

Evans, K. and Brodie, B.B., 1980, The origin and distribution of the golden nematode and its potential in the USA, Amer. Potato J., 57:79.

Fushtey, S.C., 1978, Oat cyst nematode, Factsheet AGDEX 11/632, Ontario Min. Agr. Fd., Ontario.

Griffin, G.D., 1982, The quantification of sugarbeet cyst nematode soil populations and their relationship to sugarbeet yields, in "Proceedings of the 5th Cyst Nematode Workshop, D.P. Schmidt, ed., N. Carolina State Univ., Raleigh.

Hafez, S.L. and Golden, A.M., 1984, First report of oat cyst nematode in eastern Washington, Plant Disease, 69:360.

Hirschmann, H. and Triantaphyllou, A.C., 1979, Morphological comparison of members of the *Heterodera trifolii* species complex, Nematologica, 25:458.

Holderman, Q.L., 1978, The oat cyst nematode, *Heterodera avenae* Wollenweber, 1924 - a root parasite of cereal crops and other grasses. II. A fact finding study, California Dept. Fd. Agri., Sacramento, 6pp.

Jensen, H.J., Eshtiaghi, H. and Koepsell, P.A., 1975, The oat cyst nematode occurs in Oregon, Pl. Dis. Reptr., 59:1.

Komm, D.A., Reilly, J.J. and Elliott, A.P., 1983, Epidemiology of a tobacco cyst nematode (*Globodera solanacearum*) in Virginia, Plant Disease, 67:1249.

Koshy, P.K., Swarup, G. and Seth, C.I., 1970, *Heterodera zeae* n.sp. (Nematoda: Heteroderidae), a cyst forming nematode on *Zea mays*, Nematologica, 16:511.

Lear, B., Sciaroni, R.H., Johnson, D.E. and Miyogawa, S.T., 1965, Reponse of Brussels sprouts to soil fumigation for control of cabbage root nematode, *Heterodera cruciferae*, Pl. Dis. Reptr., 49:903.

Lear, B., Miyagawa, S.T., Johnson, D.E. and Atlee, C.B., Jnr., 1966, The sugarbeet cyst nematode associated with reduced yields of cauliflower and other vegetable crops, Pl. Dis. Reptr., 50:611.

Lilly, C.E., Harper, A.M. and Hawn, E.J., 1964, Discovery of the sugarbeet nematode in Western Canada, Can. Pl. Dis. Survey, 41:288.

Lownsbery, B.F. and Lownsbery, J.W., 1954, *Heterodera tabacum* new species, a parasite of solanaceous plants in Connecticut, Proc. helminthol. Soc. Wash., 21:42.

Mai, W.F., 1977, Worldwide distribution of potato cyst nematodes and their importance in crop production, J. Nematol., 9:30.

Martens, J.W., Seamna, W.L. and Atkinson, T.G., 1984, "Diseases of Field Crops in Canada", Canad. Phytopath. Soc., Harrow, 160pp.

Meredith-C, J.A., 1976, Estudio de la reproduccion de *Heterodera rostochiensis* Wollenweber, 1923, en different condiciones ambientales sobre varias solananaceas. Ph.D. thesis, Univ. Central de Venezuela.

Miller, L.I. and Gray, B.J., 1972, *Heterodera solanacearum* n.sp., a parasite of solanaceous plants, Nematologica. 18:404.

Moore, W.F. (ed.), 1984, Soybean cyst nematode, Misc. publ., Mississippi Coop. Ext. Service, 23 pp.

Morris, R.F., 1971, Distribution and biology of the golden nematode, *Heterodera rostochiensis*, in Newfoundland, Nematologica, 17:320.

Mulrooney, R.P., 1984, Southern United States soybean disease loss estimates for 1984, Proc. South Soybean Dis. Wkrs. Ann. Met., 12:16.

Noel, G.R., 1982, Relating numbers of soybean cyst nematode to crop damage, in "Proceedings of the 5th Cyst Nematode Workshop", D.P. Schmidt, ed., N. Carolina State Univ., Raleigh, 17-19.

Noel, G.R., Jacobsen, B.J. and Leeper, C.D., 1982, Soybean cyst nematode in commercial snapbeans, Plant Disease, 66:520.

Norton, D.C. (ed.), 1984, Distribution of plant parasitic nematode species in North America, Nematode Geog. Distr. Cmttee., Soc. Nematologists, Ames., 250pp.

Norton, D.C. and Isley, D., 1967, Cyst production of *Heterodera trifolii* on some leguminosae, Pl. Dis. Reptr., 51:1017.

Orchard, W.R., 1965, Occurrence of the golden nematode on Vancouver Island, British Columbia, Can. Pl. Dis. Survey, 45:89.

Rhoades, H.L., 1971, Pathogenicity and control of a Florida population of the sugarbeet cyst nematode, *Heterodera schachtii*, on cabbage, Fla. State Hort. Soc., 84:139.

Riggs, R.D., 1977, Worldwide distribution of soybean cyst nematode and its economic importance, J. Nematol., 9:34.

Roberts, P.A., Thomason, I.J. and McKinney, H.E., 1981, Influence of non-hosts, crucifers and fungal parasites on field populations of *Heterodera schachtii*, J. Nematol., 13:164.

Sardanelli, S., Krusberg, L.R. and Golden, A.M., 1981, Corn cyst nematode, *Heterodera zeae*, in the United States, Plant Disease, 65:622.

Siddiqui, I.A., Sher, S.A. and French, A.M., 1973, Distribution of plant parasitic nematodes in California, Calf. Dept. Fd. Agric, Sacramento, ?? pp.

Singh, N.D. and North, D.C., 1970, Variability in host-parasite relationships of *Heterodera trifolii*, Phytopathology, 60:1834.

Srivastana, A.N. and Swarup, G., 1975, Preliminary studies on graminaceous plants for their susceptibility to the maize cyst nematode *Heterodera zeae* Koshy et al., 1970, Indian J. Nematol., 5:257.

Townshend, J.L. and Olthof, T.H.A,, 1967, The sugarbeet nematode, *Heterodera schachtii* Schmidt, and other plant-parasitic nematodes on rhubarb in Ontario, Can. Plant Dis. Survey., 47:14.

Wrather, J.A., Anand, S.C. and Dropkin, V.H., 1984, Soybean cyst nematode control, Plant Disease, 68:829.

OCCURRENCE OF CYST NEMATODES IN PORTUGAL

Luís G. L. Reis

Departamento de Fitopatologia, Estacao Agronomica Nacional
2780 Oeiras
Portugal

D'Oliveira (1943) was the first to report the presence of a cyst nematode, *Heterodera schachtii* Schmidt, in Portugal. Later, it was reported from Arazede on sugar beet, bean and turnip (Abrantes et al., 1978) and from Alvalade and Pombal on sugar beet (Pereira, 1977). In 1956, M. Neves (personal communication) found *Globodera rostochiensis* (Woll.) Behr. in a traditional potato growing region in the Nogueira mountains, near Braganca, northern Portugal, and this was followed by further reports of its occurrence in other potato growing areas (Caupers, 1959; Louro and Raposo, 1975). In 1962, Macara (1962) reported the occurrence of *H. goettingiana* Liebscher on peas and broad beans, *H. avenae* Woll. on wheat, *H. cruciferae* Franklin on cabbage, and *G. rostochiensis* on potatoes. Macara (1963) made a detailed investigation of the morphology and biology of the Portuguese populations of *H. goettingiana* and *G. rostochiensis*. However, regular surveys of potato growing areas for the presence of potato cyst nematodes did not begin until 1974, and work on the characterisation of *Globodera* pathotypes in Portugal has only recently been initiated.

A general survey of plant parasitic nematodes associated with vegetables and other crops was started in 1978. So far about 900 soil samples have been collected throughout Portugal. Identification of species has not been completed but some information has already emerged from the survey. Cyst nematodes occurred in about 14% of the samples; *Globodera* populations occur most frequently, and as expected in the potato growing areas, and occurrence is not related to soil type as high densities occur in granitic or schistic soils in inland areas as well as in sandy soils in the coastal regions.

In 1983, *H. carotae* Jones was found for the first time in carrot crops at Valado dos Frades, near Nazaré (unpublished information). In that area, carrots had been grown continuously for more than 15 years, and in recent years crop yields and quality had noticeably declined. I have advised farmers to rotate their carrot crops with French beans, maize, cabbage and groundnuts and in 1985 I began trials to study the effect of soil solarisation on the population densities of the nematode. *H. carotae* has not been found in other carrot growing areas of Portugal.

Further records (unpublished information) of cyst nematodes include *Heterodera* sp. of the *H. trifolii* complex on curled dock (*Rumex crispus*) at Oeiras and Almeirim. Experimentally it was shown that the species infest sugar beet and *Rumux* spp. but not cabbage. Cyst nematodes, with very long juveniles, were also found on grasses at Montalegre, near Vila Verde, northern Portugal.

REFERENCES

Abrantes, I.M.O., Morais, M.M.N. and Santos, M.S.N.de A., 1978, Nemátodos e plantas hospedeiras identificads em Coimbra, Portugal, durante 1972-1977, Ciênc.Biol. Portugal, 4:23.

Caupers, E., 1959, Generalidades sobre nemátodos e fumigantes do solo, Bol.Agric. Shell Port., 54:217.

D'Oliveira, A.B., 1943, Nota sobre alguns nemátodos de importância agrícola, Res.Comun.Apres. I Congr.Nac.Ci.Agrárias, Lisboa 1943 (summary 388) 117.

Louro, J.M.V. and Raposo, M.L.B., 1975, Prospecção preliminar da infestação de *Heterodera rostochiensis* Woll. em campos de cultura de batata-semente, Informação Técnica Fitossan, 6:1.

Macara, A.M., 1962, Contribucão para o estudo de algumas espécies do genero *Heterodera* Schmidt, 1871 encontradas em Portugal, Rel.Final Curso Eng. Agrón. Univ. Técn. Lisboa 126 pp (mimeo).

Macara, A.M., 1963, Contribuição para o estudo morfológico e biológico dos nemátodos *Heterodera goettingiana* Liebscher, 1892 e *H. rostochiensis* Wollenweber, 1923 encontrados em Portugal, Broteria, 32:25.

Pereira, M.J., 1976, *Heterodera schachtii* Schmidt, Identificacao e caracterização de populaçoes no solo, Rel.Final Curso Eng.Agrón. Univ.Técn. Lisboa 57 pp (mimeo).

CYST FORMING NEMATODES IN LIBYA

Ezarug A. Edongali
Plant Protection Department
Faculty of Agriculture
University of Al-Fateh, Tripoli
Libya

Plant nematology only became established in Libya in the late 1970s but surveys and identifications of several plant parasitic nematodes associated with different crops in Libya were made by foreign scientists, mainly Italians, before then. Carraro (1959) recorded the beet cyst nematode, *Heterodera schachtii* Schmidt on sugar beet at Sidi El-Mesri Agricultural Station. Pucci (1965) listed all the common diseases encountered in the Tripolitania region but the only cyst nematode he recorded was *H. goettingiana* Liebscher on broad bean (*Vicia faba*) at Tripoli and Ben-Ghasheer. Later, Pucci (1969) reported the presence of *H. goettingiana* and *H. avenae* Woll. on barley at Azizia.

More recently Ben Saad et al. (1981) recorded *Globodera rostochiensis* (Woll.) Behr. on potato and a survey by Edongali and Dabaj (1982) in the Tripolitania area revealed *H. cruciferae* Franklin on cabbage, cauliflower and other cruciferous plants. W. Ismail (personal communication) found a cyst nematode on alfalfa in the El-Zawia region which is probably *H. trifolii* Goffart.

REFERENCES

Ben Saad, A., Khalil, J.A., Farag, I., Abugnia, A. and Saleh, A., 1981, Diseases and pests of crop plants of Libya (in Arabic), Rpt. Univ. Al-Fateh.

Carraro, G., 1959, Contributo alla conscenza dei nematodi della piante coltivate in Tripolitania, Min. Agric. Rpt. 15 p.

Edongali, E.A., and Dabaj, M., 1982, Cauliflower (*Brassica oleracea* cv. *capitata*) cyst nematode in Libya, Libya J. agric., 11:205.

Pucci, E., 1965, Lista preliminare delle mallatie delle piante osservate in Tripolitania dal 1959 al 1964. Sintomi Danni, Lotta. Instituto Agronomico per L'ottremare Firenze.

Pucci, E., 1969, List of plant diseases recorded in Libya, Proc. FAO Meeting, Tripoli 1968, 24 p.

HOST STATUS OF CYST NEMATODES IN IRAQ

Zuhair A. Stephan

Department of Plant Protection
Abu-Ghraib, Baghdad
Iraq

The occurrence of a cyst forming nematode *Heterodera mothi* Khan & Husain was reported for the first time in Iraq by Husain et al. (1978a). *Heterodera* spp. were recorded from the roots of grapevine, almond, cherry and date palm in Ninevah Province, North Iraq (Husain et al., 1978b; Stephan et al., 1985). The main crops grown in Ninevah Province are wheat, tobacco, sugar beet, tomatoes and potatoes and the climate generally favours the activity and survival of cyst nematodes.

In the regions with a temperate climate, including the north region of Iraq, *H. schachtii* Schmidt is of economic importance on sugar beet, cabbage and rape, and *H. avenae* Woll. on wheat, oats, barley and rye. For the past 10 years, potatoes have been imported from Europe and grown extensively in the north region. Cysts of *Globodera rostochiensis* (Woll.) Behr. have probably been imported occasionally with the seed tubers but so far the species has not been recorded.

REFERENCES

Husain, S.I., Al-Zarari, A.J. and Younis, H., 1978a, Occurrence of a cyst nematode, *Heterodera mothi* Khan & Husain, 1965 in Iraq, Mesopotamia J.Agric., 13:143.

Husain, S.I., Al-Zarari, A.J., Younis, H. and Younis, M., 1978b. Report on the distribution and frequency of occurrence of stylet bearing nematodes found in the rhizosphere of fruit crops in the Ninevah Province of Iraq, Mesopotamia J.Agric., 13:129.

Stephan, Z.A., Alwan, A.H. and Antoone, B.G., 1985, Occurrence of plant parasitic nematodes in vineyard soils in Iraq, Nematol.medit., 13: (in press).

CYST NEMATODES IN INDIA

Irfan Ahmad

Department of Zoology, Aligarh Muslim University
Aligarh-202001
India

Nineteen species of cyst nematodes have been recorded from many areas in India since 1958 when the first species, *Heterodera avenae* Woll. was reported (Vasudeva, 1958). The golden cyst nematode, *Globodera rostochiensis* (Woll.) Behr., was first discovered by Jones (1961) and since then both *G. rostochiensis* and *G. pallida* (Stone) Behr. have been found extensively in potato crops in southern India (Seshadri & Sivakumar, 1962; Thangaraju, 1983). The other seventeen species have been found in association with major crops: wheat, barley, ragi, pulses, rice, sugarcane, maize, and others (Table 1). Cyst nematodes have been recorded from 13 states and the union territory of Delhi. In eleven states two or more species have been found, while in seven states three or more species have been recorded. The highest number of seven species from Delhi is probably accounted for by the greater number of surveys made there than other parts of the country. Cyst nematodes are also likely to be present in the rest of the states but few surveys have been carried out to detect them.

During the past twenty years nearly 100 research papers on cyst nematodes have been published in India, over a third of them on occurrence and distribution of various species. In recent years, awareness of the economic importance of cyst nematodes has resulted in more attention being given to the identification of biotypes and to control measures. This applies particularly to the cereal cyst nematode, *Heterodera avenae* Woll. which causes "Molya disease" in north-west India (Prasad et al., 1959; Swarup and Singh, 1961) and to the potato cyst nematode in southern India (Jones, 1961;Seshadri and Sivakumar, 1962). Several races of *H. avenae* have been identified (Mathur et al.. 1974; Swarup et al., 1980) and resistant cultivars of barley have already been recommended for cultivation in Haryana (Bhatti et al., 1976). The potato cyst nematodes have been controlled successfully by chemical means (Menon et al., 1969; Anon, 1980). In addition cultural control measures such as a three or four year rotation of potatoes with beans and peas reduced nematode populations and increased yields significantly (Menon and Thangaraju, 1973). More recently, potato cultivars have been screened for resistance (Howard et al., 1977) and cultivars resistant to *G. rostochiensis* and *G. pallida* have now been developed (Anon, 1980; Thangaraju and Sundaram, 1981).

Table 1 Cyst nematode species found in India

Species	Host	Reference
Heterodera avenae Woll.	Cereals	Vasudeva (1958)
H. cacti Filipjev & Schur. Stekh.	Cactus	Kumar (1964)
H. cajani Koshy	Pulses	Koshy (1967)
H. carotae Jones	Vegetables[1]	Swarup, Pradas & Raski (1964)
*H. chaubattia** Gupta & Edward	Apple	Gupta & Edward (1973)
H. cyperi Golden, Rau & Cobb	Grasses	Kumar (1980)
H. delvii Jairajpuri, Khan, Setty & Govindu	Ragi	Jairajpuri, Khan, Setty & Govindu (1970)
H. galeopsides Goffart	Vegetables[1]	Swarup, Prasad & Raski (1964)
H. gambiensis Merny & Netscher	Millet and weed	Krishnaprasad & Krishnappa (1983)
H. graminis Stynes	Grass	Sharma & Swarup (1983)
H. mothi Khan & Hussain	Weeds	Khan & Hussain (1965)
H. oryzicola Rao & Jayaprakash	Rice	Rao & Jayaprakash (1978)
H. sacchari Luc & Merny	Sugarcane	Swarup, Prasad & Raski (1964)
H. sorghi Jain, Sethi, Swarup & Srivastava	Sorghum	Jain, Sethi, Swarup & Srivastava (1982)
H. trifolii Goffart	Grasses	Sen (1963)
*H. vigni*** Edward & Misra	Pulses	Edward & Misra (1968)
H. zeae Koshy, Sethi & Swarup	Maize	Koshy, Sethi & Swarup (1970)
Globodera pallida (Stone) Behr.	Potato	Thangaraju & Sundaram (1981)
G. rostochiensis (Woll.) Behr.	Potato	Jones (1961)

* Considered as species inquirenda by Hesling (1978) and Sharma & Swarup (1983)

**Considered as species inquirenda by Hesling (1978)

[1] Recovered from soil where tomato was the previous crop.

REFERENCES

Anon, 1980, Annual report for the year 1979, Central Potato Research Institute of India, 192pp.

Bhatti, D.S., Dahiya, R.S., Dalal, M.R. and Dhawan, S.C., 1976, Resistant barley varieties for the control of *Heterodera avenae* Wollenweber, 1924, Curr. Sci., 45:676.

Edward, J.C. and Misra, S.L., 1968, *Heterodera vigni* n. sp. and second stage larva of *Heterodera* sp. in the Uttar Pradesh, India, The Allahabad Farmer, 42:155.

Gupta, P. and Edward, J.C., 1973, A new record of a cyst forming nematode, *Heterodera chaubattia* n. sp. from the hills of Uttar Pradesh, Curr. Sci., 42:618.

Howard, H.H., Krishananda, N., Singh, N. and Abraham, H.J., 1977, Potato cyst nematode species in the Nilgiri Hills, South India, J. Indian Potato Assoc., 4:68.

Jain, K.K., Sethi, C.L., Swarup, G. and Srivastava, A.N., 1982, *Heterodera sorghi* sp. n. a new cyst forming nematode parasitising sorghum, Revue Nematol., 5:201.

Jairajpuri, M.S., Khan, W.U., Setty, K.G.H. and Govindu, H.C., 1979, *Heterodera delvii* n. sp. (Nematoda: Heteroderidae) a parasite of ragi (*Elusine coracana*) in Bangalore, India, Revue Nematol., 2:3.

Jones, F.G.W., 1961, The potato cyst eelworm *Heterodera rostochiensis* Woll. in India, Curr. Sci., 30:187.

Khan, A.M. and Husain, S.I., 1965, *Heterodera mothi* n. sp. (Tylenchida: Heteroderidae) parasitising *Cyperus rotundus* L. at Aligarh, U.P., India, Nematologica, 11:167.

Koshy, P.K., 1967, A new species of *Heterodera* from India, Indian Phytopath., 20:272.

Koshy, P.K., Swarup, G. and Sethi, C.L., 1970, *Heterodera zeae* n. sp. (Nematoda: Heteroderidae) a cyst forming nematode on Nematologica, 16:511.

Krishnaprasad, K.S. and Krishnappa, K., 1983, Occurrence of *Heterodera gambiensis* Merny & Netscher, 1976 from India, Indian J. Nematol., 12:405.

Kumar, A.C., 1980, Studies on nematodes in coffee soils in south India. 4. Occurrence of *Heterodera cacti*, J. Coffee Res., 10:77.

Mathur, R.B., Arya, H.C., Mathur, R.L. and Handa, D.K., 1974, The occurrence of biotypes of the cereal cyst nematode (*Heterodera avenae*) in the light soils of Rajasthan and Haryana, India, Nematologica, 20:19.

Menon, P.P.V. and Thangaraju, D., 1973, Effect of crop rotation on the population of the golden nematode (*Heterodera rostochiensis* Woll.) in the Nilgiris, Indian J. Agric. Sci., 43:304.

Menon, P.P.V., Logiswaran, G., Seshadri, A.R. and Thangaraju, D., 1969, Observations on the control of the golden nematode (*Heterodera rostochiensis* Woll. 1924) of potatoes in Nilgiris, All India Nematology Symposium, New Delhi, 1969, [abstract].

Prasad, N., Mathur, R.L. and Sehgal, S.P., 1959, Molya disease of wheat and barley in Rajasthan, Curr. Sci., 28:453.

Rao, Y.S. and Jayaprakash, A., 1978, *Heterodera oryzicola* n. sp. (Nematoda: Heteroderidae) a cyst nematode on rice (*Oryza sativa* L.) from Kerala State, India, Nematologica, 24:341.

Sen, A.K., 1963, Three unlisted curcubit host plants of *Heterodera trifolii* Pl. Dis. Reptr., 47:942.

Seshadri, A.R. and Sivakumar, A.C., 1962, The golden nematode of potatoes (*Heterodera rostochiensis* Woll., 1924) - a threat to potato cultivation in the Nilgiris (Madras State), Madras Agric. J., 49:281.

Sharma, S.B. and Swarup, G., 1983, Identification scheme for *Heterodera* species in India, Indian J. Nematol., 13:171.

Swarup, G. and Singh, K., 1961, Molya disease of wheat and barley in Rajasthan, Indian Phytopath., 14:127.

Swarup, G., Prasad, S.K. and Raski, D.J., 1964, Some *Heterodera* species from India, Plant Dis. Reptr., 48:235.

Swarup, G., Sethi, C.L., Seshadri, A.R. and Kaushal, K.K., 1980, On the biotypes of *Heterodera avenae*, the causal organism of Molya disease of wheat and barley in India, Indian J. Nematol., 9:164.

Thangaraju, D., 1983, Distribution of potato cyst nematodes in Kodaikanal hills, Madurai district, Tamil Nadu, Indian J. Nematol., 13:222.

Thangaraju, D. and Sundaram, R., 1981, Studies on *Solanum vernei* 62-63-3 and hybrids of *S. vernei* 62-33-3 x Kufri Jyoti for resistance against potato cyst nematodes in Nilgiris, (Abstract), All India Nematology Symposium, coimbatoire, 1981, p.124.

Vasudeva, R.S., 1958, Annual Report, Division of Mycology and Plant Pathology, IARI, New Delhi.

CYST NEMATODES IN MEXICO, CENTRAL AND SOUTH AMERICA*

Carlos Sosa-Moss

Centro de Fitopatologia, Colegio de Postgraduados
Chapingo-Montechillos
Mexico

Although Mexico is geographically part of North America, its climatological features are more similar to Central and South America as are its agricultural activities and phytosanitary problems.

Most of the countries which constitute Latin America are considered to be tropical but the Andes of South America and the mountain ranges through Central America and Mexico are colder areas where potatoes, cereals and other temperate crops are grown. Consequently the cyst nematodes that occur include both the so-called temperate and tropical species.

The first report of a cyst nematode in Latin America was that of the golden nematode (*Globodera rostochiensis* (Woll.) Behr.) in Peru in 1952, but it was some years before it was recorded in other South American and Central American countries - Panama, Costa Rica, Chile, Venezuela, Colombia and Mexico; it has not yet been recorded in Argentina (for references see main paper). *G. rostochiensis* together with *G. pallida* (Stone) Behr. are the most common cyst nematodes in the temperate areas of Latin America.

In Mexico the potato crop occupies 70,000 hectares, and in Ecuador 33,500 hectares, and very few fields have been found to be free from the golden nematode. Yield losses can be considerable but only in a few areas is the use of chemicals considered to be econimically profitable. Crop rotation and the use of resistant cultivars are being investigated in Chile, Peru, Colombia, Ecuador, Venezuela, Panama and Mexico. Research on biological control has just begun, using the fungus *Paecilomices lilacinus*.

A species morphologically similar to the golden nematode has been found in Mexico and described by Campos-Vela in 1967 as *Heterodera (G.) mexicana*. However, due to inadequate description it remains as "inquerenda". *G. tabacum tabacum* (Lownsbery & Lownsbery) together with *G.t. solanacearum* (Miller & Gray) and *G.t. virginiae* (Miller & Gray) also frequently occur with *G. rostochiensis*

*Due to reasons beyond the control of the author the text was not submitted in time for inclusion in the book. However, it has been possible to include a summary of the paper which will be published in full in Nematologia mediterranea.

The Mexican corn cyst nematode, *Punctodera chalcoensis* Stone, Sosa-Moss & Mulvey is widely distributed in the upland valleys of Central Mexico and causes considerable yield losses in corn (maize) crops. *P. matadorensis* Mulvey & Stone has also been found in Mexico.

Heterodera schachtii Schmidt was discovered in 1969 infecting sugar beet in the Chalco area, near Mexico City. It was probably introduced with beet seeds imported from Europe and so far populations are low and localised. *H. cyperi* Golden, Rau and Cobb has recently been recorded on *Cyperus* spp in two states of Mexico.

Cactodera amaranthii (Stoyanov) has been found on spinach and other crop plants, and on weeds of the families Chenopodiaceae and Amaranthaceae in central Mexico. *C. cactii* (Filip'jev & Schuurmans Stekhoven) has also been recorded.

MEASURING FIELD POPULATIONS OF CYST NEMATODES[1)]

J. Willem Seinhorst

Formerly: Instituut voor Plantenziektenkundig Onderzoek
Wageningen
The Netherlands

INTRODUCTION

The density of field populations of cyst nematodes is estimated by counting the numbers of individuals in samples of soil. This raises two questions, 1) how to take a sample and of what size, and 2) how to arrive at a representative count of the individuals in the sample (extraction method, biotest). Here only the first question is discussed. Once the sample is in the laboratory an accurate count can be made of both the cysts and eggs contained therein. Whatever subsampling is done (investigating part of the field sample after thorough mixing, taking aliquots from egg suspensions) is supposed to result in variances according to binomial or Poisson distributions. This stage of perfection may not have been attained in all laboratories (Müller, 1983), but it certainly is not impossible to approach it. An extensive discussion of laboratory methods is to be found in Seinhorst (1986).

Sampling is done for two purposes: 1) to estimate population densities in plots in field experiments, 2) to estimate population densities in farmers' fields to serve as a basis for advice on cropping and nematode control. In the first case the sampling error must be known in order to determine the sample size required and the number of replicates needed to keep observations in experiments within certain limits of confidence, and in the second case because without a known measure of reliability advice to farmers is not trustworthy. Nevertheless, nematologists and agricultural advisers have in many cases virtually tried to do without this knowledge. Experimental errors were also taken as they came, with the result that observations in many field experiments were too inaccurate to draw clear conclusions.

Only a bare minimum of investigations on sampling error have been made, almost all of them on the error of numbers of nematodes counted in samples from whole fields (Jones, 1945; Anscombe, 1950; Raeuber and Stelter, 1963; Goodall and Ferris, 1980; McSorley and Parrado, 1982). In all cases single cores were taken from several points in the fields to be investigated,

---

[1)]This paper is an extensive summary of J. W. Seinhorst (1987), The estimation of densities of nematode populations in field plots and whole fields. II Sampling and sampling error (in preparation).

distributed according to a predetermined pattern (stratified sampling). Numbers of cysts or active nematodes in the separate cores were then assessed and efforts were made in most cases to detect general patterns in the frequency distributions of these numbers. In several cases negative binomial distributions could be fitted to the numbers counted. The numbers of cores required to guarantee that variances of estimated nematode densities will not exceed certain limits were derived from the distribution parameters. However, this procedure is based on two tacit but mostly wrong assumptions. The first is that nematode distribution within and in the immediate vicinity of each core is random (as seems to be the case with soil arthropods). At least, this could be the only reason why the authors did not pay attention to a possible relation between core size and sampling error apart from the effect of numbers of animals per core. Fenwick (1962) does not mention core size at all, but states that differences in cyst density between adjacent cores are larger than according to a random distribution of the cysts. Anscombe (1950) leaves the choice between 100 g and 200 g soil per core. Goodall and Ferris (1980) mention core size but, obviously, do not consider it to be of great importance.

The second (wrong) assumption is that each core is to be considered to have been chosen from between all possible cores, as is done by Goodall and Ferris (1980) and McSorley and Parrado (1982), to calculate the numbers of cores required to obtain a certain degree of accuracy. This denies the existance of large scale variations in nematode density over distances larger than those between sample cores. Anscombe (1950) mentions increase of core to core variability with increase of distance between cores, and because of which calculation of sampling errors must be based on distribution parameters derived from differences in nematode numbers in pairs of neighbouring cores.

CONDITIONS FOR PREDICTABILITY OF SAMPLING ERRORS

In the distribution of (cyst) nematodes in fields very small (cm), small, medium and large scale variations can be distinguished, each with its own standard deviation. Obviously medium and large scale variations must be expected to be unpredictable, unless they can be derived from previous observations, e.g. on place and size of patches of poor growth of a host. Medium and large scale variation may have become unimportant because the infestations are long standing ones and, therefore, have spread more or less uniformly over the whole field, as the potato cyst nematodes in most of the fields investigated by Anscombe (1950). However, local differences in soil condition within fields may maintain large scale irregularity and possibly more so for beet than for potato cyst nematodes. On the other hand the spread of new pathotypes after growing resistant potato cultivars for some time may give rise to mostly elongate patches of infestation by potato cyst nematodes, which have a more or less predictable size and shape (Seinhorst, 1982). Sampling fields with such patches results in small cyst counts with very large errors.

The predictability of very small and small scale variability of population density is of great importance for the planning of field experiments. It depends on whether there is a minimum area in which the distribution of cysts and eggs per cyst is always (more or less) the same, e.g. because it is determined by the distribution of the root system of host plants. Then the sampling error for such an area is known beforehand once it has been determined and sample size and numbers of replicates in field experiments can be based on it. If there is no such predictable sampling error, no field experiment can be planned properly without a previous assessment of the sampling error to be expected for the field plots to be used, e.g. by investigating pairs of samples from a sufficiently large number of plots.

EFFECT OF SUBSAMPLING ON SAMPLING ERROR

In several investigations on sampling error the nematodes counted were extracted from a subsample taken from a well mixed bulk sample. If the number of nematodes or cysts in a bulk sample is m and a proportion a of the bulk sample is investigated, then the variation coefficient of the average number of nematodes or cysts in the subsample, am, is according to a binomial distribution of the numbers of nematodes or cysts between the subsample and the remaining part of the bulk sample:

$$\{a(1 - a)m(am)^{-2}\}^{\frac{1}{2}} = \{(1 - a)(am^{-1}\}^{\frac{1}{2}}.$$

If the numbers of nematodes in bulk samples, obtained by repeated sampling of the same area, have a negative binomial distribution with mean m and "coefficient of aggregation" k then the variance of m is

$$s_m^{\,2} = (m^{-1} + k^{-1})m^2 \quad .... \ (1).$$

That of the expected number of nematodes or cysts am in a subsample taken after thorough mixing of the bulk sample then is

$$s_a^{\,2} = am^2\{m^{-1} + k^{-1} + a(1 - a)m(am)^{-2}\}$$

$$(am)^2\{(am)^{-1} + k^{-1}\} \quad .... \ (2)$$

The coefficient of variation $s_a(am)^{-1}$ of the number of nematodes to be expected in the subsample is determined only by k for the bulk sample and the number of nematodes finally counted. Therefore, subsampling can be ignored in the following discussion.

Subsampling after thorough mixing of a bulk sample composed of borings with a non-random distribution of nematodes or cysts between borings results in a smaller variance of the numbers of individuals counted than taking a bulk sample of the size of the subsample from which all nematodes or cysts are extracted and counted. If $\{s(am)^{-1}\}^2$ for the smaller bulk sample = $(am)^{-1} + k_{am}^{\;-1}$, then k_{am} is smaller than k. Most probably $k_{am} = ak$.

There are limits, however, to the possibilities of mixing, depending on the sizes of unbroken lumps of soil and, with active nematodes, on the numbers killed by the mixing. Therefore, it is advisable to investigate as large a subsample as can be handled. With cysts the numbers required for a certain degree of accuracy mostly preclude the use of small subsamples.

SMALL SCALE DISTRIBUTION OF CYSTS IN FIELD SOIL

There are no investigations on the very small scale distribution of cysts in field soil. Seinhorst (1987) compared the numbers of *Rotylenchus uniformis* Thorne in samples from square metre plots in an experiment of Seinhorst (1973), consisting of sixty cores, 1 cm diameter x 20 cm long (20 g per core) with those in samples from the same plots composed of five cores, 5 cm diameter x 20 cm long (400 g per core). The coefficient of variation of nematode numbers in the sixty core samples was known to be 17 per cent, from which that for the five core samples was calculated to be 11.5 per cent. Sampling error could, therefore, be considered to be inversely proportional to square root sample weight, irrespective of the

Table 1 Various observations on sampling errors for different nematode species

Author(s)	Nematode species		Number and size of plots	Number of samples per plot	Number of cores per sample
Den Ouden (1957)	*G. rostochiensis*	cysts / eggs	$4x1m^2$	10	80
Widdowson (1967)	*G. rostochiensis*	cysts	$364x1m^2$	2	1
Mugniéry & Zaouchi (1976)	*G. pallida*	cysts	$29x30m^2$	2	40
		cysts	78x10 to $20m^2$	2	40
		cysts / eggs	$20m^2$	40	40
		cysts / eggs	$30m^2$	40	40
		cysts / eggs	$45m^2$	40	40
	H. schachtii	cysts	$30x30m^2$	2	40
		cysts / eggs	$20m^2$	40	40
		cysts / eggs	$30m^2$	40	40
		cysts / eggs	$45m^2$	40	40

Depth of sampling cm	Weight of core g	Number of nematodes counted/ sample	Variation coeffi- cient %	*k* for 1.5 kg soil	*k* eggs	V.c. eggs/ cyst
20	20	42-108	6-19	72,81, 88		
		>200	20-27		21	16
?	240?	av.130	30	76?		
20	35	700-4200	6.1	361		
20	35	280-2200	14	55		
20	35	2200	11.5	81		
		44600	24		19	21
20	35	2640	8	166		
		47500	16		43	14
20	35	2040	30	12		
		66700	41		6	36
20	35	140-560	13.5	70		
20	35	320	16	44		
		530	138		0.56	
20	35	360	13	64		
		900	196		0.55	
20	35	300	25	16		
		430	121		0.26	

number of cores the sample was composed of if only five or more. It apparently did not make any difference whether 400 g soil was collected from 0.2 m^2 as a single 5 cm diameter cylinder or as twenty cores of 1 cm diameter well distributed over the whole area of 0.2 m^2. The variability distinctly exceeded that of a randomly distributed population. Much of the non-randomness of the distribution of the nematodes in the top 20 cm of an area of 0.2 m^2 apparently was already contained in a 5 cm diameter core. As the sampling error of numbers of nematodes in samples from small areas depends more on the quantity of soil collected than on the number of cores comprising the samples, this sampling error cannot be reduced by taking more but smaller cores, and keeping the total quantity of soil the same. This is only effective if the distribution of animals within and in the immediate vicinity of the core is a random one, as for soil arthropods (Gérard and Berthet, 1971).

To cope with the problem of reducing sampling error at small cyst densities of *Globodera rostochiensis* (Woll.), Seinhorst and den Ouden (1981) collected 5 kg soil from square metre plots by taking seven lots of about 700 g. With 50 to 100 cysts per sample this resulted in about the same sampling error as with 1.5 kg samples composed of eighty 1 cm diameter cores, also containing 50 to 100 cysts per sample. This also suggests that for cysts the number of cores is less important than the quantity of soil collected from a small area. The only reason to take eighty 1 cm diameter cores per (small) plot in field experiments when a 1.5 kg sample suffices is that it can be done faster than taking five 5 cm diameter cores.

SAMPLING ERRORS OF COUNTS OF *GLOBODERA* CYSTS IN SAMPLES FROM SMALL PLOTS

Only three investigations were found in the literature on the sampling error of counts of cysts in samples from small plots, apart from the more extensive investigations of Seinhorst (1982). The results of these three investigations are brought together in Table 1. Cyst densities were medium to high in all three and it was not possible to extract information on underlying distributions of the cysts in the soil and of the resulting numbers in the samples, except that the first were not random but also not very irregular.

The nature of the distribution of numbers of cysts in samples from square metre plots at densities between 2 and 3,000 cysts per kg soil was derived from cyst counts by den Ouden in samples from over 400 plots in nine fields and eight years by Seinhorst (1982). Rows of these plots were situated parallel to the long axes of elongate patches of infestations with *G. rostochiensis*. Apart from an area in the centre of certain patches, where densities were the same in a number of adjacent plots, log cyst densities along the rows could be considered to be linearly related to the distance of the centre of the plots from a certain point in the row at least for eight to sixteen adjacent plots (Fig. 1). Contrary to the data in Table 1, which are based on repeated sampling of the same plots, deviations of actual counts from densities expected according to these linear regressions of log cyst density upon distance from an arbitrary point in the row were considered to be a measure of the sampling error. As other disturbances of the linearity of the sampling error are to be expected the estimates obtained in this way are exaggerated. They do allow an assessment, however, of the field to field variability of the estimates and of their relation to cyst density. Field to field variability was non significant. Further, the distribution of the deviations of the counts from those expected according to the above mentioned linear relationships were, over the density range of 2 to 3,000 cysts per kg soil, in reasonable accordance with frequency

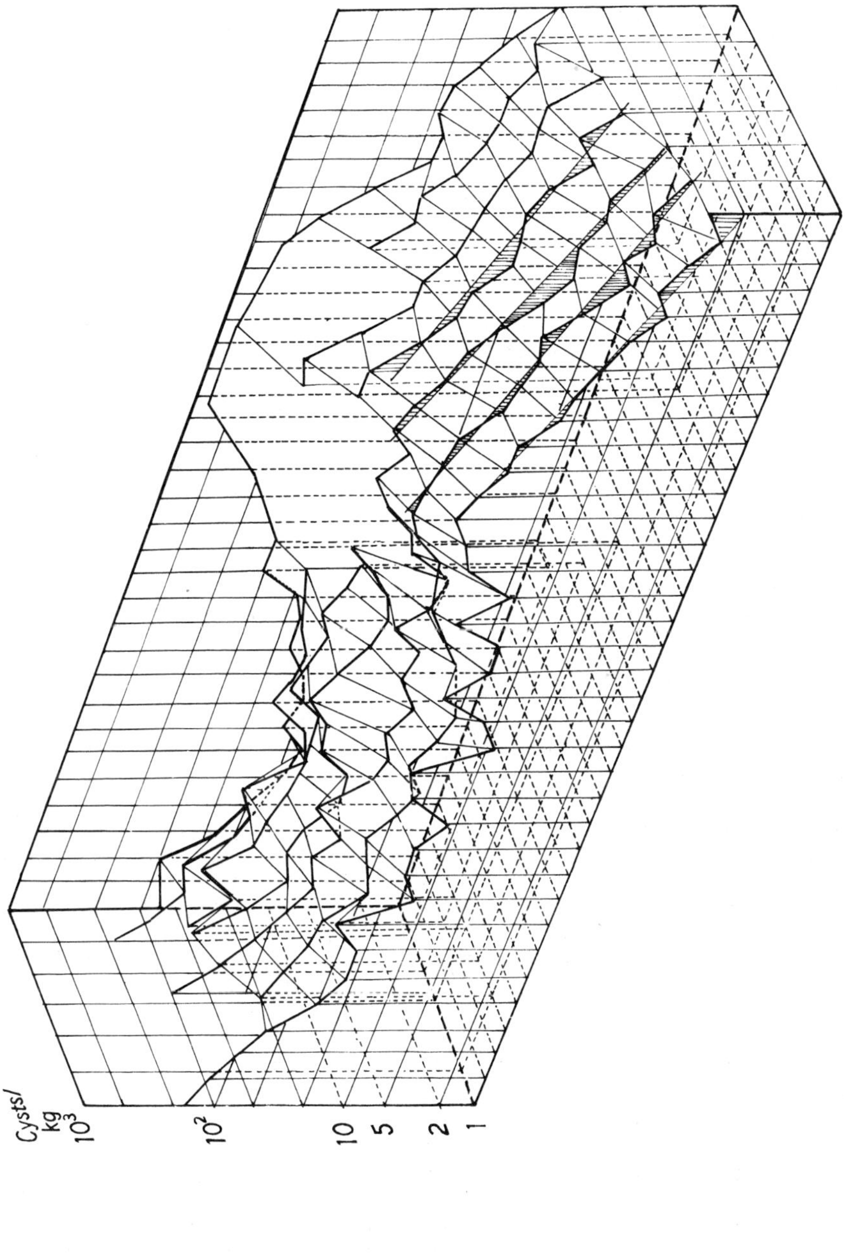

Fig. 1 Densities of *Globodera rostochiensis* cysts in square metre plots in part of the field used by Seinhorst and den Ouden (1981). Hatched areas in four rows in the right hand part of the diagram: deviations of numbers of cysts counted from densities according to a linear relation between log cyst density and distance of the center of the plots from a fixed point in the row.

distributions of cyst counts in samples according to negative binomial distributions with the same value of the "aggregation coefficient" k at all densities. The coefficient of variation of the cyst counts relative to linear log density gradients was, therefore, $sm^{-1} = (m^{-1} + k^{-1})^{\frac{1}{2}}$, in which m is the expected cyst count and s its standard deviation. The estimated value of k was 44 for 1.3 kg soil samples, which, with k proportional to sample size, amounts to 50 for 1.5 kg samples (80 cores, 1 cm diameter x 20 cm long). However, the distribution of deviations of log cyst counts from expected log cyst densities, although identical in the ranges 20 to 100 and 100 to 1,000 cysts expected per kg soil, showed an excess of observations as compared with a normal distribution, in the ranges of deviations 0 to s_{log} and > $2s_{log}$ (s_{log} = the standard deviation of log cyst counts). This could be interpreted as the result of a superposition of deviations of the true cyst densities from the assumed ones and sampling error. The value of k for sampling error would then be larger than 50 for 1.5 kg samples (Seinhorst, 1987).

Assuming sampling errors to be according to negative binomial distributions of cysts in the samples, values of k were calculated according to equation (2) from the coefficients of variation in Table 1. All values except that for *G. pallida* (Stone) in a 45 m^2 plot were well above 50 (55 to 166) for 1.5 kg samples. Seinhorst (1982) derived a value for k of 53 for 1.5 kg samples from Seinhorst and den Ouden's (1980) observations, but a reconsideration of deviations of counts in only the 2 kg samples in their experiment from values according to linear regressions of log cyst density upon distance of plots from an arbitrary point in the row considered suggested 70 as a better estimate. The investigations made so far then suggest that as a guidance for planning field experiments, sampling errors of counts of *Globodera* cysts in samples from small plots (one m^2 or perhaps a little larger) can be assumed to be largely independent of field, year and soil type and can be considered to be in accordance with a negative binomial distribution of cyst counts in samples with a value of k of 70 or more for 1.5 kg samples.

SAMPLING ERROR OF COUNTS OF *HETERODERA SCHACHTII* CYSTS IN SAMPLES FROM SMALL PLOTS

Values of k of 44 to 70 were derived for counts of *Heterodera schachtii* (Schmidt) cysts in 1.5 kg samples from plots of 30 m^2 or smaller from data by Mugniéry and Zaouchi (1976) (Table 1). Unpublished observations by Kaai (personal communication) suggest values of k of about 80 for 1.5 kg samples from square metre plots.

VARIATION OF AVERAGE EGG CONTENT OF CYSTS IN SAMPLES FROM SMALL PLOTS

So far the nature of the variability of average numbers of eggs in cysts from samples taken in small plots has been totally neglected and probably its importance has been underestimated.

Potato cyst nematodes

Den Ouden (1957) found variation coefficients of 12 to 19 per cent for average numbers of eggs (60 to 110) per cyst (40 to 200 cysts per sample) in 1.5 kg samples from square metre plots, and Mugniéry and Zaouchi (1976) of 21, 14 and 36 per cent for about 2,500 cysts per 1.4 kg soil sample from 20 m^2, 30 m^2 and 45 m^2 plots (average numbers of eggs per cyst 20, 18 and 33) respectively.

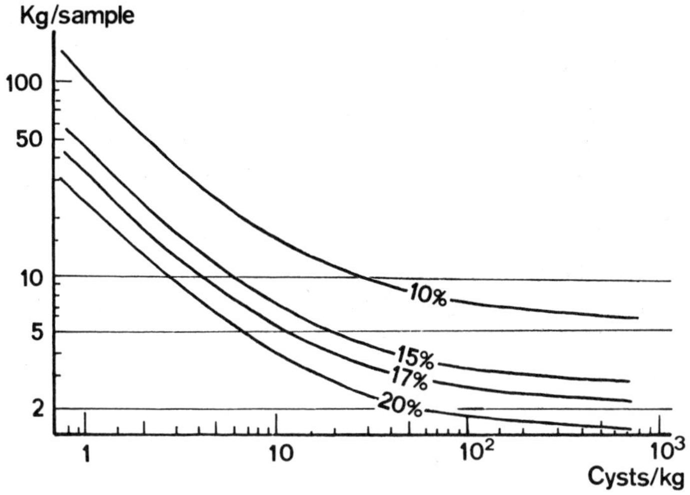

Fig. 2 Relation between population density of cysts of *Globodera rostochiensis* and *G. pallida* and sample size required to obtain variation coefficients of 10, 15, 17 and 20 per cent of estimated egg densities, assuming a negative binomial distribution of cyst counts with a value of k of 70 for 1.5 kg samples and a variation coefficient of average numbers of eggs per cyst in 1.5 kg samples of 16 per cent (k_{eggs} = 25).

Coefficients of plot to plot variation of average numbers of eggs per cyst in both adjacent plots and plots with about the same cyst densities in the same field were 13 to 21 per cent with an average of 18 per cent irrespective of cyst density, if between 35 and 700 per kg soil in the nine field experiments used by Seinhorst (1982) to determine sampling errors of cyst counts. The observations suggest a decrease at > 700 cysts per kg. The similarity of the between-plot and the within-plot (Table 1) variability of average egg contents of cysts and the independence of the first from cyst density suggests that both are in fact the same within-plot variability. If it is to be reduced this can only be done by taking larger samples.

If the variation coefficient of average numbers of eggs per cyst is assumed to be 16 per cent (the 18 per cent mentioned above probably contains some plot to plot variability) the coefficient of variation of egg densities (cv_{eggs}) in samples from small plots can be calculated as follows:

$$(cv_{eggs})^2 = m^{-1} + k_{eggs}^{-1} \quad .... \quad (3)$$

in which m is the expected number of cysts in the (sub) sample (estimated by the number counted) and

$$(k_{eggs})^{-1} = (k_{cysts})^{-1} + 0.16^2 \quad .... \quad (4)$$

For k_{cysts} = 70, k_{eggs} becomes 25 and for k_{cysts} = 50 it is 21. Apparently the large variability of average numbers of eggs per cyst reduces the importance of possible variation of k_{cysts}. Relations between cyst density and sample size required to obtain certain variation coefficients of estimates of egg densities in small plots are given in Fig. 2. The variation coefficient of average numbers of eggs per cyst of 16 per cent will only apply if percentages of empty cysts are small and, therefore, average numbers of eggs per cyst at least fairly large. Further investigations on the variability of numbers of eggs per cyst are very much needed.

Beet cyst nematodes

Populations of beet cyst nematodes tend to contain much larger percentages of empty cysts than those of potato cyst nematodes, probably at least partly because more than one generation is produced per season. There is a positive linear regression of average number of eggs per cyst upon percentage of cysts with eggs (Kaai, in litt). This results in very large sampling errors for small egg densities, while those of cyst densities still are reasonably small (Mugniéry and Zaouchi, 1976), (Table 1).

SAMPLING FIELDS WITH IRREGULAR NEMATODE DISTRIBUTION

The distribution of nematodes in a field is called irregular if there are considerable differences between population densities in small areas in different parts of the field and the areas with the different densities are not distributed according to a regular and predictable pattern. Fig. 1 depicts part of such an irregular distribution as far as the locations of the areas with large densities are concerned. However, the density gradients around these areas vary only within narrow limits as was also found for other patches of infestation with potato cyst nematodes (Seinhorst, 1982, 1987). If a distribution is called uniform here, this does not mean that the variance of the density estimate is equal to the mean, indicating a totally random distribution of the nematodes, but only that densities in the small areas mentioned above are more of less the same.

Three components can be distinguished which contribute to the variability of nematode density estimates derived from numbers of individuals counted in soil samples from fields with irregular distributions of these nematodes: the predictable small scale variation in density, discussed in the previous section, the differences between average densities in the areas (n^{-1} times the total area of the field, if a sample is compossed of n cores) from which a single core is taken (large scale variation) and the differences in densities between the small areas (one to a few square metres) within these larger areas to which the small scale variability applies (small scale variation). The first two components cause most if not almost all of the variability of the density estimates: that reflected by the sum of the variances, due to the small scale variation of the numbers of nematodes in the separate cores. The third component largely depends on the occurrence of density gradients as depicted in Fig. 1. The small scale variation and, therefore, also the sum of variances due to it of numbers of individuals in single cores, is reduced by taking larger cores (or more than one core in the immediate vicinity of each other). The variance due to the third component is reduced by taking more cores even if total sampling weight is not increased. As a result, total sampling error can only be reduced by increasing sample size by increasing the number of cores or both the number and the size of the cores.

Calculations of sampling errors (as due to the small and large scale variation of nematode density) for simple distributions (an elongate and a round patch (Seinhorst, 1982)) demonstrate positive correlations with the ratio between areas with small densities and those with large ones. Even if the average density is the same, and also with the ratio between the largest and smallest densities still occurring in sizeable proportions of the field. However, there are mahy combinations of areas with different densities resulting in the same sampling errors.

To determine the best sampling strategy for farmers' fields with generally unknown nematode distributions it is necessary to sample these distributions by determining them in a sufficiently large number of fields to be representative of those on which information could be asked in the future, calculate associated sampling errors and determine sample size and number of cores required to obtain a certain degree of accuracy of nematode density estimates for the most irregular distributions that still occur in sizeable proportion of these fields and for the smallest average density that still requires control before a host crop can be grown economically. A second factor determining the size of the sample is the accuracy required. If the crop yield without or after control measures is to be predicted, the coefficient of variation of the nematode density estimate should not be larger than about 15 per cent (true density within a range of 0.7 to 1.4 times the counted number of individuals in 95 per cent of the cases). This keeps the variation of the expected loss of yield between 0 and 10 per cent of the yield in the absence of nematodes if a nematode density twice the tolerance limit (T) is measured or can be expected after the proper control measures.

Nematode densities do not exceed certain maxima: 200 to 400 eggs of *G. rostochiensis*, fewer than 100 eggs of *H. schachtii* and of *H. avenae* Woll. per g soil. Therefore, with such large average densities the distribution could never be very irregular. However, even at an average nematode density half the expected maximum density sampling error can already be much larger than without large scale variation in density. Moreover, this knowledge is of little help before a density estimate is available.

An accurate measurement of average nematode density in a field is not enough to advise adequate control measures if nematode distribution is very irregular. If 20 per cent of the area of a field is occupied by (an) elongate patch(es), as that of Fig. 9 of Seinhorst (1986) infested with potato cyst nematodes, a third of which with the maximum density (about 350 eggs per g soil) then the average density in the field is 25 eggs per g soil, the expected yield loss would be 10 per cent of the yield in the absence of nematodes, if potatoes were grown in the field. If the degree of patchiness of the infestation and the places of the patches are not known, the average density could only be reduced by a treatment of the whole field, which would be highly uneconomical. However, if the infestation would have been uniform the expected yield loss would have been about 36 per cent and control would be highly economic. Therefore, measures should be taken to detect developing patches (as of infestations with new pathotypes of potato cyst nematodes) as early as possible and with other means than taking soil samples e.g. by inspecting plants in small patches or poor growth.

REFERENCES

Anscombe, F.J., 1950, Soil sampling for potato root eelworm cysts, Ann. appl Biol., 37:206.

Fenwick, D.W., 1961, Estimation of Field Populations of Cyst Forming Nematodes of the Genus *Heterodera*, J. Helminthol., R.T. Leiper Suppl., 63.

Gerard, G. and Berthet, P., 1971, Sampling strategy in censuring patchy populations, in "Statistical Ecology Vol. I: Spatial Patterns and Statistical Distributions", G.P. Patil, E.C. Pielou and W.E. Waters. eds., University Park and London.

Goodall, P. and Ferris, H., 1980, Plant Parasitic Nematode Distributions in an Alfalfa Field, J. Nematology., 12:136.

Jones, F.G.W., 1945, Soil populations of beet eelworm (*Heterodera schachtii*) in relation to cropping, Ann. appl. Biol., 32:357.

McSorley, M. and Parrado, J.I., 1982, Estimating relative error in nematode numbers from single soil samples composed of multiple cores, J. Nematology., 14:522.

Müller, J., 1983, Zur Problematik der quantitativen Erfassung von *Heterodera schachtii* mit Hilfe von Bondenuntersuchungen, I. Ermittlung des Nematodebesatzes in Mischproben; II. Ermittlung des Nematoden-vesatzes in Feldproben; III. Einfluss von Bearbeiter und Extraktions-methodik, NachrBl.dt. PflSchutzdienst (Braunschweig), 35:132; 35:150; 35:168.

Mugniéry, D. and Zaouchi, H., 1976, Etude critique de la méthode d'echantillonnage de nématodes a kystes dans le sol et de la validité des differentes critères d'estimation des population, Ann. de Zool. et Ecologie animales, 8:177.

Raeuber, A. and Stelter, H., 1963, Die Sicherheit bei der Bestimmung der Verseuchung eines Feldes mit dem Kartoffelnematoden (*Heterodera rostochiensis* Woll.), Biol. Zentralbl., 82:455.

Seinhorst, J.W., 1973, The combined effects of dichloropropane-dichloro-propene mixture injected into the soil and of dazomet applied to the soil surface, Netherl. J. Pl. Path., 79:194.

Seinhorst, J.W., 1981, The distribution of cysts of *Globodera rostochiensis* in small plots and the resulting sampling error, Nematologica, 28:285.

Seinhorst, J.W., 1987, The estimation of densities of nematode populations in field plots and whole fields, I. The extraction of nematodes from soil and plant tissue; II. Sampling and sampling error, (in preparation).

Seinhorst, J.W. and Ouden, H. den, 1980, The effect of different dosages of aldicarb on the multiplication at small population densities of *Globodera rostochiensis* on potato in rotavated and non rotavated plots, Netherl. J. Pl. Path., 86:147.

Ouden, H. den, 1957, Betrouwbarrheid van bemonstering, Jaarverslag 1956, Instituut voor Plantenziektenkundig Onderzoek, Wageningen, p. 121.

Widdowson, E., 1962, The estimation of soil populations of *Heterodera rostochiensis* Woll., in "Progress in Soil Zoology", P.W. Murphy, ed, London, pp. 59-64.

DISCUSSION

Responding to queries and viewpoints about the numbers and sizes of samples necessary to provide a basis for advice to farmers, Seinhorst observed that it is generally accepted that every potato field in Europe could be infected with PCN and he would propose to investigate a large number of fields but because of the limit to the number of samples that could be taken these should be taken in a way to provide advice about the worst fields. Müller referred to field sampling in Germany whereby 50 soil cores are taken along a transect to provide 6 kg soil, from which a 500 g sub-sample is taken; this is repeated 20 times to establish the error; the cyst content error; the error in counting the eggs in suspension after their extraction; the sampling error in the field. The sampling technique for the sugar beet cyst nematode had been investigated and it was concluded that more work was required in sampling and extraction methods.

Jacobsen asked if some risk equations could be developed as a basis for advice to farmers. Seinhorst replied that accuracy in sampling, extraction etc. was essential as the farmer is likely to lose either by using too much nematicide or will suffer a loss in yield because no action is taken against the nematode.

CHEMICAL AND INTEGRATED CONTROL OF CYST NEMATODES

Alan G. Whitehead

Rothamsted Experimental Station
Harpenden, Herts, AL5 2JQ
England

INTRODUCTION

Cyst-nematodes, *Heterodera* spp. and *Globodera* spp., can be controlled by crop rotation, use of resistant cultivars or by nematicides. Biological control may also be feasible in the future. None of these methods is effective by itself in all situations and control is best when measures are combined. The use of resistant cultivars may lead to selection of virulent pathotypes (Triantaphyllou, 1975; Turner et al., 1983; Young, 1984) or may entail a yield or quality penalty. If the speed of decline of a cyst nematode population in the soil is slow, control by crop rotation may entail a long and unacceptable interval between host crops. Nematicides are expensive and usually toxic, require careful handling and appropriate use, and may not give complete control when used alone. Nevertheless, nematicides are likely to be needed for the foreseeable future as one of the elements in the integrated management of cyst nematodes.

In the USA, nematicides are used on about 700,000 ha, less than one per cent of the cultivated land, mostly on cotton, soya bean, maize and tobacco (Johnson and Fassuliotis, 1984). In the UK, nematicides are used on about 52,500 ha, mostly on potatoes, onions and sugar beet, which is about 0.7 per cent of the arable land. In South Australia, however, more than 2 million ha of wheat are infested with the cereal cyst nematode, *H. avenae* Woll., annual losses are estimated at US $70 million (Brown, 1984a) and nematicide use is increasing rapidly (100,000 ha in 1983, probably 250,000 ha in 1984) (Brown, 1984b).

Uncertainty of the future size of the nematicide market and increasing concern for environmental and human safety, resulting in the withdrawal from general use of DBCP and EDB soil fumigants, may be discouraging investment in the discovery and registration of new nematicides. Nevertheless, new nematicides are needed, in particular toxicologically safer and rather more persistent compounds, which will be effective in a wide range of situations and climates. Until they materialise, we have to make the best possible use of the nematicides we have.

NEMATICIDES AVAILABLE

Currently available nematicides belong to four groups of pesticides, two being fumigants and two non-fumigants:-

Fumigants: (i) halogenated aliphatic hydrocarbons - methyl bromide, ethylene dibromide (EDB)* and 1,3-dichloropropene (1,3-D) mixtures (e.g. D-D, Telone II).

(ii) methyl isothiocyanate precursor compounds - metham-sodium and dazomet, and methyl isothiocyanate mixtures, including solutions in xylol or in dichloropropene mixtures, with or without added cloropicrin.

Non-Fumigants: (i) Organophosphates - ethoprophos, fenamiphos, fensulphothion, thionazin.

(ii) Carbamates - carbofuran, aldicarb and oxamyl.

*Use now prohibited in USA

Some commonly used nematicides are in Table 1. Further details are in Whitehead and Bridge (1985).

The soil fumigants have the advantages of killing nematodes deep in the soil (Harrison et al., 1963; Taconis and Kuiper, 1964; Whitehead et al., 1970) and spreading themselves through the soil. The disadvantages are phytotoxicity, which entails delay between application and planting; and rapid escape from the soil surface and relatively low toxicity to nematodes, which necessitates large dosages. The non-fumigants are usually non-phytotoxic, more toxic to nematodes and not usually very persistent in the soil but they are very toxic to vertebrates and usually must be mixed with the soil to be effective against cyst-nematodes.

MODES OF ACTION

Fumigants diffuse through the pore system of the soil, small amounts dissolving in the water films around soil particles and penetrating the cuticle of all stages of nematodes. At first the alkyl halide fumigants induce a period of hyperactivity, then a reversible narcosis and finally paralysis, which is not at first fatal but eventually the nematodes are killed (Marks et al., 1968; Evans and Thomason, 1971; Van Gundy et al., 1972).

The non-fumigant nematicides, dissolving in the soil water, may be toxic to eggs, juveniles in the soil and in roots, and males in the soil.

Effects on unhatched juveniles

Hatching of *Globodera rostochiensis* (Woll.) Behr. juveniles was inhibited or retarded by aldicarb at 5 ppm of soil, but not at 1 ppm (Hague and Pain, 1970), by diffusate from the roots of potato plants sprayed with oxamyl (Harrison, 1971) and by large amounts of ammonia in the soil following partial soil sterilisation or treatment of the soil with ammonium carbonate (Johnson and Townsend, 1949). In more recent work (Hague and Pain, 1973) aldicarb at 5 ppm (11.2 kg ha^{-1}) drenched on to soil in pots delayed emergence slightly but at 6 weeks 90% of the eggs had hatched in untreated and in treated soil. At 2 or 8 ppm, oxamyl or carbofuran delayed hatching slightly, whereas fenamiphos markedly inhibited hatching (Hague, 1975). Hatching of cereal cyst-nematodes, *H. avenae*, was progressively reduced by fenamiphos at dosages between 20 and 160 ppm (Sultan and Singh, 1984).

Table 1. Widely available nematicides

	SOIL FUMIGANTS			ORGANOPHOSPHATES		CARBAMATES		
Active Ingredients	Methyl bromide	1,3-dichloropropene mixtures	Dazomet	Fenamiphos	Ethoprophos	Aldicarb	Oxamyl	Carbofuran
Vapour Pressure (mm Hg)	1420 (20°C)	55 (20°C)	21 (20°C)	1×10^{-6} (30°C)	3.5×10^{-1} (26°C)	1×10^{-4} (25°C)	2.3×10^{-4} (25°C)	2×10^{-5} (33°C)
Mammalian Acute Oral Toxicity LD_{50} or LC_{50}	514 (ppm)	140-500 (mg kg^{-1})	500 (mg kg^{-1})	15 (mg kg^{-1})	62 (mg kg^{-1})	0.9 (mg kg^{-1})	5.4 (mg kg^{-1})	8-14 (mg kg^{-1})
Effective dosages (kg ha^{-1})	400-1500	156-1153	165-528	3-20	10-20	3-12	3-12	3-12

Hatching of beet cyst-nematode, *H. schachtii* Schmidt, was stimulated by aldicarb at 0.01 or 0.05 ppm (Steele and Hodges, 1975a) and by aldicarb or carbofuran at 1 ppm (Steele, 1983), probably as a result of hyper-activity with induced stylet thrusting in the unhatched juvenile (Nelmes, 1970). However, at 0.5 ppm aldicarb did not affect hatching, whereas 5 ppm completely suppressed it (Hough and Thomason, 1975), 1 ppm in the soil retarded hatching (Steudel, 1972; Steele and Hodges, 1975) and 5 500 ppm inhibited hatching (Steele and Hodges, 1975). Aldicarb, aldicarb sulphoxide and aldicarb sulphone, all at 1, 10, 50 or 100 ppm, reduced but did not stop the hatching of *H. schachtii*, even at the highest concentration (Steele, 1977). Hatching was also significantly retarded by 5 ppm or more of aldicarb, carbofuran or fensulphothion and inhibited by fenamiphos at 1 ppm (Steele, 1983). Inhibition of hatching due to aldicarb or carbofuran is reversible except at high concentrations e.g. 20 ppm for 12 weeks for potato cyst-nematode (Osborne, 1973).

As non-fumigant nematicides are rarely used above 2.5 ppm (5 kg ha^{-1}) in the soil, their effects on hatching of cyst-nematodes are probably small in commercial practice.

Effects on nematodes in soil and roots

Aldicarb at 5 ppm killed hatched *G. rostochiensis* juveniles and 1 ppm prevented them from invading potato roots (Hague and Pain, 1970). At 5 ppm in soil (11.2 kg ha^{-1}) second stage juveniles accumulated in soil in pots, unable to penetrate the roots of Majestic potatoes (Hague and Pain, 1973). In-vitro, aldicarb inhibits locomotion of *G. rostochiensis* juveniles but stimulates stylet thrusting. The effects of sub-lethal concentrations are reversible when the juveniles are transferred to water (Nelmes, 1970). Aldicarb is more toxic to the juveniles than aldicarb sulphoxide, which in turn is more toxic to them than aldicarb sulphone (Nelmes, 1970).

Migration of second stage juveniles of *H. schachtii* in sand was unaffected by aldicarb at 0.5 ppm but was inhibited by exposure to 1 ppm. Similarly, infection of sugar beet roots was inhibited at 1 ppm but not at 0.1 or 0.01 ppm (Hough and Thomason, 1975). Development of beet cyst-nematode juveniles was affected when they were exposed to aldicarb solutions before being used to inoculate beet plants (Steele and Hodges, 1975). At 1 or 0.01 ppm, aldicarb stopped males of *H. schachtii* in sand being attracted to females and in a field experiment aldicarb at 2.5 or 5 kg ha^{-1} reduced the percentage of females fertilised from 69% in untreated plots to 49 and 45% respectively (Hough and Thomason, 1975).

Roots of potato plants grown in solutions of aldicarb or fenamiphos repelled potato cyst-nematode juveniles before or soon after they had penetrated the roots, whereas the development of the nematode was unaffected when roots were treated 11 days after they had been invaded (den Ouden, 1971). Potato roots grown for four weeks in soils treated with large amounts of soil fumigants or small amounts of organophosphate or oximecarbamate nematicides may contain 0.1% or less of the juveniles found in roots grown in untreated soils (e.g. Whitehead et al., 1975b). Orientation of second stage juveniles towards roots prior to their invasion was impaired by oxamyl at 0.5 ppm and 0.2, and 0.8 ppm prevented invasion of roots by juveniles; orientation of males towards females was similarly impaired at 1 ppm. Development of juveniles in roots was inhibited at 2 ppm (Evans and Wright, 1982). Oxamyl, not its breakdown products, is effective as a nematicide (Bromilow, 1973).

The main activity of organophosphate and carbamate nematicides on cyst-nematodes would seem to be against the hatched juveniles in the soil and perhaps in roots which have absorbed them before they are invaded by the nematodes. The possibility of disrupting reproduction by disorientation of males needs further investigation.

EFFICACY OF AVAILABLE NEMATICIDES

The efficacy of fumigant and non-fumigant nematicides in controlling cyst-nematodes has been reviewed by Whitehead (1973). The efficacy of many organophosphorus, carbamate, benzimidazole and other compounds in controlling *G. rostochiensis* has been reported by Whitehead et al., (1985). Large amounts of methyl bromide, 1,3-dichloropropene, dazomet, metham-sodium or methyl isothiocyanate are needed for partial or full control of cyst-nematodes; in organic soils control is poor. Much smaller amounts of organophosphorus compounds and carbofuran are effective in soils without much organic matter. The most effective compounds in a wide range of temperate soils are the oximecarbamates, particularly aldicarb and oxamyl; in very alkaline or warm soils their persistence is greatly reduced.

AIMS OF CONTROL

The objectives of nematode control are a) to prevent injury to the crop plant, which may result in loss of yield and quality of the produce, and b) to limit or prevent population increase, so that the susceptible crop can be grown at the desired frequency.

Control of damage

Referring to the sigmoid curve (Fig. 1) relating total plant yield (Y) to numbers of nematodes in the soil at planting (P_i) in which

$$Y = Y_{min} + (1-Y_{min})cz^{P_i} \quad \text{(Seinhorst, 1965)}$$

the aim is to reduce P_i to P_t, the threshold for damage, or if possible to P_x, at which yield may exceed that in uninfested soil. To do this, a percentage (K), of the nematodes must be killed or be immobilised.

$$\text{(i)} \qquad K = \frac{100(P_i-P_t)}{P_i}$$

Control of nematode increase

Nematode increase is density dependent and inversely proportional to nematode numbers (Fig. 2), so that if P_i is reduced to P_t or P_x it approaches or reaches a maximum, maximum crop growth providing a maximum of feeding sites in the large root system. Nematode increase is prevented (full control) when P_f (eggs after harvest) = P_i (eggs before planting). Empirically, 'kill' (K) depends on nematode increase (a) for those that hatch i.e.

$$\text{(ii)} \qquad K = 100(1 - \frac{1}{a})$$

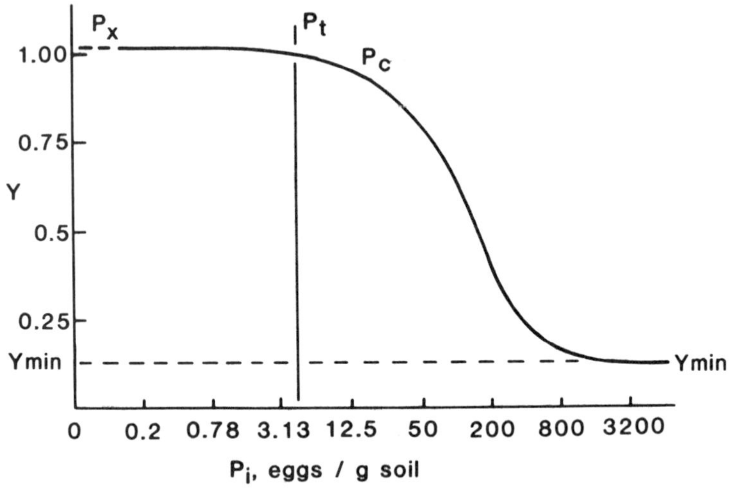

Fig. 1. Relationship between total plant yield (Y) and the number of cyst nematodes in the soil at planting (P_i)

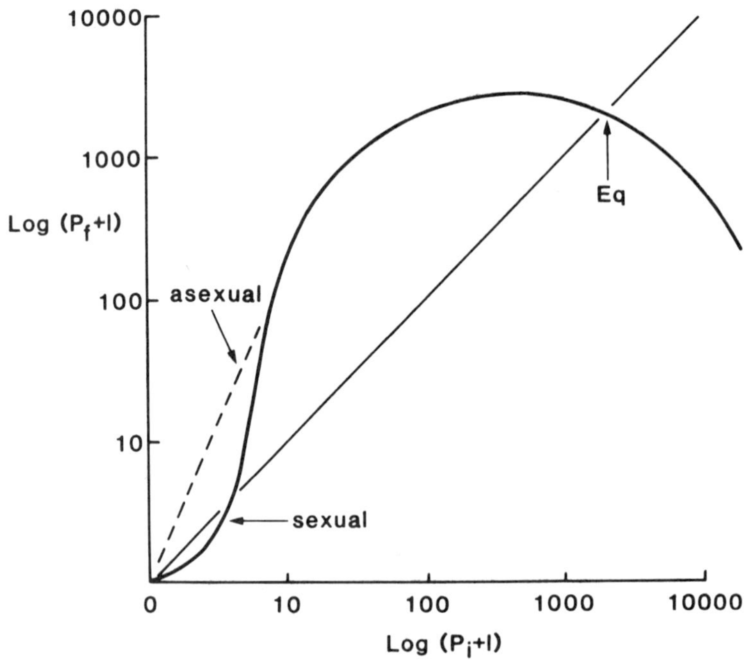

Fig. 2. Final number of cyst nematodes (P_f) in relation to initial numbers (P_i).

Actual multiplication is

(iii)
$$\frac{P_f - P_u}{P_n}$$

where P_u is the number of viable nematodes which do not hatch when the host plant is grown, and P_n is the number of nematodes attacking the plant after using a nematicide. P_n is easily determined only when all the eggs are retained in the cyst. In very lightly infested soils, potato cyst-nematode may increase 150-fold (den Ouden and van der Veer, 1977) but it usually does not multiply more than 50-fold in infested soils. Other cyst nematodes may pass more than one generation on a susceptible host crop and second or subsequent generations may not be controlled, even if the first generation is well controlled (Table 2).

Equations applicable to nematicides used to control cyst-nematodes were derived by Jones and Perry (1978) from their model of the dynamics of cyst-nematode populations. For soil populations well below the equilibrium density (i.e. most field situations) the 'kill' needed to keep $P_f = P_i$ (K_{min}) is given by the equations:-

(iv) soil fumigants

$$K_{min} = 1 - \frac{2}{a(1-C_p) + 2C_p}$$

and

(v) non-fumigants (nematostats)

$$K_{min} = 1 - \frac{2}{a}$$

Table 2. Number of generations of some cyst-nematodes

Cereal (*H. avenae*, Wollenweber)	1
Potato (*Globodera spp.*)	usually 1
Beet (*H. schachtii*, Schmidt)	2-5
Soya bean (*H. glycines*)	3-5
Pea (*H. goettingiana*, Leibscher)	1
Brassica (*H. cruciferae*, Franklin)	1-3

So if a = 50 and C_p (proportion of unhatched eggs) = 0.3

$$(\%)\ K_{min}\ (\text{fumigants}) = 94.4$$

$$(\%)\ K_{min}\ (\text{non fumigants}) = 96.0$$

When P_i approaches or exceeds the equilibrium density, K_{min} is given by more complex equations requiring knowledge of the equilibrium density of the cyst-nematode and plant yield at the equilibrium density as a fraction of that in the absence of nematodes (see Whitehead and Bridge, 1985). The empirical equation (ii) may be used to obtain a rather over-estimated required 'kill' for a cyst-nematode.

It is usually easier to prevent damage due to cyst-nematodes than to control their increase fully, especially when P_i is small, requiring only a small K value to prevent damage but a large K value to prevent increase. Usually K values large enough to prevent damage (except at very large P_i values) but not large enough to prevent nematode increase can be obtained in commercial practice. Usually nematode increase can be reduced so that partial control is achieved.

The exact shape of the sigmoid curve relating yield loss to P_i will depend on soil moisture stress, cultivar tolerance, cyst population virulence, interaction with other soil organisms and probably other unidentified factors, so that the yield benefits of nematicide use can also vary considerably.

FACTORS AFFECTING CONTROL OF CYST-NEMATODES BY NEMATICIDES

Factors important in the control of cyst-nematodes with nematicides were reviewed by Whitehead (1973). Hague (1980) and Whitehead and Bridge (1985) also reviewed factors affecting nematicides generally, so they need only brief mention here. The maximum multiplication rate and the number of generations passed on a host crop, as already mentioned, may limit control by nematicides.

In annual crops, young plants are usually more injured by cyst-nematode attack than older plants, so that controlling the nematodes in the topsoil or in part of it may prevent most yield loss. However, potato and beet cyst-nematodes may also occur in large numbers in the subsoil (Whitehead, 1977) and may cause some yield loss. They are difficult and expensive to control with nematicides but may be controlled by the use of a resistant cultivar.

Soil fumigants do not work well in organic soils, being strongly adsorbed onto the organic matter fraction. In other types of soil their efficacy is much affected by the soil moisture content. Ideally, the soil should be in seed-bed condition and well drained, so that the gases can diffuse through all the soil spaces occupied by the nematodes. Loss of gas at the soil surface can be rapid. 'Sealing' the surface by rolling, watering, using a chemical 'seal' or covering with polyethylene film may enhance the activity of the fumigant in the soil. Although soil temperature affects the speed of gaseous diffusion it is less important than correct soil moisture, provided it is not too low. Cyst-nematode kill is related to the effective fumigant concentration X time product (CTP).

Bromilow (1980) showed that the organophosphate nematicides, ethoprophos and fenamiphos, carbofuran and the oximecarbamates, aldicarb and oxamyl, only diffused in the soil water and not in the soil air, so that their efficacy depended on adequate incorporation in the soil and any subsequent transfer within the soil by mass flow following rain or irrigation water. Bromilow (1973, 1980) also showed that the mobility of a non-fumigant nematicide in the soil is a function of the partition coefficient Q.

$$Q = \frac{\text{Chemical concentration in soil organic matter}}{\text{Chemical concentration in the soil water}}$$

Q is determined by the polarity of the molecule i.e. degree of charge separation in the molecule. Small values of Q are therefore associated with mobility. The value of Q can be obtained directly by measuring the sorption of the nematicide in soils of known organic matter contents, or, more conveniently, Q can be estimated from the partition coefficient (P) of the compound between octan-1-ol and water using the equation of Briggs (1973):-

$$\log Q = 0.52 \log P + 0.62$$

P can often be estimated accurately from the chemical structure. Aldicarb and oxamyl, the most effective nematicides in a wide range of temperate soils, have small values of Q, carbofuran is intermediate and the less polar organophosphates have large values of Q and are much less effective in organic soils.

Soil moisture level in the soil after incorporation of a non-fumigant nematicide may affect nematode control, for Griffin (1977) found aldicarb more effective in preventing invasion of beet seedlings by *H. schachtii* at 80-100% of field capacity than at 20-30% field capacity.

The persistence of organophosphate and oximecarbamate nematicides is reduced in very alkaline and in warm soils.

Finally, increase of *G. pallida* (Stone) Behr. on potatoes in some irrigated organic soils was not affected by oxamyl (Whitehead et al., 1984) or aldicarb (Whitehead unpublished data) at commercial rates, which control *G. rostochiensis* increase. At present, the most likely explanation is either a) the extended period of hatching of *G. pallida* in moist soil or b) migration of juveniles into the treated soil from below, when the activity of the nematicide in the soil above has ceased. In other soils, there seemed to be no differential control of the two *Globodera* spp. by aldicarb or oxamyl (Moss et al., 1975, 1976; Brown, 1983; Trudgill et al., 1983).

On tomatoes under glass, *G. pallida* was fully controlled by treating the soil with 11.2 kg oxamyl ha^{-1} or large amounts of methyl bromide, dazomet, Telone or Di-Trapex CP (Whitehead et al., 1975b; Whitehead et al., 1979c).

APPLICATION OF NEMATICIDES FOR CYST-NEMATODE CONTROL

Treatment of seeds and setts

In Australia, oxamyl at 360 g or carbofuran at 630 g per 100 kg of wheat seed is permitted seed dressing for control of cereal cyst-nematode (Brown, 1984a). Cysts of potato cyst-nematode on seed tubers are killed without harm to the tubers by dipping in sodium hypochlorite (1% available chlorine) for 2 hours (Wood and Foot, 1975). By contrast, Brodie (1983) found that soaking cut potato tuber pieces (standard planting material in USA) only significantly reduced cyst numbers on roots of the potato at phytotoxic dosages. Although seed or sett treatment with nematicides may increase crop yield, the effect is too localised for effective cyst nematode population control.

Seed furrow and row treatments

Damage to annual crops may largely be prevented by applying an effective nematicide to the seed furrows during sowing or to the bands or ridges of soil in which the young plants are grown. Placing aldicarb or oxamyl in the seed furrows at sowing time does not control potato or beet cyst-nematode increase on susceptible crops because the roots soon grow away from the narrow treated band of soil and the nematicides are not translocated towards the root tips. Applied in 10 cm wide bands over the seed furrows at planting, aldicarb, oxamyl, carbofuran, fenamiphos and ethoprophos significantly reduced the increase of *G. rostochiensis* on susceptible potatoes (Brodie, 1983). Seed-furrow treatments have been more successful in closely-spaced crop rows. Brown (1973) found that 2.2 kg aldicarb ha^{-1} drilled with wheat seed, increased yield as much and controlled *H avenae* as well as did 9 kg ha^{-1} incorporated in the top 10 cm of the soil. Oxamyl applied in the seed furrows controlled pea cyst-nematode as well in one experiment as did oxamyl incorporated in the seed bed before sowing peas; in other experiments such seed-furrow treatments were less successful, probably because of insufficient rain to leach the nematicide downwards with the extending roots (Whitehead et al., 1979b). In South Australia, as few as 2 cereal cyst-nematode eggs g^{-1} soil depresses wheat yield by 20%, wheat yields are in general small, and only an 11-13% yield increase is needed to cover the cost of current nematicide treatments. Apart from the seed dressings already mentioned, these are seed furrow treatments of EDB 3.7 l ha^{-1}, terbufos 4 kg ha^{-1}, aldicarb 2 kg ha^{-1} or carbofuran 1.1 l ha^{-1}. Increases of 30% or more in wheat yields following these treatments are common (Brown, 1984a). However, King et al. (1982) showed that grain yield increases due to aldicarb which were statistically significant in 12 of 20 field experiments, could not be correlated with nematode egg density in the soil, length of seminal root axes, or severity of 'knotting' of the roots due to nematode attack. An ultra-low-volume applicator was developed for the application of small volumes of liquid nematicides in the seed furrows during sowing (Gurner et al., 1980), Using this equipment, small volumes of EDB or dichloropropene similarly applied did not improve the yield of susceptible oats in sandy loam infested with *H. avenae* in England, probably because of phytotoxicity, whereas 1.5 kg oxamyl ha^{-1} in the seed furrows greatly increased oat grain yield (Whitehead et al., 1982). In four field experiments in Victoria, Australia, comparing small amounts of aldicarb, oxamyl, DBCP, EDB and terbufos applied to the seed furrows during sowing, cysts per plant were consistently and significantly reduced by EDB at 3.7 - 11.1 l ha^{-1} and aldicarb at 2 or 4 kg ha^{-1} and yields were consistently and significantly increased (Brown et al., 1982).

Topsoil and subsoil treatments

In widely spaced crop rows, cyst-nematodes are usually fully or partially controlled only when effective nematicides are distributed throughout the topsoil (Lear et al., 1961; Steudel and Thielemann, 1967; Whitehead and Tite, 1968; Hague and Pain, 1970). However, this can be difficult to achieve, because a large volume of soil has to be treated (2.5×10^3 m^3 ha^{-1} in the top 25 cm of the soil). Fumigants can be applied under the plough sole during ploughing, through blade coulters 20-30 cm apart on a tractor drawn toolbar, or by under-soil spraying behind A-blade shares. The soil crumbler and powered flat roller of the Rumpstad fumigant injector makes a particularly smooth and compacted soil surface behind the under-soil sprayer, so that nematode kill is increased near the soil surface where nematode kill is often poor.

Non-volatile nematicides must usually be mixed well with the top 15 cm of the soil to control cyst-nematode populations effectively. This can be done very effectively by rotary cultivation (rotavation) but it is slow, for the rotor speed must be fast in relation to forward speed of the tractor and the hood must be lowered to shatter clods and to aid remixing. Rotary cultivators with L-shaped tines may 'smear' damp soils containing significant amounts of clay or silt. Spike tines are preferable on such soils but even they are unacceptable on poorly structured silt soils. Rotary cultivators also mix weathered surface soil with unweathered soil from below, which is harmful to crops grown in level seedbeds and in the heavier soils. Nevertheless, treating the soil, which is then made up into potato ridges, with effective nematicides, partially or fully controlled *G. rostochiensis* on potatoes and often greatly increased tuber yields in infested soils (Whitehead et al., 1973a,b,c,d,e,f; Moss et al., 1975, 1976; Kalyviotis-Gazelas, 1982; Brown 1983; Trudgill et al., 1983). The problems of rotary cultivating granules into the topsoil can be avoided by blowing them into vertical bands 20 cm apart in the top 12-15 cm of the soil and then mixing them laterally by rotary harrowing, using for example the Lely 'Roterra' (Whitehead et al., 1981). Nematode control is as good as that achieved by rotary cultivating nematicide granules into the soil and it is faster and safer. Granules applied to the soil surface and harrowed in are not mixed uniformly with the soil, some 60-80% of them usually remaining in the top 5 cm of the soil (Whitehead et al., 1975a).

The beet cyst nematode, *H. schachtii*, was fully controlled in heavily infested soils (50, 51, 91, 93 eggs g^{-1} soil) by incorporating aldicarb or oxamyl into the topsoil before sugar beet seeds were sown, but in a less heavily infested soil (30 eggs g^{-1} soil) control was partial (Whitehead et al., 1979a). Even in the untreated plots of this last soil, the nematodes increased only 2.3-fold and nematophagous fungi found in cysts may have lessened final numbers of eggs. As little as 2 kg a.i. ha^{-1} doubled sugar yields in heavily infested soil. Similar results were obtained by sowing sugar beet seeds in rows 48.3 cm apart in the middle of bands of soil 15 cm wide and 15 cm deep in which aldicarb had been incorporated by blowing the granules into pairs of vertical bands in the soil 48.3 cm apart and 7.6 cm in between adjacent bands and mixing them laterally with the soil by a reciprocating harrow. In very heavily infested soil, 1.4 kg aldicarb ha^{-1} increased sugar yield from 2 to 7.5 t ha^{-1} whereas 8 kg aldicarb ha^{-1} increased sugar yield to 10 t ha^{-1}. In less heavily infested soil, beet yields were also greatly increased by similar treatments (Whitehead, unpublished data).

1,3-dichloropropene mixtures, injected into the topsoil before sowing, partially controlled beet cyst-nematode (Altman and Fitzgerald, 1960; Radewald et al., 1971) and aldicarb at up to 13.5 kg ha^{-1} in three applications to sugar beet gave partial or full control (Steudel and Thielemann, 1967, 1968). The longer the period between planting and the application of aldicarb as a sidedressing treatment irrigated into the sugar beet rows in infested soil, the smaller the increase in sugar beet yield (Griffin, 1978). In two fields, D-D at 200 l ha^{-1} and Telone at 175 l ha^{-1} injected into the soil in autumn reduced, by the following April, viable eggs of *H. schachtii* to 8 and 23% (D-D) and 13% (Telone in one field) of the pretreatment values.

Temik at 1.35 g m^{-1} row reduced the numbers of juveniles entering the roots compared with those in fumigated plots, whereas 2.25 g m^{-1} row totally prevented juveniles from entering (Thielemann, 1981). Best results may therefore be obtained by the combined use of a fumigant in autumn and aldicarb in spring. Combined treatments with different non-fumigant nematicides or repeated treatments of one of them are better than a single application for beet cyst-nematode control (Hafez and Gillian, 1983), which is not surprising for a pest which passes several generations on its host plant.

Foliar applications

Systemically translocated nematicides which could be sprayed safely on the foliage of nematode-infested plants would be valuable. To date, no very effective or safe compound has been found. Spraying crops with oxamyl reduced the invasion of roots by nematodes but nematodes developing in the roots were unaffected (Potter and Marks, 1971). Harrison (1971) also found *G rostochiensis* juveniles were able to develop in roots of potato plants sprayed with oxamyl. Dipping shoots of susceptible young potato plants in aqueous oxamyl (2000 ppm) did not lessen nematode increase (Whitehead et al., 1973c). Up to six sprays (each of 1, 2, or 4 kg oxamyl ha^{-1}) at about 3 weekly intervals to potato foliage reduced the increase of *G rostochiensis* on susceptible potatoes to about half. More than 6 kg oxamyl ha^{-1} left unacceptable residues in the tubers (Whitehead et al., 1979d). Griffin (1975) found that foliar applications of fenamiphos or oxamyl retarded development of *H. schachtii* on sugar beet plants in a glasshouse. However, the best results were obtained when treatments preceded inoculation or were made at the time of inoculation.

INTEGRATED CHEMICAL CONTROL OF CYST-NEMATODES

Strict quarantine measures, involving prohibition of potato growing, dosage of foci of infestation with soil fumigants and the use of *G. rostochiensis* resistant cultivars, have prevented the spread of potato cyst-nematodes on the US mainland (Evans and Brodie, 1980). In France, early potatoes may not now be grown where the soil is infested with potato cyst-nematodes. However, two potato trap crops in successive years, together with 10 kg ethoprophos ha^{-1} spread on the soil surface on the day of planting, either in the first or second year, reduced soil populations by about 96.5% (P_i 2 eggs g^{-1} soil) and > 99.5% (P_i 50 eggs g^{-1} soil). The average reduction of 98.5% would have taken about 21 years to achieve with non-host crops, because the decline rate of the nematode in these trials (*G. pallida* Pa 3) was only 18.2% p.a. Each trap crop appeared to reduce soil populations by 80%. In addition to the speed of reduction in the soil population, this integrated approach lessened the risk of spread of cysts from infested to uninfested fields (Mugniery and Balandras, 1984). In the Netherlands, an integrated system of controlling potato cyst nematode has been operating in the starch potato industry for many years. This involves growing potatoes and non-host crops in alternate years, a resistant potato cultivar in alternate potato years and soil fumigation with 1,3 dichloropropene (Telone II) in the autumn following susceptible potato crops (Nollen and Mulder, 1970). After November 15, fumigation is forbidden and granular nematicides must be used in the spring before the next potato crop. The rotation is only 2 years and the system relies heavily on the control achieved by the resistant cultivar and the fumigant. Where *G. pallida* occurs and there is no suitable resistant cultivar, and where the fumigant gives less than the expected 80% kill, control is inadequate. In a three year rotation, *G. rostochiensis* was controlled both by Maris Piper potatoes, which are resistant to *G. rostochiensis* Ro 1, and by oxamyl or dazomet (Whitehead et al., 1980). Mugniery (1982) has pointed out that because of the variation in population decline rates under non-host crops, hatching, virulence and speed of nematode development in potatoes, which occur in populations of *G. rostochiensis* and *G. pallida* in different geographical areas, integrated control procedures must be modified accordingly. It may be possible in the future to control beet cyst-nematode by the combined use of a granular nematicide, applied as previously described to bands of soil in which the host crop is grown, and nemataphagous fungi which control population increase.

The integrated approach to cyst-nematode control has several strengths. It does not rely exclusively on one method of control, so there is less risk of very rapid selection of virulent pathotypes on resistant cultivars, less risk of nematicide resistance developing in the nematodes, as has already occurred in animal parasitic nematodes with repeated use of anthelmintics (Kelly, 1981) and rotations can be shortened considerably to allow intensive production of host crops on the most suitable soils. Assuming the control measures are independent of each other, the fraction of the soil population left after an integrated control programme is the product of the fractions left by the individual measures i.e.,

$$P_f = P_i(F_o)^a(F_n)^b(F_r)^c$$

where

F_o = fraction left after one non-host crop,

F_n = fraction left after nematicide use,

F_r = fraction left after a resistant cultivar,

a,b,c, = number of times each measure used.

REFERENCES

Altman, J. and Fitzgerald, B.J., 1960, Late fall application of fumigants for the control of sugar beet nematodes, certain soil fungi, and weeds, Plant Dis.Reptr., 44:868.

Briggs, G.G., 1973, A simple relationship between soil absorption of organic chemicals and their octanol/water partition coefficients, Proc.7th Brit.Insect.Fung.Conf.1973, 83.

Brodie, B.B., 1983, Control of *Globodera rostochiensis* in relation to method of applying nematicides, J.Nematol., 15:491.

Bromilow, R.H., 1973, Breakdown and fate of oximecarbamate nematicides in crops and soils, Ann.appl.Biol., 73:473.

Bromilow, R.H., 1980, Behaviour of nematicides in soils and plants, Workshop Manual,Ass.appl.Biol.meeting Rothamsted Exp.Sta. 5-6 June,1980, 87.

Brown, E.B., 1983, The relationship of potato yield with and without nematicide to density of potato cyst-nematodes, *Globodera rostochiensis* and *G. pallida*, Ann.appl.Biol., 103:471.

Brown, R.H., 1973, Chemical control of the cereal cyst nematode (*Heterodera avenae*) - a comparison of methods and rates of application of two systemic nematicides, Austr.J.Exp.Agric. and Anim.Husb., 13:587.

Brown, R.H., 1984a, Cereal cyst nematode and its chemical control in Australia, Pl.Disease, 68:922.

Brown, R.H., 1984b, Ecology and control of cereal cyst nematode (*Heterodera avenae*) in Southern Australia, J.Nematol., 16:216.

Brown, R.H., Pye, D.L. and Stratford, G.T., 1982, A comparison of low volume in-row applications of nematicides at seeding for control of the cereal cyst nematode (*Heterodera avenae*) in wheat, Nematol.Medit., 10:9.

Den Ouden, H., 1971, De ontwikkeling van *Heterodera rostochiensis* larven in aardappelwortel bij behandeling met systemische nematiciden in het bijzonder aldicarb, Meded.Rijksfaculteit Landb.wetensch.,Gent, 36:889.

Den Ouden, H. and van der Veer, R.F., 1977, The effect of some systemic nematicides on the control of *Heterodera rostochiensis* in the field, Neth.J.Pl.Path., 83:129.

Evans, A.A.F. and Thomason, I.J., 1971, Ethylene dibromide toxicity to adults., larvae and moulting stages in *Aphelenchus avenae*, Nematologica, 17:243.

Evans K and Brodie, B.B., 1980, The origin and distribution of the golden nematode and its potential in the USA, Amer.Potato J., 57:79.

Evans, S.G. and Wright, D.J., 1982, Effects of the nematicide oxamyl on life cycle stages of *Globodera rostochiensis*, Ann.appl.Biol., 100:511.

Griffin, G.D., 1975, Control of *Heterodera schachtii* with foliar application of nematicides, J.Nematol., 7:347.

Griffin, G.D., 1977, Effect of soil moisture on control of *Heterodera schachtii* with aldicarb, J.Nematol., 9:211.

Griffin, G.D., 1978, Aldicarb post-plant control of the sugar beet cyst nematode, *Heterodera schachtii*, Pl.Dis.Reptr., 62:1027.

Gurner, P.S., Dube, A.J. and Fisher, J.M., 1980, Chemical control of cereal cyst-nematode (*Heterodera avenae*) on wheat by a new low-volume applicator, Nematologica, 26:448.

Hafez, S.L. and Gillian, 1983, Sugar beet cyst nematode control with combinations of systemic nematicide - insecticides, J.Nematol., 15:480.

Hague, N.G.M., 1975, Nematicides - past and present. Proceedings 8th British Insecticide and Fungicide Conference, 837.

Hague, N.G.M., 1980, The control of plant parasitic nematodes by volatile chemicals, Workshop Manual Ass.appl.Biol.meeting Rothamsted Exp.Sta.5-6 June,1980, 19.

Hague, N.G.M. and Pain, B.F., 1970, Some observations on the effect of 'Temik' on the potato cyst eelworm, *Heterodera rostochiensis* Woll, Pl.Path., 19:69.

Hague, N.G.M. and Pain, B.F., 1973, The effect of organophosphorous compounds and carbamates on the potato cyst-nematode *Heterodera rostochiensis* Woll., Pestic.Sci., 4:459.

Harrison, M.B., 1971, Evidence of the mode of action of a systemic nematicide, [Abstract] J.Nematol., 3:311.

Harrison, B.D., Peachey, J.E. and Winslow, R.D., 1963, The use of nematicides to control the spread of arabis mosaic virus by *Xiphinema diversicandatum* (Micol.), Ann.appl.Biol., 52:243.

Hough, S.A. and Thomason, I.J., 1975, Effects of aldicarb on the behaviour of *Heterodera schachtii* and *Meloidogyne javanica*, J.Nematol., 7:221.

Hough, A., Thomason, I.J. and Farmer, W.J., 1975, Behaviour of aldicarb in soil relative to control of *Heterodera schachtii*, J.Nematol., 7:241.

Johnson, A.W. and Fassulotis, G., 1984, Nematode parasites of vegetable crops, in "Plant and Insect Nematodes", W.R. Nickle, ed., Dekker, New York and Basel.

Johnson, L.R. and Townsend, W.N., 1949, The inhibition of hatching of potato root eelworm (*Heterodera rostochiensis* Woll.) in partially sterilised soil, Ann.appl.Biol., 36:504.

Jones, F.G.W. and Perry, J.N., 1978, Modelling populations of cyst-nematodes (Nematoda: Heteroderidae), J.appl.Ecol., 15:349.

Kalyviotis-Gazelas, C., 1982, Control of the golden nematode *Globodera rostochiensis* with granular nematicides, Ann.Instit.Phytopath.Benaki, 13:145.

Kelly, J.D., 1981, Animal nematodes: growing resistance to anthelmintics, Span, 24:123.

King, P.M., Rovira, A.D., Brisbane, P.G, Simon, A. and Brown, R.H., 1982, Population estimates of cereal cyst nematode and response of wheat to granular nematicide, Austr.J.Exp.Agric. and Anim.Husb., 22:209.

Lear, B., Sciaroni, R.H., Atlee, C.B., Jr., and Hart, W.H., 1961, Yield response of Brussels Sprouts to soil fumigation for control of sugar beet nematode, *Heterodera schachtii*, Pl.Dis.Reptr., 45:739.

Marks, C.F., Thomason, I.J. and Castro, C.E., 1968, Dynamics of the permeation of nematodes by water, nematicides and other substances, Exp.Parasitol., 22:321.

Moss, S.R., Crump, D. and Whitehead, A.G., 1975, Control of potato cyst-nematodes, *Heterodera rostochiensis* and *H. pallida*, in sandy peaty and silt loam soils by oximecarbamate and organophosphate nematicides, Ann.appl.Biol., 81:359.

Moss, S.R., Crump, D. and Whitehead, A.G., 1976, Control of potato cyst-nematodes, *Globodera rostochiensis* and *G. pallida*, in different soils by small amounts of oxamyl or aldicarb, Ann.appl.Biol., 84:355.

Mugniery, D., 1982, Diversités régionales d'application de lutte intégré en culture de pommes de terre en fonction de la variabilité des populations de nématodes à Kystes, Agronomie, 2:629.

Mugniery, D and Balandras, C., 1984, Examen des possibilités d'éradication du nématode à Kystes, *Globodera pallida* Stone, Agronomie, 4:773.

Nelmes, A.J., 1970, Behavioural responses of *Heterodera rostochiensis* larvae to aldicarb and its sulphoxide and sulphone, J.Nematol., 2:223.

Nollen, H.M. and Mulder, A.A., 1970, A practical method for economic control of potato cyst nematode (*Heterodera rostochiensis*), Proc.5th Brit. Insect.Fung.Conf., 671.

Osborne, P., 1973, The effect of aldicarb on the hatching of *Heterodera rostochiensis* larvae, Nematologica, 19:7.

Porter, J.W. and Marks, C.F., 1971, Effect of Du Pont 1410-X on rate of development of *Heterodera schachtii* on cabbage, J.Nematol., 3:325.

Radewald, J.D., Hall, B.J., Shibuya, F. and Nelson, J., 1971, Results of preplant fumigation trial for the control of sugar beet nematode on cabbage, Pl.Dis.Reptr., 55:841.

Seinhorst, J.W., 1965, The relation between nematode density and damage to plants, Nematologica, 11:137.

Steele, A.E., 1977, Effects of selected carbamate and organophosphate nematicides on hatching and emergence of *Heterodera schachtii*, J.Nematol., 9:149.

Steele, A.E., 1983, Effects of selected nematicides on hatching of *Heterodera schachtii*, J.Nematol., 15:467.

Steele, A.E. and Hodges, L.R., 1975, In vitro and in vivo effects of aldicarb on survival and development of *Heterodera schachtii*, J. Nematol., 7:305.

Steudel, W., 1972, Versuche zum Einfluss von Aldicarb auf den Schlupfvorgang bei Zysten von *Heterodera schachtii* nach längerer Einwirkung, Nematologica, 18: 270.

Steudel, W and Thielemann, R., 1967, Zur Frage der Wirkung eines Carbamolyloxim-granulates auf die Vermehrung des Rübennematoden (*Heterodera schachtii* Schmidt) und den Ertrag von Zuckerrüben, Nachr. Bl.dt.-Pfl.Schutzdienst.,Stuttgart, 19:129.

Steudel, W. and Thielemann, R., 1968, Versuche zur Frage der Empfindlichkeit von Zuckerruben gegen den Rübennematoden (*Heterodera schachtii* Schmidt), Meded.fakul.Landbou.,Gent, 33:707.

Sultan, M.S. and Singh, I., 1984, Effect of nemacur on the hatching of cereal cyst nematode, *Heterodera avenae*, Ind.J.Nematol., 14:60.

Taconis, P.J. and Kuiper, K., 1964, Overdracht van het *Nicotiana* virus 5 door aaltjes van het geslacht *Trichodorus* in zanilingen van 5 gewassen, Versl.Meded.Plziektenk.,Dienst,Wageningen, 141:177.

Thielemann, R., 1981, Verminderung des *Heterodera schachtii* - Befalls in Zuckerrüben feldern durch Flächen-und Reihenbehandlung mit Nematiziden, Zuckerindustrie, 106:445.

Triantaphyllou, A.C., 1975, Genetic structure of races of *Heterodera glycines* and inheritance of ability to reproduce on resistant soybeans, J.Nematol., 7:356.

Trudgill, D.L., Holliday, J.M., Mathias, P.L., French, N., Mackintosh, G.M. and Tones, S.J., 1983, Effect of the nematicide oxamyl on the multiplication of *Globodera rostochiensis* and *G. pallida* and on the haulm growth and yield of six potato cultivars with different levels of resistance and tolerance, Ann.appl.Biol., 103:477.

Turner, S.J., Stone, A.R. and Perry, J.N., 1983, Selection of potato cyst-nematodes on resistant *Solanum vernei* hybrids, Euphytica, 32:911.

Van Gundy, S.D., Munnecke, D., Bricker, J. and Minteer, R., 1972, Response of *Meloidogyne incognita, Xiphinema index* and *Dorylaimus* sp. to methyl bromide fumigation, Phytopathology, 62:191.

Whitehead, A.G., 1973, Control of cyst-nematodes, (*Heterodera* spp.) by organophosphates, oximecarbamates and soil fumigants, Ann.appl.Biol., 75:439.

Whitehead, A.G., 1977, Vertical distribution of potato, beet and pea cyst nematodes in some heavily infested soils, Pl.Pathol., 26:85.

Whitehead, A.G. and Bridge, J., 1985, Control of plant nematodes, in "Pesticide application: principles and practice", P.T. Haskill, ed., Oxford University Press.

Whitehead, A.G. and Tite, D.J., 1968, Rep.Rothamsted Exp.Sta. 1967, 154.

Whitehead, A.G., Fraser, J.E. and Greet, D.N., 1970, The effect of D-D chloropicrin and previous crops on numbers of migratory root-parasitic nematodes and on the growth of sugar beet and barley, Ann.appl.Biol., 65:351.

Whitehead, A.G., Tite, D.J., Fraser, J.E. and French, E.M., 1973a, Control of potato cyst-nematode, *Heterodera rostochiensis*, in silt and peat loams by ten pesticides applied to the soil at planting time, Ann. appl.Biol., 73:197.

Whitehead, A.G., Tite, D.J., Fraser, J.E. and French, E.M., 1973b, Treating potato ridges in spring with aldicarb, D-D or dazomet to control potato cyst-nematode, *Heterodera rostochiensis*, in sandy clay and peat loam soils, Ann.appl.Biol., 73:203.

Whitehead, A.G., Tite, D.J. and Fraser, J.E., 1973c, Control of potato cyst-nematode, *Heterodera rostochiensis*, in sandy loam, by Du Pont 140 (S-methyl 1-(dimethyl-carbamoyl) -N- [(methylcarbamoyl) oxy] thioformimidate) applied to the soil at planting time, Ann.appl.Biol., 73:325.

Whitehead, A.G., Tite, D.J., Fraser, J.E. and French, E.M., 1973d, Effects of D-D, Telone or dazomet applied to potato ridges in spring on potato cyst-nematode, *Heterodera rostochiensis*, in sandy loam and silt loam soils, Ann.appl.Biol., 74:105.

Whitehead, A.G., Tite, D.J., Fraser, J.E. and French, E.M., 1973e, Control of potato cyst-nematode, *Heterodera rostochiensis*, in three soils by small amounts of aldicarb, Du Pont 1410 or Nemacur applied to the soil at planting time, Ann.appl.Biol., 74:113.

Whitehead, A.G., Tite, D.J., Fraser, J.E. and French, E.M., 1973f, Control of potato cyst-nematode, *Heterodera rostochiensis*, in sandy, peaty and silt loam soils by dazomet and Telone applied in different ways, Ann.appl.Biol., 75:257.

Whitehead, A.G., Bromilow, R.H., Lord, K.A., Moss, S.R, and Smith, J., 1975a, Incorporating granular nematicides in soil, Proc.8th Brit.Insect. Fung.Conf., 133.

Whitehead, A.G., Fraser, J.E., French, E.M. and Wright, S.M., 1975b, Chemical control of potato cyst-nematode, *Heterodera pallida*, on tomatoes grown under glass, Ann.appl.Biol., 80:75.

Whitehead, A.G., Tite, D.J., Fraser, J.E. and French, E.M., 1975c, Incorporating granular nematicides in soil to control potato cyst-nematode *Heterodera rostochiensis*, Ann.appl.Biol., 80)85.

Whitehead, A.G., Tite, D.J., Finch, P.H., Fraser, J.E. and French, E.M., 1979a, Chemical control of beet cyst-nematode, *Heterodera schachtii*, in some peaty loam soils, Ann.appl.Biol., 92:73.

Whitehead, A.G., Bromilow, R.H., Tite, D.J., Finch, P.H., Fraser, J.E. and French, E.M., 1979, Incorporation of granular nematicides in soil to control pea cyst-nematode, *Heterodera goettingiana*, Ann.appl.Biol., 92:81.

Whitehead, A.G., Fraser, J.E. and French, E.M., 1979, Control of potato cyst-nematode, *Globodera pallida*, on tomatoes grown under glass, by applying steam or chemical nematicides to the soil, Ann.appl.Biol., 92:275.

Whitehead, A.G., Tite, D.J., Fraser, J.E. and French, E.M., 1979, Rep. Rothamsted Exp.Sta.1978, 170.

Whitehead, A.G., Tite, D.J., Fraser, J.E. and French, E.M., 1980, Control of potato cyst-nematode, *Globodera rostochiensis*, in a three-course rotation, J.Agric.Sci., 95:293.

Whitehead, A.G., Tite, D.J. and Bromilow, R.H., 1981, Techniques for distributing non-fumigant nematicides in soil to control potato cyst-nematodes, *Globodera rostochiensis* and *G. pallida*, Ann.appl.Biol., 93:311.

Whitehead, A.G., Tite, D.J., Fraser, J.E. and Nichols, A.J.F., 1982, Rep. Rothamsted Exp.Sta.1981, 158.

Whitehead, A.G., Tite, D.J., Fraser, J.E. and Nichols, A.J.F., 1984, Differential control of potato cyst-nematodes, *Globodera rostochiensis* and *G. pallida* by oxamyl and the yields of resistant and susceptible potatoes in treated and untreated soils, Ann.appl.Biol., 105:231.

Whitehead, A.G., Bromilow, R.H., Fraser, J.E. and Nichols, A.J.F., 1985, Control of potato cyst-nematodes, *Globodera rostochiensis*, and root-knot nematode, *Meloidogyne incognita*, by organophosphorous, carbamate, benzimidazole and other compounds, Ann.appl.Biol., 106:489.

Wood, F.H. and Foot, M.A., 1975, Treatment of potato tubers to destroy cysts of potato cyst nematode; a note, N.Z.J.Exp.Agric., 3:349.

Young, L.D., 1984, Effects of continuous culture of resistant soybean cultivars on soybean cyst nematode reproduction, Pl.Disease, 68:237.

DISCUSSION

Whitehead explained that in experiments on control using aldicarb and oxamyl, the effectiveness of a treatment in terms of nematode mortality, was measured by sampling roots to estimate juvenile invasion. Recent work had concentrated on P_f/P_i as an indicator of control, with the final population regarded as the important parameter. Sikora observed that in Germany preplant treatment of sugar beet with aldicarb generally had failed to reduce final populations of *H. schachtii* and sought an explanation in view of aldicarb supposedly remaining effective for 6 weeks. Whitehead responded that in his experiments the multiplication of *H. schachtii* in untreated plots was low with P_f/P_i 2.4 to 2.5, suggesting that natural antagonists may have limited multiplication of the second generation (the experiment was on a site with a long history of sugar beet cultivation). Incorporation of aldicarb around the rows was considered to have largely prevented juvenile invasion of the roots, consequently reducing the inoculum for the second generation.

In a discussion on the use of nematostatic chemicals, Whitehead referred to the effective control of aphids with aldicarb because it is translocated to the leaves (oxamyl is less effective); to the leaching of aldicarb from light sandy soils in very heavy rain; and to the fact that neither residues of aldicarb or oxamyl, nor their breakdown products, had been detected above their permitted levels in potato crops. Reference was also made to the approval of carbofuran for potatoes; also to the phytotoxic effects associated with the use of EDB in UK trials, in contrast to its use in Australia, although it was pointed out that in the UK dichlorethylene was used as a solvent for EDB to lower the freezing point and hence the treatments were different. Asked if there was renewed interest in metham-sodium in Europe, as there is in the USA, Whitehead replied that persistence may be a problem in the cold UK soils. Trudgill added that metham sodium is used in the Netherlands to control potato cyst nematodes in potato crops for starch production where it can be applied in late summer before sowing cereals.

Wyss asked if avermectins might be used in the future. Whitehead said that Avermectin B2A had behaved poorly in organic soil because of high adsorption. Roberts added that there is some evidence from small scale field trials in the USA that root-knot nematodes could be controlled successfully with avermectins.

Stone referred to integrated control using resistant cultivars of potato and the selection for *G. pallida* which had occurred in southern England as a result of the over-use of Maris Piper with H_1 resistance gene. He asked if it is not a matter of concern that cultivars becoming available with resistance to *G. pallida* derive this resistance from only two genetic sources. Whitehead agreed that this was of concern and thought that rotation with susceptible cultivars would be necessary to minimise selection for virulence.

AN ASSESSMENT OF THE ROLE OF PARASITES AND PREDATORS IN THE REGULATION OF CYST NEMATODE POPULATIONS

Brian Kerry

Rothamsted Experimental Station
Harpenden, Herts, AL5 2JQ
England

INTRODUCTION

Cyst nematodes have four distinct phases during their life cycle and at each stage they are subject to predation or parasitism by several soil organisms (Table 1). Although a number of predators have been observed to feed on cyst nematode eggs, second-stage juveniles, males and females, they tend to be non-specific and attack all nematodes in soil. Many of these organisms frequently occur together, are common in agricultural soils and are often numerous, but there is little evidence that they have any significant effect on the numbers of a particular pest species. Such non-specific control is difficult to measure and exploit and consequently little research has been done to demonstrate the effects of predation on nematode populations (Esser and Sobers, 1964; Sayre, 1971; Mankau, 1980). Arthropod predators tend to be confined to the upper layers of soil and are too large to pursue nematodes through narrow soil pores. More information is needed on factors affecting the activity of predators in soil if only to ensure that agricultural practices do not remove any beneficial effects that may be occurring. Research on the biological control of cyst nematodes, as with other plant parasitic genera, mainly concerns microbial agents.

Rickettsia-like organisms were first observed in females of *Heterodera goettingiana* Liebscher and *Globodera rostochiensis* (Woll.) Behr. (Shepherd et al., 1973) and more recently in *H. glycines* Ichinohe (Endo, 1978; Noel and Wilson, 1981). It was not clear whether these organisms were parasitic or symbiotic, but they were found in all tissues of adult nematodes and second-stage juveniles except for the genital primordia (Walsh et al., 1983). Noel and Wilson (1981) noted that infected *H. glycines* Ichinohe juveniles failed to develop normally and the numbers of females produced on soybeans were markedly smaller in comparison with an uninfected population. *G. rostochiensis* juveniles containing rickettsias were shorter lived than if uninfected but fecundity was not consistently reduced (Walsh, 1981). It will remain difficult to prove the effects of these organisms on nematode populations until infection can be achieved in the laboratory.

Spores of the bacterium *Pasteuria penetrans* have been observed adhering to second-stage juveniles of several species of cyst nematode isolated from soil (Esser, 1980; Sturhan, 1985). Most studies of the life cycle of this organism have been with *Meloidogyne* spp. but Page and Bridge (1985) have observed *Heterodera* females on sorghum that were filled with spores and

Table 1 Natural enemies of cyst nematodes

Phase	Stages	Natural enemies	Reference
Resting stages in soil	Encysted eggs	Endoparasitic fungi	Tribe (1977)
Active stages in soil	Infective juveniles and adult males	Predatory nematodes	Thorne (1972)
		Theratromyxa weberi (amoeba)	Weber et al. (1952)
		Mites	Doncaster (1962)
		Pasteuria penetrans (bacterium)	Esser (1980)
		Nematode trapping fungi	Duddington (1960)
		Endoparasitic fungi	Evans (1969)
Developing stages in roots	Second, third and fourth-stage juveniles	Root colonising fungi	James (1968)
		Rickettsias	Noel and Wilson (1981)
Egg production on root surface	Maturing females	Collembola	Murphy and Doncaster (1952)
		Symphylids	Doncaster (1962)
		Mites	Doncaster (1962)
		Staphilinid beetles	Doncaster (1962)
		Centipedes	Doncaster (1962)
		Enchytraeids	Doncaster (1962)
		Predatory nematodes	Doncaster (1962)
		P. penetrans	Page and Bridge (1985)
		Endoparasitic fungi	Kerry and Crump (1977)

failed to form cysts; presumably the developmental stages are similar in both nematode genera. The taxonomic status of *P. penetrans* and past research has been recently reviewed (Kerry, 1986). The lack of suitable methods to grow this organism in vitro has prevented its commercial development as a biological control agent but an in vivo method of spore production in *Meloidogyne* females (Stirling and Wachtel, 1980) has enabled its potential to be evaluated in small plots. *P. penetrans* was successfully introduced into soil and effectively controlled *M. javanica* on tomatoes (Stirling, 1984) but the numbers of nematodes killed depended upon their period of activity in soil and the concentration and distribution of the non-motile spores. Although many first generation juveniles were infected as they migrated towards roots, those from subsequent generations that hatched from eggs on the root surface and moved only short distances in soil were more difficult to parasitise. It may prove difficult on a field scale to ensure that the spores are thoroughly incorporated in soil and that many nematodes are infected. Cyst nematodes that complete only one generation in a growing season are more likely to be effectively controlled than those that have several. A proper assessment of the potential of *P. penetrans* will only become apparent when sufficient inoculum can be produced for field trials on a range of crops.
P. penetrans was associated with small populations of *H. elachista* Oshima under intensive rice cropping in Japan and controlled the nematode after introduction into soil in pot tests (Nishizawa, 1984); the first time this organism has been used to control a cyst nematode. Our scanty knowledge

of the potential and principles of microbial control of nematodes is derived largely from research with nematophagous fungi.

The stage in the life cycle that is attacked by a biological control agent will have an effect on nematode multiplication and the damage caused to crops. An organism that destroys second-stage juveniles may reduce plant damage but because of the pyramidal age structure of the population (Jones and Kempton, 1978) it will tend merely to remove those individuals surplus to the carrying capacity of the host and have little effect on numbers of nematodes at the end of the season. Those organisms that attack females on roots and prevent the production of healthy eggs are more likely to reduce nematode multiplication than those agents infecting juvenile stages.

Biological control may be induced when an organism is introduced by man to control a pest species or be natural when the organism(s) increase in soil to densities that effectively limit nematode multiplication without being specifically introduced. In general, natural control develops where there has been a long association between a nematode pest and its parasites and predators and consequently is associated with perennial crops or those grown in monocultures. Agents that provide effective natural control in the field may not necessarily have the attributes necessary to induce control after introduction into soil where they are absent (Kerry, 1986). So far there have been a number of cases where natural control of cyst nematodes has been reported but as yet predictable induced control has not been achieved at rates of application that would be practical for a farmer.

NATURAL CONTROL

Populations of several cyst nematode species have been observed to decline under prolonged monocultures of susceptible crops: *G. rostochiensis* (Goswami & Rumpenhorst, 1978), *H. avenae* Woll. (Gair et al., 1969), *H. cruciferae* Franklin (Doncaster, 1962), *H. glycines* (Hartwig, 1981; Carris et al., 1984) and *H. schachtii* Schmidt (Crump & Kerry, 1983; Heijbroek, 1983; Thielmann and Steudel, 1973). In most cases, endoparasitic fungi that attacked females and eggs were implicated as causal agents of the suppression of nematode multiplication but only those studies on *H. avenae* were supported by data on infection rates in the field (Kerry et al., 1982). The development of the natural control of *H. avenae* populations in cereal monocultures (Gair et al., 1969) has the characteristics if an induced suppression (Deacon, 1983; Kerry et al., 1986a) which are summarised below (Table 2).

The decline of *H. avenae* populations in the UK is associated with high rates of infection of females and eggs by the endoparasitic fungi, *Nematophthora gynophila* and *Verticillium* Chlamydosporium. These fungi are widespread in cyst nematode infested soils (Kerry et al., 1986a). They are often present in small numbers in soil where *H. avenae* populations increase but are more numerous where the nematode fails to multiply on susceptible crops. During the early years of monoculture of such crops these fungi increase on the many females that develop; this induction phase may last five years during which severe yield losses may occur (Collingwood, 1962). In general, the density of *N. gynophila* spores in soil is more dependent on the presence of nematode hosts than that of *V. chlamydosporium* (Kerry & Andersson 1983). The microbial origin of the suppression of *H. avenae* was demonstrated by treatment with the partial sterilant, formalin (38 per cent formaldehyde), as a soil drench that resulted in marked increases in nematode multiplication (Kerry et al., 1980, 1982). Formalin was an

Table 2 Some characteristics of a decline phenomenon (induced suppression) (After Deacon, 1983).

1. Control agents initially present in soil.
2. Pathogen favours multiplication of control agent.
3. Severe damage develops in the early years of monoculture.
4. Soil fails to support nematode multiplication when inoculated with pathogen and sown with a susceptible crop.
5. Suppressiveness is biological in origin and removed by biocidal treatments.
6. Suppressiveness is often specific to particular pathogens.
7. Suppressiveness can be transferred to conducive soils.

effective fungicide but poor nematicide and had no effect on *H. avenae* multiplication in soils where *N. gynophila* and *V. chlamydosporium* were absent. Biological control of *H. avenae* can be transferred from a suppressive soil to one that is normally conducive i.e. where the nematode normally causes damage. About 95 per cent fewer *H. avenae* females survived on wheat plants in pots in which the soil had been mixed with a soil fraction containing spores of *N. gynophila* and *N. chlamydosporium* washed from a suppressive soil (Stirling and Kerry, 1983). Although these two fungi parasitise a number of cyst nematode species and are widespread, the numbers of spores in most soils are few and as yet there has been no evidence that they cause effective natural control of any other species than *H. avenae*.

Assessing the nature and extent of natural control

Natural control is more frequently observed in soils where the nematode pest is present but not causing obvious disease symptoms in susceptible crops (Baker and Cook, 1974). A nematode population may not increase to densities that cause damage because the physical environment is unsuitable. To determine whether suppression was due to biological agents, two experimental approaches have been used that have proved successful in studies of soils suppressive to fungal pathogens (Lindeman et al., 1983). Firstly, selective biocidal treatments have been applied to soils and the response of the nematode population assessed. The partial soil sterilant, formalin, and the fungicide captafol (Ortho Difolotan 80W) have both resulted in marked increases in the multiplication of cereal and beet cyst-nematodes on susceptible crops (Williams, 1969; Kerry, 1984b). Also, small quantities or selected aliquots of a suppressive soil have been added to conducive soils and reduced nematode multiplication. Suppressiveness is often eliminated by heat treatment at 55-60°C for 30 minutes and transference of such soil can be used as a control treatment. All cases of natural control of nematodes so far observed have involved a suppression of nematode multiplication and not simply a suppression of disease symptoms. In the case of *H. avenae* decline in the UK it is fungal parasitism of developing females and their eggs from May to August that prevents populations increasing on susceptible crops. Studies on the biological nature of nematode-suppressive soils must be supported by observations on parasitism.

Isolation of microbial parasites of cyst nematodes

Both direct extraction of parasitised nematodes from soil and the baiting of soil with disease-free juveniles have been used to isolate microbial parasites. The incidence of *P. penetrans* was determined by extracting *Meloidogyne* second-stage juveniles and by adding healthy nematodes from cultures to small soil samples and agitating in a water suspension; juveniles were examined for the presence of spores adhering to their cuticles (Slana and Sayre, 1981; Stirling and White, 1982). Such techniques could also be used for the isolation of endoparasitic fungi that infect second-stage juveniles of cyst nematodes. Infected nematodes are less mobile than those that are healthy and so they should be extracted from soil using a method such as sugar flotation (Caveness and Jensen, 1955) that does not rely on the activity of nematodes for efficient recovery (Jaffee and Zehr, 1982; Stirling and Kerry, 1983). Baiting soil samples with healthy nematodes was the most sensitive method of detecting the presence of *P. penetrans* (Stirling and White, 1982) and could be used to determine differences in spore densities in soil (Dutky and Sayre, 1978).

Several methods have been developed for the isolation of nematode trapping fungi from soil but most are time consuming and necessitate observations over several weeks. Also, there is often considerable variation in the fungi isolated on replicate agar plates and at best the methods are only semi-quantitative. Hence, it has proved difficult to study factors affecting the activity and density of these fungi in soil (Barron, 1982; Kerry, 1984a).

Methods for the isolation of fungi from cyst nematode females and eggs have been recently reviewed (Kerry et al., 1986b). Standard mycological techniques can be used to isolate fungi from females and eggs and reliably estimate parasitism. If host or parasite densities in soil are low, second-stage juveniles should be added to increase the density of females that develop and the chances of their becoming infected. Levels of fungal parasitism, particularly of those species that were infected by motile zoospores, often fluctuated markedly from week to week and were much affected by soil moisture (Kerry et al., 1982). Hence, sampling on more than one occasion is often necessary to isolate all but the most common parasities. Females that are totally destroyed by fungi such as *N. gynophila* were difficult to extract from soil and frequent sampling was essential to determine levels of parasitism. To overcome some of these problems Crump and Kerry (1981) developed a method of recovering spores of *N. gynophila* and *V. chlamydosporium* from soil and demonstrated that spore densities were greater in suppressive soils than where *H. avenae* caused damage. The presence in soil of actively growing fungal parasites of nematode eggs was determined by burying discs of water agar containing eggs in soil for 10 days (Culbreath et al., 1984). Microscopical examination of the eggs revealed a number of fungal parasites whose activity in soil could be increased by the addition of various amendments. However, the authors stressed that the agar may select and increase the activity of particular fungal species. Simple observation chambers have been used to study the effects of fungal parasitism on female survival on roots in undisturbed conditions (Kerry and Crump, 1977; LaMondia and Brodie, 1984).

Approximately 140 species of fungi have so far been found colonising eight species of cyst nematode females, eggs or cysts. However, only 40 per cent of those fungi isolated from cysts also colonised eggs and few have been tested for their parasitic status. Gintis et al. (1983) demonstrated that in a population of *H. glycines* the proportion of

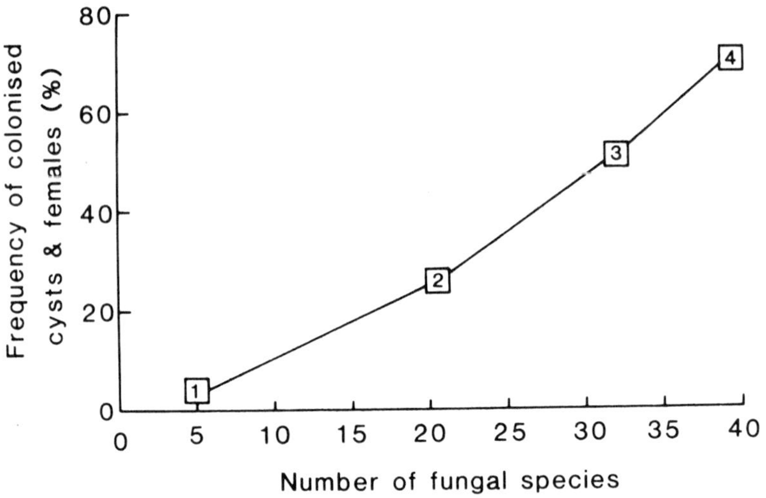

Fig. 1. The relationship between different developmental stages of *H. glycines*, frequency of fungal colonisation and the number of fungal species isolated. (After Gintis et al., 1983).

[1] = immature females; [2] = mature females; [3] = new cysts; [4] = cysts in soil.

individuals colonised, and the number of fungal species isolated, increased with maturation of the female (Fig. 1). Some fungi were associated with particular developmental stages of the nematode and colonisation of the cyst continued in soil. More eggs were parasitised in *H. avenae* cysts than in females (Kerry et al., 1982) although *H. schachtii* encysted eggs were more heavily infected on roots than in soil (Tribe, 1979; Nigh et al., 1980). The methods used to isolate fungi will greatly affect the number and composition of species found. In general, those surveys based on the isolation of fungi from cysts have found a much greater range of fungi than those from females and eggs (Table 3). Fungi in cysts may be confined to the mucilage surrounding the eggs which themselves may not be colonised. Such saprophytes are likely to be faster growing than many parasitic species and be preferentially isolated particularly if rich media are used. Parasitic fungi may be more readily isolated if the female tissues and eggs are spread on water agar plates (Kerry et al., 1986b). However, fungi in the mucilage may affect the development and hatching of eggs without penetrating the eggshell (Morgan-Jones and Rodriguez-Kabana, 1985) and often fungi normally considered as soil saprophytes can be found in eggs (Kerry et al., 1982). Whatever method of isolation is used it is essential to determine the parasitic status of the fungus.

Some fungi have been tested for their pathogenicity in vitro to nematode eggs (Table 4). Tests on agar provide a simple screening method

Table 3. Some recent surveys of endoparasitic fungi in cyst nematodes

Nematode species	Stages	No. of sites	No. of fungal spp. isolated	Reference
H. schachtii	eggs	7	19	Bursnall and Tribe (1974)
	eggs	32	9	Nigh et al. (1980)
	cysts and eggs	2	35	Vinduska (1982)
H. avenae	females and eggs	3	15	Kerry et al. (1982)
	females and eggs	4	15	Dackman and Nordbring-Hertz (1985)
H. glycines	cysts, females and eggs	1	65	Gintis et al. (1983)
	cysts and eggs	2	50	Carris et al. (1984)
G. rostochiensis	cysts and eggs	5	38	Clovis and Nollen (1983)

for determining parasitism but fungi that were highly pathogenic may not be virulent in soil where the ability to survive and spread will also determine the number of eggs killed (see below). Immature eggs were more readily infected than those that contained second-stage juveniles suggesting that the colonisation of young females on roots would result in greater parasitism of eggs, and possibly reduced fecundity, than if young cysts were colonised when most of the eggs had completed their development. Different isolates of the same fungal species from the same soil sample varied greatly in their ability to colonise nematode eggs.

Estimation of natural control in soil

If biological factors in soil are responsible for the suppression of a nematode population, and some potential causal agents have been identified, the next stage is to determine whether the organisms account for the control observed. Three types of experiment have been made to quantify the effects of microbial parasites on populations of nematodes (Kerry, 1984a) and are dealt with here only briefly:

a) Collection of parasitised nematodes through the growing season. The estimation of control based on the number of diseased nematodes is the most obvious method of assessing the role of parasites in suppression. However, such assessments are very time consuming because frequent sampling is necessary; the proportion of individuals killed on one sampling occasion is not a reliable estimate of the total parasitised throughout a growing season.

b) Use of treatments to suppress the activity of biological agents. Such treatments have proved very useful in identifying nematode suppressive soils but it is impossible to apply treatments and only affect particular organisms, and so there are frequently side effects that make the results of experiments difficult to interpret. Formalin proved to be an effective fungicide but only a weak

Table 4. Some fungi tested in vitro for their ability to parasitise cyst nematode eggs

Fungal species	Host	Parasitic status	Reference
Chaetomium indicum	*H. glycines*	-	Godoy et al. (1982)
Codinea heteroderae		+	
Stagonospora heteroderae		+	
Thielavia terricola		-	
Verticillium leptobactrum		+	
V. lamellicola		+	
Fusarium solani		-	
Neocosmospora vasinfecta		-	
Phoma macrostoma		+	
P. multirostrata		+	
F. oxysporum		-	
F. oxysporum	*H. schachtii*	+	Nigh et al. (1980)
Acremonium strictum		+	
Dactylella oviparasitica		+	Stirling & Mankau (1979)
Paecilomyces lilacinus	*G. pallida*	+	Jatala et al. (1979)
V. chlamydosporium	*H. avenae*; *H. schachtii*	+	Kerry et al. (1985)

nematicide and enabled the quantification of the role of nematophagous fungi in *H. avenae* suppressive soils. In a series of experiments at Rothamsted (Kerry et al., 1980; 1982) nematode multiplication was consistently increased in formalin treated soil and four results indicated that the major effect of treatment was on the nematophagous fungi:

i) Similar numbers of females were produced in treated and untreated soil.

ii) The numbers of diseased females and eggs were much reduced in formalin treated soil.

iii) There was no increase in nematode multiplication in treated soils where the fungi were absent.

iv) The response to formalin applications increased with increased densities of nematophagous fungi in soil.

Other biocidal treatments with more selective modes of action may also prove effective.

c) The addition of biological control agents to soil. The importance of agents considered responsible for producing natural control could be demonstrated by adding similar numbers of propagules to a soil where they are absent and inducing control of nematode multiplication. However, such experiments rely on being able to produce sufficient inoculum, and are subject to the difficulties involved in inducing biological control in soil (see below).

Factors affecting the use and expression of natural control

Natural control has become established only in soil where crops have been grown in monocultures in the presence of cyst nematode pests. High densities of cyst nematodes were necessary for obligate parasites such as *N. gynophila* to increase (Kerry and Andersson, 1983) whereas facultative agents such as *V. chlamydosporium* and other egg parasites were less dependent on numbers of nematodes and probably survived saprophytically in soil or on other hosts.

There is little information on factors affecting the activities of nematophagous fungi but, as might be expected, temperature, moisture and soil texture can be important in determining levels of parasitism (see Kerry, 1984a). In general, those organisms that grow in soil are more likely to be affected by soil conditions than those that remain dormant as spores germinating only in the presence of a host. The activity of trapping fungi was much affected by the nutritional status of organic amendments in soil (Cooke, 1963).

The addition of organic amendments to soil to enhance the natural control caused by trapping fungi has rarely reduced cyst nematode damage (Duddington and Wyborn, 1972). The increase in trapping activity that resulted from applications of organic matter was short-lived and difficult to coincide with the period of invasion of the nematode (Hams and Wilkin, 1961). Also, the addition of large amounts of organic matter to soil can produce yield benefits and reduce nematode populations without affecting the activity of nematophagous fungi. Some amendments may selectively increase nematophagous fungi in soil (Mian et al., 1982) but there has been no evidence that this results in predictable or longlasting control and the large amounts of material necessary would prevent their use on field crops.

H. avenae suppressive soils are widespread in Europe (Gair et al., 1969; Jakobsen, 1974; Ohnesorge et al., 1974); Stoyanov, 1982)). Although the nematophagous fungi that control *H. avenae* kill a number of other important cyst nematodes, little is known whether the spread and establishment of these other pests is affected. At Rothamsted Experimental Station, *H. schachtii* introduced into soil naturally suppressive to *H. avenae* rapidly established a heavy infestation that caused significant damage to sugar beet even in a three year rotation with cereals; presumably the former species required higher densities of spores in soil to prevent multiplication and even in this situation natural control was slow to establish.

The slow rate at which antagonists appear to increase naturally in soils means that other control measures must be used to reduce nematode damage during the first few years of monoculture. Any agent added to soil would need to be compatible with these control measures and other agronomic practices. Biological control is unlikely to be effective unless incorporated into an integrated management strategy that would probably need expert supervision; rapid control of nematodes in heavily infested soil is unlikely to occur.

INDUCED CONTROL

Soil organisms have two basic survival strategies that should be considered when attempting to apply biological control agents. Firstly, those organisms such as *P. penetrans* and *N. gynophila* that are obligate parasites produce resistant resting structures and do not grow vegetatively in soil. Spores of other endoparasites may be short-lived if they have a wide host range. These organisms may not need an energy source to aid their establishment but because the soil is not actively colonised, spread is likely to be slow and large numbers of spores would need to be introduced and be extremely well mixed to ensure that a large proportion of nematodes became infected. Such thorough incorporation may be difficult on a field scale. Obligate parasites that have narrow host ranges are more likely to respond to changes in population densities of their hosts. However, unless resting stages are produced such parasites may be more difficult to maintain in soil at effective densities. A narrow host range that limits the market for a biological control agent would militate against commercial development. The second survival strategy is that followed by the nematode trapping fungi and the parasites of eggs. Both are facultative parasites that do not depend upon nematodes for their nutrition and can usually be cultured on standard laboratory media. These fungi produce a mycelium in soil and so an energy source is required to aid colonisation; the type and form of that substrate will affect the subsequent growth and development of the introduced organism in soil. A number of nematophagous fungi have proved sensitive to soil mycostatic factors and have failed to become established after introduction (Mankau, 1962; Cooke and Satchathananthavale, 1968; Eren and Pramer, 1968; Giuma and Cooke, 1974). The use of precolonised substrates can enable an organism to withstand competition from the residual soil microflora and establish itself in soil (Lewis and Papavizas, 1984). Those fungi that colonise the rhizosphere (Mankau, 1964; Stirling, 1979; Kerry et al., 1984) may enable applications with the planting material and greatly reduce the total amount of inoculum required.

Most studies on induced biological control have been empirical. Once pathogenicity to nematodes had been shown, and in vitro inoculum production established, the organism was added to soil with little consideration of the factors that affected its growth and survival. Egg parasitic fungi could be most conveniently added to soil when the crop is planted or drilled and so would have to survive for several weeks in sufficient numbers to kill a large proportion of the new generation of cyst nematode females and eggs. *V. chlamydosporium* introduced on inert clay granules survived in soil for at least six months but when added on barley grain survival of some isolates was reduced to only a few days. Presumably, the introduced organism was unable to withstand the saprophytic competition from the residual microflora that rapidly developed on the grain. Until we understand much more of the factors that affect the growth and survival of fungi introduced into soil it seems unlikely that biological control of cyst nematodes will be predictably induced by amounts of inoculum that could be practically applied by a grower.

Little research has been done for the past 25 years on the use of nematode trapping fungi against cyst nematodes. Although some developments have more recently occurred in the control of root knot nematodes by these fungi, it remains difficult to obtain predictable benefits and these organisms have few characteristics of effective biological control agents (Duddington and Wyborn, 1972; Mankau, 1980; Tribe, 1980; Kerry, 1984a, 1986). Few strains have been found that combine saprophytic competitiveness with pathogenicity to nematodes and we know little of how to prolong trapping activity in soil. Usually these fungi have been added with large amounts of organic matter to facilitate establihsment in soil but a method of producing a dried product is essential for use on field crops. These non-selective fungi that trap most nematodes in soil are likely to provide only partial control of cyst nematode second-stage juveniles and consequently have little effect on multiplication. Most recent research is concentrated on those organisms that attack nematode females and eggs that have been demonstrated to provide long term natural control in the field.

Research with endoparasitic fungi has largely been confined to surveys and studies of their role in the natural control of cyst nematodes. Only *Paecilomyces lilacinus* and *V. chlamydosporium* have been added to soil to induce biological control. *P. lilacinus* appeared to survive better than *V. chlamydosporium* (Godoy et al., 1983) but the fungus produced variable results in field trials with potato cyst nematodes (Lay et al., 1982; Davide and Zorilla, 1983). The isolate of *V. chlamydosporium* and the method of incorporation greatly affected the control of cereal cyst nematode on wheat grown in pots (Kerry et al., 1984). Although one isolate reduced post harvest populations by 80 per cent compared with untreated soil, the amount of inoculum required was too large to be considered practical. Coosemans (personal communication) reduced the inoculum rate and controlled *H. schachtii* by applying chlamydospores of *V. chlamydosporium* around the roots of sugar beet grown in an organic soil prior to transplanting into infested soil in the field. In glasshouse tests, where beet cyst nematodes have much greater reproductive rates than in the field, applying the fungus around the roots of sugar beet grown in peat blocks did not protect the plants after transplanting into *H. schachtii* infested soil (Kerry, unpublished data). Although fungi parasitic on cyst nematode eggs show some promise as biological control agents, much research is required on the selection of aggressive isolates and the development of soil incorporation methods before the potential of these organisms can be properly assessed.

CONCLUDING REMARKS

Effective induced biological control requires a detailed knowledge of the biology of both pest and parasite and is more likely to be affected by environmental conditions than other methods of nematode control. More studies of the development of nematode suppressive soils are necessary to provide information on the factors that affect biological control agents in soil. The demise of some nematicides and the continuing demands of growers to intensify production have increased interest in alternative methods of control. However, it is doubtful whether biological agents will be developed commercially unless they can be effective in controlling nematode populations in the year of application even if long-term control is assured. Although natural biological control is widespread in some cyst nematode infested soils we are clearly a long way from being able to predictably induce such control in the field.

REFERENCES

Baker, K.F. and Cook, R.J., 1974, Biological control of plant pathogens. San Francisco, Freeman & Co.

Barron, G.L. (1982), Nematode destroying fungi, in "Experimental microbial ecology", R. G. Burns and H. H. Slater, eds, Blackwell Scientific Publications, Oxford.

Bursnall, L.A. and Tribe, H.T., 1974, Fungal parasitism in cysts of *Heterodera*. II. Egg parasites of *H. schachtii*, Trans. Br. mycol. Soc., 62:595.

Carris, L.M., Glawe, D.A. and Edwards, D.I., 1984, A comparison of fungi associated with *Heterodera glycines* cysts in two Illinois USA soybean fields during 1983, Phytopathology, 74:830.

Caveness, F.E. and Jensen, H.J., 1955, Modification of the centrifugal-flotation technique for the isolation and concentration of nematodes and their eggs from soil and plant tissue, Proc. helminth. Soc. Wash., 22:87.

Clovis, C.J. and Nolan, R.A., 1983, Fungi associated with cysts, eggs and juveniles of the golden nematode (*Globodera rostochiensis*) in Newfoundland, Nematologica, 29:346.

Collingwood, C.A., 1962, Continuous corn growing and cereal root eelworm in the south-west, NAAS Quarterly Review, 58:70.

Cooke, R.C., 1963, The predaceous activity of nematode-trapping fungi added to soil, Ann. appl. Biol., 51:295.

Cooke, R.C. and Satchuthananthavale, V.E., 1968, Sensitivity to mycostasis of nematode-trapping Hyphomycetes, Trans. Br. mycol. Soc., 51:555.

Crump, D.H. and Kerry, B.R., 1981, A quantitative method for extracting resting spores of two nematode parasitic fungi, *Nematophthora gynophila* and *Verticillium chlamydosporium* from soil, Nematologica, 27:330.

Crump, D.H. and Kerry, B.R., 1983, Possibilities for biological control of beet cyst-nematode with parasitic fungi, Aspects Appl. Biol. 2, 59.

Culbreath, A.K., Rodriguez-Kabana, R. and Morgan-Jones, G., 1984, An agar disc method for isolation of fungi colonizing nematode eggs, Nematropica, 14:145.

Dackman, C. and Nordbring-Hertz, B., 1985, Fungal parasites of the cereal cyst nematode *Heterodera avenae* in Southern Sweden, J. Nematol., 17:50.

Davide, R.G. and Zorilla, R.A., 1983, Evaluation of a fungus, *Paecilomyces lilacinus* (Thom.) Samson for the biological control of the potato cyst nematode, *Globodera rostochiensis* Woll., as compared with some nematicides, Phil. Agr., 66:397.

Deacon, J.W., 1983, Microbial control of plant pests and diseases, Van Nostrand Reinhold, UK.

Doncaster, C.C., 1962, Naturliche Fiende wirtschaftlich bedeutsamer Schadnematoden, Die Umschau, 14:443.

Duddington, C.L., 1960, Biological control - predaceous fungi, in "Nematology, fundamentals and recent advances with emphasis on plant parasitic and soil forms", J. N. Sasser and W. R. Jenkins, eds, Chapel Hill, Univ. N. Carolina Press.

Duddington, C.L. and Wyborn, C.H.E., 1972, Recent research on the nematophagous Hyphomycetes, Bot. Rev., 38:545.

Dutky, E.M. and Sayre, R.M., 1978, Some factors affecting infection of nematodes by the bacterial spore parasite, *Bacillus penetrans*, J. Nematol., 10:285.

Endo, B.Y., 1978, Morphology and interaction of an intracellular micro-organism in tissues of the soybean cyst nematode, *Heterodera glycines*, J. Nematol., 10:286.

Eren, J. and Pramer, D., 1968, Use of a flourescent brightener as aid to studies of fungistasis and nematophagous fungi in soil, Phytopathology, 58:644.

Esser, R.P., 1980, A bacterial spore parasite of nematodes, Fla. Dept. Agric. & Consumer Services, Nematology Circular, No. 63, 3 pp.

Esser, R.P. and Sobers, E.K., 1964, Natural enemies of nematodes, Proc. Soil Crop Sci. Soc. Fla., 24:326.

Evans, K., 1969, Influence of some factors on the reproduction of *Heterodera rostochiensis*, Ph.D. Thesis Univ. London.

Gair, R., Mathias, P.L. and Harvey, P.N., 1969, Studies of cereal nematode populations and cereal yields under continuous or intensive culture, Ann. appl. Biol., 63:503.

Gintis, B.O., Morgan-Jones, G. and Rodriguez-Kabana, R., 1983, Fungi associated with several developmental stages of *Heterodera glycines*, from an Alabama soybean field soil, Nematropica, 13:181.

Giuma, A.Y. and Cooke, R.C., 1974, Potential of Nematoctonus conidia for biological control of soil-borne phytonematodes, Soil Biol. Biochem., 6:217.

Godoy, G., Rodriguez-Kabana, R. and Morgan-Jones, G., 1982, Parasitism of eggs of *Heterodera glycines* and *Meloidogyne arenaria* by fungi isolated from cysts of *H. glycines*, Nematropica, 12:111.

Godoy, G., Rodriguez-Kabana, R. and Morgan-Jones, G., 1983, Fungal parasites of *Meloidogyne arenaria* eggs in an Alabama soil. A mycological survey and greenhouse studies, Nematropica, 13:201.

Goswami, B.K. and Rumpenhorst, H.J., 1978, Association of an unknown fungus with potato cyst nematodes, *Globodera rostochiensis* and *G. pallida*, Nematologica, 24:251.

Hams, A.F. and Wilkin, G.D., 1961, Observations on the use of predacious fungi for the control of *Heterodera* spp., Ann. appl. Biol., 49:515.

Hartwig, E.G., 1981, Breeding productive soybean cultivars resistant to soybean cyst-nematode for the Southern United States, Pl. Dis., 65:303.

Heijbroek, W., 1983, Some effects of fungal parasites on the population development of the beet cyst nematode (*Heterodera schachtii* Schm.), Med. Fac. Landbouww. Rijksuniv. Gent, 48:433.

Jaffee, B.A. and Zehr, E.I., 1982, Parasitism of the nematode *Criconemella xenoplax* by the fungus *Hirsutella rhossiliensis*, Phytopathology, 72:1378.

Jakobsen, J., 1974, The importance of monocultures of various host plants for the population density of *Heterodera avenae*, Tidsskrift for Planteavl., 78:697.

James, G.L., 1968, The interrelationships of the causal fungus of brown root rot of tomatoes and potato root eelworm, *Heterodera rostochiensis* Woll., Ann. appl. Biol., 61:503.

Jatala, P., Kaltenbach, R. and Bocangel, M., 1979, Biological control of *Meloidogyne incognita acrita* and *Globodera pallida* on potatoes, J. Nematol., 11:303.

Jones, F.G.W. and Kempton, R.A., 1978, Population dynamics, population models and integrated control, in "Plant Nematology", J. F. Southey, ed. HMSO, London.

Kerry, B.R., 1984a, Nematophagous fungi and the regulation of nematode populations in soil, Helm. Abs. Series B., 53:1.

Kerry, B.R., 1984b, Report of a meeting of the IOBC/WPRS Integrated control of soil pests working group. Pathogens of nematodes sub group, Pedobiologia, 27:151.

Kerry, B.R., 1986, Biological control, in "Principles and practice of nematode control in crops", R. H. Brown and B. R. Kerry, eds, Academic Press, Sydney, (in press).

Kerry, B.R. and Andersson, S., 1983, *Nematophthora gynophila* och *Verticillium chlamydosporium*, svampparasiter pa cystnematoder, vanliga i svenska jordar med förekomst au strasadescystnematoder, Vaxtskyddsnotiser, 47:5.

Kerry, B.R. and Crump, D.H., 1977, Observations on fungal parasites of females and eggs of the cereal cyst nematode, *Heterodera avenae*, and other cyst nematodes, Nematologica, 23:193.

Kerry, B.R., Crump, D.H. and Mullen, L.A., 1980, Parasitic fungi, soil moisture and multiplication of the cereal cyst nematode, *Heterodera avenae*, Nematologica, 26:57.

Kerry, B.R., Crump, D.H. and Mullen, L.A., 1982, Studies of the cereal cyst-nematode, *Heterodera avenae* under continuous cereals, 1975-1978. II. Fungal parasitism of nematode females and eggs, Ann. appl. Biol., 100:489.

Kerry, B.R., Simon, A. and Rovira, A.D., 1984, Observations on the introduction of *Verticillium chlamydosporium* and other parasitic fungi into soil for control of the cereal cyst-nematode *Heterodera avenae*, Ann. appl. Biol., 105:509.

Kerry, B.R., Crump, D.H. and Irving, F., 1986a, Some aspects of biological control of cyst-nematodes, Proc. 1st Int. Workshop on biological control of nematodes, P. Jatala and R. Rodriguez-Kabana, eds, Lima, Peru, (in press).

Kerry, B.R., Crump, D.H. and Hornsey, J.C., 1986b, Techniques for isolating nematophagous fungi from cyst-nematodes and soil, Proc. 1st Int. Workshop on biological control of nematodes, P. Jatala and R. Rodriguez-Kabana, eds, Lima, Peru, (in press).

LaMondia, J.A. and Brodie, B.B., 1984, An observation chamber technique for evaluating potential biocontrol agents of *Globodera rostochiensis*, J. Nematol., 16:112.

Lay, E.C., Rodriguez, R. and Jatala, P., 1982, Preliminary evaluation of *Paecilomyces lilacinus* as a biological control of the potato cyst nematode, Nematropica, 12:154.

Lewis, J.A. and Papavizas, G.C., 1984, A new approach to stimulate population proliferation of *Trichoderma* species and other potential biocontrol fungi introduced into natural soils, Phytopathology, 74:1240.

Lindermann, R.G., Moore, L.W., Baker, K.F. and Cooksey, D.A., 1983, Strategies for detecting and characterizing systems for biological control of soilborne plant pathogens, Pl. Dis., 67:1058.

Mankau, R., 1962, Soil fungistasis and nematophagous fungi, Phytopathology, 52:611.

Mankau, R., 1964, Ecological relationships of predacious fungi associated with the citrus nematode, Phytopathology, 54:1435.

Mankau, R., 1980, Biological control of nematode pests by natural enemies, Ann. Rev. Phytopathol., 18:415.

Mian, I.H., Godoy, G., Shelby, R.A., Rodriguez-Kabana, R. and Morgan-Jones, G., 1982, Chitin amendments for control of *Meloidogyne arenaria* in infested soil, Nematropica, 12:71.

Morgan-Jones, G. and Rodriguez-Kabana, R., 1985, Phytonematode pathology: fungal modes of action. A perspective, Nematropica, 15:107.

Murphy, P.W. and Doncaster, C.C., 1957, A culture method for soil meiofauna and its application to the study of nematode predators, Nematologicam 2:202.

Nigh, E.A., Thomason, I.J. and Van Gundy, S.D., 1980, Identification and distribution of fungal parasites of *Heterodera schachtii* eggs in California, USA, Phytopathology, 70:884.

Nishizawa, T., 1984, Effects of two isolates of *Bacillus penetrans* for control of root-knot nematodes and cyst nematodes, Proc. First Intern. Congr. Nematology, Guelph, Canada, 60.

Noel, G.R. and Wilson, H.J., 1981, Association of a rickettsia-like organism with reduced *Heterodera glycines* populations, J. Nematol., 13:452.

Ohnesorge, B., Freidel, J. and Oesterlin, U., 1974, Investigations on the distribution pattern of *Heterodera avenae* and its changes in a field under continuous cereal cultivation, Z. Pflanzenkr. Pflanzen pathol. Pflanzenschutz, 81:356.

Page, S.L.J. and Bridge, J., 1985, Observations on *Bacillus penetrans* as a as a parasite of *Meloidogyne acronea*, Nematologica, 31: (in press).

Sayre, R.M., 1971, Biotic influence in soil environment, in "Plant Parasitic Nematodes", B. M. Zuckerman, W. F. Mai and R. A. Rohde, eds, Academic Press, New York.

Shepherd, A.M., Clark, S.A. and Kempton, A., 1973, An intercellular micro-organism associated with tissues of *Heterodera* spp., Nematologica, 19:31.

Slana, L.J. and Sayre, R.M., 1981, A method for measuring incidence of *Bacillus penetrans* spore attachment to the second-stage larvae of *Meloidogyne* spp., J. Nematol., 13:461.

Stirling, G.R., 1979, Techniques for detecting *Dactylaria oviparasitica* and evaluating its significance in field soils, J. Nematol., 11:99.

Stirling, G.R., 1984, Biological control of *Meloidogyne javanica* with *Bacillus penetrans*, Phytopathology, 74:55.

Stirling, G.R. and Kerry, B.R., 1983, Antagonists of the cereal cyst nematode, *Heterodera avenae* Woll. in Australian soils, Aust. J. Exp. Agric. Anim. Husb., 23:318.

Stirling, G.R. and Mankau R., 1979, Mode of parasitism of *Meloidogyne* and other nematode eggs by *Dactylella oviparasitica*, J. Nematol., 11:282.

Stirling, G.R. and Wachtel, M.F., 1980, Mass production of *Bacillus penetrans* for the biological control of root knot nematodes, *Meloidogyne*, Nematologica, 26:308.

Stirling, G.R. and White, A.M., 1982, Distribution of a parasite of root-knot nematodes in south Australian vineyards, Pl. Dis., 66:52.

Stoyanov, D., 1982, Cyst-forming nematodes on cereals in Bulgaria, EPPO Bull., 12:341.

Sturhan, D., 1985, Untersuchungen über Verbreitung und Wirte des Nematodenparasiten *Bacillus penetrans*, Mitt. biol. BundAnst. Ld-u. Forstw., 226:75.

Thielemann, R. and Steudel, W., 1973, Nine years experience with monocultures of sugar beet in soil infested with *Heterodera schachtii* (Schmidt), Nachr.Bl. dt. Pflschutzdienst, 25:145.

Thorne, G., 1927, The life history, habits, and economic importance of some Mononchs, J. Agric. Res., 34:265.

Tribe, H.T., 1977, Pathology of cyst-nematodes, Biol. Rev., 52:477.

Tribe, H.T., 1979, Extent of disease in populations of *Heterodera*, with especial reference to*H. schachtii* , Ann. appl. Biol., 92:61.

Tribe, H.T., 1980, Prospects for the biological control of plant parasitic nematodes, Parasitology, 81:619.

Vinduska, L., 1982, Parazitace hub na cystach hadatka repreho, *Heterodera schachtii*, Rostlinna Vyroba, 27:257.

Walsh, J.A., 1981, The distribution and significance of rickettsia-like organisms in cyst nematodes, Ph.D. Thesis, Univ. Leeds.

Walsh, J.A., Shepherd, A.M. and Lee, D.L., 1983, The distribution and effect of intracellular rickettsia-like micro-organisms infecting second-stage juveniles of the potato cyst-nematode *Globodera rostochiensis*, J. Zool., Lond., 199:395.

Weber, A.P., Zwillenberg, L.C. and Van der Laan, P.A., 1952, A predacious amoeboid organism destroying larvae of the potato root eelworm and other nematodes, Nature, Lond., 169:834.

Williams, T.D., 1969, The effects of formalin, nabam, irrigation and nitrogen on *Heterodera avenae* Woll., *Ophiobolus graminis* Sacc. and the growth of spring wheat, Ann. appl. Biol., 64:325.

DISCUSSION

Responding to a question about the activity of single spore isolates, Kerry said that cultures seem to be stable for at least five years, the longest time so far tested. At present there was no information on mutations. Stone observed that some fungi are pathogenic to man and asked if there was any information about precautions that should be observed in working with such organisms in the laboratory and in field trials, and eventually by farmers who may use them. Barnes responded that at the University of Arkansas there were investigations on toxicology using mice but he did not think that any protocol had so far been established.

Asked by Barnes if any promising bacterial isolates had been identified, Kerry replied that he had not made any extensive search for such. Noel referred to the Rickettsia-like organisms found in cyst nematode populations and their possible damage to humans.

Roberts asked about the response of repeated applications of biological control organisms to the soil, as opposed to the idea that there needs to be a survival strategy. Kerry replied that the organisms are so slow to build up that repeated applications will be necessary in commercial practice.

QUARANTINE AND LEGISLATION

Charles E. Taylor

Scottish Crop Research Institute
Invergowrie, Dundee, DD2 5DA
Scotland

INTRODUCTION

Almost all countries have statutory regulations which restrict the entry of plants, plant products and commodities. Their aim is to protect agriculture and the environment from avoidable damage by pests and diseases that may be introduced by man. Although many pests and diseases, including nematodes, are widespread they have not necessarily reached the potential of their biological distribution and may be absent from a country or geographical zone. Plant quarantine regulations therefore provide the means of delaying, if not preventing, their introduction. A country such as Australia can impose particularly stringent regulations because it is self-sufficient in the majority of agricultural products (Morschel, 1971) and has the advantage of being isolated by the sea from other land masses. The surrounding sea combined with a national policy of isolation during the Tokugawa era (1639-1859) provided an effective quarantine for Japan but then the development and expansion of international trading resulted in the introduction of several pests and diseases that were a threat to agricultural production e.g. woolly aphid, arrowhead scale and bacterial crown gall (Anon, 1975). The virtually unrestricted transport of plants and agricultural produce led to phytosanitary problems in many countries throughout the world, the response to which was usually the introduction of legislation to deal with a particular problem, although often embodied in a broadly based statute. For example, in Great Britain the Destructive Insects Act of 1877 referred only to preventing the entry and establishment of the Colorado beetle, and the following Destructive Insects and Pests Act of 1907 was intended mainly to prevent the entry of American gooseberry mildew on imported plants.

INTERNATIONAL COOPERATION

International cooperation in securing common and effective plant quarantine is relatively recent and stems from the increasing awareness through scientific investigation of the distribution of non-indigenous pests and diseases with plants and plant products transported between countries. A significant step forward was the recognition of the Food and Agriculture Organisation's (FAO) International Plant Protection Convention (IPPC) in Rome, 1951 (Chock, 1979). The aim of the IPPC is the strengthening of international efforts to combat important pests affecting plants or plant products and to prevent their spread across international boundaries (a pest as defined by FAO is any form of animal or plant life, or any pathogenic agent, injurious or potentially injurious to plants or plant products). The Convention has 15 articles and the most important provisions include the requirements for each contracting party to establish an official phytosanitary organisation which can carry out surveys and the inspection of growing plants; inspect consignments of plants and plant products moving in international trade and undertake disinfection or disinfestation as required; issue phytosanitary certifices conforming to the model FAO pattern; report outbreaks and the spread of plant pests; publish its quarantine regulations.

The IPPC foresaw the desirability of establishing regional organisations with agreements to safeguard the interests of neighbouring countries with similar phytosanitary problems. Regional action is particularly desirable to prevent a pest that is absent from a whole area from being introduced into any part of the area, as its entry into one territory would immediately endanger neighbouring countries. The regionalisation concept has been discussed by Mathys and Smith (1984) who considered that many barriers such as the sea, mountains, and deserts, remain effective obstacles to pest dissemination as long as trade exchanges are properly regulated. Further, that once agreement has been reached on quarantine measures in a bio-geographical region, understanding at a global level will become easier and more relevant. Mathys (1977) proposed eight biogeographical regions which relate to the eight international plant protection organisations that have now been established (Smith, 1979). These include the Inter African Phytosanitary Commission (IAPSC) which has 48 member states and covers the whole of Africa except the most southern areas; the European Plant Protection Organisation (EPPO) with 35 member governments in Europe and the Meditteranean area (including north African and near East countries); the North American Plant Protection Organisation (NAPPO) with three members, Canada, Mexico and USA; the Organismo International Regional de Sanidad Agropecuaria (OIRSA) for several Central American countries; the Plant Protection Committee for South East Asia and the Pacific Region (PPC/SEAP); the Comite Inter Americano de Protecciòn Agricola (CIPA); the Caribbean Plant Protection Committee; the Near East Plant Protection committee (NEPPC).

PEST RISK ANALYSIS

Quarantine measures for a country or biogeographical region are generally based on an assessment of the pest risks associated with the importation of plants and plant products. EPPO collected information on 342 pathogenic agents and weeds, including 15 nematode species, and recommended quarantine measures based on the extent to which the pests were considered to have an economic impact on agriculture and foresty. Lists of pests and pathogens are published periodically in special issues of the EPPO Bulletin (Mathys, 1977). The A1 list includes exotic organisms which are absent from all EPPO countries and for which a nil tolerance is required in quarantine regulations. The A2 list includes organisms that are present in some EPPO countries but not all, e.g. *Radopholus similis* among the five nematodes listed (Smith, 1979). A country free from an A2 organism would normally impose a nil tolerance, but a small tolerance might be accepted if the risk of establishment were slight. The A1 and A2 lists are under continuing revision by the EPPO Working Party on Phytosanitary Regulations as the criteria on which the organisms are regulated are constantly changing as more becomes known about the biology of the organisms, and as new crops are grown or old ones abandoned.

In the USA, doubts about the efficacy of efforts to prevent the entry of exotic pests and diseases in the face of a rising volume of international trade prompted an in-depth study and an assembly of information on 1,330 organisms considered to be a threat to the country's agriculture and forestry (Mathys, 1975, 1977). A computer analysis of the information produced a list of the most dangerous exotic pests and diseases and classified into lists (A, B, C), similar to the EPPO categories.

In most other territories a first essential for the establishment of a plant quarantine service is to survey the plant pest and disease situation both within the country and throughout the world and by deductions determine the threats to the agricultural economy. An illustration of the approach that might be taken, with particular reference to Britain, is provided by Baker and Bailey (1979). They point out that the host range of pest or pathogen is one of the few pieces of knowledge that is available for most species, although sometimes alarmingly incomplete. The starting point is to select out only those species known to attack plants grown as economic or amenity crops in that country. The world distribution of the species of concern is then considered in relation to three arbitrary climatic zones (Fig. 1). Species from Zone 1 are expected to be capable of surviving outdoors in Britain, where their host crops are grown, and of causing economic damage both outdoors and under glass. Few species from Zone 2 seem likely to be able to live outdoors in Britain except possibly in a few specially favourable localities in southern England. Zone 3 comprises the tropics and subtropics from which it is most unlikely that many species could survive outdoors in Britain, although there are several species that could probably become pests under glass. The picture illustrated in the Fig. 1 leaves many questions unresolved, e.g. the potential of species from mountain areas in tropical or sub-tropical latitudes. Following an assessment based on geographical distribution the next step would be to estimate the timing of the annual life cycle in Britain, because this determines the stage that is exposed to particular weather conditions in the field. Together with an estimate of reproductive potential, survival factors etc., an assessment is arrived at of the existing capability of an alien species to survive and cause damage in Britain (although this does not take into account the ability of a pest to adapt to new conditions).

Khan (1983) depicts the concept of pest risk analysis as the interaction of biological, economic and political factors in quarantine situations, although with the actual or perceived risk involving the biological consequences of moving plants and plant products along man-made pathways. Taking all pest risk factors into account leads to the development of a phytosanitary control strategy, albeit with many questions of a scientific nature probably remaining unanswered as yet (Southey, 1979). Preventative activity can be required to start at source in the exporting country and can extend along the trade route to the point at which the pest reaches a new environment at the destination. Short of prohibiting the import of host plants, which is justified in some cases, the aim is to impose conditions at the points where they will be most effective. The key to achieving good safeguards lies in the careful integration of measures, e.g. inspections, treatment, quarantine etc., in such a way as to maximise their effect.

NEMATODES

Plant parasitic nematodes are found in all the agricultural areas of the world in virtually every crop. Many species are indigenous or long established, having moved from wild hosts to cultivated crops and some of them have been dispersed unknowingly from their native habitat. Over short distances nematodes may enter new areas on soil adhering to farm implements or on the feet of birds, in flood and irrigation water, and occasionally in wind-blown soil or seeds. Such short-distance dispersal is extremely difficult to prevent, as evidenced by the rapid infestation of new polders (land reclaimed from the sea) in the Netherlands (Kuiper, 1962) or the spread of sugar beet cyst nematode *Heterodera schachtii* in eastern England (Southey et al., 1982). Over long distances many species have been dispersed unknowingly as stowaways on plants or plant products, and their influx into new areas has increased in recent times with the expansion of international trade despite awareness of the problem and the introduction of phytosanitary regulations. Examples of recent introductions are *Globodera rostochiensis* into Japan, New Zealand, India and the USA; *H. glycines* into Egypt and the USA, and *Aphelenchoides besseyi* into East and Central Africa. It would, however, seem unduly pessimistic to accept the view of Jones (1969) that quarantine services everywhere are fighting losing battles against noxious alien species. Nevertheless, it must be recognised that regulatory control of nematodes poses special problems, particularly because their small size makes them impossible to detect by normal inspection procedures. Any quarantine measures against nematode pests therefore are likely to be applied indirectly through restrictions on the importation of their known host plants and particularly by the prohibition of the import of soil which is invariably a source of nematodes. Safeguards which are stipulated by importing countries usually include (a) a permit authorising importation from a specific geographic area (b) a phytosanitary certificate issued by the quarantine services of the exporting country and based on the IPPC model (c) inspection upon entry, and (d) depending on whether pests are present, treatment of the consignment if available. The phytosanitary certificate states that the plants or plant products concerned "have been inspected according to appropriate procedures and found to be free from quarantine pests, and practically free from other injurious pests" and confirms that the importing country's requirements have been complied with. Domestic regulations imposed by the exporting country provide a further safeguard with regard to plant health.

Domestic plant health regulations may be introduced when it is in the national interest to eradicate or limit the spread of a recently discovered or introduced pest or pathogen, or to control an established one that threatens crop production or the export of plant material. The action taken by US Department of Agriculture following the discovery of the potato cyst eelworm *G. rostochiensis*, on Long Island, New York State, in 1941 (Chitwood et al., 1942) illustrates the economic cost of attempting to eradicate a soil-borne pathogen. Infested fields were removed from host crop production, the sale or shipment of potatoes, all movement of plants and soil on Long Island were subject to quarantine regulations, and a massive exercise of fumigation with D-D of infested soil was undertaken (Spears, 1969). An expenditure of several million dollars with a strict enforcement of quarantine measures restrained the development and spread of the nematode on Long Island but could not prevent its eventual appearance in several other parts of North America near both the Atlantic and Pacific coasts (Anon, 1962; Orchard, 1965; Spears, 1969). Nevertheless, the USA continues to prevent *G. rostochiensis* from having other than a tentative foothold restricting imports or potatoes and other root crops through its Federal Quarantine legislation.

Potato cyst nematodes, *G. rostochiensis* and *G. pallida*, have been too long established in the traditional potato growing areas of Europe for eradication to be considered as a practical proposition. Nevertheless, because of their serious economic effects most European countries legislate against them with the aim of preventing their spread to apparently clean areas and to prevent their multiplication to population densities which cause serious loss of yield. Usually the regulations prescribe crop rotation, sometimes with chemical fumigation of the soil as practised in the Netherlands, and special attention is generally paid to seed potatoes to ensure that they are produced only on "eelworm-free" land.

The Potato cyst Eelworm (Great Britain) Order 1973 complies with a Directive of the European Economic Community (EEC) on the control of the nematode and is a result of the United Kingdom entry to the EEC in that year. The Order is intended particularly to provide safeguards against the spread of potato cyst nematodes with seed potatoes. It requires that seed potatoes intended for marketing are produced only on land on which no potato cyst nematode has been found by official soil examination. It also prohibits the planting, except under licence, of any potatoes on, or removal of plants for transplanting from, any land officially declared to be infested. Regulations referring to potato cyst nematodes appear in the phytosanitary statutes of many countries and are aimed at either prevention of entry or the control, and possible elimination, of indigenous infestations.

The history of the beet cyst nematode, *H. schachtii*, in Britain is an example of the relaxation of quarantine regulations as more information on its biology became available. The beet cyst nematode was first recorded in Britain in 1919 but the first confirmed report of damage to sugar beet was from Cambridgeshire in 1934 (Whiteway et al., 1982). There followed a rapid increase in the number and extent of infestations throughout the principal sugar beet growing areas, particularly in the black peat soils of the East Anglian fens. Control by crop rotation was enforced from the 1940s until 1983 through growers' contracts with the British Sugar Corporation and subsequently reinforced by successive Beet Eelworm Orders, 1943-1962 (Southey et al., 1982, Anon, 1984). When the first Beet Eelworm Order was promulgated, it was thought that the nematode had a limited distribution in Britain and part of the purpose of legislation was to contain its spread. It has since become evident that *H. schachtii* is established in all sugar beet growing areas and is probably indigenous. Further, that crop losses due to the nematode have remained small. Thus it was concluded that the Beet Eelworm Orders, which restricted the frequency of growing any host crops of the nematode in a 'scheduled area' of the fens of eastern England and in all other fields known to be infested, were no longer justified on biological or economic grounds. Consequently the Beet Cyst Nematode Order 1977 was introduced, which revoked the earlier regulations and provides reserve powers to reintroduce control of cropping and other measures if and where there is evidence that infestations are increasing to a serious degree. In order to detect local signs of such increases an official series of surveys was begun in 1977 to monitor the incidence of infestations in the areas believed to be most at risk. The survey data have been computer mapped to define the cumulative distribution of the beet cyst nematode in Britain and to provide a sensitive and accurate method for detecting changes (Southey et al., 1982 ; Whiteway et al., 1982).

Most of the cyst nematode problems associated with international trade in plant and plant products can be obviated by ensuring that all consignments are free from soil. Nevertheless, many countries wish to reduce the risk of importing cyst nematodes to a minimum by placing restrictions on imports of host plants and may require an inspection and sampling of the crop in the field to ensure freedom from infestation. Many countries, e.g. USA, Canada, accept imported planting material that has originated only from fields "believed to be free" from potato cyst nematodes as found by soil-sampling tests using officially approved methods. Southey (1974) and

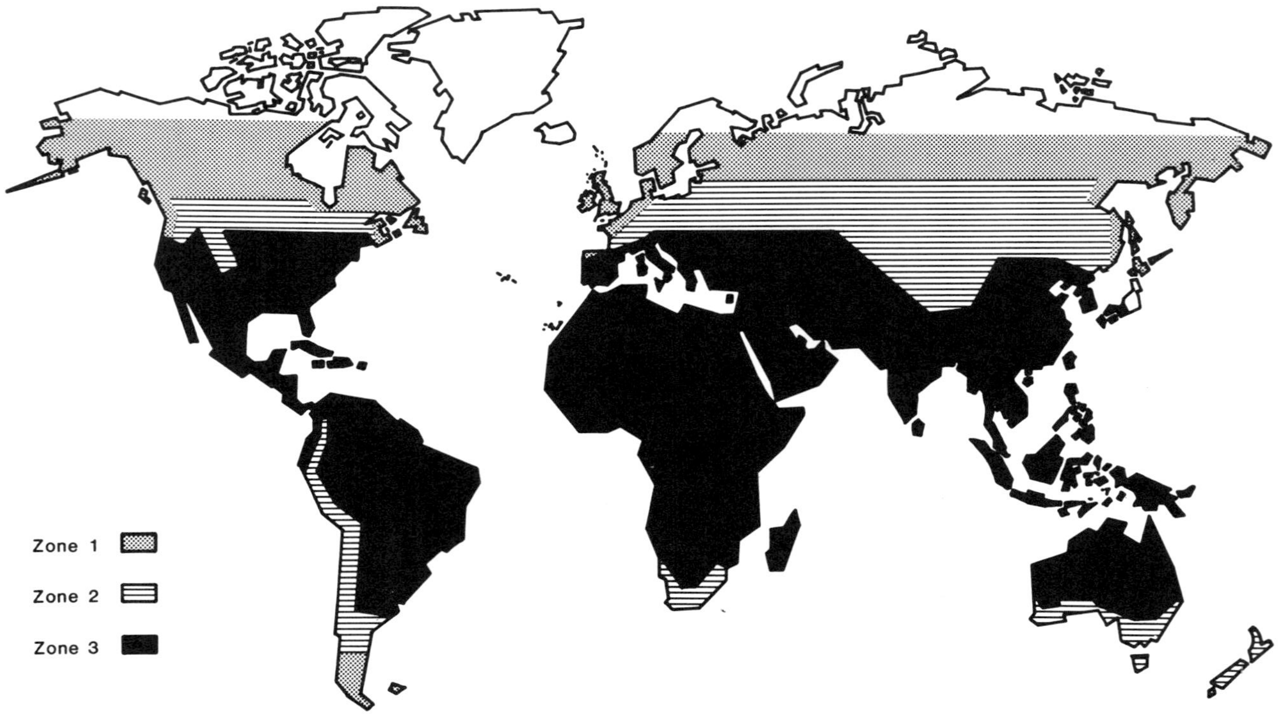

Fig. 1. Diagrammatic map of world climate zones used in assessing the potential of alien pests and diseases for survival in Britain. For explanation of zones 1-3 see text. (Redrawn after Baker and Bailey, 1979).

Jones (1969) refer to the severe limitations on undertaking adequate soil sampling to provide a reasonable level of assurance that no potato cyst nematodes are present in a growing crop. Indeed, Southey concludes that soil sampling does not in itself provide an effective safeguard against transport of cysts with plant material and considers it more important to have effective systems of control on potato growing land at source and to minimise the amount of soil that accompanies exported plants and seed potatoes. Detection and estimation of cyst nematodes from soil samples is dealt with elsewhere in this volume.

CONCLUSIONS

In its strict sense 'plant quarantine' refers to the holding of plants in isolation until they are considered to be healthy, but common usage has broadened the meaning to include all aspects of the regulation of the movement of plants between territories or parts thereof. Plant quarantine legislation varies from country to country but in most cases restrictions are imposed on the movement or entry of commodities, and associated materials such as soil with plants or packaging, to prevent or delay the establishment of plant pests in areas where they are not known to occur. Any legislation introduced for plant quarantine purposes must be based on fundamental prerequisites which include (a) the quarantine measure must be based on sound biological grounds; the pest must pose a threat to major crops or forests (b) the quarantine measure must be established only to control or prevent entry of a particular pest and not for the furtherance of trade or the attainment of some other objective (c) as conditions change or further information becomes available, quarantine should be modified accordingly (d) the proposed quarantine must represent a necessary or desirable measure for which no other suitable substitute is available, the objective of the measure introduced must be considered reasonable to achieve, and the legislation should be clear and consistent with the objectives to be attained.

Quarantine regulations should be regarded as a means of reducing the frequency of introduction of noxious pests and diseases rather than their complete exclusion from a particular area. The measures that are taken are sometimes costly but to delay or lessen spread or to reduce infestation to less harmful levels is invariably better than doing nothing.

REFERENCES

Anon, 1962, Outbreaks and new records. Canada. Occurrence of potato root eelworlm, Pl.Prot.Bull.FAO., 10:115.

Anon, 1975, Plant quarantine in Japan, Minist.Agric.For.Toyko 36 pp.

Anon, 1984, Beet cyst nematode, Leaflet 233, Minist.Agric.Fish.Fd., 6 pp.

Baker, C.R.B., and Bailey, A.G., 1979, Assessing the threat to British crops from alien diseases and pests, in "Plant Health", D.L. Ebbels and J.E. King, eds, Blackwell, Oxford.

Chitwood, B.G., Clement, R.L., Morgan, R., and Tank, R., 1942, *Heterodera rostochiensis*, the golden nematode of potatoes, in New York State, Pl.Dis.Reptr, 26:390.

Chock, A.K., 1979, The International Plant Protection Convention, in "Plant Health", D.L. Ebbels and J.E. King, eds, Blackwell, Oxford.

Jones, F.G.W., 1969, Some reflections on quarantine, distribution and control of plant nematodes, in "Nematodes of Tropical Crops", J.E. Peachey, ed, Commonw.Bur.Helminth, Farnham Royal.

Khan, R.P., 1983, Safeguarding the international exchange of plant germplasm, Proc.10th Int.Congr.Pl.Prot., 2:866.

Kuiper, K., 1962, Dissemination of plant parasitic nematodes with new polder soil, Nematologica 7:16.

Mathys, G., 1975, Thoughts on quarantine problems, EPPO Bull., 5:55.

Mathys, G., 1977, Phytosanitary regulations and the transfer of genetic resources, in "Plant Health and Quarantine in International Transfer of Genetic Resources", W.B. Hewitt and L. Chiarappa, eds, CRC Press, Cleveland.

Mathys, G., and Smith, I.M., 1984, Regional and global plant quarantine strategies with special reference to developments with EPPO, EPPO Bull., 14:83.

Morschel, J.R., 1971, Introduction to plant quarantine, Australian Govt. Publ.Service, Canberra, 71 pp.

Orchard, W.R., 1965, Occurrence of the golden nematode on Vancouver Island, British Columbia, Can.Pl.Dis.Surv., 45:89.

Smith, I.M., 1979, EPPO : the work of a regional plant protection organisation with particular reference to phytosanitary regulations, in "Plant Health", D.L. Ebbels and J.E. King, eds, Blackwell, Oxford.

Southey, J.F., 1974, Methods for detection of potato cyst nematodes, EPPO Bull., 4:463.

Southey, J.F., 1979, Preventing the entry of alien diseases and pests into Great Britain, in "Plant Health", D.L. Ebbels and J.E. King, eds, Blackwell, Oxford.

Southey, J.F., Alphey, T.J.W., Cooke, D.A., Whiteway, J.A., and Mathias, P.L., 1982, Surveys of beet cyst nematode in England 1977-80, Pl.Path., 31:163.

Spears, J.F., 1969, Organised nematode control programmes in the USA, in "Nematodes of Tropical Crops", J.E. Peachey, ed., Commonw.Bur. Helminth., Farnham Royal.

Whiteway, J.A., Alphey, T.J.W., Mathias, P.L., and Southey, J.F., 1982, Computer mapping of records of beet cyst nematode (*Heterodera schachtii*), 1928-77, Pl.Path., 31:157.

DISCUSSION

Sikora referred to the likelihood of introducing pests into a country or geographical area when food such as potatoes, grain, etc. is brought in to relieve conditions resulting from national disasters e.g. floods, drought. Taylor said that under such conditions it would be difficult to impose quarantine, except to a limited degree, but whilst pests might be more extensively introduced than normally there would be less chance of their becoming established because of the absence of host plants in the disaster areas.

Several participants discussed the cleaning of seed potatoes to eliminate any cysts of *Globodera* spp. that might be present on them. Chlorox had been used to good effect in New Zealand but Stone reported that the station undertaking the treatment had recently been closed because of the high running costs. Trudgill added that drying of the tubers was a problem after the dipping treatment and increased rotting resulted. Müller reported that chlorox had been tried in West Germany but he was not aware of the results. Mabbot said that a dipping treatment used in Scotland incorporated an improved method of drying the tubers.

It was agreed that European countries should jointly develop more effective ways of assessing the presence of potato cyst nematodes in cargoes of potatoes and further, that measures should be taken to prevent any such infestations.

PARTICIPANTS

LAURA AMBROGIONI - Istituto Sperimentale per la Zoologia Agraria, Cascine del Riccio - 50100 Firenze, Italy.
KIRSTEN ANDERSEN - Dept. Crop Husbandry and Plant Breeding, Thorwaldsensvej 40 - DK 1871 Copenhagen V, Denmark.
SIGURD ANDERSEN - Dept. Crop Husbandry and Plant Breeding, Thorwaldsensvej 40 - DK 1871 Copenhagen V, Denmark.
STIG ANDERSSON - Swedish University of Agric. Sci., Dept. Plant and Forest Protection, Box 44 - S 230 53 Alnarp, Sweden.
MARIA ANTONIOU - Benaki Phytopathological Institute, 14561 Kiphissia, Athens, Greece.
VINCENT AUBERT - Station Fédérale de Recherches Agronomiques de Changins, Nématologie - CH 1260 Nyon, Switzerland.
CATHERINE BALANDRAS - INRA, Laboratoire de Recherches de la Chaire de Zoologie, Domaine de la Motte-au-Vicomte - Le Rheu 35650, France.
JAMES BALDWIN - Department of Nematology, University of California - Riverside CA 92521, USA.
JOHN M. BARNES - USDA, Cooperative State Research Service, Plant and Animal Sciences - Washington D.C. 20251, USA.
M. MOULDI B'CHIR - I.N.A.T., 43, Avenue Charles Nicolle - Tunis, Tunisie.
SVEIN M. BIRKENES - Norwegian Plant Protection Institute - N 1432 AS-NLH, Norway.
TERESA BLEVE ZACHEO - Istituto di Nematologia Agraria CNR, Via G. Amendola 165/A - 70126 Bari, Italy.
MICHEL BOSSIS - INRA, Laboratoire de Recherches de la Chaire de Zoologie, Domaine de la Motte-au-Vicomte - Le Rheu 35650, France.
ANGELA CAPUSSO TOSI - Istituto di Nematologia Agraria CNR, Via G. Amendola 165/A - 70126 Bari, Italy.
EDWARD CASWELL - Department of Nematology, University of California, Riverside CA 92521, USA.
AURELIO CIANCIO - Istituto di Nematologia Agraria CNR, Via G. Amendola 165/A - 70126 Bari, Italy.
MAGHESPREN CHINAPPEN - Plant Pathology Division, Ministry of Agriculture, Reduit, Mauritius.
RENATO CROZZOLI - UCV, Fac. Agronomia, Inst. Zoologia Agricola, Apdo 4579 Maracay, E. Aragua, Venezuela.
FRANCESCO P. D'ERRICO - Istituto di Entomologia Agraria, Università, 80055 Portici, Italy.
MAURO DI VITO - Istituto di Nematologia Agraria CNR, Via G. Amendola 165/A, 70126 Bari, Italy.
MARIA M. TEIXEIRA DUARTE - Escola Superior Agraria de Santarem, Sao Pedro, 2000 Santarem, Portugal.
BURTON Y. ENDO - USDA, ARS, Northeastern Region, Plant Protection Institute, Beltsville MD 20705, USA.
EZARUG A. EDONGALI - Plant Protection Department, El-Fateh University, P.O. Box 13274 - Tripoli, Lybia.
FRANCO ELIA - Istituto di Nematologia Agraria CNR, Via G. Amendola 165/A, 70126 Bari, Italy.

JAVIER FRANCO - Dept. Nematology and Entomology, International Potato Center, Apdo 5969, Lima, Peru.
A. MORGAN GOLDEN - USDA, ARS, BARC-W, Nematology Lab., PPI, Beltsville MD 20705, USA.
NICOLA GRECO - Istituto di Nematologia Agraria CNR, Via G. Amendola 165/A, 70126 Bari, Italy.
JASPER DE HAAN -H.L. Hilbrands Laboratory for Soilborne Diseases, P.O. Box 323 - 9400 AH Assen, The Netherlands.
LARS M. HANSEN - Institute for Plant Protection, Zoological Department - DK 2800 Lyngby, Denmark.
DIETER HEINICKE - Pflanzenschutzamt Hannover, Wunstorfer Landstrasse 9 - D 3000 Hannover, W. Germany.
I.K.A. IBRAHIM - Dept. Plant Pathology, College of Agriculture, Alexandria University - Alexandria, Egypt.
ANITA IREHOLM - Swedish University of Agric. Sci., Dept. Plant and Forest Protection, Box 44 - S 230 53 Alnarp, Sweden.
BARRY J. JACOBSEN - Dept. Plant Pathology (N-533b), University of Illinois, 1102 South Goodwin Avenue, Urbana IL 61820, USA.
PARVIZ JATALA - Dept. Nematology and Entomology, International Potato Center, Apdo 5969 - Lima, Peru.
JABR A. KHALIL - Plant Protection Department, El-Fateh University, P.O. Box 13274 - Tripoli, Lybia.
BRIAN KERRY - Dept. Nematology, Rothamsted Experimental Station - Harpenden, Herts AL5 2JQ, United Kingdom.
CHRISTOF KüNAST - BASF Aktiengesellschaft, Landwirtschaftliche Versuchs-station - D 6703 Limburgerhof, Germany.
FRANCO LAMBERTI - Istituto di Nematologia Agraria CNR, Via G. Amendola 165/A - 70126 Bari, Italy.
MICHEL LUC - Muséum National d'Histoire Naturelle, Laboratoire des Vers, 61, Rue de Buffon - 75231 Paris Cedex 05, France.
THOMAS W. MABBOT - Agricultural Scientific Services Ltd., Dept. Agric.and Fisheries for Scotland, East Craigs - Edinburgh EH18 8NJ, United Kingdom.
ANNA MARINARI PALMISANO - Istituto Sperimentale per la Zoologia Agraria, Cascine del Riccio - 50100 Firenze, Italy.
JULIA A. MEREDITH - UCV, Fac. Agronomia, Inst. Zoologia Agricola, Apdo 4579 - Maracay, E. Aragua, Venezuela.
LAWRENCE I. MILLER - Dept. Plant Pathology and Physiology, Virginia Polytechnic Institute and State University - Blacksburg VA 24061, USA.
DIDIER MUGNIERY - INRA, Laboratoire de Recherches de la Chaire de Zoologie, Domaine de la Motte-au-Vicomte - Le Rheu 35650, France.
JOACHIM MüLLER - Biologische Bundesanstalt fur Land-und Forstwirtschaft, Institut für Nematologie, Toppheideweg 88 - D 4400 Münster, Westf., Germany.
GREGORY R. NOEL - USDA, ARS, Midwest Area, Dept. Plant Pathology, W-321 South Goodwin Avenue - Urbana IL 61801, USA.
ROLAND N. PERRY - Dept. Nematology, Rothamsted Experimental Station - Harpenden, Herts AL5 2JQ, United Kingdom.
LUIS G.L. REIS - Departamento de Fitopatologia, Estacao Agronomica Nacional - 2780 Oeiras, Portugal.
PHIL ROBERTS - Department of Nematology, University of California - Riverside CA 92521, USA.
M. DOLORES ROMERO DUQUE - Instituto de Edafologia y Biologia Vegetal, C.S.I.C., Serrano 115 dpdo - 28006 Madrid, Spain.
J. WILLEM SEINHORST - Middelberglaan 6 - 6711 JH Ede, The Netherlands.
RICHARD A. SIKORA - Institut für Pflanzenkrankheiten, Nussallee 9 - D 5300 Bonn, Germany.
CARLOS SOSA-MOSS - Centro de Fitopatologia, Colegio de Postgraduados - Chapingo 56230, Mexico.

CORNELIA STENDER - Botanische Institut der Tierärztlichen Hochschule, Bünteweg 17d - D 3000 Hannover 71, W. Germany.
ALAN R. STONE - Dept. Nematology, Rothamsted Experimental Station - Harpenden, Herts AL5 2JQ, United Kingdom.
CHARLES E. TAYLOR - Scottish Crop Research Institute - Invergowrie, Dundee DD2 5DA, United Kingdom.
DAVID L. TRUDGILL - Scottish Crop Research Institute - Invergowrie, Dundee DD2 5DA, United Kingdom.
ROLAND VALLOTTON - Station Fédérale de Recherches Agronomiques de Changins, Nématologie - CH 1260 Nyon, Switzerland.
EVANGELOS VLACHOPOULOS - Benaki Phytopathological Institute, 14561 Kiphissia - Athens, Greece.
NICOLA VOVLAS - Istituto di Nematologia Agraria CNR, Via G. Amendola 165/A-70126 Bari, Italy.
ALAN G. WHITEHEAD - Dept. Nematology, Rothamsted Experimental Station - Harpenden, Herts AL5 2JQ, United Kingdom.
URS WYSS - Institut für Phytopathologie der Universitat, Olshausenstrasse 40-60, Haus S 24 c - D 2300 Kiel, Germany.
GIOVANNI ZACCHEO - Istituto di Nematologia Agraria CNR, Via G. Amendola 165/A - 70126 Bari, Italy.
GIUSEPPE ZACHEO - Istituto di Nematologia Agraria CNR, Via G. Amendola 165/A - 70126 Bari, Italy.
ULRICH ZUNKE - Institut für Phytopathologie der Universitat, Olshausenstrasse 40-60, Haus S 24 c - D 2300 Kiel, Germany.

INDEX